THE TAXONOMY,
MORPHOLOGY AND
ECOLOGY OF
RECENT OSTRACODA

Professor H. V. Howe

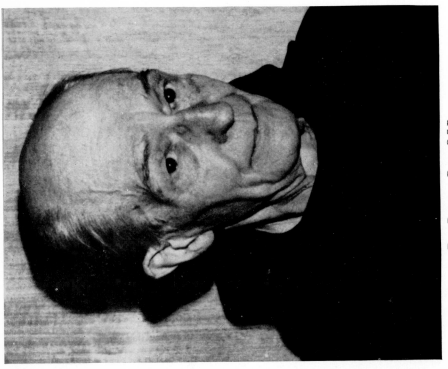

Dom Remacle Rome O.S.B.

THE TAXONOMY, MORPHOLOGY AND ECOLOGY OF RECENT OSTRACODA

Edited by

JOHN W. NEALE

Department of Geology
University of Hull

OLIVER & BOYD · EDINBURGH

Dedicated to
Professor H. V. Howe
and
Dom Remacle Rome, O.S.B.

OLIVER AND BOYD LTD
Tweeddale Court, Edinburgh 1

05 001794 2
First published 1969
© 1969 The Authors
All rights reserved

Printed in Great Britain by
Robert Cunningham and Son Ltd, Alva.

CONTENTS

IV. LOCALISATION OF MATERIAL

FOREWORD

In common with other branches of science, the study of micropalaeontology, and in particular of the Ostracoda, has witnessed a phenomenal expansion during the last two decades. In 1963 Dr Harbans Puri organized the memorable Naples Symposium and yet the advent of new work and workers in the last four years has been such that it is already time to take stock and review the progress that has been made in the study of the taxonomy, morphology and ecology of Recent Ostracoda.

For this end a Symposium was arranged in Hull from 10th to 14th July 1967, and was preceded by a Collecting Excursion covering the North of England and South of Scotland. Forty-six people took part in the Symposium, and twenty-nine contributions from nine different countries were read and discussed during the formal part of the proceedings.

We were very fortunate in having with us Professor H. V. Howe and Dom Remacle Rome, both of whom have devoted a lifetime to the study of Ostracoda and to whom all other workers are greatly indebted. Professor Howe's standard works on Ostracod Taxonomy and (with Dr Laurencich) on Cretaceous Ostracoda are the basic requisites of anyone who attempts to work on these small crustacea, and Dom Rome's brilliant researches into the morphology and taxonomy of Ostracoda, especially his elucidation of the Cyprididae, are too well known to need comment. It gives me and all my colleagues and fellow-workers particular pleasure to dedicate this volume to these two fine gentlemen.

This volume contains original papers and the accounts of discussion sessions dealing with various aspects of the Taxonomy, Morphology and Ecology of Recent Ostracoda, with occasional excursions into the palaeontological field where it was felt that this threw some light on common problems. The order of the papers in this volume differs somewhat from the order in which they were read and the volume falls roughly into four parts, beginning with Taxonomy, continuing with detailed Morphology and Function, and thence by way of Ecology to Localisation of Material. It is hoped that this book will prove useful to other workers in indicating the breadth and scope of work at present in progress, the different approaches of different workers to the same problem, the varying techniques employed, and the areas where major problems still remain and much work still needs to be done.

I am most grateful for the help of Dr O. Elofson, M. N. Grekoff, Dr J. P. Harding, Professor H. V. Howe, Dr H. S. Puri, Dr I. G. Sohn and Professor F. M. Swain who chaired the various sessions of the Symposium. The chairmen of the Discussion Sessions kindly agreed to prepare shortened and edited versions of these sessions which are included here. It is a great pleasure to acknowledge my indebtedness to Mrs P. M. Neale and Mrs L. F. Penny and thank them for their help in connection with the programme for accompanying families during the Symposium, and similarly to thank Dr J. E. Robinson for his invaluable help during the preceding Field Collecting

Excursion, and Dr S. M. Marshall, F.R.S., of the Millport Marine Station, Mr A. M. Tynan of the Hancock Museum, Newcastle-upon-Tyne, and Mr and Mrs J. T. Dealtry of Howden for their kindness and hospitality. It is a pleasure to thank Professor House and the University of Hull for their encouragement and kindness in making the full facilities of the Department of Geology available during the Symposium, Miss H. Sanderson and Messrs. R. G. Clements and A. R. Lord for their help during the latter and last, but by no means least, my especial thanks are due to the British Petroleum Company for a generous grant to help with the organization of the Symposium.

JOHN W. NEALE
Hull

AUTHORS

Dr F. J. Adamczak (Sweden)
Dr R. H. Benson (U.S.A.)
Dr G. Bonaduce (Italy)
Dr G. Carbonnel (France)
Dr D. L. Danielopol (Romania)
Dr C. D. Evenson (U.S.A.)
† Professor E. Ferguson (U.S.A.)
Dr A. M. Gervasio (Italy)
Mrs J. Gilby (England)
C. W. Hart, Jr., (U.S.A.)
Mrs D. G. Hart (U.S.A.)
Professor H. V. Howe (U.S.A.)
Dr N. C. Hulings (U.S.A.)
Dr R. L. Kaesler (U.S.A.)
T. I. Kilenyi, Esq. (England)

Dr L. S. Kornicker (U.S.A.)
D. L. McGregor, (U.S.A.)
Dr K. G. McKenzie (Australia)
Dr M. Masoli (Italy)
Dr J. W. Neale (England)
Dr T. Petkovski (Yugoslavia)
Dr H. S. Puri (U.S.A.)
Dr S. Reys (France)
Dr J. E. Robinson (England)
D. R. Rome, O.S.B. (Belgium)
Dr I. G. Sohn (U.S.A.)
Professor F. M. Swain (U.S.A.)
D. R. Wall, Esq. (Wales)
Dr R. C. Whatley (Wales)
R. Williams, Esq. (Wales)

PARTICIPANTS

Dr F. J. Adamczak (Sweden)
Dr P. Ascoli (Italy)
Dr R. H. Bate (England)
Dr R. H. Benson (U.S.A.)
Professor W. A. van den Bold (U.S.A.)
Dr G. Carbonnel (France)
R. G. Clements, Esq. (England)
Dr D. L. Danielopol (Romania)
J. T. Dealtry, Esq. (England)
Dr L. D. Delorme (Canada)
Dr O. Elofson (Sweden)
Dr C. D. Evenson (U.S.A.)
† Professor E. Ferguson (U.S.A.)
Dr G. Fryer (England)
Mrs J. Gilby (England)
Dr M-C. Guillaume (France)
M. N. Grekoff (France)
Dr L. Hagerman (Sweden)
M. J. Hardenbol (France)
Dr J. P. Harding (England)
C. W. Hart, Jr., (U.S.A.)
Mrs D. G. Hart (U.S.A.)
Dr C. W. Haskins (England)

Professor M. R. House (England)
Professor H. V. Howe (U.S.A.)
Dr N. C. Hulings (U.S.A.)
Dr R. L. Kaesler (U.S.A.)
T. I. Kilenyi, Esq. (England)
Dr L. S. Kornicker (U.S.A.)
Dr S. A. Levinson (U.S.A.)
A. R. Lord, Esq. (England)
D. L. McGregor, (U.S.A.)
Dr K. G. McKenzie (Australia)
Dr J. W. Neale (England)
Dr H. J. Oertli (France)
L. F. Penny, Esq. (England)
Dr T. Petkovski (Yugoslavia)
Dr H. S. Puri (U.S.A.)
Dr J. E. Robinson (England)
D. R. Rome, O.S.B. (Belgium)
Dr I. G. Sohn (U.S.A.)
Professor F. M. Swain (U.S.A.)
Professor P. C. Sylvester-Bradley (England)
D. R. Wall, Esq. (Wales)
Dr R. C. Whatley (Wales)
R. Williams, Esq. (Wales)

† We regret to record the death of Professor Ferguson on 12th June 1968.

I. TAXONOMY

HOMONYMY AND OTHER MAN-CREATED TAXONOMIC AFFLICTIONS OF RECENT OSTRACODA

HENRY V. HOWE
Louisiana State University, Baton Rouge, U.S.A.

SUMMARY

The indexing of the very valuable proceedings of the 1963 Naples Symposium on ostracod ecology published in 1965 turned up a number of taxonomic problems. These are summarized briefly in the ensuing pages in the hope that they will make the volume of greater use to later workers. They are treated under the headings: same species under different names; failure to give author and/or date; same species in checklist under different names; failure to recognize first description of a species; misspelled names; different classification in different parts of same paper; and a list is given of homonyms encountered.

INTRODUCTION

Ostracoda are extremely interesting creatures in their own right and as such are wel worthy of study. Time has shown that in many parts of the world they are suitable as stratigraphic markers. Few organisms that can be used in the interpretation of sediments encountered in the drilling of oil wells live in as diverse habitats as do the Ostracoda. Some are planktonic, some occur in brackish and in fresh waters, and some even are terrestrial. Most species, however, are benthonic, selective in their habitat, and are not world-wide in their distribution.

Oil companies have found that conditions of sedimentation in particular basins govern the manner and places of petroleum accumulation (Weeks, 1958). To properly evaluate past environments, one must know where and how present organisms live. A great start was made in that direction with the holding of the Symposium at Naples in June, 1963. More than 20 authors contributed to the success of the published volume (Symposium, 1965), and following each article are the comments of the members of the Symposium.

Unfortunately I was in Africa on a collecting trip at the time of the Symposium, but after returning home I indexed the genera and species which had been reported upon. Some papers were easy to index, some were quite difficult, and a number of taxonomic problems presented themselves. Undoubtedly it was a realization that only a start had been made on the ecology of Ostracoda that was the motive force behind the present Symposium. I therefore felt that a brief summary of taxonomic problems presented in the first Symposium would be helpful to later workers using that most valuable volume and might be helpful in the preparation of subsequent volumes.

Ecology is only as good as taxonomy permits. As Hartmann (Symposium 1965, p. 591) says: 'The main unit we have in zoology is the species.' Tribes, subgenera,

3

genera, subfamilies, families, superfamilies, and higher categories in any system of classification are all artificial means of cataloguing species so that other workers may be able to understand how you think the species you have under consideration is related to other species. Just as cataloguers in libraries differ somewhat in the catalogue numbers they place on books in a library, so no two ostracod workers can agree on where to place the majority of species. We have flexible guide lines to follow which are outlined in the Code of the International Commission, but most workers depart from these at one time or another. This is not all bad, as classification is artificial and if the worker has a good enough cross-index, he should be able to understand the concepts of fellow workers. If, however, a species is reported under the wrong specific name, everyone suffers. Again from Hartmann (Symposium 1965, p. 591): 'We should, therefore, be very careful in considering the species, and *identify only if the identification seems to be sure.*' Neale (Symposium 1965, p. 258) properly goes even further: 'In studies of ecology and distribution accurate taxonomy is a *sine qua non,* and in this respect the species is the most significant and important unit.'

With the above thoughts in mind, I have, in indexing the Naples Symposium, made note of the following items which I hope will be of help to succeeding users of this most valuable volume.

1. SAME SPECIES UNDER DIFFERENT NAMES IN DIFFERENT LISTS

Mutilus convexus (Baird) 1850 and *Mutilus punctatus* (Münster) are given as separate species by Ruggieri (Symposium 1965, p. 525). These were earlier considered to be the same species by Ruggieri (1959, p. 200). Same or not, the species reported by Ruggieri as *convexus* is given under the name *Aurila convexa* (Baird) by Neale (Symposium 1965, pp. 229, 264, 272) and by Elofson (Symposium 1965, pp. 483, 484, 486, 487), both of whom give excellent ecologic data concerning the species. Rome (Symposium 1965, p. 206) lists *A. convexa* from Monaco, and Puri, Bonaduce and Malloy (Symposium 1965, pp. 113, 117, 168, 169, Fig. 22) give *Aurila* aff. *A. convexa* from Naples.

Here we are presented with two problems. First, are we dealing with two separate species? Secondly, are we dealing with two separate genera?

The older species *Cythere punctata* Münster (1830, p. 62) was described from the Pliocene of Castell'arquato, Italy. The younger, *Cythere convexa* Baird (1850, p. 174) was described from recent sand at Torquay on the south coast of England. I have good material from the south coast of England and also from Castell'arquato and I had so much difficulty in trying to separate specimens from the two places, I followed Ruggieri when preparing material for the Treatise and considered them synonyms despite the fact that some of the variants from Italy can be separated from some of the variants from England. It might be well for the assembled group to examine specimens from the two localities to see if agreement is possible, as *Cythere convexa* Baird was designated the type species of *Aurila* by Pokorný (1955, p. 17).

The genus *Mutilus* dates from Neviani's (1928, pp. 72, 90) 'Gruppo' [= subgenus] of *Cythereis,* under which he listed several species, from which Ruggieri (1956, p. 167) designated *Cythereis (Mutilus) laticancellata* Neviani, n.sp. to be the type, but placed it in the synonymy of *Cythere retiformis* Terquem (1878, p. 116). Again we are confronted with species problems. I have specimens from Brindisi, sent me by Ruggieri which correspond well with his 1956 figures. I also visited Rhodes and collected specimens which correspond well with Terquem's figures. They certainly are not the same species. Those from Rhodes have much thicker ribs forming the reticulations, and have a rounded postero-dorsal region. The question of whether *Aurila* is synonymous with *Mutilus*, a subgenus of *Mutilus*, or is a separate genus obviously is a matter of personal opinion. The genotype of *Mutilus*, however, I feel must be Neviani's species, rather than Terquem's.

2. FAILURE TO GIVE DATE OF AUTHORSHIP IN FAUNAL LISTS

Mutilus marsupius (Neviani) is listed by Ruggieri (Symposium 1965, p. 525) and is presumably the same species he (1952, p. 125) had listed previously as *Hemicythere marsupia* (Neviani). In both cases the listing was without dates or bibliographi, reference. I assume Ruggieri probably refers to *Cythere marsupia* Capeder (1902, p. 13c Fig. 25) as Capeder's figure appears to be that of a hemicytherid species that might well be referred to *Mutilus*, but am unable to prove it. Cross-checking by others is possible if dates of authorship are given in lists as done by Benson (Symposium 1965, pp. 397–417).

3. FAILURE IN LISTS TO GIVE EITHER AUTHOR OR DATE

An illustration is the list given by McKenzie (Symposium 1965, pp. 448–53) in his very thorough study of the ecology of Oyster Harbour. In this paper McKenzie discusses in the text some of the species listed, but some like *Bairdia milne edwardsi* var. *dentata*, I have been unable to find. I assume that he does not refer to *Bairdia dentata* Issler (1908, p. 95), a Liassic species, but I have no other reference to '*dentata*' in my index. '?*Aglaiella setigera*' is given by Benson (Symposium 1965, p. 399) under the original name *Macrocypris setigera* Brady, 1880, and *Loxoconcha honoluliensis* is given by Benson (Symposium 1965, p. 416) properly under the generic name *Loxoconchella*, as it is the type species of that genus. Other species in the list have had their generic name changed so that the reader must necessarily guess to what genus they were originally referred. *Ambostracon pumila* and *Callistocythere crenata* are illustrations. Another complication arises when the regional distribution of a species is greatly extended without explanation. *Cytheridea exilis* of McKenzie's list I assume refers to *Cytheridea exilis* Seguenza (1880, pp. 194, 290, 364, Pl. 17, Fig. 49) which in my index has been cited by authors only as fossil from Italy or Albania.

4. SAME SPECIES REPORTED IN CHECKLIST UNDER MORE THAN
ONE GENERIC AND SPECIFIC NAME

Benson (Symposium 1965, p. 397) reports *Cythere truncata* Thomson, 1879, and on the next page he reports *Hemicythere innominata* Brady, 1879 (sic). The specific name *truncata* had been used in conjunction with the generic name *Cythere* several times prior to Thomson. *Cythere truncata* was used by Bosquet (1847, p. 357, Pl. 1, Figs 2a–e) for a Cretaceous species. *Cythere* (*Bairdia*) *truncata* was used by Kirkby (1858, p. 433, Pl. 11, Figs 3, 3a) for a Permian species, and *Cythere truncata* Sars (1866, p. 117) was a recent species. This was recognized by Brady (1898, [*non* 1879], p. 443, Pl. 46, Figs 1, 2) who renamed the species *Cythere innominata* Brady, *nom. nov.*

5. FAILURE TO RECOGNIZE THE FIRST DESCRIPTION OF A SPECIES

As most workers realize, Terquem's (1878) figures of Pliocene species from Rhodes are not all easy to recognize, and I understand that his types may be lost. Nevertheless the figures of *Cythere irregularis* Terquem (1878, p. 101, Pl. 11, Figs 10a–c) appear to be similar to those of *Loxocythere angulosa* (Seguenza), which species was reported by Ruggieri (Symposium 1965, p. 524). I have material from the Pliocene of Italy and also from Rhodes and cannot separate the species when laid side by side. Both show the same sexual dimorphism. *Cythere irregularis* Terquem clearly has priority over *Cytheridea angulosa* Seguenza (1880, p. 363, Pl. 17, Fig. 47), but unfortunately it is not clear whether it has priority over *Cythere irregularis* Miller (1878, p. 106, Pl. 3, Figs 7, 7a), a species described during the same year from the Ordovician of Ohio. The *Journal of the Cincinnati Society of Natural History*, volume 1, carries on the title page, 'April, 1878, to January, 1879'. I have not been able to ascertain the month in which Terquem's monograph was published, and, therefore, have not changed the generic assignment of Terquem's species.

6. MISSPELLED NAMES

Misspelled names are usually occasioned by typographical errors and do not often cause trouble for others who may keep an index. As an illustration, Hartmann's (Symposium 1965, p. 564) spelling *Caspiolina* Mandelstam, 1957 falls close alphabetically to the true name *Caspiollina* Mandelstam (1957, p. 169). However, Hartmann's (Symposium 1965, p. 564) spelling *Taminocypris* Zalányi, 1944 is widely separated in an index from the true name *Thaminocypris* (Zalányi, 1944, pp. 49, 165). I am sure that all active ostracod workers must have at one time or another been guilty of a misspelling, and it is one of the most difficult errors to catch in reading proof, but every effort should be made to avoid it.

7. DIFFERENT CLASSIFICATION IN DIFFERENT PARTS OF A PAPER

There is no ostracod worker who does not change his ideas as to the proper placement of genera and species as his knowledge grows. It does not help a beginning student, however, when in reading Hartmann's paper (Symposium 1965, p. 573) he finds the precise placement of *Paleomonsmirabilia* Apostolescu, 1955 in the tribe CYTHERIDAE-CYTHERIDEINAE-CUNEOCYTHERINI, and later in the same paper (p. 586) he finds it 'CYTHERIDAE of uncertain subfamilies'.

8. HOMONYMS

No attempt was made to identify all homonyms in the Naples Symposium. First will be from Benson's (Symposium 1965, pp. 397–417) compiled list of Pacific Ostracoda. This excellent paper is the first summary of the ecologic distribution of Ostracoda in the Pacific region since John Murray (1896) compared the Kerguelen Antarctic fauna with that of the south Atlantic and Pacific. Benson (p. 387) states: 'No attempt has been made to rework and modernize the classification of most of the older species.' However, not all of them were given under their original names and in the following list of homonyms, I shall give the page in Benson's report, the name used by Benson in quotes, followed by the data on homonymy.

Symposium page:
397: '*Cythere truncata* Thomson, 1879' (*non* Bosquet, 1847; *non* Kirkby, 1858, *non* Sars, 1866) = *Cythere innominata* Brady (1898, *nom. nov.*, p. 445, Pl. 46, Figs 1, 2); and given by Benson (Symposium 1965, p. 398) as '*Hemicythere innominata* Brady, 1879' (sic).

400: '*Bairdia foveolata* Brady, 1868' is not *Bairdia foveolata* Bosquet (1852, p. 21, Pl. 1, Figs 5a–d) from the Eocene of France.

400: '*Cythere caudata* Brady, 1890' is not *Cythere* (*Bairdia*) *plebeia caudata* Kirkby (1860, p. 145, Pl. 9, Figs 9, 10, 12, 12a) from the Permian of England.

400: '*Cythere cuneolus* Brady, 1890' is not *Cythere cuneola* Jones and Kirkby (1867, p. 223, *nom. nud.*) nor *Cythere cuneola* Jones and Kirkby, in Vine (1883, p. 234, Pl. 12, Figs 6, 6a, 7), a species later referred to *Bythocypris*? by Jones and Kirkby (1886, p. 250, Pl. 6, Figs 3–7), and to several other genera by later authors. It came from the Carboniferous of Scotland. It may be reasoned that the difference in ending, and the fact that neither species is a *Cythere*, may justify saving Brady's specific name. Brady, however, indicated that his species might be the young of some other species, and hence hardly worth saving.

401: '*Cythere labiata* Brady, 1890' is not *Cythere labiata* Terquem (1878, p. 105, Pl. 12, Figs 5a–c) from the Pliocene of Rhodes.

401: '*Cythere ovalis* Brady, 1880' is not *Cythere ovalis* Stoddart (1861, p. 489, Pl. 18, Figs 5, 5a, 5b) from the Carboniferous of England.

401: '*Cytherella truncata* Brady, 1880'. This species was first described by Brady (1869, p. 154, Pl. 19, Figs 3, 4) from Colon, Panama. The form reported by Benson (Brady, 1880, p. 174, Pl. 36, Figs 3a–d) was from Torres Strait and may not be the same. At any rate, it is a junior homonym of *Cythere truncata* Bosquet (1847, p. 7, Pl. 1, Figs 2a–c) [= *Cytherella truncata* by Bosquet (1854, p. 333, list), not *Cythere* (*Cytherella*) *truncata* (Bosquet) of Jones (1849, p. 30, Pl. 7, Figs 25a–e)]. Bosquet's species is a well recognized *Cytherella* from the Maestrichtian of Maestricht. Both of the above species are senior to *Cytherella truncata* LeRoy (1939, p. 274, Pl. 12, Figs 1–2) from the Miocene of central Sumatra.

403: '*Loxoconcha tumida* Chapman, 1902' (p. 428, Pl. 37, Figs 5a–c) was described from Funafuti, and is not *Loxoconcha tumida* Brady (1869, p. 48, Pl. 8, Figs 11, 12), a species Brady described from the eastern Mediterranean.

406: '*Cytherura quadrata* Hanai, 1957'. Hanai's species is a junior homonym of *Cytherura quadrata* Norman (1869, p. 292), where it was described but not figured. It was figured, however, by Brady and Robertson (1872, p. 55, Pl. 1, Figs 10–11). Hanai (1961, p. 358, text-figs 2a, b) changed the generic assignment of his species to *Semicytherura*. It still needs a new specific name.

409: '*Paradoxostoma cuneata* Lucas, 1931' from Vancouver Island is not *Paradoxostoma cuneatum* Brady and Robertson (1874, p. 117, Pl. 5, Figs 6, 7), a species described from Grimsby Harbour, Scilly Islands, 10–15 faths.

412: '*Hemicytheria pokornyi* Hartmann, 1962'. The name *Hemicytheria pokornyi, sp. nov.* appears in a paper by Sheremeta (1959, p. 86). It was reported from the Pannonian of the Ukraine, and presumably is described elsewhere, but no bibliography accompanied this paper by Sheremeta.

414: '*Argilloecia affinis* Chapman, 1919'. This name has been used several times. According to my records the first was by Chapman (1902, p. 419, Pl. 37, Figs 1a–c) off Funafuti at a depth of 1489 fathoms, though listed by Benson as characteristic of the Antarctic Realm. The second record was *Argilloecia affinis*, n. sp. by Brady (1903, p. 99) in a list, but not described from 29 miles east of Alnmouth, England, at a depth of 59 fathoms. The third was *Argilloecia affinis* Brady (1911, p. 396, Pl. 20, Figs 9, 10), described as new, from off Madeira at a depth of 70 fathoms.

B

415: '*Cytheropteron foveolata* (Brady), 1880'. This was originally described as *Cythere foveolata* Brady (1880, p. 75, Pl. 13, Figs 5a–h) from Kerguelen and Heard Island. It is not *Cythere foveolata* Seguenza (1880, p. 75, Pl. 13, Figs 5a–h) from the Quaternary of Italy. Brady's species was renamed *Cythereis exfoveolata* nom. nov. by Neviani (1928, p. 106). Triebel (1958, p. 111, Pl. 3, Figs 15a, b) later referred Brady's species to the genus *Uroleberis*. Perhaps the correct name should be *Uroleberis exfoveolata* (Neviani), 1928.

415: '*Paradoxostoma antarcticum* Müller, 1908' [1909 on title page] is apparently senior to *Paradoxostoma antarcticum* Scott (1912, p. 585, Pl. 14, Figs 27, 28). Scott's new species was described from Scotia Bay, South Orkneys.

415: '*Bairdia attenuata* Brady, 1880'. The name *Bairdia attenuata* was first used by Reuss (1854, p. 140, Pl. 27, Figs 3a–c) for a species Reuss (1846, p. 104, Pl. 24, Figs 15a, b) had originally described as *Cytherina attenuata* from Turonian and Cenomanian of Bohemia, and which has since been referred to the genus *Paracyprideis* by Triebel (1941, p. 160, Pl. 3, Figs 17–25). Both the above species are senior, however, to *Bairdia attenuata* Girty (1910, p. 237, no Figs) which was described from the Mississippian of Arkansas.

415: '*Cythere circumdentata* Brady, 1880' is a senior homonym of *Cythere circumdentata* Seguenza (1884, p. 180, Pl. 2, Fig. 4) from the Quaternary of Rizzolo, Italy.

416: '*Cythere reussi* Brady, 1869' was described from Colon, Panama and Brady's original figures do not look like the specimens he later figured from the Pacific. It certainly is not *Cythere* (*Bairdia*) *reussi* Speyer (1863, p. 45, Pl. 1, Figs 7a–c). It is, however, senior to *Cythere reussi* Procházka (1893, p. 56, Pl. 2, Figs 1a–b), a Miocene species from Moravia.

416: '*Cytherella punctata* Brady, 1880'. This species was described by Brady (1866, p. 362, Pl. 57, Figs 2a–b) from sponge sand in the Levant, not from the Pacific. It is a senior homonym of *Cytherella punctata* Batalina (1941, pp. 300, 308, Pl. 2, Fig. 4) from the Devonian of the U.S.S.R.

The remaining homonyms are from Hartmann (Symposium, 1965).

570: '*Cypridella* Vavra, 1895'. Vávra's genus is a junior homonym of *Cypridella* Koninck (1841, p. 20), a Carboniferous genus.

571: '*Kassina* Mandelstam, 1956', [sic = *Kassinia* Mandelstam (1956, p. 118)]. The generic name was preoccupied by Khabakov, 1937, and was renamed *Kassinina* by Mandelstam (1961, *Basic Paleontology*, p. Q361).

The above observations should make it evident that precise ecologic study in the field of Ostracoda requires the worker to have available a comprehensive index of the species and genera that have been previously described. It is not enough to have just an index of living Ostracoda, or of Mesozoic Ostracoda. In several of the homonyms given above the senior species was described from the Palaeozoic. Unfortunately, I suspect that no one has a complete index. My personal index, which I have worked on assiduously for more than 40 years, contains in excess of 32,000 cards on which are perhaps a half-million entries. It is not complete and at the rate of publication for the past ten years, I see no possible chance that I can bring it up to date.

Many described species have been assigned to six or seven genera since they were erected. If an author insists on using subgenera the number of cards must be doubled. They must be doubled again if an author uses subspecies, as subspecies have equal taxonomic rank with species under the rules. I fully realize the importance of showing variation within a species, but letters or symbols could well be given to the variants, rather than varietal names, as later authors frequently have different concepts from these of the original author, and varieties tend to be raised to specific rank by later workers.

The problem of keeping up with changing ostracod taxonomy may be illustrated best by the number of families, subfamilies, genera and subgenera that have been erected since the Naples Symposium was held. My index contains cards for more than 300 such new taxonomic categories, and I know that it is far from complete.

In conclusion, I feel that I must again emphasize the importance of Hartmann's remarks to the effect that for ecologic studies to be of value, the species is of utmost importance and should not be identified unless the identification appears to be certain.

REFERENCES

BAIRD, W. 1850. *The Natural History of the British Entomostraca.* London, The Ray Society. i-viii, 1-364, Pls 1-36 (Ostracoda 138-182, Pls 18-23).

BASIC PALEONTOLOGY. 1960. Arthropoda Volume, Y. A. Orlov, director; E. N. Tschernysheva, ed. 1-514, pls 1-18. (Ostracoda 264-421, Figs 600-1238, Pl 17). Moscow.

BATALINA, M. A. 1941. Ostracods of the Main Devonian Field, in Fauna of the Main Devonian Field. 1. *Izv. Akad. Nauk. SSSR*, 285-306 (Russ.), 307-14 (English), 2 Pls.

BOSQUET, J. 1847. Description des Entomostracés fossiles de la craie de Maestricht. *Mém. Soc. r. Sci. Liège*, 4, 353-78, Pls 1-4.

——. 1852. Description des Entomostracés fossiles des terrains tertiaires de la France et de la Belgique. *Mém. cour. Mém. Sav. étr. Acad. r. Sci. Belg.*, 24, 1-142, Pls 1-6.

BRADY, G. S. 1866. On new or imperfectly known species of Marine Ostracoda. *Trans. zool. Soc. Lond.*, 5, 359-93, Pls 57-62.

——. 1868. La mer à Noumea. *Les Fonds de la Mer.* 1, chap. XIII, 54-59, Pl. 7, Figs 4-6. Folin et Perier, Paris.

——. 1869. Colon-Aspinwall (supplement). *Les Fonds de la Mer.* 1, chap. XXX, 152-5, Pls 18, 19.

——. 1869. Contributions to the study of the Entomostraca No. IV. Ostracoda from the River Scheldt and the Grecian Archipelago. *Ann. Mag. nat. Hist.*, (4), 3, 45-50, Pls 7-8.

——. 1880. Report on the Ostracoda dredged by *H.M.S. Challenger* during the years 1873-1876. *Report of the Scientific Results of the Voyage of H.M.S. Challenger*, 1873-76, Zoology, 1, Pt. 3, 1-184, Pls 1-44.

——. 1898. On new or imperfectly-known species of Ostracoda, chiefly from New Zealand. *Trans. zool. Soc. Lond.*, 14, 429-52, Pls 43-47.

——. 1903. Report on Dredging and other Marine Research off the North East Coast of England in 1901. *Trans. nat. Hist. Soc. Northumb.*, 14, 87-101.

——. 1911. Notes on Marine Ostracoda from Madeira. *Proc. zool. Soc. Lond.*, 595-601, Pls 20-22.

—— AND ROBERTSON, D. 1872. Contribution to the Study of the Entomostraca. No. VI. On the Distribution of the British Ostracoda. *Ann. Mag. nat. Hist.*, (4) 9, 48-72, Pls 1, 2.

—— AND ROBERTSON, D. 1874. Contributions to the Study of the Entomostraca. No. IX. On the Ostracoda taken amongst the Scilly Islands, and on the Anatomy of *Darwinella Stephensoni*. *Ann. Mag. nat. Hist.*, (4), 13, 114-18, Pls 4, 5.

CHAPMAN, F. 1902. On some Ostracoda from Funafuti. *J. Linn. Soc.*, Zoology, 28, 417-33, Pl. 37.

HANAI, T. 1957. Studies on the Ostracoda from Japan. III. Subfamilies Cytherurinae G. W. Müller (emend. G. O. Sars, 1925) and Cytheropterinae n. subfam. *J. Fac. Sci. Tokyo Univ.*, Sect. 2, 11, Pt. 1, 11-36, Pls 2-4.

ISSLER, A. 1908. Beiträge zur Stratigraphie und Mikrofauna des Lias in Schwaben. *Palaeontographica*, 55, 1-104, Pls 1-7.

JONES, T. R. 1849. A monograph of the Entomostraca of the Cretaceous Formation of England. *Palaeontogr. Soc.*, 1-40, Pls 1-7.

—— AND KIRKBY, J. W. 1867. On the Entomostraca of the Carboniferous Rocks of Scotland. *Trans. geol. Soc. Glasg.*, 2, 213-28.

—— AND KIRKBY, J. W. 1886. Notes on the Palaeozoic Bivalved Entomostraca, No. XXII. On some undescribed species of British Carboniferous Ostracoda. *Ann. Mag. nat. Hist.*, (5), 16, 249-69, Pls 6-9.

KIRKBY, J. W. 1858. On Permian Entomostraca from the Fossiliferous Limestone of Durham. *Ann. Mag. nat. Hist.*, (3), 2, 317-30, Pls 10, 11.

——. 1860. On Permian Entomostraca from the Shell-Limestone of Durham. *Trans. Tyneside Nat. Field Club*, 2, 122-71, Pls 8-11.

KONINCK, L. DE. 1841. Mémoire sur les Crustacés fossiles de Belgique. *Mém. Acad. r. Sci. Lett. Belg.*, **14**, 1-20, plate.

LEROY, L. W. 1939. Some small foraminifera, ostracoda, and otoliths from the Neogene ('Miocene') of the Rokan-Tapanoeli Area, Central Sumatra. *Natuurk. Tijdschr. Ned.-Indië*, **99**, 272-77, Pls 10-12.

LUCAS, V. Z. 1931. Some Ostracoda of the Vancouver Island Region. *Contr. Can. Biol. Fish.*, N.S. **6**, 397-416, 6 Figs.

MANDELSTAM, M. I. AND OTHERS. 1956. Ostracoda. In *Material of Paleontology*, n. ser., **12**, Paleontology, New Families, New Genera. *Trudȳ vses. nauchno-issled. geol. Inst.*, Moscow. 1-356, Pls 1-43.

MANDELSTAM, M. I., SCHNEIDER, G. F., KUZNETSOVA, Z. V. AND KATZ, F. I. 1957. New genera of Ostracoda in the families Cypridae and Cytheridae. *Ezheg. vses. paleont. Obshch.*, **16**, 165-93, Pls 1-4.

MILLER, S. A. 1878. Description of a new genus and eleven new species of fossils, with Remarks upon others well known from the Cincinnati Group. *J. Cincinn. Soc. nat. Hist.*, **1** (1878-9), 100-8, Pl. 3.

MÜLLER, G. W. 1909. Die Ostracoden der deutschen Südpolar-Expedition 1901-1903. *Wiss. Ergebn. dt. Südpolarexped.*, **10**, zool. II Bd., 51-181, Pls 4-19.

MÜNSTER, G. 1830. Ueber einige fossile Arten Cypris (Müller, Lamk.) und Cythere (Müller, Latreille, Desmarest). *Neues Jb. Miner. Geol. Paläont.*, 60-67.

MURRAY, J. 1896. On the Deep and Shallow-water Fauna of the Kerguelen Region of the Great Southern Ocean. *Trans. R. Soc. Edinb.*, **38**, 343-500.

NEVIANI, A. 1928. Ostracodi fossili d'Italia. I. Vallebiaja (Calabriano). *Memorie Accad. pont. Nuovi Lincei*, Ser. 2, **11**, 1-120, 2 Pls.

NORMAN, A. M. 1869. Shetland Final Dredging Report—Pt. II. On the Crustacea, Tunicata, Polyzoa, Echinodermata, Actinozoa, Hydrozoa and Porifera. *Rep. Br. Ass. Advmt Sci.*, **38**, Norwich, 247-336, Suppl. 341-2.

POKORNÝ, V. 1955. Contribution to the morphology and taxionomy of the subfamily Hemicytherinae Puri 1953 (Crust., Ostrac.). *Acta Univ. Carol.*, Geologica, **3**, 1-35.

PROCHÁZKA, V. J. 1893. Das Miocaen von Seelowitz in Mähren und dessen Fauna. *Rozpr. české Akad.*, (2), **2**, Heft 24, 1-90, Pls 1-3.

REUSS, A. E. 1846. *Die Versteinerungen der böhmischen Kreide Formation.* Schweizerbart, Stuttgart, **2**, 58-148, Pls 14-51.

——. 1854. Beiträge zur Charakteristik der Kreideschichten in den Ostalpen, besonders in Gosauthale und am Wolfgangsee. *Denkschr. Akad. Wiss., Wien*, Math. Phys. Kl., **7**, 1-156.

RUGGIERI, G. 1959. Enumerazione degli Ostracodi marini del Neogene, Quaternario e Recente italiani descritti elencati nell'ultimo decennio. *Atti. Soc ital. Sci. nat.*, **98**, 183-208.

SCOTT, T. 1912. The Entomostraca of the Scottish National Antarctic Expedition 1902-1904. *Trans. R. Soc. Edinb.*, **48**, 521-99, Pls 1-14.

SEGUENZA, G. 1880. Le formazioni terziarie nella provincia di Reggio (Calabria). *Atti Accad. naz. Lincei Memorie*, (3), **6**, 1-446, Pls 4-17.

——. 1884. Il. Quaternario di Rizzolo. *Naturalista sicil.*, Anno Terzo, 16-22, 48-51, 67-71, 115, 118, 141-5, 179-83, 223-7, 262-6, 287-91, 308-11, 349-52, Pls 1-4.

SHEREMETA, V. G. 1959. Important Ostracods in the Stratigraphic Separation of Sedimentary Deposits of the Pannonian Basin in Zakarpat. *Trudȳ III Sess. vses. paleont. Obshch.*, 83-88.

SPEYER, O. 1863. Die Ostracoden der Casseler Tertiärbildungen. *Ber. Ver. Naturk. Cassel*, **13**, 1-62, Pls 1-4.

STODDART, W. W. 1861. On a Microzoal Bed in the Carboniferous Limestone of Clifton, near Bristol. *Ann. Mag. nat. Hist.*, (3), **8**, 486-90, Pl. 18.

SYMPOSIUM. 1965. Ostracods as Ecological and Paleoecological Indicators. (Ed. H. S. Puri). *Pubbl. Staz. zool. Napoli* [1964]. **33** (suppl.), 1-612, plates, and papers by P. ASCOLI, R. H. BENSON, G. BONADUCE, J. M. GILBY, J. P. HARDING, G. HARTMANN, N. C. HULINGS, L. S. KORNICKER, J. MALLOY, K. G. McKENZIE, J. W. NEALE, V. POKORNÝ, H. S. PURI, A. RITTMANN, D. R. ROME, G. RUGGIERI, P. SANDBERG, I. G. SOHN, F. M. SWAIN, and C. W. WAGNER.

TERQUEM, M. O. 1878. Les foraminifères et les entomostracés-ostracodes du Pliocène supérieur de l'Ile de Rhodes. *Mém. Soc. géol. Fr.*, **3**, 81-135, Pls 10-14.

THOMSON, G. M. 1879. On the New Zealand Entomostraca. *Trans. Proc. N.Z. Inst.*, **11**, 251-63, Pl. 11.

TRIEBEL, E. 1941. Fossile Arten der Ostracoden-Gattung *Paracyprideis* Klie. *Senckenberg. leth.*, **23**, 153-64, Pls 1-3.

——. 1958. Zwei neue Ostracoden-Gattungen aus dem Lutet des Pariser Beckens. *Senckenberg. leth.*, **39**, 105-17, Pls 1-3.

VÁVRA, V. 1895. Die von Dr. F. Stuhlmann gesammelten Süsswasser-Ostracoden Zanzibar's. *Jb. hamb. wiss. Anst.*, **12**, 1-23, 9 text Figs.

VINE, G. R. 1883. Notes on the Carboniferous Entomostraca and Foraminifera of the North Yorkshire shales. *Proc. Yorks. geol. polytech. Soc.*, N.S. **8** (1882-4), 226-39, Pl. 12.

WEEKS, L. G. 1958. *Habitat of Oil*. A symposium conducted by the American Association of Petroleum Geologists. Tulsa, 1384 pp.

ZALÁNYI, B. [DR. TORDAI ZALÁNYI BELA]. 1944. Magyarorszagi Neogén Ostracodák. *Geologica hung.*, Ser. Palaeontologica, Fasc., **5**, 1-153, Pls 1-4.

DISCUSSION

SOHN: There are several points that I would like to make. I wish Professor Howe that you would emphasize the importance of using parentheses. The transfer of species from one genus to another gets lost unless it is indicated by parentheses. If genus A and species B are cited with an author's name without putting the parentheses around it, how are we to know the binomen under which it was originally described?

HOWE: That is of course correct, and it is also most desirable to put subgenera in parentheses. This was not done at first by Jones and others. This has caused some difficulties in species described as *Cythere*—subgenus *Bairdia*, particularly in the Palaeozoic, which really does not belong here, but I am going to hand you a sheet of corrections.

SOHN: The other point I would like to make is that you are quite correct that homonyms should be renamed, but this is dangerous because when one finds a homonym one is seldom sure whether there might be an available name for the homonym. I am afraid that at times more damage than good is done by renaming a homonym without doing the proper homework.

HOWE: It should never be done without that. Those of you that have used the Cretaceous book by Dr Laurencich and myself must have noted that we did not rename the homonyms, except where I had the actual materials. Those I renamed were based on material supplied me by Van Veen. Now the list I read from the Pacific came from Benson's compilation. He is the man who has been doing the work in that area and I think you should all agree that he is the one who should check these species, particularly those of Brady 1880, and should be the one to rename them.

NEALE: In agreeing with the foregoing remarks, it might be useful to point out that *Paradoxostoma antarcticum* Scott, which Professor Howe has drawn attention to as being a junior homonym of *P. antarcticum* Müller, has recently been renamed *P. scotti* by Hartmann (1962, p. 216, *Mitt. hamb. zool. Mus. Inst.*, **60**).

SWAIN: I should like to recommend that the list of 300 entries Professor Howe has assembled since the Naples symposium be made available to ostracod workers and other palaeontologists through publication in the *Journal of Paleontology* and, furthermore, that Stuart Levinson be encouraged to continue his annual bibliography of ostracod publications. Although Ephraim Gerry has made an excellent start at this annual listing, he has been dependent on irregular contributions of information and not on anything systematic.

HOWE: If you will pardon me, this is not a complete list at all. It merely gives the name of the genus, the author and the date, no page reference at all.

MCKENZIE: I hope that the accompanying written contribution regularizes the names in my 1965 paper to Professor Howe's satisfaction.

Corrected nomenclature of species list in McKenzie 1965 pp. 448-53

Bairdia villosa	= *Bairdia* cf. *B. villosa* Brady, 1880
Bairdia cf. *B. angulata* (not *angulata* Brady 1870)	= *Bairdia* sp. 3
Bairdia sp. 1	= *Bairdia* sp. 1
Bairdia sp. 2	= *Bairdia* sp. 2
Bairdia milne-edwardsi var. *dentata* (not *dentata* Issler 1908)	= *Bairdia* sp. 4
Genus A sp. A	= *Australoecia* sp.
? *Algaiella setigera*	= ? *Aglaiella setigera* (Brady, 1880)
Phlyctenophora sp.	= *Phlyctenophora* sp.
Propontocypris sp.	= *Propontocypris* sp.

? *Loxocythere scaphoides*	= '*Cytheropteron*' cf. *C. scaphoides* Brady, 1880
Cytheridea exilis	= '*Cytheridea*' cf. *C. exilis* (Brady, 1880)
Cyprideis sp.	= *Cyprideis* sp.
? *Perissocytheridea* sp.	= *Loxocythere* cf. *L. hornibrooki* McKenzie, 1967
Parakrithella sp.	= *Parakrithella australis* McKenzie, 1967
'*Cythere*' sp.	= '*Cythere*' sp.
Cytheropteron wellingtoniense	= *Cytheropteron wellingtoniense* Brady, 1880
Hemicytherura sp.	= *Hemicytherura* sp.
Semicytherura sp.	= *Semicytherura* sp.
Hemicythere kerguelenensis	= *Hemicythere kerguelenensis* Brady, 1880
Ambostracon pumila	= '*Ambostracon*' *pumila* (Brady, 1866)
Hemicytheridea sp.	= '*Hemicytheridea*' *portjacksonensis* McKenzie, 1967
Callistocythere crenata (not *crenata* Brady, 1880)	= *Callistocythere purii* McKenzie, 1967
Callistocythere sp.	= *Callistocythere* sp.
Loxoconcha sp. 1	= *Loxoconcha australis* Brady, 1880
Loxoconcha avellana (not *avellana* Brady, 1866)	= *Loxoconcha* cf. *L. australis* Brady, 1880
Loxoconcha honoluluensis? (not *honoluluensis* Brady, 1880)	= indet. valve
Loxoconcha sp. 2	= *Loxoconcha* sp.
Trachyleberis melobesioides (juveniles)	= *Trachyleberis* aff. *T. melobesioides* (Brady, 1869)
Trachyleberis militaris	= *Ponticocythereis* cf. *P. militaris* (Brady, 1866)
Archicythereis sp.	= *Ponticocythereis* cf. *P. militaris* (Brady, 1866)
? *Henryhowella* sp. (not *Henryhowella*)	= Trachyleberidid sp. 1
? *Orionina* sp. (not *Orionina*)	= Trachyleberidid sp. 2
Quadracythere prava subsp. (not *Quadracythere*)	= *Jugosocythereis prava* (Baird, 1850) subsp.
Xestoleberis depressa (not *depressa* Sars, 1865)	= *Xestoleberis* sp. 1
Xestoleberis granulosa	= *Xestoleberis granulosa* Brady, 1880
Xestoleberis margaritea (not *margaritea* Brady, 1866)	= *Xestoleberis* sp. 1
Xestoleberis nana? (not *nana* Brady, 1880)	= *Xestoleberis* sp. 1
Xestoleberis variegata	= *Xestoleberis* aff. *X. variegata* Brady, 1880
? *Microxestoleberis* sp. (not *Microxestoleberis*)	= *Xestoleberis* sp. 2
Hemicytherid sp.	= Hemicytherid sp.

Notes: '*Cythere*' sp. resembles forms referred by Brady (1880, Pl. 14, Figs 9d, e) to *Cythere cancellata* Brady (1868, p. 62, Pl. 7, Figs 9-11) but evidently not that species when compared against type material (Hancock Museum, Newcastle-upon-Tyne collection).

'*Hemicytheridea*' *portjacksonensis* McKenzie, 1967, belongs in a new generic category foreshadowed by Teeter (1966).

'*Ambostracon*' *pumila* (Brady, 1866) may well belong in a new generic or subgeneric category (McKenzie 1967, p. 93).

REFERENCES

BRADY, G. S. 1868. Ostracoda. In *Les Fonds de la Mer*. **1**, 114 pp., Folin et Perier, Paris.

BRADY, G. S. 1880. Report on the Ostracoda dredged by *H.M.S. Challenger* during the years 1873-1876. *Report of the Scientific Results of the Voyage of H.M.S. Challenger, 1873-76. Zoology*, **1**, Pt. 3, 1-184, Pls 1-44.

McKenzie, K. G. 1965. The ecologic associations of an ostracode fauna from Oyster Harbour, a marginal marine environment near Albany, Western Australia. *Pubbl. Staz. zool. Napoli*, [1964], **33** (suppl.), 421-61, 18 Figs, 2 tables.

McKenzie, K. G. 1967. Recent Ostracoda from Port Phillip Bay, Victoria. *Proc. R. Soc. Vict.*, **80**, (1), 61-106, Pls 11-13, 9 Figs.

Teeter, J. W. 1966. *The distribution of Recent marine Ostracoda from British Honduras*, Ph.D. thesis, Rice University, Houston, Texas, 212 pp., 19 Pls.

THE HISTORY OF *BERNIX TATEI* JONES 1884

J. E. ROBINSON
University College, London, England

SUMMARY

Since its establishment in 1884, the genus *Bernix* Jones, has acquired a unique character in the literature of Palaeozoic ostracods, and has been interpreted as a Palaeozoic myodocopid on grounds of the ornament depicted in the original figures. In view of the shortcomings of the available specimens deposited in the British Museum, topotype material has been collected and studied, resulting in some clarification as to the nature and affinities of *Bernix*.

The name of George Tate figures frequently in early accounts of the geology of the North of England, particularly in the recording of fossils from the Carboniferous of Northumberland. His collecting included several microfossils, many of which were described by T. R. Jones, and H. B. Brady.

In 1857 while preparing an account of the Geology of the Roman Wall, he collected a carbonaceous shale containing ostracods from Brunton in the North Tyne Valley, later submitting this fauna to T. R. Jones for description. The specimens were duly described in 1864 (*Proc. Berwickshire Nat. Club*), when the similarity in shape to Canadian ostracods being studied at the same time by Jones and Holl, caused Jones to term his new species *Beyrichia tatei*. Perhaps the most significant feature of the description is the following sentence: 'In one specimen the pyritous cast seems to show that radiating, sinuous tapering furrows existed on the inside of the shell as in *Leperditia*.'

In the following twenty-five years, no further records of the genus were made in the considerable Carboniferous faunas described by Jones and Kirkby, but in 1882 casts from the Devonian 'Lower Culm' of Torquay were described as being reminiscent of the Brunton specimens, which were in the same paper renamed *Primitia tatei*, the shape and form being judged more primitiid than beyrichiid. Even this revision, however, did not satisfy, and Jones proceeded to restudy the materials of Tate's collection supplemented by topotype material collected by G. A. Lebour, and redescribed the form as *Bernix tatei* (1884, *Proc. Berwick Nat. Club*). In this revised account, Jones entered a full and reasoned assessment of the affinities of *Bernix*, commenting on some features akin to *Leperditia*, others to *Primitia*, and still others to *Carbonia* (sic). Points of particular interest, however, refer to sulcation and surface ornament, and are quoted in full: 'The surface is smooth to the eye, but, when magnified, shows a radiating reticulation, which varies in intensity as different individuals; it is also impressed with a shallow furrow in the middle third of the dorsal region, reaching to the centre of the valve; and it bears a small submedian tubercle in front of the sulcus accompanied by some small irregular depressions of the surface ... again reference is made to the

14

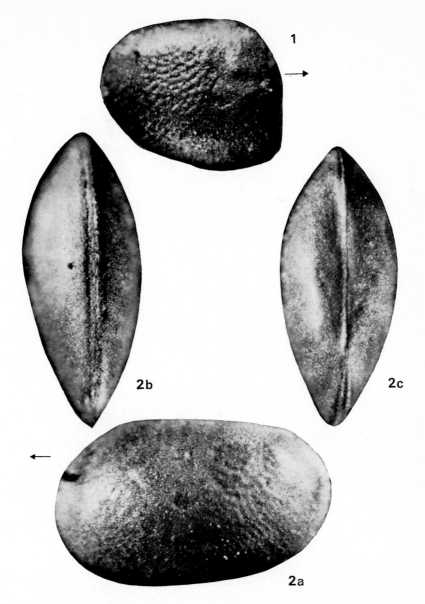

PLATE I.
Fig. 1. Right aspect of immature moult stage showing typical *Bernix* ornamentation (B.M.N.H.)
Io700. × 65 approx.
Fig. 2. (a) Left aspect of adult steinkern (Io701). × 60; (b) Ventral aspect of 2a, showing grooving
of the steinkern ventral edge representing a thickened valve margin (Io701); (c) dorsal aspect, showing
median channelling of the hingeline (Io701).

PLATE II.

Fig. 1. Right aspect of the smallest moult found (Io702). × 65.

Fig. 2. Left aspect of immature moult: strength of sulcus is exaggerated in the illumination of this specimen (Io702). × 65.

Fig. 3. Left aspect of immature moult (Io702). × 65.

Fig. 4. Dorsal aspect of specimen retaining some original shell at the hinge line (Io704). × 65.

Fig. 5. Right aspect of large adult steinkern showing slight suggestion of S_1 and S_2 (Io703). × 60.

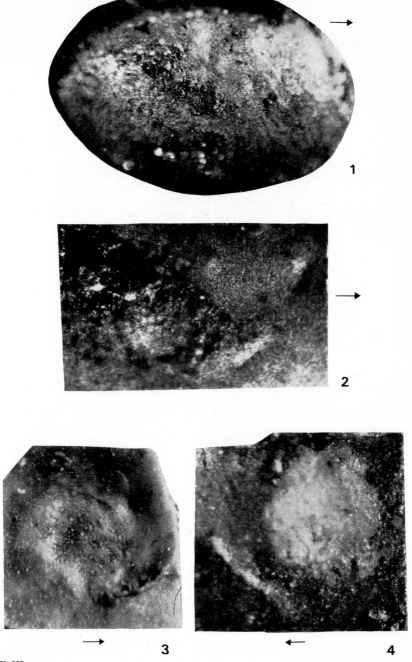

PLATE III.

Fig. 1. Right aspect of large adult, showing muscle scar position. Out of focus in this figure, is the hinge line which retains shell material conveying a paraparchitid outline (Io705). × 65.

Fig. 2. Enlargement of muscle scar of Fig. 1.

Fig. 3. Muscle scar pattern of *Paraparchites* sp, Tournaisian Lower Limestone Shales, Forest of Dean. × 100.

Fig. 4. Muscle scar pattern of *Paraparchites* cf. *P. inornatus* (McCoy), Visean Scremerston Coal Group, Warksburn, Northumberland.

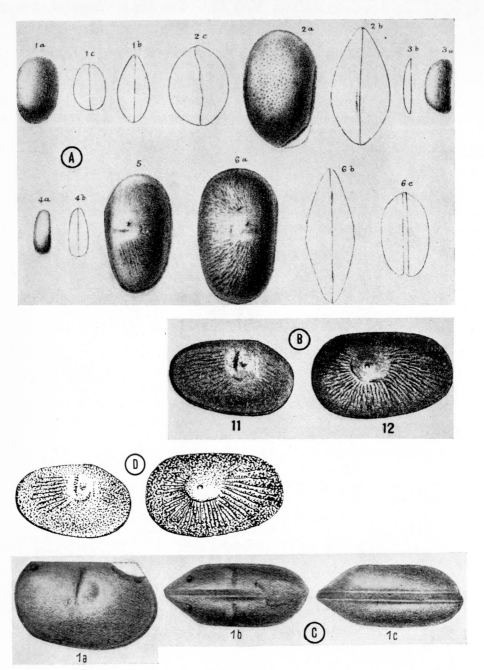

PLATE IV. Published illustrations of 'Bernix'.

A. Plate II of Jones 1884 'Notes on the Late Mr. George Tate's Specimens of Lower Carboniferous Entomostraca from Berwickshire and Northumberland' in *The Berwickshire Naturalists Club*. Figs 5 and 6, *Bernix tatei*, Fig. 6 being the specimen from Walwick Grange, near Brunton, considered to be the type in the B.M.N.H. collection. The specimen, I 2550, is decomposed beyond use.

Fig. 2 of this same Plate II, '*Carbonia fabulina* J & K *var*', according to Jones, and from 'the Mountain Limestone near Bellingham', is a form which fulfils the present author's conception of *Bernix* with its valves restored to the steinkern. Fig. 2. (and *Bernix*) would be recognized as a paraparchitid.

B. Figs 11 and 12 of the Bibliographic Index of Bassler and Kellett, 1934.

C. Figs 1a ,b and c of Tafel III, Kummerow, 1939. *Bernix venulosa*.

D. Fig. 1030 Pokorny, redrawn from Bassler and Kellett 1934.

radiate patterns upon the valve flanks . . . the radiate ornament on the valves from Northumberland reminds us of similar markings sometimes visible on the outside of *Leperditia* (*L. artica*), *and always on the inside or cast*. But the broad, definite, central muscle spot, and the escutcheoned eye spot of *Leperditia* are both wanting.'

Jones was at pains to correct an impression created by his woodcut Fig. 3 of the 1864 paper, in which he considered that his dorsal line was too straight, '. . . giving the valve too much of a Leperditioid appearance', yet such is the outline of Professor Lebours' specimen figured as Fig. 5 of Plate II, 1884.

Bernix was next figured in 1934 by Bassler and Kellett in the *Bibliographic Index of Paleozoic Ostracoda*, the drawings numbered 11 and 12 of their figure 12, being Figs 5 and 6A of Jones, although in the process of reproduction the faint radial surface ornament of the originals has become appreciably coarsened. In the text, *Bernix* is considered a genus of the family Kloedenellidae, with the diagnosis, 'Carapace equivalved, moderately convex with straight hinge line. Surface with a shallow furrow from hinge to center of valve and with radiating reticulations.'

In 1939, Kummerow described *Bernix venulosa*, from the Lower Tournaisian of Insemont in Belgium, and Ratingen in the Rhineland; the generic diagnosis (p. 25) was precisely that given by Bassler and Kellett, and again the taxonomic allocation was to the family Kloedenellidae. This species is distinct from *B. tatei* in its weaker patterns of radial ornamentation and its more rectangular outline. A third point might be added that from the dimensions given by Kummerow (l 2·71 mm, h 1·6 mm, t 1·18 mm), *B. venulosa* is double the size of *B. tatei*, for which the only measurement given by Jones is that it was 'about $\frac{1}{20}$ of an inch long' (1·25 mm) (1864, p. 88).

In 1941, Triebel made reference to *Bernix* in the well-known paper, 'Morphologie und Ökologie der fossilen ostrakoden. . .'. Commenting upon the branching pattern of 'canals' within the shell of certain recent ostracods, and similar patterns in *Leperditia* from the Lower Palaeozoic, he seems to imply first that the radiate patterning of *Bernix* is homologous to the 'blutekanale' of *Cypridina*, and second that possessing such a vascular system, *Bernix* should properly belong to a primitive suborder of ostracods. Following from this last point, rather than place the genus within the Kloedenellidae, the opinion of previous authors, Triebel preferred to regard it as a member of the Myodocopa, but lacking the well-defined rostral incision of that group (Triebel, 1941, p. 331).

It was in this spirit it seems, that Pokorný considered the genus; for in his *Grundzüge der Zoologischen Mikropaläontologie, Band II*, *Bernix* was referred to as of uncertain relationship, but placed for convenience with the Thaumatocypridinae, and within the Myodocopida (Pokorný 1958, p. 309). The accompanying Fig. 1030 was acknowledged to Bassler and Kellett, but again the process of reproduction seems to have caused the radial ornamentation to be over-emphasized.

The same year, Mertens in a tabular statement on ostracod classification, considered *Bernix* as *genus incerta familiae*, but attached the genus to the margin of his Table 7, devoted to the Beyrichiacea (1958).

From a review of references in the literature to the genus *Bernix*, it is evident that some remarks stem from the description and Plate II of the 1884 paper, but probably most are based upon the diagnosis of Bassler and Kellett and the modified figures which they reproduced.

In each of these instances of citation of *Bernix*, it seems reasonable to assume that it was the original illustration of 1884, or more probably that of Bassler and Kellett, 1933, which was the standard of reference. If deposited type material had been sought in the British Museum, it is almost certain that it would have been of little help, because of the instability of the pyritised steinkerns when stored untreated in collections. In the re-ordered collections of the Museum, there are several pieces of carbonaceous shale labelled as *Bernix tatei*, including specimens purchased from T. R. Jones, 27th June 1891 (I 2550), and several specimens of rock, part of the Birley Bequest (I 8052–63), but all surfaces are decomposed to such an extent as to make the fossils unrecognizable. In some cases, the decomposition following oxidization of the pyrite, seems to have completely penetrated the shale flakes. From observation of collected material, moist shale can undergo such change in a matter of months, and the museum specimens were probably so reduced by the beginning of this century by the same token.

Fortunately, the locality and the mode of occurrence were fully described by Jones, and it is possible to locate the abandoned coal tips at Brunton Bank, and obtain topotype material. The most profitable source is a tip some 60 yards north of the abandoned incline to Cocklaw Quarry, the horizon being presumably only a short distance stratigraphically below the base of the Great Limestone worked in that quarry (Horizon: Lower Namurian, Lower *Eumorphoceras* subzone).

In the past the rock containing *Bernix*, has been termed 'black band ironstone', 'canneloid shale', 'carbonaceous shale', and 'a real oil shale'.

The tip material is a fissile carbonaceous shale, interbedded in which are lenses of more cannel-like character, poorly-bedded and breaking with a conchoidal fracture. (I am indebted to Dr D. G. Murchison of King's College, Newcastle, for the opinion that these beds are of detrital origin, and that the rock term carbargillite would be appropriate.) Within the rock there is an abundance of pyrite, both as discrete pellets, and as replacements of fossils. Of this fauna, apart from *Bernix*, other ostracods include *Carbonita fabulina*, *Carbonita pungens*, and *Darwinula berniciana*, all in the form of internal moulds. These pyrite moulds of the inner surface of the original shells retain many details such as traces of rudimentary duplicatures, outlines of muscle-fields, and in the case of *Bernix*, the radiate markings which have received attention in the literature. The fact that *Bernix* was an internal mould or steinkern was appreciated by Jones, as may be evident in the quotations given above.

Admittedly too little is known of the replacement processes by which pyrites substitutes for shell material, but invariably the process seems to relate to the principal surfaces of the shell, external or internal. Pyrite moulds then reflect the original form reasonably closely. In the process, however, it is clear that minor irregularities either of ornament, or crystal surfaces may become accentuated beyond their normal appearance. In the case of *Bernix*, it is the inner face of the valve surface which is so preserved.

How far '*Bernix*' can be related to known ostracod genera remains to be discussed.

In lateral outline, the better preserved specimens prove to have a straight hinge line, the ventral margin being evenly convex. As internal moulds it is conceivable that actual external shell outline would be more angular, and this is indeed the case in specimens in shale matrix, the pyrite core being fringed dorsally by a moat-like cavity to give an outline similar to Jones' Fig. 5 of 1884. Some specimens freed from the shale matrix show corrosion of the cardinal angles, and conform to the more ovate outline of Jones' Fig. 6 of 1884. As mentioned above, Jones considered his first drawings (Fig. 3, 1884) gave the genus too 'Leperditioid' an outline, and preferred to stress the ovate outline in a later description; topotype material however confirms the former outline to be more accurate. In recent descriptive terminology the outline could be called preplete.

All the larger moulds are lightly sulcate, the print of S_2 being weak at the dorsum, then constricted, finally expanding to a circular print in front of mid-length and at mid-height. In front of the weak lobe bounding this sulcus, a faint parallel groove represents S_1. In the smaller moult stages, S_2 is more prominent, with a keyhole outline, which weakens with growth. In all cases it must be stressed, this 'sulcation' is weak and possibly too weak to justify description in these terms.

Shaver, commenting on *Bernix*, mentioned affinities with *Hypotetragona* (1961, Q 412), but both outline and sulcation would seem to rule out reference to this genus, or to *Geisina*, as it is understood in Britain, these being more rectangular, and their sulcate character being still further emphasized in internal moulds. The genus *Sansabella* likewise, although weakly sulcate externally, may be markedly sulcate internally. Particularly, *Bernix* differs from all three genera mentioned above, in respect of sex-dimorphism, for while *Hypotetragona*, *Geisina* and *Sansabella* show marked kloedenellid dimorphism, in all the specimens of *Bernix* studied, no posterior inflation is evident. Invariably the moulds show a mid-length position of greatest thickness, but in the larger adult forms there is a suggestion in some of an increased convexity below mid-height. This is a condition recognized in *Paraparchites*, otherwise often regarded as a parthenogenetic genus (Scott, 1961 Treatise, Q 86) without dimorphism.

This comparison is strengthened by features of muscle-scar pattern, both in general plan, and in certain details. The muscle scar in *Bernix* is a slightly depressed area, circular in outline, sometimes patterned by clusters of small pits. Below this area, and extending obliquely upwards toward the higher anterior margin, one or two linear scars may occur. All these features occur as incisions upon the mould surface, and naturally represent positive, raised features of the inner shell surface. The pattern is strongly reminiscent of the central muscle field of *Paraparchites*, together with its mandibular scars associated with the mouthparts. In front of the main sulcus, a small round tubercle occurs in some specimens close to the narrowed part of the sulcus. Should this be an eye tubercle, this would confirm the orientation suggested by the mandibular scar mentioned.

Generally it is difficult to determine the nature of valve hingement and overlap from the moulds, except to note that the even continuity of the grooves about the margin suggests a thickened shell margin. In cross-section, the moulds are almost symmetrical showing the position of greatest width about mid-height, rather than below as is the case in kloedenellids. A variety of kloedenellid carapaces have been etched, and their valves prised off to reveal the mould-form; such preparations were only partly successful, those most like *Bernix* were *Paraparchites* specimens.

A final support to this conclusion might be called palaeoecological. *Bernix* is by far the most abundant ostracod form in the fauna, specimens of *Carbonita* being the only other ostracods and totalling, at the most, 15% of the fauna. Experience of Carboniferous faunas such as that of the Viséan Scremerston Coal Group of Northumberland, shows that assemblages dominated by a single species of a single genus are quite common in the carbonaceous facies, and then most often the dominance is of *Paraparchites* or *Sansabella*. Compared with the greater generic and specific variety to be found in nearby calcareous shales of undoubted marine character, such 'monotypic' faunas could best represent the restricted fauna of brackish-water environment as documented diagrammatically by Remane in 1934, or more recently by Wagner. Such a palaeoecological conclusion based upon the microfauna, would be appropriate for the formation of canneloid shale, and for the Brunton carbargillite.

Undoubtedly, the one single character which differentiated *Bernix* from other ostracods to Jones, and probably prevented him regarding it as *Paraparchites* [*Leperditia*], was the surface ornament of delicate raised lines, and it remains to be asked what the pattern might represent in shell morphology, and whether it is in fact unique to *Bernix*.

Striking patterns of radial anastomosing ridges are well known from illustrations of *Leperditia baltica* (Hisinger) var. *arctica* Salter 1852, from *Leperditia gigantea* Roemer 1858, from *Leperditia* (*Isochilina?*) *formosa* and *Leperditia solitaria* Barrande 1872. For the latter, Barrande suggested that these were traces 'd'origine vasculaire'. A similar concrete suggestion, that in the markings of *Bernix* were a trace of a vascular system, was made by Triebel, with the further implication that this was cause to group the genus together with heart-bearing Myodocopida without rostral incision. Apart from the species of *Leperditia* already mentioned, however, markings of similar character have been illustrated in specimens of a beyrichiid, *Mastigobolbina incipiens* Ulrich and Bassler (1923, Plate LIII, Fig. 12), and have been observed by the author upon a mould of *Glyptopleura plicata* (Jones and Kirkby). Specimens of *Bernix venulosa* Kummerow, have not been seen, but from the illustrations given, it seems probable that the type was an internal mould, possibly of *Beyrichiopsis* or a kloedenellid. At present, too few internal moulds have been figured in the literature to extend these comments, but '*Bernix*'-patterns appear to cut across several classificatory groups, suggesting a cause more universal than the possession of a well developed vascular system. Such a cause may be a shell structure in living ostracods described by Müller in his classic work on the fauna of the Gulf of Naples (1894). In a section concerned with the 'chitinous foundation of the calcified layer', he described a variety of patterns of solid chitin rods (*chitinbalken*) forming a network system at the base of the shell layer (1894, pp. 96-97). In plan, Fig. 32 of Plate 36 shows a polygonal mosaic in *Cythereis prava*, the chitin rods grouped about the pore canals, the whole network focused upon the central muscle scar. As it is situated upon the innermost surface of the calcified shell, the 'network' could be expected to form raised features upon internal mould surfaces. It is for this reason that this shell character seems to answer the problem posed in '*Bernix*'.

SUMMARY

It is thought that *Bernix* Jones 1884, is in all probability the steinkern of a paraparchitid ostracod, although not necessarily a synonym of *Paraparchites* Ulrich and Bassler 1906. As a genus, *Paraparchites* has suffered a very broad interpretation since 1906, and many species so named depart far from the character of *Paraparchites humerosus*, the genotype, recently re-described by Scott. Subject to future revision of British forms, *Bernix* as a steinkern could relate to members of the *okeni-*, or *inornatus-*group of ostracods currently referred to the genus *Paraparchites*.

The interesting surface ornamentation of *Bernix*, the cause of its separation, and much speculation, may represent a fundamental shell structure, and not a vascular system with the taxonomic implications sometimes implied. Details of the muscle scar pattern of *Paraparchites* (s.1.), shared by *Bernix*, are illustrated.

These photographs (Plates I-IV) were taken using a Zeiss Photomicroscope, using several light sources from low incident positions in order to emphasize the low relief ornamentation of the pyritized steinkerns. Among other results this tends to suggest stronger sulcation than is evident in normal viewing of the specimens. The numbers are from the Register of the Collections of the British Museum (Natural History).

Deposition of Materials

Three specimens in the mounted slide collection of the British Museum (Natural History), Slide I 1744, show *Bernix* characters; those illustrated have been added to the collection under number Io700–4 (those figured here), together with additional specimens, numbered Io705–8.

Slide Io709 contains valve-internal surface fragments of *Paraparchites* cf. *inornatus* (McCoy) from the Scremerston Coal Group of Kingwater, Cumberland, which show typical, paraparchitid muscle scars for comparison with '*Bernix*'.

Slide Io710 contains available steinkerns of Carboniferous genera, for comparison with the '*Bernix*' form.

Specimens of the lithologies of the carbonaceous shale are available both in its decomposed and weathered form (the type specimen Io2550, and the Birley Collection), and in its fresh, little altered state.

REFERENCES

BASSLER, R. S. AND KELLETT, B. 1934. Bibliographic Index of Paleozoic Ostracoda. *Spec. Pap. geol. Soc. Am.*, **1**, 500 pp. (p. 30, Figs 11-12).
JONES, T. R. in TATE, G. 1864. Description of Entomostraca from the Mountain Limestone of Berwickshire and Northumberland. *Hist. Berwicksh. Nat. Club*, **5**, 83-89.
JONES, T. R. 1882. Notes on some Palaeozoic bivalved Entomostraca, No. 15. *Ann. Mag. nat. Hist.* (5) **10**, 358-60, Figs 1a, b.
JONES, T. R. 1884. Notes on the late Mr George Tate's specimens of Lower Carboniferous Entomostraca from Berwickshire and Northumberland. *Hist. Berwicksh. Nat. Club*, **10**, 312-26, Pl. 11, Figs. 1-6.
JONES, T. R., AND KIRKBY, J. W. 1886. Notes on the distribution of the Ostracoda of the Carboniferous formations of the British Isles. *Q. J. geol. Soc. Lond.*, **42**, 496-514 (pp. 496, 513 table).
KUMMEROW, E. H. 1939. Die Ostrakoden und Phyllopoden des deutschen Unterkarbons. *Abh. preuss. geol. Landesanst.*, Heft, **194**, 1-107, 7 Pls (p. 26, Pl. 3 Figs 1a-c).
MERTENS, E. 1958. Zur Kenntnis der Ordnung Ostracoda (Crustacea). *Geol. Jb.*, **75**, 311-18, Tables 1-16.
MÜLLER, G. W. 1894. Die Ostracoden des Golfes von Neapel und der angrenzenden Meeres-Abschnitte. *Fauna Flora Golf. Neapel.* Mongr. **21**, i-viii, 1-404, Pls 1-40. (pp. 96-98, Pl. 36, Fig. 32).
POKORNÝ, V. 1958. *Grundzüge der Zoologischen Mikropaläontologie.* **2**, Deutsch. Verlag der Wiss., Berlin, 453 pp., Figs 550-1077 (p. 309).
REMANE, A. 1934. Die Brackwasserfauna (mit besonderer Berucksichtigung der Ostsee). *Verh. dt. zool. Ges.*, **36**, 34-74.
REMANE, A., AND SCHLIEPER, C. 1958. *Die Biologie des Brackwassers. Die Binnengewasser.*, **22** (p. 21).
SCOTT, H. W. 1959. Type species of *Paraparchites* Ulrich and Bassler. *J. Paleont.*, **33**, 670-4, Pl. 87.
SHAVER, R. H. 1961. Nomina Dubia. Q.412-Q.414. In *Treatise on Invertebrate Paleontology*, R. C. Moore, ed., Pt. Q., Arthropoda 3, Crustacea: Ostracoda. 442 pp., 334 Figs. University of Kansas Press, Lawrence, Kansas.
TRIEBEL, E. 1941. Morphologie und Ökologie der fossilen Ostrakoden mit beschriebung einiger neuer Gattungen und Arten. *Senckenbergiana*, **23**, 294-400, 15 Pls (pp. 331-2).
ULRICH, E. O., AND BASSLER, R. S. 1906. New American Paleozoic Ostracoda. Notes and descriptions of Upper Carboniferous genera and species. *Proc. U.S. natn. Mus.*, **30**, 149-64, Pl. 11.
ULRICH, E. O., AND BASSLER, R. S. 1923. Silurian Volume. Systematic Paleontology, Ostracoda. *Md. geol. Surv. stratigr. Mem.*, 500-704, Pls 36-65 (pp. 632-33, Pl. 53, Fig. 12).
WAGNER, C. W. 1957. *Sur les Ostracodes du Quaternaire Récent des Pays-Bas et leur utilization dans l'étude géologique des dépôts holocènes.* Mouton and Co., The Hague, 260 pp. (p. 108, Fig. V).

DISCUSSION

BATE: Do you consider that this species and genus is marine, or fresh-water or brackish in habit?

ROBINSON: Following from my last comment about the palaeoecology, I think that my opinion would be that it was brackish. The additional data which I can give in the paper include the fact that *Bernix* is associated with lamellibranchs in this carbonaceous shale, fossils which of course themselves are controversial, but forms which would generally be regarded as non-marine. Other than these, nothing more than a few fish-scales occur. The shale fits into a cyclothemic sequence of sedimentation, with very close below, a coal seam, while above progressively one gets shales which lead one within five or ten feet into a calcareous shale beneath the Great Limestone, thickest of our Yoredale limestones. These shales yield a great variety of ostracods and foramini-fera which would of course in any estimation represent marine conditions. In contrast to this last, the carbonaceous canneloid shale must represent something very different, and while one wouldn't go as far as to say fresh-water, certainly a low salinity would seem a reasonable judgement. Just one final point, I am very anxious that specimens of *Bernix* should be as widely distributed as possible, and I would be very glad if anyone who thinks they know of any 'niche' into which *Bernix* could be fitted for reference purposes, would ask for topotype specimens.

ADDENDUM: In 1886, Jones and Kirkby in another tucked-away reference to *Bernix*, said of its occurrence: 'In some estuarine beds of the Mountain Limestone series of North-umberland.'

REFERENCE

JONES, T. R. AND KIRKBY, J. W. 1886. A list of the genera and species of bivalved Entomostraca found in the Carboniferous Formations of Great Britain and Ireland, with notes on the genera, and their distribution. *Proc. Geol. Ass.*, 9, 505.

NUMERICAL TAXONOMY OF SELECTED RECENT BRITISH OSTRACODA

ROGER L. KAESLER
University of Kansas, Lawrence, Kansas, U.S.A.

SUMMARY

As the only extant group of arthropods well represented in the fossil record, ostracods possess unique attributes for furthering the development of numerical taxonomy. Numerical taxonomic methods, in turn, can provide much information about ostracods masked by current, phylogenetic classifications.

A numerical taxonomic pilot study was made of fourteen specimens representing thirteen species and twelve genera of seven families of Recent podocopid ostracods from Britain. Both correlation coefficients and average distance coefficients were computed, several methods of cluster analysis were used, and phenograms were presented. Cophenetic values from the phenogram prepared from distance coefficients clustered by the unweighted pair-group method using arithmetic averages gave highest correlation with the original distance matrix, a value of 0·857. The phenogram contained considerably more structure than did a tree based on the *Treatise* classification of ostracods.

Fruitful areas of further investigation include: (*i*) tests of congruence of numerically derived classifications using characters from adults with those based on characters from larvae; (*ii*) tests of congruence of numerical classifications based on characters from carapaces only with those based on appendage morphology; (*iii*) numerical taxonomic studies of large genera (e.g., *Loxoconcha*) and entire families; (*iv*) construction of hierarchies based on characters from carapaces and appendages of adults and all instars; (*v*) factor analytic approach to continuously evolving lineages; (*vi*) automatic data collection by computer operated optical scanning devices; and (*vii*) R-type analyses to determine correlation among characters.

INTRODUCTION

Numerical taxonomy, from its inception in the late 1950s, has been the subject of much testing, primarily by its proponents, to determine the utility of numerical taxonomic methods in classification of organisms, both real and hypothetical. Lamentable are the facts that, (*i*) with a few notable exceptions (Rowell, 1967, p. 115), development of numerical taxonomy of Recent invertebrate animals has been left to entomologists; (*ii*) collections of fossil organisms have been largely neglected (but see Rowell, 1967); and (*iii*) the more vigorous opponents of the new method have bemoaned its shortcomings and pontificated against it but have not actually *done* numerical taxonomy and thereby tested it with the organisms on which they specialize.

Numerical taxonomy was conceived primarily by entomologists and microbiologists, both of whom deal with vast groups of organisms notoriously deficient in their fossil

21

records. One is not surprised, therefore, that further development of a purely phenetic taxonomic system is viewed with somewhat more urgency by entomologists than by many other invertebrate zoologists.

Other reasons further account for the interest of entomologists in phenetic classification. First, insects grow by ecdysis. Assuming that the specialist can distinguish adults from instars, he encounters few of the difficulties met by specialists in groups that grow by other means, e.g., by continuous growth, by chamber addition, or in colonies (Kaesler, 1967).

Second, and largely as a result of the mode of growth of insects, entomologists *generally* have less difficulty identifying homologues than do specialists in other groups. Noteworthy cases in point might be Porifera and, again, chamber-building organisms.

Finally, insects are complex, highly organized, often highly specialized organisms from which a great number of taxonomic characters are available. This is certainly not the case with many groups of organisms, e.g., *Orbulina*, subfossil pteropods, and fossil scaphopods.

But it is not my purpose here to consider at length the mutual advantages that obtain between entomology and numerical taxonomy. Instead, the purpose of this paper is to present results of a pilot numerical taxonomic study of some ostracod species. The preceding discussion was presented to emphasize the applicability of numerical taxonomy to the classification of arthropods. It is hoped that these methods, when applied to the study of ostracods, will provide the same objectivity, repeatability, and stability that they have for other groups of arthropods, notably the insects.

Furthermore, ostracod specialists have much to contribute to numerical taxonomy. As the only extant group of arthropods with an extensive fossil record, the ostracods occupy a position of singular importance for further development of numerical taxonomy. For example, at present little is known of the structure of a matrix of similarity coefficients among species in a continuously evolving sequence. If such a matrix is not hierarchically structured, as one intuitively suspects that it is not, the use of phenograms may be less appropriate than factor analytic techniques.

THE BASIS OF CLASSIFICATION

Numerical taxonomy is empirical and phenetic (Sokal and Sneath, 1963; Sokal and Camin, 1965). Evaluation and comparison of the philosophies of phenetic and phylogenetic taxonomy are the subject of an extensive literature. Neither time and space permit nor our purposes here require that I thoroughly review or add to that great volume of work. Examples of the discourse on the basis of classification include works by Simpson (1961), Sokal and Sneath (1963), Mayr (1965), and Sokal and Camin (1965); a symposium held by the Systematics Association (Heywood and McNeil, 1964); and summaries in the palaeontological literature by Rowell (1967) and Kaesler (1967). Nevertheless, before proceeding to the results of the study at hand, we should, perhaps, consider briefly phenetic and phylogenetic taxonomy and some of the advantages of basing classifications on phenetics.

Re-evaluation of the tenets of phylogenetic taxonomy has shown that fundamental concepts of classification have been overlooked. Two of the most basic of these, modified somewhat from Gilmour (1962), are discussed below.

1. A classification is appropriate only when developed for a specific purpose. This point has been made repeatedly and was emphasized by Gilmour (1937, 1940, 1951, 1961, 1966), Sokal and Sneath (1963), and Edwards and Cavalli-Sforza, who held

(1964, p. 68) 'that the most essential prerequisite of a classification is its purpose, clearly defined. . .'. The alternative to this point of view, labelled 'absolutist' by Gilmour (1966, p. 30), is that classification is an end in itself.

In spite of the emphasis placed on the need for purpose in classification, the point has not been universally accepted. Thus Mayr (1965, p. 76) believed that 'a classification of organisms that deliberately ignores their historical information content is prone to be misleading or at best inefficient and uneconomical'. But efficiency is not an intrinsic property; it must be gauged in terms of some purpose for which the classification was proposed. Thus a classification of organisms by biogeographic realm, such as the one attempted by Benson (1964) for the Pacific Ocean ostracod faunule, need take into account neither history of migrations nor evolution of endemic species. Edwards and Cavalli-Sforza (1964, pp. 67–68) expressed this important concept as follows:

> A classification created with no particular purpose in mind is likely to be logically indefensible and partially useless; for when it is put to a specific use, for which it was not, of course, designed, it will most probably be inefficient.

2. A classification cannot be expected to serve several diverse purposes well. As was pointed out by Gilmour (1962), this point follows from the first. It is, in fact, almost the converse of the first: a classification requires a specific purpose; conversely, a specific purpose *may* require a classification. The present taxonomic system, although not originally designed to serve all the purposes for which it is now used, is an example of a system that does not serve diverse purposes well (Sokal and Sneath, 1963, p. 6). It simultaneously classifies, names, conveys an estimate of resemblance, and shows phylogenetic relationship. But every taxonomist knows of many specific instances in which it has inadequately served one or more of these purposes.

Some confusion on this point has been perpetrated by insistance on an inherent distinction between general- and special-purpose classifications. For example, Sokal and Sneath (1963, p. 12) wrote:

> We could arrange living creatures in many ways, but we choose one way because we think it is best for some purpose. If the purpose is restricted, then the classification is a special classification, often called 'arbitrary'. Such a classification conveys less information than a general or 'natural' one.

Even Gilmour (1966) has urged that classification of organisms into taxa is general-purpose classification and has stressed that other, special-purpose classifications may be erected, usually based on fewer characters chosen for specific purposes.

In opposition to this view, Edwards and Cavalli-Sforza (1964, p. 71) have written:

> But if we are asked to classify . . . animals 'for general purposes' we must insist on knowing what purposes, though, of course, there is nothing wrong in admitting that the animals seem at first sight to be divisible into . . . homogeneous classes, and that this classification will no doubt serve the majority of purposes included under the title 'general'. *But* this fact derives from the particular data we are using, and cannot be used as a justification for a taxonomic procedure whose logical structure must be independent of the organisms being considered. The fact that different characters and different taxonomic methods lead to the same segregation of the animals into [groups] tells us something about [the groups], but nothing about the logical basis of taxonomy.

A part of the basis for this disagreement is semantic. Both Gilmour (1938, 1940, 1951, 1961) and Sokal and Sneath (1963) have equated the terms *general-purpose* and *natural* in reference to classification. This usage is consistent with the views of

C

nineteenth-century philosophers of classification, as was pointed out by Gilmour (1962). Natural (general-purpose or scientific) classification is distinguished from artificial (special-purpose or non-scientific) classification in that the former is based on degree of similarity of a large number of characters rather than on a small number or only one. Of course, the two kinds of classification merge, but they are distinct in their extreme forms.

The term *natural* is to be preferred to *general-purpose*. This is especially true in light of apparent agreement among phenetic and phylogenetic taxonomists that natural classifications have higher predictive value than artificial ones (Sneath, 1957; Sokal and Sneath, 1963; Mayr, 1965). Furthermore, Gilmour (1966) has shown that 'classification based on evolutionary history should be regarded as special-purpose classification constructed for studying one or more particular aspects of the phylogeny of the group concerned.'

The rise of phenetic taxonomy and the associated taxonomic revolution have, of course, been characterized by much more than a reconsideration of the bases of classification. In addition, the principles of phylogenetic taxonomy have been aired, particularly those related to the so-called 'New Systematics'.

Blackwelder (1964, pp. 18–19) has listed five 'acceptances' of evolutionary taxonomy that are 'relevant to taxonomy', none of which 'correctly represents taxonomic practice, either actual or potential'. These acceptances are:

> 1. that taxonomy in the past was exclusively 'morphological' and that the New Systematics differs in being 'biological';
> 2. that a so-called biological species concept is superior in taxonomic work and is now in wide use in zoology;
> 3. that species are different in nature from other taxa;
> 4. that there can be a direct basis of classification in phylogeny, or that the aim of classification is to reflect phylogeny, or that taxonomists must study the origin of the taxa which they distinguish and define; and
> 5. that only a phylogenetic classification is a natural one.

It is statements such as these, acceptable in theory but impossible to realize in practice, that numerical taxonomists seek to expose, overcome, and replace in order to give taxonomy an operational basis. Only with such a sound basis can taxonomy become a precise science rather than an art.

Walters (1962) has pointed out that most taxonomists still consider *natural* and *phylogenetic* classifications as equivalent terms (see, e.g., Simpson, 1961, pp. 54–57). Nevertheless, at least some leading evolutionary taxonomists have followed the lead of Gilmour and the pheneticists. Thus, Mayr (1965 p. 77), has acknowledged that the 'classification which has the highest predictive value is the best, the "most natural" '. It is a matter of belief, not of record, that 'that classification has the greatest predictive value which reflects . . . past evolutionary history' (Mayr, 1965, p. 77).

Rowell (1967, p. 113) has interpreted this apparent agreement as an indication of essential coincidence of the points of view of the two schools of thought on this matter. He has pointed out (p. 113) that 'the gulf separating the two positions lies in the choice of method whereby a classification with this attribute may be obtained'. This interpretation is correct, I believe; but one should not be misled into thinking that a unified approach to biological classification is imminent. Sokal, *et al.* (1965) have emphasized that views on many of the points of dispute between the two schools are matters of personal philosophy on classification. Differences of such a fundamental nature are not likely to be resolved quickly or easily.

The basic difference between the two schools of taxonomy, then, as Rowell (1967) has mentioned, is in the choice of method. The phylogenetic school bases its classification on phylogeny, which, as Sokal and Camin (1965) have emphasized, has three components: phenetics, cladistics, and chronistics. The phenetic or numerical taxonomic school is content with a classification based on phenetics alone.

In the case of organisms with a very poor fossil record, neither cladistic nor chronistic relationships can ever be known. Thus, phylogenetic relationships must be inferred from phenetics. The reason for rejecting phylogenetic classification in this case has been concisely expressed by Sokal, *et al.* (1965, p. 239):

> Numerical taxonomists *do not disparage* interpretation or speculation or the inductive-deductive method in science. They simply feel that the process of constructing classifications should be as free from such inferences as possible.

But what about groups of organisms with good fossil records? Can we determine phylogenies from their preserved remains? Neontologists seem to have as much faith in the ability of palaeontologists to determine exact phylogenies as palaeontologists have in neontologists' ability to use breeding criteria to determine biological species (see Kaesler, 1967). But Edwards and Cavalli-Sforza (1964) have shown that 'there is no substantial logical difference about estimating the course of evolution with or without fossil evidence'. It is true that study of the fossil record provides chronistic information, but chronistic information about what? There is no assurance that forms being arranged in an evolutionary sequence or phylogram by a palaeontologist are geneologically (i.e., cladistically) related. *All* such interpretation must be based on phenetic information.

In the final analysis, then, the only information available for classification is phenetic information. Even where breeding tests are performed, it is phenetic information that suggests to the taxonomist which organisms to use in his tests. And only phenetic information permits him to apply results of the tests to other organisms not so tested.

The leaders of the opposition to numerical taxonomy have failed to understand the numericists' point of view in at least two important respects, thus contributing to further misunderstanding. First, they have not realized that the purpose of phenetic taxonomy is not to show phylogeny but to show phenetic similarity. Second, since Darwin demonstrated the theoretical reason for phenetic similarity of organisms and groups of organisms, they have assumed that this reason must be made the theoretical basis of classification, ignoring the necessity of purpose in classification.

RESULTS OF NUMERICAL TAXONOMIC STUDY

Material and Characters

Material selected for study (Table 1) includes fourteen specimens of Recent podocopid ostracods from Britain collected at the localities shown in Fig. 1. These specimens represent thirteen species, twelve genera, and seven families according to the classification presented in the *Treatise on Invertebrate Paleontology* (Moore, 1961). Characters measured are listed in Table 2. All characters were taken from the right valve of the specimens with terminal elements of the hinge oriented horizontally, and most of them are geometrical characters or measurements of position of carapace features rather than being characters traditionally regarded as important in ostracod taxonomy. Material used in the study was deposited in the University of Kansas Museum of Invertebrate Paleontology as specimens 1,004,703 to 1,004,716.

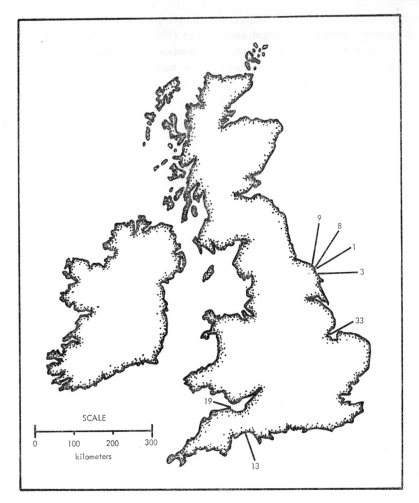

Fig. 1. Map of United Kingdom and Eire showing location of samples.

TABLE 1. Species, location of collection station, and specimen number in the University of Kansas Museum of Invertebrate Paleontology

Species	Location	Museum Numbers
Loxoconcha tamarindus	Ravenscar, Yorkshire	1,004,703
Loxoconcha tamarindus	Whitby, Yorkshire	1,004,704
Cythere lutea	Whitby, Yorkshire	1,004,705
Paradoxostoma variabile	Whitby, Yorkshire	1,004,706
Cytheromorpha robertsoni	Whitby, Yorkshire	1,004,707
Hirschmannia viridis	Robin Hood's Bay, Yorkshire	1,004,708
Hemicytherura cellulosa	Whitby, Yorkshire	1,004,709
Cytherura sella	Scarborough, Yorkshire	1,004,710
Heterocythereis albomaculata	Ravenscar, Yorkshire	1,004,711
Cytheropteron nodosum	Lyme Regis, Dorset	1,004,712
Xestoleberis aurantia	Ravenscar, Yorkshire	1,004,713
Aurila convexa	Lyme Regis, Dorset	1,004,714
Leptocythere castanea	Minehead, Devonshire	1,004,715
Leptocythere pellucida	The Wash, Lincolnshire	1,004,716

TABLE 2. Taxonomic characters used in the study, all from right valve

---

Length of right valve
Height of right valve
Width of right valve
Distance selvage to anterior extremity
Width anterior vestibule at anterior extremity
Distance anterior edge of eyespot to anterior extremity of carapace
Width of eyespot
Distance anterior of antennal muscle scar pattern (msp) to anterior of eyespot
Distance anterior of antennal msp to posterior of adductor msp
Width of adductor msp
Distance posterior of adductor msp to anterior mandibular msp
Width of mandibular msp
Width of antennal msp
Antennal msp: 0 = circular or subcircular
               1 = heart-shaped or lobed
               2 = multiple
Width posterior vestibule at posterior extremity of carapace
Width posterior duplicature at posterior extremity of carapace
Width anterior duplicature at anterior extremity of carapace
Distance selvage to posterior extremity
Depth of ventral fold
Distance venter to ventral selvage in ventral fold
Width of ventral duplicature below antennal msp
Distance anterior extremity of carapace to anterior hinge element
Length of hinge
Distance dorsum to ventral edge of terminal hinge elements
Distance dorsal edge of eyespot to dorsal edge of antennal msp
Distance dorsal edge of eyespot to dorsal edge of adductor msp
Height of antennal msp
Height of adductor msp
Distance dorsal edge of eyespot to dorsal edge of mandibular msp
Height of mandibular msp
Distance dorsal edge of eyespot to venter
Position of posterior extremity: 1 = in dorsal one third of height
                                  2 = in middle one third of height
                                  3 = in ventral one third of height
Caudal process: 1 = posterior evenly rounded
                2 = posterior slightly caudate
                3 = posterior highly caudate
Lateral distance anterior hinge element to anterior edge of eyespot
Height of eyespot
Surface ornament: 0 = smooth
                  1 = pitted
                  2 = coarsely pitted
                  3 = reticulate
                  4 = ribbed, highly ornamented
Number of radial pore canals anterior to antennal msp
Number of radial pore canals posterior to antennal msp
Number of normal pore canals
Vertical distance line of terminal hinge elements to anterior
  extremity of carapace
Vertical distance line of terminal hinge elements to posterior
  extremity of carapace
Hinge type: 1 = adont
            2 = prionodont
            3 = lophodont
            4 = merodont
            5 = entomodont, lobodont, or amphidont
            6 = schizodont or gongylodont
If hinge merodont: 0 = median element smooth
                   1 = median element crenulate

TABLE 2 (*cont.*)

If hinge merodont: 0 = median element depressed
                    1 = median element elevated
If hinge is type 5 above: 1 = entomodont
                           2 = lobodont
                           3 = paramphidont
                           4 = hemiamphidont
                           5 = holamphidont
If hinge is type 6 above: 0 = schizodont
                           1 = gongylodont
Hinge robustness: 0 = weak
                   1 = moderate
                   2 = strong, highly robust
Radial pore canals: 0 = simple
                     1 = curved
                     2 = branched, curved
Hinge shape: 0 = straight
              1 = simple arch
              2 = complex arch with horizontal terminal elements
Eye tubercle development: 0 = faint, weakly developed
                           1 = moderately developed
                           2 = distinct, strongly developed

Conventional Classification

Cain (1962, p. 226) believed that a phylogenetic classification 'will demonstrate for us the actual course of evolution'. He is not alone in this belief; this point of view has been widely adhered to by systematists in spite of the low content of phylogenetic information in many classifications. Simpson (1961), however, has shown that, *given a phylogram* (a most unlikely circumstance, to say the least), many different classifications can be made from it. Some of these will be largely vertical, others largely horizontal, and most a mixture of the two modes. But all of them may be consistent with phylogeny. Simpson (1961, p. 113) defined a consistent evolutionary classification as 'one whose implications, drawn according to stated criteria of such a classification, do not contradict the classifer's views as to the phylogeny of the group'.

The conventional classification of the ostracods used in this study (Fig. 2), taken from the *Treatise on Invertebrate Paleontology* (Moore, 1961), is presumably intended to be based on, or at least consistent with, phylogeny. Study of this figure, however, shows that the classification gives little information regarding the phylogenetic relationships of the organisms under study, thus bearing out Simpson's statement.

Perhaps more information would be conveyed if a more complete classification were presented. We cannot, however, assume that the present classification, or any classification, is ever a complete one. Therefore, it is likely that little important bias is introduced by examining only a segment of the classification of the Cytheracea.

It is also likely that inclusion of chronistic information would aid in interpreting phylogeny from the classification given in Fig. 2. But chronistics, although an important element of phylogeny, are not a part of the consistent classification. As is apparent, then, one must have a phylogeny to obtain a phylogenetic classification and not the reverse.

Fig. 2 shows a marked paucity of detailed structure. By the very nature of the diagram, all families branch from the superfamily stem at the same level, as does each subsequent category from its stem. Thus one has, for example, no indication whether the Leptocytheridae are more similar to the Xestoleberididae than are the Hemicytheridae. The three families are indicated merely as subdivisions of the superfamily

Cytheracea. Thus instead of demonstrating the course of evolution, a classification may give a highly distorted impression of that course.

By contrast, numerical taxonomic phenograms, such as the ones presented below, are replete with useful phenetic information at all levels of similarity.

Fig. 2. Graph of conventional classification of species included in this study from the *Treatise on Invertebrate Paleontology* (Moore, ed., 1961).

Methods, Assumptions and Results

Of the three types of coefficients commonly employed in numerical taxonomy, coefficients of association, correlation, and distance (Sokal, 1961), only the last two were used in this study. Coefficients of association are usually computed from two-state characters. Because only five of the fifty characters used in this study were of this kind, coefficients of association were not appropriate without extensive recoding of data.

Coefficients of correlation and distance have been discussed and compared by Sokal (1961) and Rohlf and Sokal (1965).

The only coefficient of correlation that has been used in numerical taxonomy is the Pearson product-moment correlation coefficient, the equation for which may be found in nearly any statistical textbook. Rohlf and Sokal (1965) have found that the correlation coefficient is a measure of similarity of shape of the organisms being studied rather than of their size. Eades (1965) rejected the use of the correlation coefficient as a measure of taxonomic resemblance. He based this conclusion on failure of the correlation coefficient to indicate close similarity among two operational taxonomic units (OTU's) that he regarded as obviously similar. However, because he gave no criterion for determining resemblance independently of any quantitative measure (except for comparison with distance coefficients), his argument loses much of its force. Furthermore, it is likely that shape is of more interest in taxonomy than size, so use of the correlation coefficient can easily be justified.

The distance coefficient, on the other hand, is a better measure of similarity in size than shape. This is particularly true, as Rohlf and Sokal (1965, p. 21) pointed out, 'if most of the characters used in a study are measurements of various parts of an organism, and if the OTU's differ much in over-all size (measured by the distance of an OTU from the origin). . .'. Thirty-four of the fifty characters used in this study are measurements of parts of organisms. With this mixture of types of characters and the fact that the species being studied do not vary greatly in size, little can be said about the relative values of correlation and distance coefficients in this study.

Data were standardized by rows (characters) before computation of correlation and distance coefficients. This procedure, which transforms measurements of each character so that it has a mean of zero and a standard deviation of one, is an important one. As Rohlf and Sokal (1965, p. 5) said: 'Standardization seems desirable since the coding of the various character states is arbitrary and consequently the number of states and the units of measurement vary from character to character.'

Standardization of characters before computation of correlation coefficients has the effect of permitting negative correlations, which are not likely to occur with un-standardized data. A more complete discussion of the effect of standardization on correlation coefficients was presented by Rohlf and Sokal (1965). Standardization of characters before computation of distance coefficients does not drastically change the resultant phenograms, but it is to be preferred on theoretical grounds since 'to use unstandardized characters is equivalent to weighting the characters in proportion to their observed standard deviations' (Rohlf and Sokal, 1965, p. 19).

Fig. 3 is a scatter diagram of correlation coefficients versus average taxonomic distance coefficients all computed from standardized data. The correlation coefficient between the two is -0.544. Rohlf and Sokal (1965) showed that standardization of data by OTU's (columns) would make distance coefficients a simple function of correlation. In such a case the distance coefficient would provide no information not contained in the correlation coefficient. Data were standardized by characters (rows), however; and Fig. 3 is a graphic expression of the difference in information content of the two coefficients.

Three assumptions in addition to the usual assumptions of numerical taxonomy (Sokal and Sneath, 1963) have been made in this study. One of these has been partially tested; the other two are essentially matters of definition and are untestable with the material at hand.

The first assumption is that variation within species is negligible when compared to variation among species. All specimens chosen for study were adults, so possible incongruence of larval and adult classifications need not be considered. This assumption does, however, apply to sexual dimorphism because only carapaces were used in the study; and sex could not be determined.

The assumption is necessary because the study is to some extent an application of the exemplar method of numerical taxonomy (Sokal and Sneath, 1963). If it were not a valid assumption, phenetic relationships among OTU's could be expected to vary markedly depending on what individual specimens were chosen.

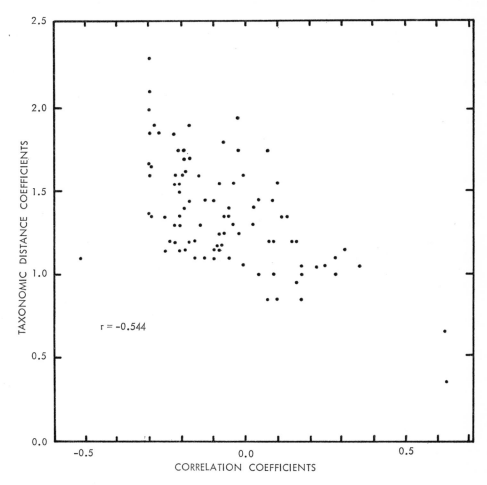

Fig. 3. Scatter diagram of correlation coefficients versus average taxonomic distance coefficients from similarity matrices of species included in this study.

The assumption was partially tested by including two specimens of *Loxoconcha tamarindus* in the study. In all phenograms these two specimens showed much greater mutual similarity than did any other specimens included in the study. The assumption was neither thoroughly nor rigorously tested, however, and such highly variable species as *Leptocythere pellucida* and *L. castanea* may show appreciable overlap.

Two other assumptions were made, both of which involve the concept of homology. The diverse approaches to this important concept have been discussed elsewhere

(Sokal, 1962; Kaesler, 1967) and need not be reviewed here. Suffice it to say that the concept of operational homology (Sokal and Sneath, 1963, pp. 69–74) was used in this study.

The second assumption is that terminal hinge elements of the right valve of the specimens studied are homologues. This assumption was necessary to provide justification for orienting all specimens in the same position for measuring, that is, with the hinge horizontal.

The third assumption is that the characters measured (Table 2) are homologous for all specimens measured. This assumption is, of course, vital because we must be sure that comparisons of organisms with each other are based on the same characters. The nature of many of the characters chosen is such that one might speak of 'geometrical homology' with an operational basis.

Matrices of both correlation and distance coefficients were clustered by several standard techniques of cluster analysis (Sokal and Sneath, 1963, pp. 178–94). These techniques included clustering by single linkage (Sneath, 1957), by complete linkage (Sørensen, 1948), and by average linkage methods (Sokal and Michener, 1958) including the unweighted pair-group method with arithmetic averages (UPGMA), the weighted pair-group method with arithmetic averages (WPGMA), and, for correlation coefficients only, Spearman's sums of variables method.

Figs 4 to 12 are phenograms that resulted from the cluster analyses listed above. Characteristics of the methods of cluster analysis given by Sokal and Sneath (1963) may be used for understanding and interpreting some of the differences among phenograms.

Each phenogram was compared with the matrix from which it was computed by using the method of cophenetic values (Rohlf and Sokal, 1962). Results of this comparison are presented in Table 3. Values in Table 3 are correlation coefficients between elements of the correlation or distance matrices and corresponding cophenetic values from the phenogram. They indicate the degree to which the similarities among species shown in the original matrix are represented by the phenogram and can be thought of as a measure of the distortion introduced by the clustering technique. A certain amount of distortion is, of course, inevitable because the phenogram is a two-dimensional representation of a multidimensional configuration.

Correlation coefficients in Table 3 between the distance matrix and cophenetic values from the distance phenograms range from 0·819 to 0·857. These correlations are of about the same magnitude as is customarily found in numerical taxonomic studies (see, e.g., Rowell, 1967). The cophenetic correlations for correlation coefficients, which range from 0·600 to 0·698, are somewhat lower than one might hope for and indicate considerable distortion in clustering.   [*Text continued* on p. 42.]

TABLE 3. Correlation coefficients between similarity matrices (correlation and distance) and cophenetic values from phenograms prepared by various clustering techniques

	Correlation Coefficient Matrix	Distance Coefficient Matrix
Single linkage method	0·600	0·833
Complete linkage method	0·647	0·819
Unweighted pair-group method with arithmetic averages	0·698	0·857
Weighted pair-group method with arithmetic averages	0·680	0·856
Spearman's sums of variables method	0·680	—

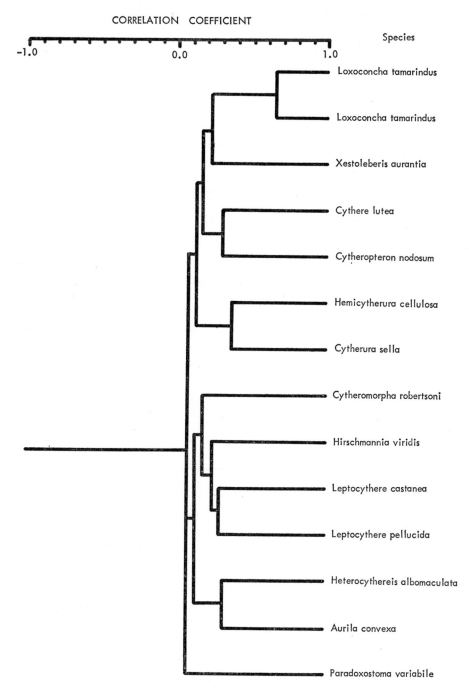

Fig. 4. Phenogram prepared from the correlation coefficient matrix by the single linkage method.

33

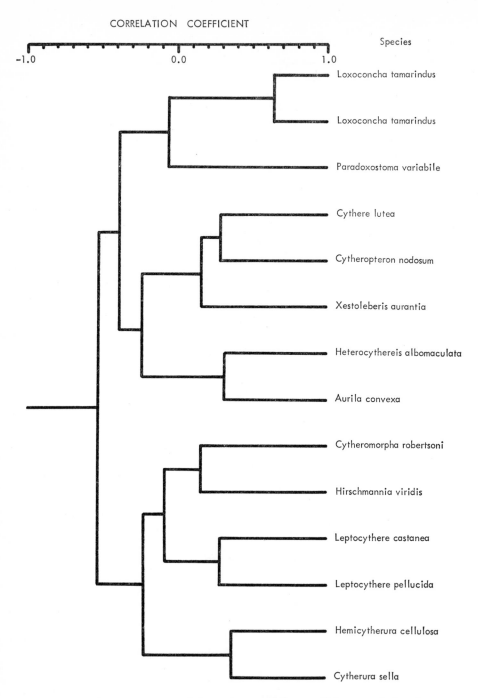

CORRELATION COEFFICIENT

Species

Loxoconcha tamarindus

Loxoconcha tamarindus

Paradoxostoma variabile

Cythere lutea

Cytheropteron nodosum

Xestoleberis aurantia

Heterocythereis albomaculata

Aurila convexa

Cytheromorpha robertsoni

Hirschmannia viridis

Leptocythere castanea

Leptocythere pellucida

Hemicytherura cellulosa

Cytherura sella

Fig. 5. Phenogram prepared from the correlation coefficient matrix by the complete linkage method.

34

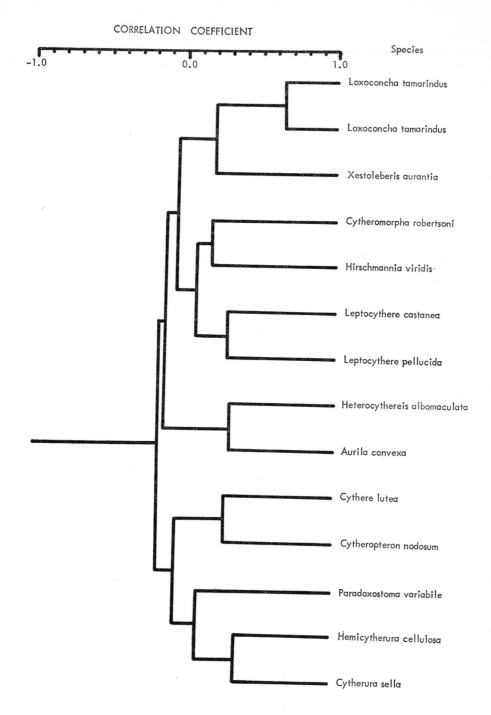

Fig. 6. Phenogram prepared from the correlation coefficient matrix by the unweighted pair-group method using arithmetic averages (UPGMA).

35

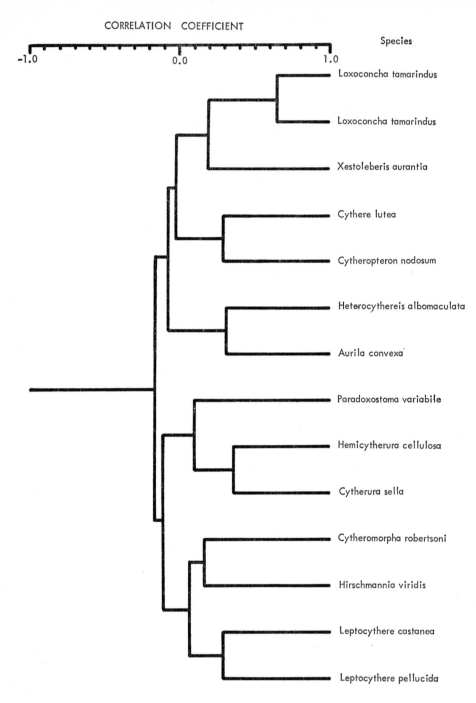

CORRELATION COEFFICIENT

Species

-1.0   0.0   1.0

Loxoconcha tamarindus

Loxoconcha tamarindus

Xestoleberis aurantia

Cythere lutea

Cytheropteron nodosum

Heterocythereis albomaculata

Aurila convexa

Paradoxostoma variabile

Hemicytherura cellulosa

Cytherura sella

Cytheromorpha robertsoni

Hirschmannia viridis

Leptocythere castanea

Leptocythere pellucida

Fig. 7. Phenogram prepared from the correlation coefficient matrix by the weighted pair-group method using arithmetic averages (WPGMA).

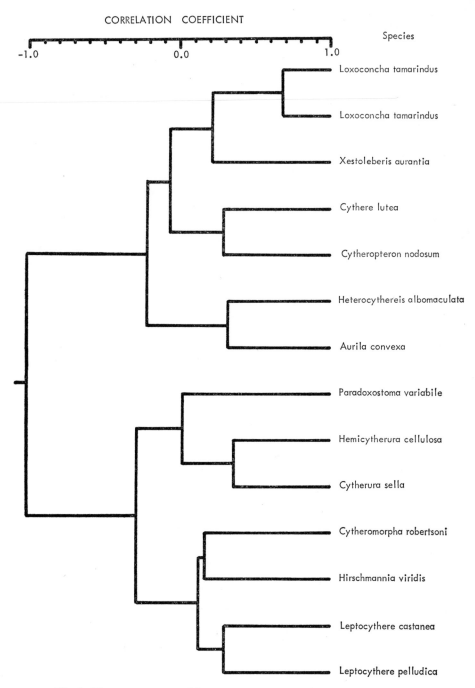

CORRELATION COEFFICIENT

Species

Loxoconcha tamarindus

Loxoconcha tamarindus

Xestoleberis aurantia

Cythere lutea

Cytheropteron nodosum

Heterocythereis albomaculata

Aurila convexa

Paradoxostoma variabile

Hemicytherura cellulosa

Cytherura sella

Cytheromorpha robertsoni

Hirschmannia viridis

Leptocythere castanea

Leptocythere pelludica

Fig. 8. Phenogram prepared from the correlation coefficient matrix by Spearman's sums of variables method.

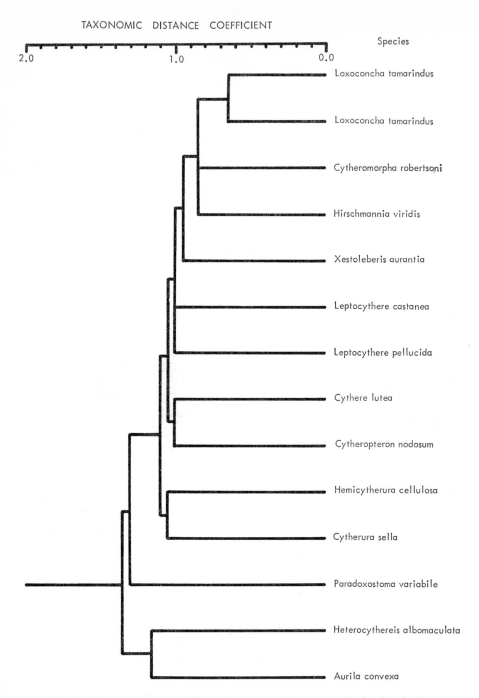

TAXONOMIC DISTANCE COEFFICIENT

Species

2.0   1.0   0.0

Loxoconcha tamarindus

Loxoconcha tamarindus

Cytheromorpha robertsoni

Hirschmannia viridis

Xestoleberis aurantia

Leptocythere castanea

Leptocythere pellucida

Cythere lutea

Cytheropteron nodosum

Hemicytherura cellulosa

Cytherura sella

Paradoxostoma variabile

Heterocythereis albomaculata

Aurila convexa

Fig. 9. Phenogram prepared from the average distance matrix by the single linkage method.

38

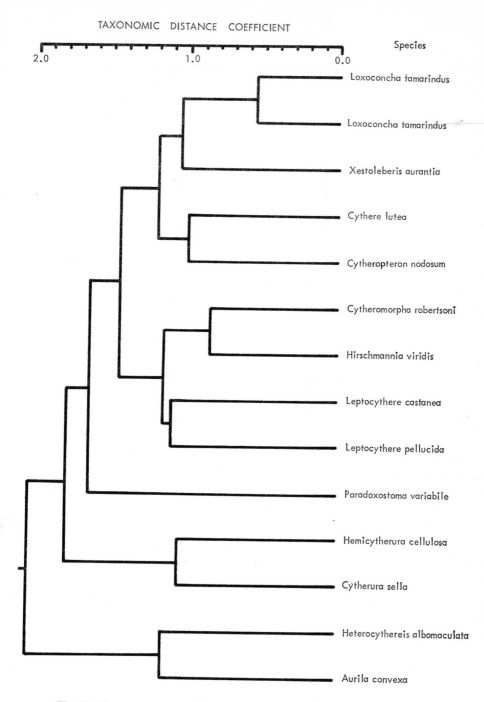

TAXONOMIC DISTANCE COEFFICIENT

Species

2.0   1.0   0.0

Loxoconcha tamarindus

Loxoconcha tamarindus

Xestoleberis aurantia

Cythere lutea

Cytheropteron nodosum

Cytheromorpha robertsoni

Hirschmannia viridis

Leptocythere castanea

Leptocythere pellucida

Paradoxostoma variabile

Hemicytherura cellulosa

Cytherura sella

Heterocythereis albomaculata

Aurila convexa

Fig. 10. Phenogram prepared from the average distance matrix by the complete linkage method.

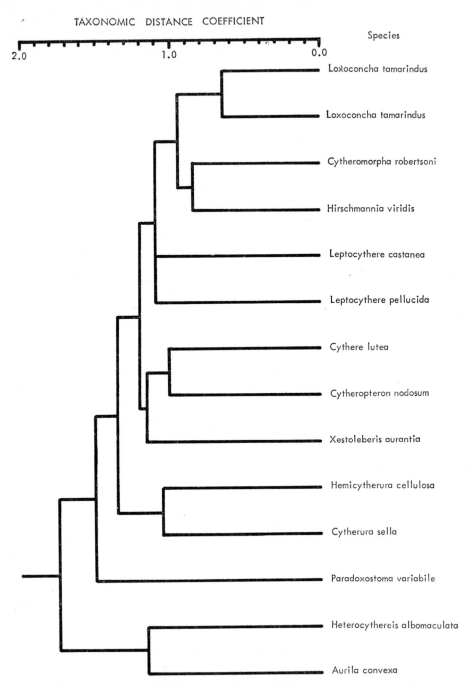

Fig. 11. Phenogram prepared from the average distance matrix by the un-weighted pair-group method using arithmetic averages (UPGMA).

40

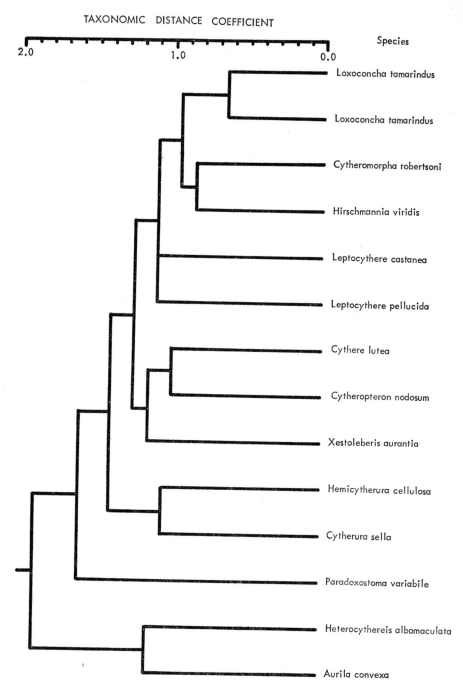

TAXONOMIC DISTANCE COEFFICIENT

Species

2.0   1.0   0.0

Loxoconcha tamarindus

Loxoconcha tamarindus

Cytheromorpha robertsoni

Hirschmannia viridis

Leptocythere castanea

Leptocythere pellucida

Cythere lutea

Cytheropteron nodosum

Xestoleberis aurantia

Hemicytherura cellulosa

Cytherura sella

Paradoxostoma variabile

Heterocythereis albomaculata

Aurila convexa

Fig. 12. Phenogram prepared from the average distance matrix by the weighted pair-group method using arithmetic averages (WPGMA).

41

The two phenograms prepared by the unweighted pair-group method with arithmetic averages have higher correlations with the matrices from which they were computed than do phenograms prepared by other methods from the same matrices. This result, which indicates that UPGMA introduced less distortion than other clustering techniques, has also been reported from most other numerical taxonomic studies in which cophenetic correlations were computed. Of the two phenograms prepared by UPGMA, the distance phenogram has by far the higher cophenetic correlation, 0·857. For this reason, we will focus our attention on Fig. 11 as the 'best' phenogram—that is, the one with the least distortion of original similarities. We could not justify using one of the other, more highly distorted phenograms without having decided *a priori* that we were interested in it specifically, regardless of the amount of distortion introduced.

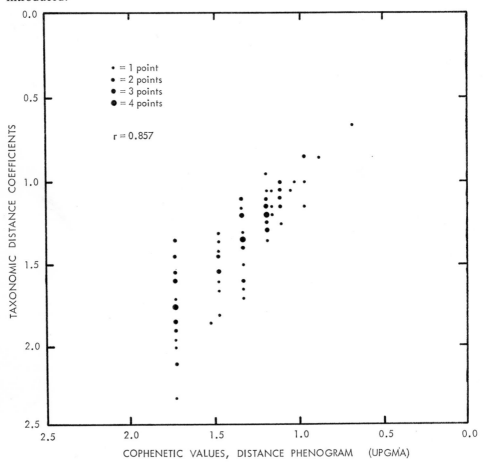

Fig. 13. Scatter diagram of cophenetic values from the average distance coefficient phenogram (UPGMA) versus average distance coefficients from the distance matrix.

Fig. 13 is a scatter diagram of cophenetic values from the UPGMA distance phenogram versus average taxonomic distance coefficients. As was previously mentioned, the scatter of points ($r = 0·857$) is a measure of distortion introduced by the clustering method. Fig. 14 is a similar scatter diagram, but it shows coded taxonomic level (1 = individual, 2 = species, 3 = genus, 4 = sub-family, 5 = family, 6 = super family) at which the branches join in Fig. 2 versus average taxonomic distance coefficient. The taxonomic level data are highly skewed because most of the species in Fig. 2 are members of different families and thus come together only at the super-family level. Nevertheless, the figure gives an estimate of the extent to which the average distance coefficient matrix portrays the *Treatise* classification. The correlation between the two is only 0·398.

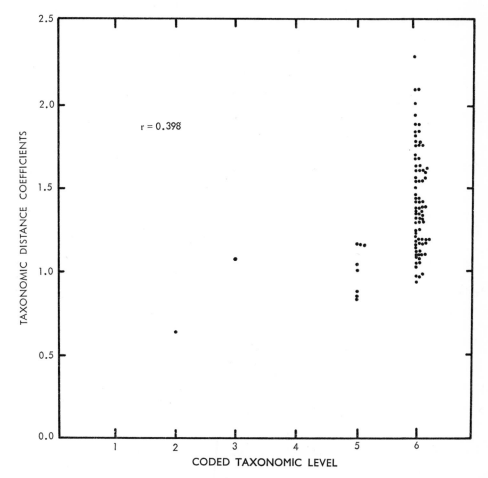

Fig. 14. Coded taxonomic level from the *Treatise* classification versus average distance coefficients from the distance matrix.

The phenogram in Fig. 11 is in close agreement with the graph of the phylogenetic classification from the *Treatise* (Fig. 2), but as mentioned before the phenogram has much more detailed structure. It must be emphasized at this point that failure of the two classifications to agree should be regarded neither as surprising nor undesirable. The ultimate purpose of the *Treatise* classification has not been stated, but one can assume that as a general classification it is intended to have maximum predictive value. This is also the purpose of the phenetic classification.

Differences between the two may have resulted from:

1. Differences in method of determining the graphs. The phenogram is based on phenetic similarity as measured by distance coefficients; the *Treatise* classification is based on speculation about phylogeny.

2. Distortion in the phenogram introduced by the clustering method.

3. Absence of intermediate and closely similar species from the phenogram.

Similarities between the two can only be regarded as a measure of the extent to which Fig. 2, the *Treatise* classification, is phenetic.

The phenogram agrees with the conventional classification in several important respects. First, the two specimens of *Loxoconcha tamarindus* clustered at a high level with a distance coefficient of less than 0·7. Second, all specimens of loxochonchids were placed in the same cluster with a distance coefficient less than 1·0. At a distance of slightly less than 1·2, the Loxoconchidae were joined by the two species of *Leptocythere*. The species of *Leptocythere*, however, did not form a distinct cluster above this level. Third, the Hemicytheridae formed a cluster with distance coefficient between 1·1 and 1·2. Fourth, *Paradoxostoma variabile*, the only representative of the Paradoxostomatidae, clustered with all species except the Hemicytheridae at a very low level, about 1·5, whereas the Hemicytheridae joined the cluster containing all other species at an even greater average taxonomic distance, between 1·7 and 1·8.

Only one major inconsistency exists between the two graphs. It is the failure of the species of Cytheruridae to cluster at a higher level than they did and the inclusion of *Cythere lutea* and *Xestoleberis aurantia* in the cluster with *Cytheropteron nodosum*. On the basis of carapace morphology alone, one would probably not object to the separation of *Cytheropteron* from the other cytherurids nor to a cluster consisting of *Cythere lutea* and species of *Cytheropteron*. However, one feels intuitively that *Xestoleberis aurantia* should have joined the other clusters more in the manner *Paradoxostoma variabile* did, rather than at a distance of only about 1·15. Choice of characters may have caused this peculiar clustering; further study is needed to determine if this is a good estimate of the overall phenetic resemblance of *Xestoleberis* species to *Cythere* and *Cytheropteron*. On the other hand, considering obvious differences in carapace morphology, it is not surprising that *Hemicytherura cellulosa* and *Cytherura sella* are separated by a taxonomic distance of 1·05.

CONCLUSIONS

I do not envisage Fig. 11 or any of the other phenograms as representing the ultimate word on the classification of the species studied. As was pointed out earlier, this was a pilot study; and it is much too early to consider determining phenons from the phenograms. The purpose was to obtain an initial estimate of the usefulness of numerical taxonomic methods in the classification of ostracods. I believe the results of the study indicate *at a minimum* that more work needs to be done. I hope they also indicate that results of such work will be fruitful both in furthering our knowledge of the ostracods and in advancing the field of numerical taxonomy.

I have identified the following as a few of the major problems that need further investigation. To do so will require co-operation from other ostracod specialists in obtaining both specimens and literature.

1. Tests of congruence of numerically derived classifications using characters from adults with those based on characters from larvae;

2. Tests of congruence of numerical classifications based on characters from carapaces only with those based solely on appendage morphology.

3. Numerical taxonomic studies of entire families and large genera with long fossil records, such as *Loxoconcha*.

4. Construction of hierarchies based on characters from both carapaces and appendages of adults and all instars.

5. Factor analytic studies of continuously evolving lineages.

6. Automatic data collection by computer operated optical scanning devices.

7. R-type analyses to determine correlations among characters.

The role of the computer in the future development of taxonomy has been universally acknowledged. At present, much of the effort in computer-based taxonomy is being directed to numerical taxonomy. We can expect this trend to continue in the future because of scientific momentum, the widespread attention and open discussion numerical taxonomic innovations are receiving, the careful thought invested in the development of each new concept, and the rigorous tests by both proponents and opponents that each development must withstand. But almost none of the numerical taxonomic literature has been concerned with palaeontological material. By participating now in this new and exciting field, ostracod specialists can both contribute to its development and help direct it in such a way as to make it of value to the entire biological community, palaeontologists as well as neontologists.

ACKNOWLEDGEMENTS

Field work during the spring and summer of 1964 was financed by a grant from The Society of the Sigma Xi, a National Science Foundation Summer Fellowship, a grant from the Department of Geology, The University of Kansas, and financial and logistic support from the Department of Geology, The University of Hull. Gratitude is expressed to Richard H. Benson and John W. Neale, who made my stay at Hull possible and who assisted me in many ways, both in the planning and early execution of my research.

Computation on the IBM 7040 computer at The University of Kansas Computation Center was made possible by a grant of computer time from that center. Preliminary computation was done using computer programs that I wrote, but most of the computation was done using the NTSYS programs written by F. James Rohlf, J. R. L. Kishpaugh, and Ronald Bartcher, all of The University of Kansas. Without the use of this valuable research tool, this study could not have been completed.

REFERENCES

BENSON, R. H. 1965. Recent marine podocopid and platycopid ostracodes of the Pacific. *Pubbl. Staz. zool. Napoli*, [1964], **33** (suppl.), 387-420.
BLACKWELDER, R. E. 1964. Phyletic and phenetic *versus* omnispective classification. *Publs Syst. Ass.*, Pub. No. 6, 17-28.
CAIN, A. J. 1962. Zoological classification. *Proc. Aslib*, **14**, (8), 226-30.
EADES, D. C. 1965. The inappropriateness of the correlation coefficient as a measure of taxonomic resemblance. *Syst. Zool.*, **14**, 98-100.
EDWARDS, A. W. F. AND CAVALLI-SFORZA, L. L. 1964. Reconstruction of evolutionary trees. *Publs Syst. Ass.*, No. 6, 67-76.
GILMOUR, J. S. L. 1937. A taxonomic problem. *Nature, Lond.*, **139**, 1040-2.
——. 1940. Taxonomy and philosophy, 461-74. In *The New Systematics*, J. S. Huxley, ed. Clarendon Press, Oxford, 583 pp.
——. 1951. The development of taxonomic theory since 1851. *Nature, Lond.*, **168**, 400-2.
——. 1961. Taxonomy, 27-45. In *Contemporary Biological Thought*. A. M. MacLeod and L. S. Cobley, eds. Oliver and Boyd, Edinburgh, and Quadrangle Books, Chicago, 197 pp.
——. 1962. Introduction. *Proc. Aslib.*, **14**, (8), 223-5.
——. 1966. Report, the classification of changing phenomena, biology. *Bull. Classif. Soc.*, **1** (2), 30-31.
HEYWOOD, V. H., AND MCNEIL, J. eds. 1964. Phenetic and phylogenetic classification. *Publs Syst. Ass.*, No. 6, 164 pp.
KAESLER, R. L. 1967. Numerical taxonomy in invertebrate paleontology, 63-81. In Essays in Paleontology and Stratigraphy, Raymond C. Moore Commemorative Volume, *Univ. Kans. Dept. Geol. Spec. Pub.* 2, Univ. Kansas Press, Lawrence, 626 pp.

MAYR, E. 1965. Numerical phenetics and taxonomic theory. *Syst. Zool.*, **14**, 73-97.

MOORE, R. C. ed. 1961. *Treatise on Invertebrate Paleontology. Pt. Q. Arthropoda*, 3. *Crustacea: Ostracoda.* 442 pp., 334 Figs. University of Kansas Press, Lawrence, Kansas.

ROHLF, F. J. AND SOKAL, R. R. 1965. Coefficients of correlation and distance in numerical taxonomy. *Kans. Univ. Sci. Bull.*, **45** (1), 3-27.

ROWELL, A. J. 1967. A numerical taxonomic study of the chonetacean brachiopods, 113-40. *In* Essays in Paleontology and Stratigraphy, Raymond C. Moore Commemorative Volume, *Univ. Kans. Dept. Geol. Spec. Pub.* **2**, Univ. Kansas Press, Lawrence, 626 pp.

SIMPSON, G. G. 1961. *Principles of animal taxonomy*, 247 pp. Columbia Univ. Press, New York.

SNEATH, P. H. A. 1957. The application of computers to taxonomy. *J. gen. Microbiol.*, **17**, 201-26.

SØRENSEN, T. 1948. A method of establishing groups of equal amplitude in plant sociology based on similarity of species content and its application to analyses of the vegetation on Danish commons. *Biol. Skr.*, **5** (4), 1-34.

SOKAL, R. R. 1961. Distance as a measure of taxonomic similarity. *Syst. Zool.*, **10**, 70-79.

——. 1962. Typology and empiricism in taxonomy. *J. Theoret. Biol.*, **3**, 230-67.

——. AND CAMIN, J. H. 1965. The two taxonomies: areas of agreement and conflict. *Syst. Zool.*, **14**, 176-95.

——. CAMIN, J. H., ROHLF, F. J. AND SNEATH, P. H. A. 1965. Numerical taxonomy: some points of view. *Syst. Zool.*, **14**, 237-43.

—— AND MICHENER, C. D. 1958. A statistical method for evaluating systematic relationships. *Kans. Univ. Sci. Bull.*, **38** (2), No. 22, 1409-38.

—— AND ROHLF, F. J. 1962. The comparison of dendrograms by objective methods. *Taxon*, **11**, 33-40.

—— AND SNEATH, P. H. A. 1963. *Principles of numerical taxonomy*. W. H. Freeman and Co., San Francisco, 359 pp.

WALTERS, S. M. 1962. Botanical classification. *Proc. Aslib.*, **14**, No. 8, 231-3.

DISCUSSION

SOHN: What characters did you measure?

KAESLER: I measured fifty characters such as length, height and width of right valve, width of the anterior vestibule at the anterior extremity, distance from the anterior edge of the eye spot to the anterior extremity of the carapace, etc.; and I included some coded data about the shape of the antennal muscle scar. These are all tabulated in my paper.

HULINGS: Will you comment further on your statement that you were not satisfied with the fifty characters that you measured?

KAESLER: Yes, I am not necessarily satisfied with these particular fifty characters. Adding others such as soft part characters might or might not change the classification appreciably and this needs to be further tested to see if these fifty characters really give a good estimate of the similarity of these groups. The test would need to be based on many other characters.

HULINGS: If you were to use both carapace and appendage morphology, have you any idea how many characters you would use?

KAESLER: Initially I think that I would want to include perhaps one hundred of each so that any additional ones would not make too much difference. Later it might be possible to reduce this somewhat.

HULINGS: There seems to be a problem of weighting these characters. Would the appendage characters and the carapace characters carry the same weight?

KAESLER: I am inclined to favour equal weighting because I cannot think of a criterion to justify unequal weighting. If you did have criteria and felt that you wanted to weight, say, the carapace characters more than the others, you could very easily do this. However, I think that it is difficult to justify weighting of any kind.

SYLVESTER-BRADLEY: In many ways numerical taxonomy is in its infancy and what it is aiming to do is to give an unprejudiced, unweighted assessment of similarity. As taxonomists most of us have grown up to realize that some characters are much more important for classification than others, in that they can discriminate between one species or one genus and another. Now, although you are introducing no weighting between the fifty characters you use, you are in fact introducing excessive weighting against the characters you omit, characters which other taxonomists may think are important. For example you have left out the appendages altogether and so you immediately add weight to the characteristics of the carapace. However, I do not think that this in any way weakens the logic of the method. I am a great believer in numerical taxonomy. We can even use it as a step towards finding out which characters are important for classification. This can be achieved by comparing an analysis of the fossil record at successive geological horizons. If in fact we can work out changes in phenetic resemblance between related species as we pass up the geological column, we may get a true picture of evolution.

KAESLER: I think that the evolutionary picture is fascinating and one of the main objects, but on the other hand I do not think that classification need be based on the evolutionary picture. It is true that in this example, the classification has been based on fifty characters. Since these are the ones included, obviously it must be weighted for them. However, this is not what is usually meant by the term weighted classification. Within the fifty characters selected there is no differential weighting, and they are all regarded as being of equal importance.

NOTES ON THE PARADOXOSTOMATIDS

K. G. McKENZIE
Monash University, Australia[1]

SUMMARY

Paradoxostomatinids evolved during the late Mesozoic, probably from small laterally compressed marine cytherid ostracods with smooth fragile carapaces which lived in warm (?) seas among calcareous algae and coarse detritus at shallow depths (< 50 m). The ancestral genus may have been *Sclerochilus* which is characterized by rounded extremities, numerous short radial pore canals and five oblique adductor muscle scars. The early forms adapted to the expanding phytal niche by developing specialized mouth parts and had spread world wide in association with marine algae and marine angiosperms by the early Miocene. Subsequently, Microcytherinae and Cytheromatinae evolved either by branching off or by parallel evolution and convergence. Although both are phytal associates, Microcytherinae also occupy a littoral interstitial niche whereas the range of some cytheromatinids includes such marginal environments as estuaries. By the late Miocene, the tropical genera *Pellucistoma* and *Javanella* had evolved thus diversifying the original stock. Microcytherinae, Cytheromatinae and Pellucistomatini all possess a prominent frontal scar (probably the attachment site for a mandibular muscle) as does *Cytherois*, whereas this scar is absent or weakly developed in genera belonging to the ancestral (?) lineage, i.e. *Sclerochilus*, *Paracytherois*, *Paradoxostoma*, *Paracythere* and *Machaerina*. Since the Pleistocene, isopod commensals have appeared, possibly evolved from *Cytherois*-like paradoxostomatinids (*Redekea*, *Laocoonella*).

The above synthesis has been formulated after morphologic comparisons of the carapaces and anatomies of most paradoxostomatid genera as the group is herein understood and after considering their environmental associations and the available fossil evidence.

INTRODUCTION

The first description of a paradoxostomatid was Baird's 1835 record of *Cythere variabilis* (cf. also Baird, 1850, p. 170, Pl. 21, Figs 10, 11). This was followed by Fischer's description of the type species (by monotypy) of the type genus *Paradoxostoma*, in diagnosing which he stressed the specialized mouth parts, still the most striking anatomical features in some members of this group of cytheraceid ostracods (Fischer 1855). By 1894, about 25 species had been described and also four genera, namely *Paradoxostoma*, *Sclerochilus* Sars 1866, *Cytherois* Müller 1884, *Machaerina* Brady and Norman 1889 (= *Xiphichilus* Brady 1870, non *Xiphocheilus* Bleeker 1856), but only the first and last of these were formally included by Brady and Norman (1889)

[1] Now at British Museum (Natural History)

in their family Paradoxostomatidae. In the Naples monograph Müller described the subfamily Paradoxostominae, including *Cytherois* and the new genus *Paracytherois* and almost as many new species as the sum of those named earlier. In the same paper, he also defined the new genera *Paracythere*, *Microcythere* and *Cytheroma* (Müller 1894). Later, *Paracytheroma* Juday 1907 was described. The group was now stabilized taxonomically for some decades and the only major change was introduced by Sars when he added *Sclerochilus* to the subfamily Paradoxostomatinae (Sars, 1928). There were also some additions to the roll of genera in *Pellucistoma* Coryell and Fields 1937, *Javanella* Kingma 1948, *Redekea* de Vos 1953, *Laocoonella* de Vos and Stock 1956 (= *Laocoon* de Vos 1953, non *Laocoon* Nierstrasz and Entz 1922), *Boldella* Keij 1957, *Cobanocythere* Hartmann 1959 and *Megacythere* Puri 1960. In 1960 and 1961 respectively, the Russian and American *Treatises* appeared (Orlov, 1960; Moore, 1961). Both re-established the family status of the group as a whole but diverged markedly in their interpretations of the associated subfamilies (Table 1). Other recent treatments (Pokorný, 1965; van Morkhoven, 1963; Hartmann, 1963) are conservative and have retained Müller's subfamilial ranking. The latest genus to be defined is *Pontocytheroma* Marinov 1963.

TABLE 1. The family Paradoxostomatidae according to the Russian and American *Treatises*

Russian *Treatise* (1960, p. 407)	American *Treatise* (1961, p. Q315)
Family Paradoxostomidae Brady et Norman, 1889	Family Paradoxostomatidae Brady and Norman, 1889
Subfamily Paradoxostominae Brady et Norman, 1889	Subfamily Paradoxostomatinae Brady and Norman, 1889
Subfamily Pseudocytherinae Schneider, subfam. nov.	Subfamily Microcytherinae Klie, 1938
Subfamily Bythocytherinae Sars, 1926	Subfamily Cytheromatinae Elofson, 1939

RELATIONSHIPS BETWEEN PARADOXOSTOMATIDAE AND BYTHOCYTHERIDAE

There is something to be said in defence of the Russian opinion, particularly as regards the general similarity in adductor muscle scar pattern between bythocytherids and the genus *Sclerochilus*. Also, at least one bythocytherid, *Pseudocythere* Sars 1866, has a smooth, fragile laterally compressed carapace. But apart from the important anatomical differences the carapace morphology of most Bythocytheridae is strikingly different to that of most Paradoxostomatidae. For instance, bythocytherid carapaces are usually strong; inflated; subquadrate to rhomboid in lateral view; often marked by a prominent mediodorsal sulcus; often strongly ornamented (exceptions *Pseudocythere*, *Jonesia* Brady 1866) by reticulations, alae, spines, tubercules; with straight radial pore canals and characterized by some type of caudal process (exception *Monoceratina* Roth 1928 s.s.). The only paradoxostomatid subfamily to possess some of these features is Microcytherinae—and Pellucistomatini have a caudal process. But, although it is true that *Microcythere* has a fairly strong, inflated carapace with a mediodorsal sulcus its shape and internal characters are quite unlike any bythocytherid and the common features also occur in ostracods belonging to other families, thus are too general to be phylogenetically significant.

The relationship between Bythocytheridae and Paradoxostomatidae becomes clearer when considering the fossil record. In the upper Mesozoic (Orlov, 1960) the representative bythocytherid genera were *Bythoceratina* Hornibrook 1952, *Monoceratina* s.l. (including *Bythocytheremorpha* Mandelstam 1958) and *Pseudocythere* whereas Paradoxostomatidae were confined to *Paradoxostoma* s.l. and *Sclerochilus*. The only record which may be incorrect is that of *Paradoxostoma*, but there is no doubt that *Sclerochilus* was present in the Barremian (Kuznetsova, 1962, p. 45, T. 7, Figs 1, 2, 3, 4—described as *Paradoxostoma pristina*). The bythocytherids in question are all very different from *Sclerochilus* and the doubtful *Paradoxostoma* species but the similarities in adductor muscle scar pattern and the broad resemblances between *Pseudocythere* and Paradoxostomatidae which have been mentioned earlier were still present. *No closer relationship was evident between these groups almost 100 million years ago than that which exists today*. I conclude that the two stocks were already distinct in the upper Mesozoic. In the Palaeozoic only *Monoceratina* and *Triceratina* Upson 1933 occur—the latter restricted to the Palaeozoic. I concede that the two families could have branched off from a *Monoceratina*-like cytheraceid ancestor in the early Mesozoic.

It is more difficult to study Mesozoic environmental associations of these groups mostly because there are so few Mesozoic records of Paradoxostomatidae (Dr P. Kaye, personal communication). In at least one Cretaceous fauna—the lower Maastrichtian of Rugen Island (Herrig, 1966)—a paradoxostomatid somewhat resembling *Sclerochilus* (described as *Paracytherois*?) occurs together with numerous bythocytherids (*Monoceratina*, *Bythoceratina* and *Pseudocythere*). Similar associations still occur in neritic environments, but it is more usual at the present time to find Paradoxostomatidae in littoral environments and at depths less than 50 m whereas Bythocytheridae usually occur in neritic environments and at greater depths. Paradoxostomatids generally have an affinity for the phytobenthos, which also implies shallow depth preferences because of the photosynthesizing requirements of marine plants, and this is not matched by most Bythocytheridae. Consequently, one may argue that even if the two groups had similar environmental associations in the Cretaceous they have tended to diverge since, in this respect.

From a biogeographic point of view both groups appear to have utilized the Tethyan corridor to achieve world-wide distribution during the upper Cretaceous and Tertiary (McKenzie, 1967a).

The stratigraphic and environmental data are interesting and, when considered together with the physical differences between bythocytherinids and paradoxostomatids some of which have already been mentioned, demand for each group a category of identical rank. Since both *Treatise* volumes have Paradoxostomatidae as a family it is consistent to consider Bythocytheridae as a family. The stratigraphic evidence is against Bythocytherinae and Pseudocytherinae as subfamilies within the family Paradoxostomatidae but supports them as subfamilies within the family Bythocytheridae. There is not sufficient anatomical disparity between Pseudocytherinae and Bythocytherinae to treat them as separate taxa *each of the same rank as* Paradoxostomatidae as is done in the Russian *Treatise*.

CARAPACE CHARACTERS

Strength of Shell

Generally these ostracods have fragile carapaces, a feature which more than any other has determined their poor fossil record. Considering present-day representatives, some free-living genera, e.g. *Microcythere* and *Paracytheroma*, do not have fragile shells. The phytobenthic parasites and the commensal and some interstitial genera, on the other hand, are characterized by thin carapaces.

Shape

Many genera are laterally compressed and most are acuminate anteriorly. Strongly compressed genera include *Machaerina*, *Paracytherois* and *Paradoxostoma* while *Sclerochilus* and *Cytherois* are not as slim. Even within Paradoxostomatinae, however, the members of one tribe (Pellucistomatini) are often ventrally inflated (e.g. some *Pellucistoma* species from the Pacific coast of Central America). *Redekea* and *Laocoonella*, the commensal genera, are slim but are rounded anteriorly (de Vos 1953). Neither Cytheromatinae nor Microcytherinae are laterally compressed. In side view the different genera have shapes which are often distinctive. Thus, *Paracytherois* is typically flexuous, *Sclerochilus* always has rounded extremities, *Machaerina* is typically pointed at both ends, *Microcythere* is strongly convex dorsally, *Pellucistoma* is sub-ovate with a caudal process and *Paracytheroma* is often oblong. *Paradoxostoma* has a wide variety of shapes some of which, e.g. wedge-shape as in *P. incongruens* and *P. versicolor*, occur only in this genus.

Ornament

Paradoxostomatids are typically smooth but *Paracytheroma* with several ribbed and reticulate species is a notable exception. Furthermore, a few species of *Paradoxostoma* and *Paracytherois* are striated lengthwise (Fig. 1), *Pellucistoma* species occasionally feature surface punctation, one *Sclerochilus* has a short ventral rib (Hartmann, 1959, Fig. 143 and text) and *Microcythere* has a dorsomedial sulcus (McKenzie, 1967b, Fig. 3h). Many species of *Paradoxostoma* and *Paracytherois* are colour-banded (usually black or brown) and I consider these pigmentation-patterns to be specific characters (Müller, 1894, Pl. 23).

Inner Lamellae

The inner lamellae of typical paradoxostomatids is very thin and is sometimes damaged when the valves of a closed specimen are separated for microscopic examination of internal shell detail. When preserved the inner margin is seen to be simple in outline. Broad anterior and posterior vestibules are features of some genera while in others, e.g. *Sclerochilus*, the line of concrescence is separated from the inner margin along its entire length.

Radial Pore Canals

These vary from short branched types (*Sclerochilus*) and short simple types (many *Paradoxostoma*) to longer and straight or wavy simple types as in several *Paracytherois* species and longer branched types (e.g. some *Pellucistoma* species). The radial pore canal pattern is sometimes difficult to see (especially in the anterior of *Cytherois* species) because individual canals are so short but in compensation appears to be species-specific in many cases. This pattern may be less useful for species determination in such genera as *Paracytheroma*, *Sclerochilus* and *Pellucistoma* although it can be used in this way (van Morkhoven, 1963, p. 435).

Normal Pore Canals

For all the groups considered here normal pore canals are simple and open, rather few and scattered. The groupings of these canals may be useful to distinguish species and

I have commented to this effect in respect of an Australian *Microcythere* (McKenzie, 5967b, p. 93).

Hingement

In Paradoxostomatinae the hingement is adont for *Paracythere* and *Sclerochilus* and possibly most *Paradoxostoma* and *Paracytherois* species, although I have examined an Australian *Paradoxostoma* (Fig. 1) and American *Paradoxostoma* and *Paracytherois* (McKenzie and Swain, 1967, text Figs 23–25) which have weakly lophodont hinges. In *Cytherois*, terminal teeth are distinctly developed in all species and in one I have seen also a weak crenulation on the posterior median element (Fig. 2). *Pellucistoma* is characterized by a strong anterior 'antislip' tooth and a weaker posterior tooth-like element. The hingement in *Microcythere* is rather similar to that in *Cytherois* except that the median element in some species is strongly crenulate (Müller, 1894, Pl. 24). The hinge of *Cobanocythere* is rectodont (Hartmann, 1963, p. 70). In Cytheromatinae, *Cytheroma* has a hinge like *Cytherois* but in *Paracytheroma* the hingement is more specialized approaching a pentodont type (Hanai, 1961, p. 361). *Redekea* and *Laocoonella* are adont.

Muscle Scars

These fall into two categories: in one a frontal scar is absent or only weakly developed. in the other it is usually prominent. The genera without a prominent frontal scar are further split into those with four or five oblique adductors (*Sclerochilus*, *Paracytherois*) and those with four vertical adductors (*Machaerina*, *Paradoxostoma*). Genera with a prominent frontal scar·include *Cytherois*, *Cytheroma*, Microcytherinae and Pellucisto-matini. Although present this scar is not as prominent in *Paracytheroma*. Unfortunately, muscle scar patterns of the Redekeini are not known.

Size

In general, Paradoxostomatidae are small-medium sized ostracods (length <0·5–0·75 mm) and only one species, *Machaerina tenuissima* (Norman), is longer than 1 mm. Height is less than half the length in most species and breadth has been considered in the paragraph on shape. Little work has been done on size trends through time within the family although van den Bold has graphed this data for several *Paracytheroma* (including *Megacythere*, Puri, 1960) species (van den Bold, 1963, p. 410) and no trend is evident. The lengths of living *Sclerochilus* species correlate rather well with temperatures expressed as latitude to show that cold water (higher latitude) species are longer than warm water (low latitude) species. Assuming a similar past distribution and if no size increase with time has taken place then the length of fossil *Sclerochilus* species could be used to interpret palaeolatitudes—and/or palaeotemperatures (Fig. 4). Other data, however, do show a trend towards size increase with time (van den Bold, 1960, p. 156; McKenzie, 1967c, p. 110) so that such interpretations, if made without qualification, may indicate a lower palaeolatitude than is correct.

ANATOMICAL CHARACTERS

First antennae

These are attenuated and six-jointed in all Paradoxostomatidae (Figs 2, 3) except *Sclerochilus* in which they are seven-jointed and less attenuated. The proximal joints

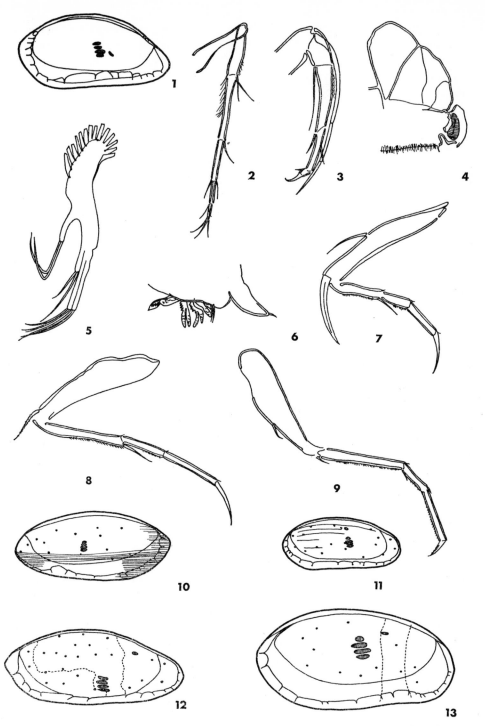

Fig. 1. 1, *Paradoxostoma* sp. 1, internal view LV; 2, 1st antenna; 3, 2nd antenna; 4, oral cone and suctorial disc; 5, maxilla; 6, posterior of body; 7, PI; 8, PII; 9, PIII; 10, *Paradoxostoma* sp. 2, internal view RV; 11, *Paradoxostoma* sp. 3, internal view LV; 12, *Paradoxostoma* sp. 4, internal view LV; 13, *Paradoxostoma* sp. 5, internal view LV. Magnifications: 1, 10-13, × 90 approx.; 2-9, × 200 approx. All species occur in Westernport Bay, Victoria. The anatomy illustrated is that of *Paradoxostoma* sp. 1 (a female).

53

are often pilose and distal joints carry a few bristles; usually three on the terminal joint (four shown for *Redekea*). In Microcytherinae the terminal joint is much reduced, and the bristles are relatively longer than in Paradoxostomatinae. The first antennae of Cytheromatinae are less attenuated and may be considered to resemble those of *Sclerochilus* except that they are wider and carry fewer and shorter bristles, including a terminal club-shaped bristle.

Second antennae

Broadest in a *Cytheroma* species (Elofson, 1938, Fig. 25) and in *Pontocytheroma* Marinov and most elongate in *Machaerina* and *Paracytherois* with *Sclerochilus* somewhere in between, the second antennae are five-jointed in Paradoxostomatinae, four- or five-jointed in Microcytherinae and four-jointed in Cytheromatinae. Endopodite joints bear several bristles and may be pilose: terminally, they have one or two claws except *Sclerochilus* which carries three (the third sometimes short and spinose— *S. ventriosus* Hartmann—at other times longer and smooth—*S. contortus* (Norman)). The exopodites (spinneret bristles) may be two-, three- or four-jointed (Marinov, 1964, Figs 8g, h; Figs 2, 3). The spinneret glands may be bilobate (*Sclerochilus*) or unilobate (*Paracytherois*) and for some *Paradoxostoma* are pipe-like.

Mandibles

Styliform coxales typify several Paradoxostomatinae but not *Sclerochilus*, *Cytherois*, *Paracythere*, Redekeini and Pellucistomatini in which coxal teeth are present; and the palp is reduced in all genera of this subfamily except in *Paracythere*, the number of epipodial bristles varying from one to three. In Microcytherinae the coxales have an elongate anterior tooth and there are either two equally long epipodial bristles (*Microcythere*) or else these are (?) absent (*Cobanocythere*). Cytheromatinae resemble Microcytherinae except that the anterior coxal tooth is less strongly developed and there are either two equally long epipodial bristles (*Paracytheroma*) or one or two long and one very short bristle (*Cytheroma*).

Maxillae

The maxilla palp and lobes in Paradoxostomatinae are reduced in number and variable with only *Cytherois* having the normal complement (three lobes and a palp). In this respect, Microcytherinae and Cytheromatinae are relatively normal, like *Cytherois*. The number of aberrant Strahlen on the maxilla epipodite also varies: *Paracythere* has three, several other Paradoxostomatinae have two, Microcytherinae and Cytheromatinae only one and Redekeini apparently none. The number of normal Strahlen is usually below ten for Redekeini, and around six in *Cobanocythere labiata*. In *Microcythere* about ten Strahlen are present while in most Paradoxostomatinae and in Cytheromatinae there are about fourteen.

Fifth, sixth and seventh limbs (PI–PIII)

The fifth limb of *Sclerochilus*, *Paradoxostoma*, *Paracytherois* and probably *Machaerina*, bears a coarse dorso-distal claw on the protopodite, but other genera (as is normal) carry two bristles here. The fifth limb of *Sclerochilus* also bears three proximo-ventral protopodite bristles, in other genera one may occur. In Cytheromatinae and in *Cytherois* the fifth limb is distinctly shorter than the sixth and seventh limbs. In

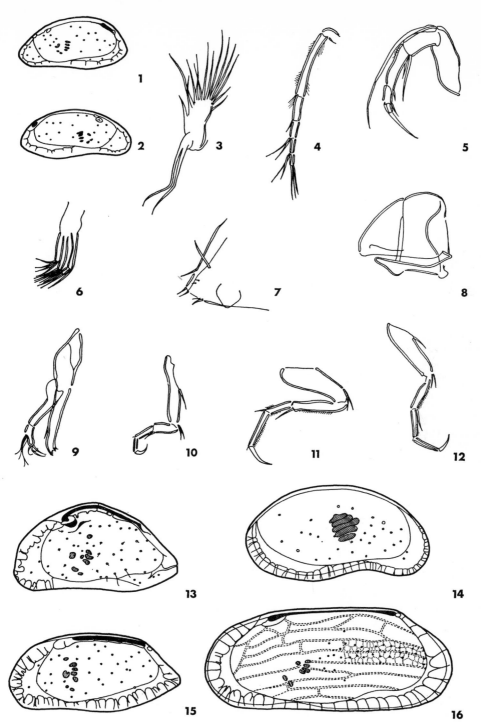

Fig. 2. 1, *Cytherois* sp., internal view RV; 2, *Cytherois* sp., internal view LV; 3, maxilla epipodite; 4, 1st antenna; 5, 2nd antenna; 6, maxilla palp and lobes; 7, posterior of body; 8, oral cone; 9, mandible; 10, PI; 11, PIII; 12, PII; 13, *Pellucistoma bensoni* McKenzie and Swain, 1967, internal view RV; 14, *Sclerochilus* sp., internal view LV; 15, *Javanella kendengensis* Kingma 1948, internal view RV of holotype, Mineralogic-Geologic Institute of the State University of Utrecht micropalaeontological collection, slide no. D 31989; 16, *Megacythere punctocostata* Swain 1967, internal view RV.

Magnifications: 1, 2, 14, 15, × 90 approx.; 3-12, × 200 approx.; 13, × 100 approx.; 16, × 120 approx.

Cytherois sp. is from Westernport Bay, Victoria. *Sclerochilus* sp. from the Tertiary of Victoria at Bells Headland, nr. Torquay. The anatomy is that of *Cytherois* sp. (a female).

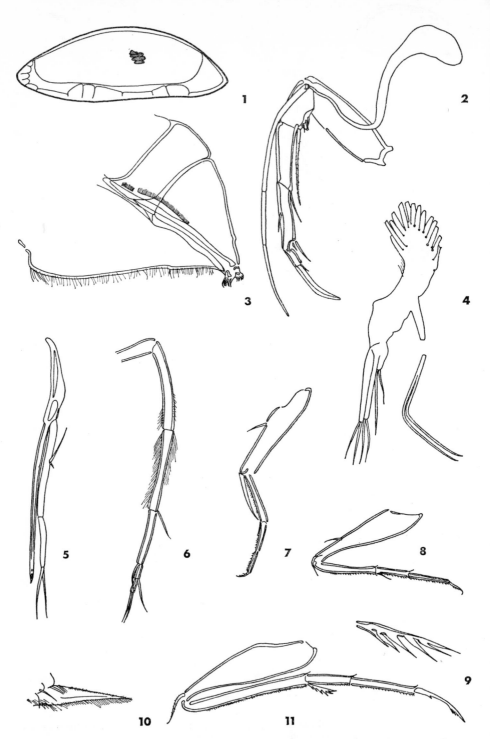

Fig. 3. 1, *Paracytherois* sp., internal view LV; 2, 2nd antenna with spinneret gland; 3, oral cone; 4, maxilla; 5, mandible; 6, 1st antenna; 7, PI; 8, PII; 9, detail of dorso-distal bristle, 1st endopodite joint of PIII; 10, pilose bristle at posterior of body; 11, PIII.

Magnifications: 1, × 90 approx.; 2-8, 10, 11, × 200 approx.; 9, × 450 approx.

Paracytherois sp. occurs in Westernport Bay, Victoria.

general, these limbs are more attentuated in Paradoxostomatinae than in Micro-cytherinae and Cytheromatinae.

Other characters

The eye is an interesting diagnostic character among Paradoxostomatinae. A single well developed eye is present in *Cytherois*, *Machaerina* and *Paradoxostoma*; *Paracytherois* is blind and *Sclerochilus* and Redekeini have confluent eyes. *Pellucistoma* bears a mediodorsal eye spot on each valve. The furca is also variable, e.g. in *Sclerochilus* it comprises two setiferous lobes, in *Paracytherois* a single long hirsute bristle (Fig. 3). The reproductive parts of *Cytheroma* include a diagnostic chitinized process (Genitalhöcker). The anterobasal part of the penis is narrow and pointed in *Microcythere*; broad and pointed in Cytheromatinae; broad and pointed in Redekeini, and relatively lobate, also pointing downwards, in other Paradoxostomatinae. Brush-shaped organs are always present, the hairs usually long. Typical Paradoxostomatinae have a more or less attentuated oral cone sometimes with a terminal suctorial disc.

ENVIRONMENTAL ASSOCIATIONS

There is a gathering body of data to indicate world wide periodic changes in the geo-chemistry of the oceans which have affected, for example, the oceanic concentrations of carbonates and sulphates (Fairbridge, 1964; Holser and Kaplan, 1966). One such change took place in the Cretaceous. During this time also marine incursions were widespread and marine sedimentation took place in relatively warm water on extensive continental shelves. Furthermore, the Tethys corridor was open for most of the Cretaceous and Tertiary up to the Vindobonian except for relatively brief regressive phases (Termier and Termier, 1960). Biological activity was favoured in the warm, shallow, carbonate-rich seas and once stocks had become established the corridor ensured that they were widely and rapidly dispersed. The Paradoxostomatidae evolved and became distributed within this general ecologic framework.

A factor of great importance is the contemporary expansion of marine angiosperms and marine algae because many paradoxostomatids are phytal associates. Marine angiosperms have been described from the Upper Cretaceous of Europe and from the Tertiary of Middle Europe, the southern Soviet Union and Japan (Gothan and Weyland, 1954; Arnold, 1947). Their present distribution is world wide in relatively shallow water (usually less than 100 m) and lies predominantly within the warm temperate to tropical zones (Good, 1964). Most groups of marine algae have been recorded world wide during the Mesozoic and Tertiary. Of these, the calcareous algae, which include such limestone-building genera as *Lithothamnion*, are highly significant (Johnson, 1961).

Turning to paradoxostomatids, the longest ranging genus, *Sclerochilus*, lives today mostly at shallow depths—exceptions are some Antarctic species (Müller, 1908). Its known temperature tolerances range from below 0°C to about 25°C, although the ranges of individual species are more restricted. The preferred substrate is either phytal (including calcareous algae) or among fine organogenic sand (Hartmann, 1964, p. 95; Reys, 1965a, p. 270) and coarse detritus (Elofson, 1941, p. 345; Hartmann, 1959, p. 228). These substrate preferences would appear to have changed little since the Cretaceous. Most other genera in Paradoxostomatinae have similar ecologic associations. Apart from their phytal preferences, some are also known to live among coarse sand and detritus (Elofson, 1941, pp. 347–53; Reys, 1965b, pp. 264–65) and on rocky littorals (Caraion, 1963, pp. 51–58), others have been found living among fine organogenic sand (Reys, 1965a, pp. 270–2). Pellucistomatini may not be so closely associated with the phytal niche but are known to occur among very fine quartz sand (McKenzie and Swain, 1967, pp. 299–300) and in sand-mud mixtures (Baker and Hulings, 1966, pp. 114–15). The Redekeini, which are commensal upon the wood-boring isopod *Limnoria*, form a special case (de Vos, 1953; de Vos and Stock, 1956).

Cytheromatinae range from Miocene to Recent. *Cytheroma* occurs at shallow depths, from the eulittoral to about 80 m; its temperature tolerances are boreal to sub-tropical; and the preferred substrate varies from phytal (*Posidonia*) to fine sands (Hartmann, 1964, p. 98) and greyish silts (Caraion, 1962, p. 113). Furthermore, at least one species is interstitial (Elofson, 1938, p. 18). *Pontocytheroma*, as yet confined to the Black Sea, lives in sublittoral sand at 10–15 m depth (Marinov, 1963, p. 559). *Paracytheroma* is a Caribbean and central Pacific coast of America genus with the best fossil record among Cytheromatinae. It prefers shallow water; tolerates temperatures ranging from tropical to temperate; and has been recorded usually from sandy and muddy sand substrates. Its known habitats include polyhaline coastal lagoons and estuaries (Hartmann, 1962, p. 217; Hulings and Puri, 1965, p. 336; McKenzie and Swain, 1967, p. 300), mangrove esteros (Hartmann, 1962, pp. 155–8) and poorly con-solidated shelly marine sublittorals (Hartmann, 1959, p. 230). The fossil records of this genus include several from inner neritic marine facies (Puri, 1953) and others which indicate brackish conditions similar to those in which it occurs today. In some sequences the faunas are mixed, e.g. the Upper Miocene Cubagua Formation of north-eastern Venezuela (van den Bold, 1966).

Microcytherinae also range from Miocene to Recent. Their environmental associa-tions usually include shallow depth; temperature tolerances in the temperate to tropical range; and phytal as well as sandy substrates. One genus, *Cobanocythere*, prefers an interstitial niche among coarse sand and shelly detritus. The type genus, *Microcythere*, lives interstitially in coarse sandy eulittorals where it apparently tolerates variable salinities (Marinov, 1962, pp. 106–7), also is associated with cal-careous algae (Puri, 1963, p. 4) and even occurs in rocky tide pools (McKenzie, 1967b, p. 93). Fossil records of *Microcythere* are few, mostly from inner neritic to brackish marginal facies in the Caribbean and Gulf Coast America (Puri, 1953, p. 298; van den Bold, 1966, p. 40).

CONCEPTS

In developing the phylogenetic relationships suggested by Fig. 5 and in documenting phylogenetically useful trends I have been guided by the following principles:

1. I have attempted to gather as much palaeontological evidence as possible on this poorly fossiliferous group of ostracods, from the literature and comparative material, and have then assumed that the oldest genus (according to the fossil record) is ancestral. This is the chronological principle and it has been equally significant in my consideration of the environmental associations of Paradoxostomatidae.

2. I have considered evolution of the group in terms of morphologically expressed adaptations by the different genera to their environments and to changes in these environments—such as the expansion of the phytal niche. I have dealt in particular with adaptations in feeding and locomotory habits and have (in this instance) neglected those in the reproductive habit although variations in penis structure have been referred to. For these purposes I have used homologies in the sense of essential similarity of position in the parts being compared.

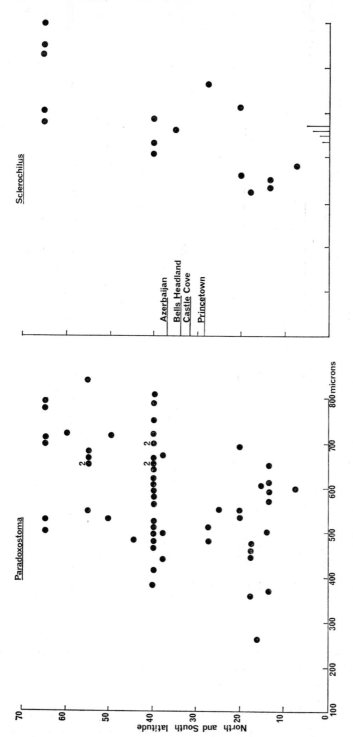

Fig. 4. Relationships between length and latitude for extant species of *Paradoxostoma* and *Sclerochilus*. Data from the literature and comparative material. Interpolation of the Victorian Tertiary species of *Sclerochilus* (Table 2) and the Russian Barremian species from Azerbaijan, *Sclerochilus pristina* (Kuznetsova), 1962, suggests that palaeolatitudes were lower and/or seas warmer when they were living.

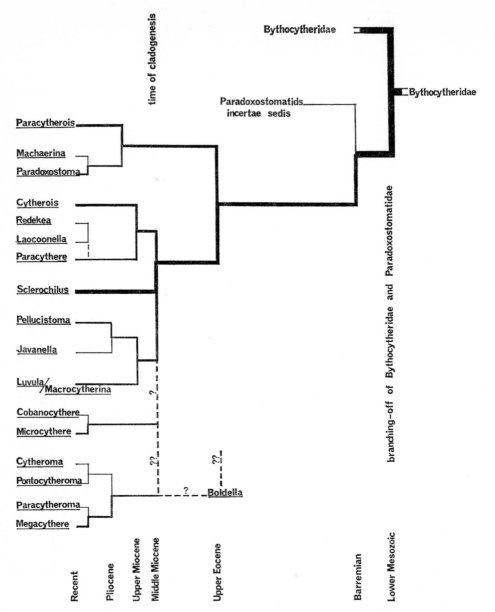

Fig. 5. Suggested phylogeny of the Family Paradoxostomatidae.

CONCLUSIONS

The relationships postulated in Fig. 5 may well record parallel evolution and convergence at the subfamily level (and even within subfamilies) but in my view cytheromatinids and microcytherinids show sufficient similarities to some Paradoxostomatinae to suggest a common origin. This opinion is based upon the fact that the range of characters expressed within the type subfamily embraces those of the other two.

Comparing carapace features, all three subfamilies have open normal pore canals, and the paradoxostomatinid *Cytherois* has a muscle scar pattern also a vestibular and radial pore canal pattern which are matched by genera in the other subfamilies.

Paracytheroma species which exhibit surface reticulation (here referred to subgenus *Megacythere*) differ in some shell characters from *Cytheroma* and *Pontocytheroma*. Thus, *Megacythere* may be an instance of convergence within a subfamily but its close anatomical similarity to other cytheromatinids tends to discount this possibility.

Turning to anatomical features, no Microcytherinae or Cytheromatinae possess mouth parts which are as specialized as in some Paradoxostomatinae. But this specialization does not occur in all Paradoxostomatinae: for example, styliform mandible coxales are not present in all genera and only *Paradoxostoma* and *Cytherois* have a terminal suctorial disc. *Paracythere* which in many anatomical features is a good paradoxostomatinid has a mandibular palp like Microcytherinae and Cytheromatinae. The second antennae in Cytheromatinae are only four-jointed but this may well represent a simple reduction in the number of endopodite segments. The fifth limbs of *Cytherois* and Cytheromatinae are similar. Futhermore, whereas in most Paradoxostomatinae the penis structure differs from that in Microcytherinae and Cytheromatinae, in Redekeini (which possess oral cones and reduced mandibular palps) it is like that in Microcytherinae.

Common environmental associations are emphasized, particularly affinities with the phytobenthos and the tendency to seek an interstitial niche. In addition, the stratigraphic record shows that when Microcytherinae and Cytheromatinae first appeared (in the Miocene) both occurred in inner neritic facies as well as nearshore in environments similar to those they favour today. Similarly, several Australian Tertiary paradoxostomatinids are associated with inner neritic and littoral facies (Table 2) although today speciation in Paradoxostomatinae probably is most diverse in littoral environments.

TABLE 2. Records of Paradoxostomatidae in the Tertiary of Victoria, Australia (McKenzie, unpublished data)

Genus	Locality	Age
Paradoxostoma spp. (2)	Creek at Forsythe's Bank on Grange Burn, nr. Hamilton	Lower Pliocene
Cytherois sp.	North side of Bunga Creek, roadcut E. of Lakes Entrance	Lower Pliocene
? *Paracythere* sp.	Rutledge Creek, shell beds at base of cliff, nr. Port Campbell	Upper Middle Miocene
Microcythere sp.	North side of Bunga Creek, roadcut E. of Lakes Entrance	Lower Pliocene
	Port Campbell Limestone, E. of Port Campbell	Upper Miocene
	Glenample Clay at Gibsons steps, E. of Port Campbell	Middle Miocene
Paracytherois sp.	Port Campbell Limestone, E. of Port Campbell	Upper Miocene
	Upper part of section, Grices Creek, nr. Mornington	Middle Miocene
	Glenample Clay at Gibsons Steps, E. of Port Campbell	Middle Miocene
	Fishing Point Marl, at Fishing Point	Lower Miocene
	Jan Juc Marl, 3'-6' below Bird Rock, nr. Torquay	Oligocene
	Browns Creek Clays, nr. Johanna River	Upper Eocene
Sclerochilus sp.	Gellibrand Clay, nr. Princetown Bridge	Lower Miocene
	Lower shell bed, at Bells Headland, W. of Torquay	Oligocene
	Glen Aire clays, at Castle Cove	Oligocene

Several phylogenetically useful trends can be pointed out.

1. Shape trends include those towards anterior rounding (e.g. commensal Redekeini) from an early pointed anterior, compression (as in phytal genera such as *Paradoxostoma* and *Paracytherois*), and ventral inflation (e.g. interstitial Microcytherinae).

2. Vestibules have developed from a condition originally like that in some *Sclerochilus* species in which the vestibule is continuous ventrally to one in which separate anterior and posterior vestibules are present.

3. In the first antennae (from seven to six) and probably also in second antennae (from five to four) there has been a reduction in the number of joints. Spinneret glands have changed also from bilobate (*Sclerochilus*) to unilobate (*Paracytherois*) and pipe-like (some *Paradoxostoma* species).

4. Muscle scar patterns have evolved from five oblique adductors (*Sclerochilus*) towards four vertical adductors with or without a weak frontal scar (*Paradoxostoma*) and then to four vertical or crescentic adductors plus a large frontal scar (*Cytherois*, Pellucistomatinae).

5. Mouth part trends include attenuation of the oral cone (*Paracytherois*), also development of suctorial discs (*Paradoxostoma*, *Cytherois*) and styliform mandible coxales (Paradoxostomatinae as understood herein).

6. Eyes originally may have been confluent (*Sclerochilus*); now some genera have single eyes (*Cytherois*), and at least one is blind (*Paracytherois*).

7. With respect to mandibular palps and maxillae the trends may have been retrogressive since 'normal' organizations characterize some younger genera and subfamilies.

8. Other trends relate to the development of a coarse dorso-distal spine on the fifth limb protopodite in some genera (Paradoxostomatinae as understood herein) which possibly expresses a grappling adaptation, and to changes in the number of aberrant Strahlen on the maxilla epipodite although the functional significance of the latter is obscure.

The best example within the family of the significance of these trends lies in the evolution of feeding adaptations in respect of which, for example, Paradoxostomatinae are uniquely adapted to the phytal environment. The important shell feature in this respect is the frontal scar. In two recent papers on Cyprididae a frontal scar has been associated with a mandibular muscle and in one of these the muscle is shown connecting a frontal scar to the mandible coxale behind and above the attachment site of the mandibular palp (Kesling, 1965, p. 17; Smith, 1965, p. 17). I am assuming that the frontal scar pattern in Cytheridae is homologous with that in Cyprididae hence infer that at least part of the frontal scar in Paradoxostomatidae provides an attachment site for one of the mandibular muscles. Significant 'soft part' characters are the mouth parts and mandibles.

Frontal scars are rarely illustrated for *Sclerochilus* which has a weak mandibular palp, a mandible coxale with an oblique and finely denticulated cutting edge and an oral cone with two lips, the lower of which is serrated. From this condition forms seem to have evolved in two main lines. One group of genera (Paradoxostomatinae) has become adapted to feed suctorially, by means of attenuated oral cones, suctorial discs (*Paradoxostoma*) or finely-toothed jaws (*Paracytherois*) and styliform mandible coxales—for piercing plant tissues—with mandibular palps poorly developed, i.e., principally by mouth part modification. In the other main evolutionary line (Microcytherinae, Cytheromatinae) mandibular parts have become strengthened so that the distal coxales are fitted for a triturating role and the palps can act as accessory feeding organs by passing food particles back to the mouth. In these groups frontal scars are prominent. The remaining subfamily (Sclerochilinae) includes genera with what appear to be intermediate features. Thus, *Redekea* has an oral cone and also powerful mandibles; *Cytherois* an imperfect sucking disc and poorly developed mandibular palps but a prominent frontal scar.

OTHER GENERA

Several genera with obscure affinities might be referable to this family. Of these, *Luvula* and *Macrocytherina* are paradoxostomatids (van den Bold, personal communication) probably related to the Pellucistomatinae although they were originally referred to Bythocytheridae (Coryell and Fields, 1937). *Boldella*, A Palaeogene European genus, is poorly known but was originally referred to Paradoxostominae (Keij, 1957, p. 163). It could be related to *Megacythere* or may be a cytherurinid. *Parvocythere* has only been recorded twice (Hartmann, 1959, p. 235; Marinov, 1962, p. 97). I believe that this genus may be related to Paradoxostomatidae with the loss of a walking limb providing the basic differentiating character (and another phylogenetic trend). Hartmann (1964, p. 99) has commented on the general difficulty in determining the affinities of interstitial genera, chiefly because of their similar shape adaptations to the environment.

ACKNOWLEDGEMENTS

This paper was written during tenure of the inaugural Shell Research Fellowship for Monash University. Some of the comparative material was picked with support from a Commonwealth Science and Industry Endowment Fund grant. Miss D. Wade was responsible for the draughting which in part included inking my original drawings. Mrs T. McConnell typed the manuscript. I am grateful to Professor F. M. Swain for providing copies of drawings (made by Mr F. J. Gunther) of the anatomy of *Pellucistoma* species, and to Mr T. Freudenthal for airmailing the types of *Javanella*. Dr G. Bonaduce, of Stazione Zoologica, very kindly provided a slide containing several Paradoxostomatidae from the Bay of Naples, supplementing my earlier collections from the Bay which were made during tenure of an Italian Government Scholarship and a Royal Society Table at the Stazione. Finally, I have been most fortunate in having been allowed access to the Brady collection from the Hancock Museum, Newcastle-upon-Tyne.

REFERENCES

ARNOLD, C. A. 1947. *Introduction to Paleobotany*. McGraw Hill, New York, 433 pp.
BAIRD, W. 1850. *The natural history of the British Entomostraca*. The Ray Soc., London. [Ostracoda: 138-82, Pls 18-23]. 364 pp.
BAKER, J. H. AND HULINGS, N. C. 1966. Recent marine ostracod assemblages of Puerto Rico. *Publs Inst. mar. Sci. Univ. Tex.*, **11**, 108-25.
BOLD, W. A. VAN DEN. 1960. Eocene and Oligocene Ostracoda of Trinidad. *Micropaleontology*, **6** (2), 145-96.
——. 1963. Upper Miocene and Pliocene Ostracoda of Trinidad. *Micropaleontology*, **9** (4), 361-424.
——. 1966. Miocene and Pliocene Ostracoda from north-eastern Venezuela. *Verh. K. ned. Akad. Wet.*, *1st Ser.*, **23** (3), 43 pp.
BRADY, G. S. 1870. Notes on Entomostraca taken chiefly in the Northumberland and Durham district (1869). *Trans. nat. Hist. Soc. Northumb.*, **3** (2), 361-73.
BRADY, G. S. AND NORMAN, A. M. 1889. A Monograph of the marine and freshwater Ostracoda of the North Atlantic and of North-Western Europe. Section I. Podocopa. *Scient. Trans. R. Dubl. Soc.*, *Ser.* 2, **4**, 63-270.

CARAION, F.-E. 1962. Cytheridae noi (Crustacea-Ostracoda) pentru fauna Pontica Romineasca. *Studii Cerc. Biol. Seria 'biologie animala'*, **14** (1), 111-21.

——. 1963. Contributii la cunoasterea faunei de ostracode petricole din lungul litoralului Rominesc (Agigea si Mangalia). *Studii Cerc. Biol. Seria 'biologie animala'*, **15** (1), 45-63.

CORYELL, H. N. AND FIELDS, S. 1937. A Gatun ostracode fauna from Cativa, Panama. *Am. Mus. Novit.*, **956**, 1-18.

ELOFSON, O. 1938. Neue und wenig bekannte Cytheriden von der Swedischen westküste. *Ark. Zool.*, **30A** (21), 1-22.

——. 1941. Zur kenntnis der marinen Ostracoden Schwedens mit besonderer berücksichtigung des Skageraks. *Zool. Bidr. Upps.*, **19**, 215-534.

FAIRBRIDGE, R. W. 1964. The importance of limestone and its Ca/Mg content to palaeoclimatology. In A. E. M. Nairn (ed.), *Problems in Palaeoclimatology*, 431-78. Interscience, London.

FISCHER, S. 1855. Beitrag zur kenntniss der Ostracoden. *Abh. bayer. Akad. Wiss. Math.-phys. Kl.*, **7**, 3, 635-66.

GOOD, R. D'O. 1964. *The Geography of Flowering Plants*. Longmans, London. 518 pp.

GOTHAN, W. AND WEYLAND, H. 1954. *Lehrbuch der Paläobotanik*. Akad. Verlag, Berlin. 535 pp.

HARTMANN, G. 1957. Zur Kenntnis des Mangrove-Estero-Gebietes von El Salvador und seiner Ostracoden-Fauna. *Kieler Meeresforsch.*, **13** (1), 134-59.

——. 1959. Zur Kenntnis der lotischen Lebensbereich der pazifischen Küste von El Salvador unter besonderer Berücksichtigung seiner Ostracodenfauna. *Kieler Meeresforsch.*, **15** (2), 187-241.

——. 1962. Zur Kenntnis des Eulitorals der chilenischen Pazifikküste und der argentinischen Küste Südpatagoniens unter besonderer Berücksichtigung der Polychaeten und Ostracoden. Teil III. Ostracoden des Eulitorals. *Mitt. hamb. zool. Mus. Inst.*, **60**, 169-270.

——. 1963. Zur Phylogenie und Systematik der Ostracoden. *Z. zool. Syst. Evolut.-forsch.* **1** (1/2), 1-154.

——. 1964. Zur Kenntnis der Ostracoden des Roten Meeres. *Kieler Meeresforsch.*, **20**, 35-127.

HERRIG, E. 1966. Ostracoden aus der Weissen schriebkreide (Unter-Maastricht) der Insel Rügen. *Paläont. Abh., Abt. A, Paläozool.*, **2** (4), 693-1024.

HOLSER, W. T. AND KAPLAN, I. R. 1966. Isotope geochemistry of sedimentary sulfates. *Chem. Geol.*, **1**, 93-135.

HULINGS, N. C. AND PURI, H. S. 1965. The ecology of shallow water ostracods of the West Coast of Florida. *Pubbl. Staz. zool. Napoli* [1964], **33** (suppl.), 308-44.

JOHNSON, J. H. 1961. *Limestone-building Algae and Algal Limestones*. Johnson Publ. Co., Boulder, Colorado. 297 pp.

JUDAY, C. 1907. Ostracoda of the San Diego region. II. Littoral forms. *Univ. Calif. Publs Zool.*, **3** (9), 135-56.

KEIJ, A. J. 1957. Eocene and Oligocene Ostracoda of Belgium. *Mém. Inst. r. Sci. nat. Belg.*, No. 136, 210 pp.

KESLING, R. V. 1965. Four reports of Ostracod investigations. 1: Anatomy and dimorphism of adult *Candona suburbana* Hoff. *N.S.F. Project GB-26*: i-iv, 1-56. University of Michigan Press.

KINGMA, J. T. 1948. *Contributions to the knowledge of the young Caenozoic Ostracoda from the Malayan region*. Kemink en Zoon N.V., Utrecht. 106 pp.

KUZNETSOVA, Z. B. 1962. Representative Ostracoda in the families Cytheridae and Paradoxostomidae of the Cretaceous and lower Tertiary deposits of Azerbaijan. *Coll. Sci.-Tech. Inf.*, **1**, 30-50 [in Russian].

MCKENZIE, K. G. 1967a. The distribution of Tertiary-Recent marine Ostracoda from the Gulf of Mexico to Australasia. In G. A. Adams and D. V. Ager (eds), *Aspects of Tethyan Biogeography*. *Publs Syst. Ass.* No. 7.

——. 1967b. Recent Ostracoda from Port Phillip Bay, Victoria. *Proc. R. Soc. Vict.*, **80** (1), 61-106.

——. 1967c. Saipanellidae: a new family of Podocopid Ostracoda. *Crustaceana*, **13** (1), 103-13.

MCKENZIE, K. G. AND SWAIN, F. M. 1967. Recent Ostracoda from Scammon Lagoon, Baja California. *J. Paleont.* **41** (2), 281-305.

MARINOV, T. 1962. On the Ostracod-fauna of the western Black Sea coast. *Bull. Inst. cent. Rech. Sci. pisc. pech, Varna*, **2**, 81-108 [in Russian].

——. 1963. *Pontocytheroma arenaria* n.g.n.sp.—eine neue Ostracode aus der Sandbiozone des Schwarzen Meeres. *C.r. Acad. bulg. Sci.*, **16** (5), 557-60.

——. 1964. Contribution on the Ostracode fauna of the Black Sea. *Acad. bulg. Sci.*, **4**, 39-60 [in Russian].

MOORE, R. C. (ed.) 1961. *Treatise on Invertebrate Paleontology. Pt. Q. Arthropoda. 3. Crustacea: Ostracoda.* 442 pp, 334 Figs. University of Kansas Press, Lawrence, Kansas.

MORKHOVEN, F. P. C. M. VAN. 1963. *Post-Palaeozoic Ostracoda their morphology, taxonomy and economic use.* II. *Generic Descriptions.* Elsevier, New York. 478 pp.

MÜLLER, G. W. 1884. Zur näheren Kenntniss der Cytheriden. *Arch. Naturgesch.* **1**, 1-18.

——. 1894. Die Ostracoden des Golfes von Neapel und der angrenzenden Meeresabschnitte. *Fauna Flora Golf. Neapel, Monog.*, **21**, i-viii, 1-404.

——. 1908. Die Ostracoden der Deutschen Südpolar-Expedition 1901-1903. *Wiss. Ergebn. dt. Sudpolarexped.*, **10**, Zool. 11 Bd., 51-181.

ORLOV, Y. A. (dir.). 1960. *Basic Paleontology: Arthropoda-Trilobitomorpha and Crustaceamorpha.* Moscow [in Russian]. 516 pp.

POKORNÝ, V. 1965. *Principles of Zoological Micropalaeontology.* **2**, Pergamon, London. 465 pp.

PURI, H. S. 1953. Contribution to the study of the Miocene of the Florida Panhandle. *Bull. Fla. St. geol. Surv.*, **36**, 215-309.

——. 1960. Recent Ostracoda from the west coast of Florida. *Trans. Gulf Cst Ass. geol. Socs.*, **10**, 107-49.

——. 1963. Preliminary notes on the Ostracoda of the Gulf of Naples. *Experientia*, **19** (368), 1-6.

REYS, S. 1965a. Note préliminaire sur les Ostracodes d'un sable fin organogène. *Recl Trav. Stn mar. Endoume*, **37** (53), 263-75.

——. 1965b. Ostracodes de la biocénose des fonds détritiques côtiers et de ses facies d'algues calcaires. *Recl Trav. Stn mar. Endoume*, **38** (54), 255-67.

SARS, G. O. 1926-8. *An Account of the Crustacea of Norway. Vol. IX. Ostracoda, Parts* 13-14, 15-16, 207-77. Bergen Museum, Bergen.

SMITH, R. N. 1965. Four reports of Ostracod investigations. 3: Musculature and muscle scars of *Chlamydotheca arcuata* (Sars) and *Cypridopsis vidua* (O. F. Müller) (Ostracoda-Cyprididae). *N.S.F. Project GB*-26, i-vi, 1-40. University of Michigan Press.

TERMIER, G. AND TERMIER, H. 1960. *Atlas de Paleogeographie.* Masson et Cie, Paris.

VOS, A. P. C. DE 1953. Three new commensal Ostracods from *Limnoria lignorum* (Rathke). *Beaufortia*, **4** (34), 21-31.

VOS, A. P. C. DE AND STOCK, J. H. 1956. On commensal Ostracods from the Wood-infesting Isopod *Limnoria*. *Beaufortia*, **5** (55), 133-9.

DISCUSSION

SOHN: Where does *Monoceratina* fit into the picture?

MCKENZIE: As I understand it, *Monoceratina sensu stricto* first occurs in the Devonian and the type species lacks a caudal process.

SOHN: What do you mean by *sensu stricto*?

MCKENZIE: In the sense of Roth (1928) and some illustrations of the type species. As far as I can tell from the latter there is no caudal process in the type species, an observation also made by van Morkhoven. Later forms referred to *Monoceratina* have more or less well developed caudal processes, as indeed do all bythocytherids. In consequence, I use *Monoceratina sensu lato* to incorporate forms generally considered by workers to be *Monoceratina* even though in a particular feature (the caudal process) they do not resemble illustrations of the type species.

SWAIN: It seems unlikely to me that there is a close relationship between the Devonian and Recent forms, even though the muscle scar patterns were found to be very similar. The scar pattern appears to be the main basis for comparing the Recent with the Devonian forms and I think that eventually the Recent ones will have to be separated into a different taxon.

MCKENZIE: Many Mesozoic *Monoceratina* s.l. have caudal processes and it is such forms that I have postulated as the immediate ancestors of Paradoxostomatidae. As a name *Monoceratina* has been used for numerous Ostracoda which range in age from Devonian to Recent.

ROBINSON: I am always worried about *Monoceratina* Roth, and the absence of characteristics which are associated with later so-called *Monoceratina*; but certainly in the lowest Carboniferous, at the very base of the Tournaisian, there are forms which have good caudal processes and the muscle-scars which have just been referred to by Professor Swain. What is more, this muscle-scar pattern is of the type which you illustrated at the beginning of your series in your outline phylogeny. Secondly, there is a tendency for the shells in well-preserved forms to show a very delicate inner 'shelf' within the margin, a duplicature one might say, and this, combined with the caudal process, does suggest some relationship to these later forms. The Roth type, which I know nothing about, has always seemed something of an oddity.

In the Upper Palaeozoic, these forms are useful, and we separate them on the basis of ornament, which can be very varied.

MCKENZIE: I remain uneasy about the referral of Cytheromatinae to Paradoxostomatidae. Here it may be germane to recall common environmental associations, particularly with regard to the interstitial niche. This habitat has been recorded for some Micro-cytherinae and Paradoxostomatinae and for a *Cytheroma* which Dr Elofson has described. Dr Hartmann has stressed previously, however, that the similar morpho-logical characteristics of some interstitial ostracods may be due to adaptive converg-ence rather than to phylogenetic relationship. The 'soft part' morphology of Cythero-matinae is not strikingly similar to that of most Paradoxostomatidae although in some respects quite close to that of *Cytherois*.

The postulated relationships between Paradoxostomatinae and the commensal genera (*Redekea, Laocoonella*) are strengthened by the recent discovery of *Paradoxostoma rostratum* Sars, 1866 as a commensal on two species of marine amphipods (Baker and Wong, 1968, in *Crustaceana* **14** (3), 307-11.

THE TYPE SPECIES OF THE GENUS
STENOCYPRIS SARS 1889
WITH DESCRIPTIONS OF TWO NEW SPECIES[1]

EDWARD FERGUSON, JR.
Lincoln University, Missouri, U.S.A.

SUMMARY

An examination of ostracods collected from Nagpur and from Ceylon by the Rev. S. Hislop and Mr A. Haly respectively reveals some interesting facts relative to the genus *Stenocypris* Sars 1889.

Ostracod specimens collected by the Rev. Hislop from Nagpur in 1845 which are now on deposit in the British Museum (Natural History) as collection 9.26.207–209 and bearing the label *Cypris cylindrica* Sowerby, and collections 9.26.210–212 and 59.61 and labelled *Cypris cylindrica* var. *major* Baird were examined. Two groups of specimens on deposit in the Hancock Museum were also examined: samples bearing the label *Cypris cylindrica* Baird Types, Nagpur, and those designated *Cypris cylindrica* Baird *malcolmsoni*. Permanent microscopic mounts were made of specimens from all of the samples.

It is generally agreed among ostracod workers that *Cypris cylindrica* Sowerby (= *Stenocypris cylindrica* Sowerby) has no representatives among extant forms. Consequently, the specimens on deposit in the British Museum (Natural History) as *Cypris cylindrica* Sowerby have been redescribed as a new species of the genus *Stenocypris*. Stained microscopic mounts of the female holotype and two female paratypes have been placed in the collection of the British Museum (Natural History). An examination of Hancock Museum specimens '*Cypris cylindrica* Baird Types Nagpur' shows structural characteristics, especially those associated with the band of radial pore canals, that indicate that it is a new species of *Stenocypris*, distinct from the British Museum form and from all other known species. Permanent microscopic mounts of the holotype female and a paratype female of the new species are on deposit in the Hancock Museum at Newcastle-upon-Tyne, England.

Comparison of the morphology of British Museum specimens '*Cypris cylindrica* var. *major* Baird' from Nagpur and of Hancock Museum specimens '*Cypris cylindrica* Baird *malcolmsoni*' from Ceylon reveals only a slight difference in size between them, the representatives from Ceylon being the smaller. The size difference does not appear to be significant and it is concluded that *Cypris cylindrica major* Baird 1859 (= *Stenocypris cylindrica major* (Baird, 1859)) and *Cypris malcolmsoni* (= *Stenocypris malcolmsoni* (Brady, 1886)) are synonyms. Under Rules *Stenocypris cylindrica major* (Baird, 1859)) being the older name takes priority over the junior synonym, and thereby becomes the valid type species for the genus *Stenocypris* Sars 1889.

[1] Supported by National Science Foundation Grants GB-1534 and GB-5553.

The status of the subfamily Stenocyprinae Ferguson 1964 is discussed briefly. Despite a contrary opinion by Danielopol (1965), the author is convinced of the validity of this taxon. A comprehensive review of cyprid ostracods possessing a wide anterior inner duplicature, asymmetrical, lamelliform furcal rami, one or both of which are either denticulated or pectinated along the dorsal margin, and having the dorsal seta absent, will doubtlessly add other genera to the subfamily Stenocyprinae.

The guessing game over which group of ostracods Sars had before him at the time he erected the genus *Stenocypris* continues at an accelerating pace. McKenzie (1965) published what he calls a frolicsome note entitled, 'The Great *Stenocypris* Muddle: A Cautionary Tale for Taxonomists.' He is of the opinion that Sars's description of the genus *Stenocypris* is based upon the structure of an Australian form which McKenzie equates with Brady's species *Cypris malcolmsoni*.

Baird (1859) described some ostracods from Nagpur and designated them as *Cypris cylindrica major*. According to Baird the chief difference between his species and the fossil species *Cypris cylindrica* Sowerby is in the relative size, *C. cylindrica major* being about twice the size of *C. cylindrica* in all dimensions. Sars (1889) referring to Brady's species *C. malcolmsoni* states: 'This beautiful species is undoubtedly identical with the form described by Baird from Nagpur, India, under the name *Cypris cylindrica* Sowerby and more especially with the figures given for his variety "major".' Brady (1886) states, 'I have been enabled to compare the Ceylon specimens with some from the Nagpur gathering, described by Dr Baird, for which specimens, as well as for some of the fossil *C. cylindrica* collected by Dr Malcomson, I am indebted to my friend, Professor T. Rupert Jones. The two series are undoubtedly identical; but I learn from my brother, Mr H. B. Brady, that those preserved in the British Museum are much larger—probably Baird's variety *major*.' Brady in a footnote asserts that he is unable to follow Baird in identifying recent specimens with Sowerby's *C. cylindrica*.

Recently I had an opportunity to examine some ostracods collected in 1845 by the Rev. S. Hislop from Nagpur and currently on deposit in the British Museum (Natural History). These specimens bear the following collection numbers: 9.26 207–209 and labelled *Cypris cylindrica* Sowerby; and 9.26 210–212 and 59.61 labelled *Cypris cylindrica* var. *major* Baird. Two groups of specimens on deposit in the Hancock Museum have also been examined, namely, samples bearing the label *Cypris cylindrica* Baird Types, Nagpur and those designated *Cypris cylindrica* Baird *malcolmsoni*. The Hancock Museum specimens were collected from Ceylon by Mr A. Haly. A careful and detailed examination of all these organisms suggests that the specimens on deposit in both the British Museum (Natural History) and the Hancock Museum as *Cypris cylindrica* Sowerby represent two previously undescribed species.

A comparison of the structure of *Stenocypris cylindrica major* (Baird, 1859) (= *Cypris cylindrica major* Baird, 1859) with that of *Stenocypris malcolmsoni* (Brady, 1886) (= *Cypris malcolmsoni* Brady, 1886) gives strong evidence that the two are conspecific.

Systematic Record
Family Cyprididae Baird, 1845
Subfamily Stenocyprinae Ferguson, 1964
Genus *Stenocypris* Sars, 1889

Stenocypris hislopi n. sp.
Fig. 1, 2–6

Female

Valves from above elliptical, left valve overlaps right; seen laterally valves reniform; surfaces with numerous puncta, but without other ornamentation; anterior and posterior margins broadly rounded; ventral margin with a weak anterior sinuation; band of radial pore canals along margins except dorsally; dorsum with a low arch sloping downward posteriorly. Inner duplicature very wide anteriorly, considerably narrowed posteriorly. Length of valves 1·38–1·50 mm; height 0·55–0·65 mm; maximum height at posterior end. Natatory setae of second antennae well developed but not reaching tips of claws. Terminal podomere of third thoracic appendage elongate, length approximately eight times least width, distal end with a digitiform projection between a long seta and a shorter claw-like denticulated one. Furcal rami asymmetrical, wider of the two pectinated along distal half of ramus; terminal and subterminal spines of both rami pectinated; length of the narrower ramus sixteen times least width, ramus smooth, ventral margin with a slight curvature anteriorly; length of wider furca approximately thirteen times least width; dorsal seta absent, terminal setae longer than terminal spines.

Male

Unknown.

Type locality

Specimens of *S. hislopi* were collected from pools in Nagpur, India, in 1845 by the Rev. S. Hislop in whose honour the new species is named. Specimens of *Cypris subglobosa* Sowerby were also present in these samples.

Type specimens

Stained microscopic mounts of the female holotype and female paratypes and numerous unstained dried specimens are on deposit in the British Museum (Natural History). The registered numbers are for the holotype B.M. 1966.10.17.1 and for the stained paratypes B.M. 1966.10.17.2 and B.M. 1966.10.17.3 respectively.

Remarks

Stenocypris hislopi shows a remarkable resemblance to *Stenocypris cylindrica major* (Baird, 1859). It is immediately obvious that despite the close resemblance between the two species, *S. hislopi* is significantly smaller than *S. cylindrica major*. The following are measurements of several mature females of *S. hislopi* and of seven mature females of *S. cylindrica major*.

Stenocypris hislopi n. sp.

Length of Valve	Height of Valve
1·38 mm	0·55 mm
1·40	0·58
1·43	0·58
1·43	0·63
1·45	0·60
1·50	0·65

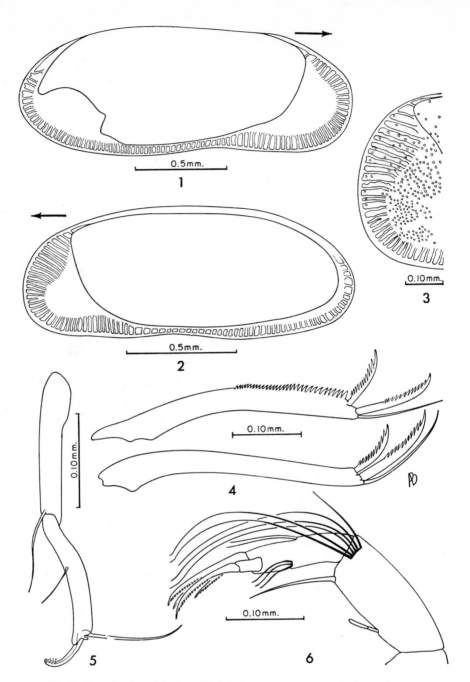

Fig. 1. Unstained mesial view of left valve of *Stenocypris malcolmsoni* (Brady, 1886) [(= *Stenocypris cylindrica major* (Baird, 1859)]. 2-6. *Stenocypris hislopi* n. sp., stained in a 1% alcoholic solution of eosin Y and mounted in Canada balsam: 2. Mesial view of right valve of female holotype; 3. Details of radial pore canals of female holotype; 4. Furcal rami of female paratype; 5. Third thoracic appendage of female paratype; 6. Second antenna of female paratype.

Stenocypris cylindrica major

Length of Valve	Height of Valve
2·00 mm	0·80 mm
2·00	0·80
2·01	0·75
2·08	0·82
2·08	0·83
2·10	0·83
2·13	0·87

There is apparently no overlapping of sizes between the two species. The maximum length recorded for *S. hislopi* is 1·50 mm; on the other hand the smallest specimen of *S. cylindrica major* is 2·00 mm. *S. hislopi* could easily be mistaken for a late instar of *S. cylindrica major*, except for the fact that the examination of a large number of valves of *S. hislopi* shows them to be completely adult shells. *S. hislopi* resembles *S. cylindrica major* in possessing a distinct band of radial pore canals along the margins of the valves except dorsally. It differs from the larger species in having a denticulated distal claw-like seta and a digitiform projection at the distal end of the third thoracic appendage. The inner duplicature of the new species is very narrow posteriorly, much wider and irregularly shaped in *S. cylindrica major*.

Stenocypris halyi n. sp.
Fig. 2, 1–4

Female

Eye prominent. Shell elliptical, ventral margin with a slight sinuation anteriorly; inner duplicature wide anteriorly, very narrow posteriorly; hairs along margins except dorsally; valve surfaces with puncta, but hairs and other ornamentations absent. Pore canals not forming a distinct striated band along margins; alternating with radial pore canals are a series of shorter canals. Length of valves 1·40–1·50 mm; maximum height 0·44–0·50 mm. Natatory setae of second antenna not quite reaching tips of claws. Dorsal margin of wider furcal ramus denticulated along distal one-half; length approximately thirteen times least width and slightly curved at anterior end; dorsal margin of narrower ramus smooth, length of ramus sixteen times least width and straight; terminal and subterminal spines denticulate; dorsal setae absent; terminal setae short with a length about one-fourth that of terminal spine.

Male

Unknown.

Type locality

Collected by Mr A. Haly from pools in Ceylon.

Type specimens

Permanent microscopic mounts of the holotype female and three female paratypes are on deposit in the Hancock Museum.

F

Remarks

Specimens of *Stenocypris halyi* n. sp. have been on deposit in the Hancock Museum
for many years, and have borne the label *Cypris cylindrica*. Assuming the validity of
the consensus that *Stenocypris cylindrica* (Sowerby) occurs only as a fossil form, it
appears that the specimens in the Hancock Museum collections are not identical with
Sowerby's species. An examination of *S. halyi* reveals that it differs significantly in
two respects from closely related species. The absence of a distinct band of radial pore
canals along the anterior, ventral, and posterior margins of the valves distinguishes
this species from *S. hislopi*. *S. halyi* resembles *S. fontinalis* Vávra, 1895 in the absence
of a distinct striated band of radial pore canals, but may be separated from the latter
species by size differences and by the structure of the narrower ramus. *S. fontinalis*
measures 1·70 mm in length, and the narrower of its furcal rami bears short denticles
at the distal end; *S. halyi* measures 1·40–1·50 mm in length, and its narrower furcal
ramus is smooth.

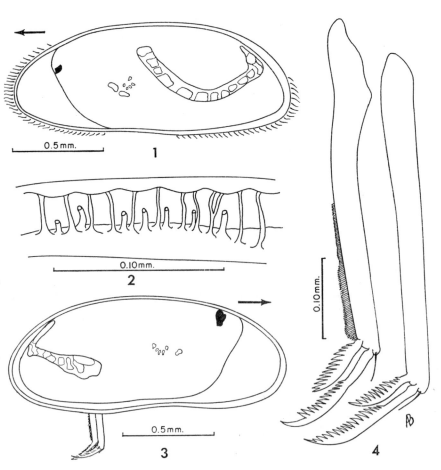

Fig. 2. 1-4. *Stenocypris halyi* n. sp: 1. Mesial view of right valve of female holotype. 2. Details of pore canals of female holotype. 3. Mesial view of left valve of
female paratype. 4. Furcal rami of female paratype.

Discussion

Sars (1889), as pointed out earlier in this paper, was somewhat uncertain about the specimens he had before him. However, I am led to the conclusion that he was examining what he considered to be *Stenocypris malcolmsoni*, but what was in fact *Stenocypris cylindrica major*. A considerable measure of support is given by Sars himself in his statement: 'This beautiful species is undoubtedly identical with the form described by Baird from Nagpur, India, under the name *Cypris cylindrica* Sowerby and more especially with the figures given for his variety "major".'

A detailed examination of specimens designated *Stenocypris cylindrica major* (Baird, 1859) and *Stenocypris malcolmsoni* (Brady, 1886) respectively, reveals no

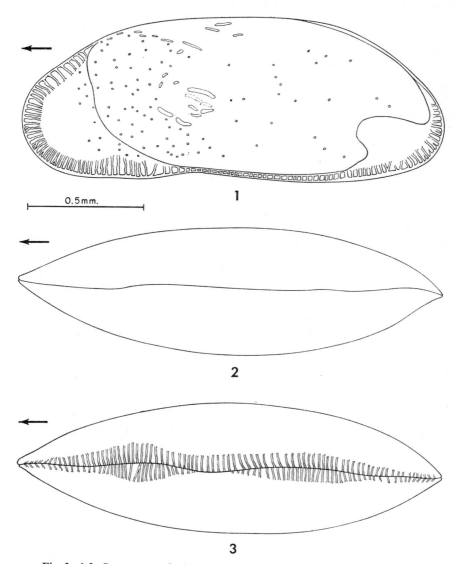

0.5mm.

1

2

3

Fig. 3. 1-3. *Stenocypris cylindrica major* (Baird, 1859): 1. Mesial view of right valve; 2. Dorsal view of right and left valves; 3. Ventral view of right and left valves.

major morphological differences between the two species. Brady (1886) in his description of *S. malcolmsoni* gives the length as $\frac{1}{12}$ inch or 2·1 mm. *S. cylindrica major* specimens from Nagpur (Fig. 3, 1–3) have a length varying from 2·00 to 2·13 mm; specimens of *S. malcolmsoni* from Ceylon (Fig. 1, 1) measure from 1·95 to 2·00 mm in length. However, these slight differences in the length of the valves when placed alongside the similarities in structure the two groups possess in common appear to be insignificant. Consequently, I agree with Howe (1962) and others who are of the opinion that *S. cylindrica major* (Baird, 1859) and *S. malcolmsoni* are synonyms. Under Rules *Stenocypris cylindrica major* being the older name takes priority over the junior synonym, *Stenocypris malcolmsoni* (Brady, 1886), and is therefore, the valid type-species of the genus *Stenocypris* Sars, 1889.

Ferguson (1964) erected the subfamily Stenocyprinae to accommodate cyprid ostracods having the following characteristics: shell elongate narrow, elliptical from above, frequently reniform, surface smooth with scattered puncta, pore canals when present restricted to anterior, posterior, and ventral margins; dorsal margin evenly arched, occasionally flattened; hairs along margins except dorsally; inner duplicature large at anterior end; length usually twice the maximum height. Natatory setae of second antennae barely reaching tips of claws. Ultimate podomere of third thoracic appendage bearing a curved claw-like seta, an extremely short one, and a very long, unreflexed seta. Furcal rami large, lamelliform, dissimilar; dorsal margin of one or both either denticulated or pectinated; dorsal seta absent; terminal and subterminal claws heavily pectinated. Males and females, with males unknown for many species.

Danielopol (1965) apparently questions the validity of the subfamily Stenocyprinae. He is of the opinion, which I share, that the genera *Stenocypris*, *Acocypris*, *Herpetocypris* and *Isocypris* show affinities which doubtlessly place them in a single phyletic line. However, he states, 'Il existe aussi une asymétrie de la furca chez *Isocypris quadrisetosa* Rome et l'absence du poil postérieur de la furca est propre aussi au genre *Acocypris*. Donc je considère qu'on ne peut pas établir une nouvelle sous-famille en partant de ces caractères.' However, it should be emphasized that the subfamily Stenocyprinae was not erected on the basis of a single morphological character. Relative to the symmetry of the furcal rami and the presence or absence of the dorsal seta, members of the subfamily Stenocyprinae possess lamelliform, asymmetrical furcae, the dorsal margin of one or both of which is denticulated or pectinated and having the dorsal seta absent. The furcal ramus of members of the genus *Herpetocypris* bears a dorsal seta. Rome (1947) in Fig. 3, L, furca: g, d. shows each furcal ramus of *Isocypris quadrisetosa* with a dorsal seta, which is only slightly removed from the subterminal spine.

It is my opinion that the subfamily Stenocyprinae is a valid taxon, and that it rests upon a solid morphological foundation. Subsequent investigations will doubtlessly add other genera to this subfamily.

ACKNOWLEDGEMENTS

The author acknowledges with thanks the assistance of Dr J. P. Harding, Mrs Patricia L. Barker, Messrs W. A. Smith and A. V. Hounsome of the British Museum (Natural History), and Mr A. M. Tynan, Curator, and Mrs O. Marshall of the Hancock Museum. He also expresses his appreciation to Miss Alice Boatright of the University of Illinois and to Mrs June Gilby for aid in the preparation of the drawings.

REFERENCES

BAIRD, W. 1859. Description of some new recent Entomostraca from Nagpur collected by the Rev. S. Hislop. *Proc. zool. Soc. Lond.*, No. 398, 231-4.
BRADY, G. S. 1886. Notes on Entomostraca collected by Mr A. Haly in Ceylon. *J. Linn. Soc.*, Zoology, **19**, 293-317.
DANIELOPOL, D. L. 1965. Nouvelles données sur les ostracodes d'eau douce de Roumanie: *Cordocythere phreaticola* n. g. n.sp., *Eucypris petkovskii* n.sp., Limnocytherinini et Metacyprini nouvelles tribus de la sous-famille Limnocytherinae Sars, 1925. *Annls Limnologie*, **1** (3), 443-68.
FERGUSON, E. JR. 1964. Stenocyprinae, a new subfamily of freshwater cyprid ostracods (Crustacea) with description of a new species from California. *Proc. biol. Soc. Wash.*, **77**, 17-24.
HOWE, H. V. 1962. *Ostracod Taxonomy*. Louisiana State University Press, Baton Rouge, La., v-xix, 366 pp.
MCKENZIE, K. G. 1965. The great *Stenocypris* muddle: a cautionary tale for taxonomists. *Aust. Soc. Limnol.*, **4** (2), 51-2.
ROME, D. R. 1947. Contribution a l'étude des ostracodes de Belgique. I.—Les ostracodes du Parc —St. Donat à Louvain. *Bull. Mus. r. Hist. nat. Belg.*, **23** (34), 1-24.
SARS, G. O. 1889. On some freshwater Ostracoda and Copepoda raised from dried Australian mud. *Forh. VidenskSelsk. Krist.* No. 8. 79 pp. 8 Pls.

DISCUSSION

MCKENZIE: We still do not know for certain the status of the original specimens which Sars described as *Stenocypris malcomsoni*. Sars stated that in his opinion the Australian material was identical with Baird's *cylindrica major* which Brady renamed *malcolmsoni*. Lately, Dr Triebel and Prof. Ferguson both have noted that *cylindrica major* and *malcolmsoni* are identical, with the former, of course, taking priority. But is Sars's Queensland material identical with this species? I hope that Dr Ferguson will get to Australia sometime to re-collect the Queensland localities and sort out this particular problem. I have a second comment to make: regarding *S. halyi*. Since it lacks radial septa this species probably belongs in *Chrissia* Hartmann. Furthermore, would Dr Ferguson agree that his subfamily Stenocyprinae now comprises the three genera *Stenocypris* Sars, *Chrissia* Hartmann and *Parastenocypris* Hartmann?

SYLVESTER-BRADLEY: The question of the type species of *Stenocypris* does not rest on whether Sars correctly identified the material or not. According to the Rules of Nomenclature it is the species he thought he was dealing with that becomes the type species and so I don't think it is necessary for Dr Ferguson to go to Australia to settle this particular problem.

ÜBER DIE NOTWENDIGKEIT EINER REVISION DER SÜSSWASSER-OSTRACODEN EUROPAS

T. PETKOVSKI
Prirodonaučen Muzej, Skopje, Jugoslawien

SUMMARY

Many of the fresh-water ostracod species of Europe, particularly those belonging to the genera *Candona* and *Potamocypris*, have been very inadequately described. Recognition of species is already difficult for zoologists. Palaeontologists, who have only the shell available for comparative purposes, find themselves in a very complicated situation.

During detailed investigations of material of *Candona neglecta* G. O. Sars from south-east Europe, the author was able to separate four new species (two of which *C. fasciolata* Petkovski and *C. altoides* Petkovski have already been described) on the basis of the genital organs in both sexes. In this case it is a question of homoeomorphic species which have an outwardly similar shell—an environmental feature, but which differ widely in the morphology of the genitalia.

A completely different state of affairs was found in a revision of species of *Candona* belonging to the *C. candida* group from Lake Ohrid in Macedonia. Over a dozen species had a single main type of male copulatory organ (*C. neglecta* type). Fortunately one finds that in the females the genitalia are already well differentiated morphologically, which enables them to be assigned to natural groups. Three such groups have been established. All these species can be separated on the outline of the shell very clearly. Such divergence in shell form perhaps arose more as adaptation to the different ecological niches of the lake bottom, and is certainly intra-lacustrine.

The example of the forms related to *Candona neglecta* quoted above shows how difficult it is, on the basis of shell outline alone, to say anything about the phylogenetic relationships. The author has not yet examined the detailed structure of the shell; it is possible that the detailed structure will offer many reliable features and will strongly reflect the true relationships. It is assumed that similar problems occur in the other groups of fresh-water ostracods. A revision of these ostracods is therefore absolutely essential.

ZUSAMMENFASSUNG

Viele Ostracodenarten des Süsswassers von Europa, insbesonders aber die aus den Gattungen *Candona* und *Potamocypris*, sind sehr dürftig beschrieben. Eine Wiedererkennung der Arten ist schon dem Zoologen schwierig. In einer viel komplizierterer Lage befinden sich die Paläontologen, denen nur die Schalen zu Vergleichszwecken dienen können. Von den Schalen aber sind meist nur unexakte Umrisse abgebildet. Beispiele dafür werden angeführt.

Während der detaillierten Untersuchungen des Materials von *Candona neglecta*

G.O. Sars aus Südost-Europa, konnte der Autor, aufgrund der Genitalorgane bei beiden Geschlechtern, vier neue Arten absondern (zwei davon wurden schon beschrieben: *fasciolata* Petkovski und *altoides* Petkovski). In diesem Falle handelt sich es um homöomorphe Arten, die äusserlich sehr ähnliche Schalen haben (milieubedingt), die aber in der Morphologie der Genitalien weit auseinander stehen.

Ein ganz umgekehrtes Verhältnis fand sich bei der Bearbeitung der *Candona*-Arten aus der *candida*-Gruppe vom Ohridsee in Mazedonien. Über ein Dutzend Arten besitzen einen einzigen Haupttypus von männlichen Kopulationsorganen (*neglecta*-Typus). Glücklicherweise findet man bei den Weibchen dagegen morphologisch schon gut differenzierte Genitalien, die eine Verteilung der Arten in natürliche Gruppen ermöglichen. Es sind drei solcher Gruppen festgestellt worden. Alle diese Arten lassen sich aber im Umriss der Schalen sehr deutlich, schon äusserlich unterscheiden. Eine solche Divergenz in der Form der Schalen ist wahrscheinlich als Anpassung an verschiedenen ökologischen Nischen des Seebodens entstanden, und zwar intralakustrisch.

Die oben angeführten Beispiele von *Candona neglecta*-Verwandten weisen darauf hin, wie schwierig es ist, nur aufgrund des Umrisses der Schalen, etwas über phylogenetische Zusammenhänge auszusagen. Der Autor hat noch nicht die Feinstruktur der Schalen näher untersucht; es ist möglich, dass der Feinbau manche zuverlässige Merkmale bietet und die richtige Verwandtschaft stärker widerspiegelt. Es ist anzunehmen, dass auch in den anderen Ostracoden-Gruppen des Süsswassers ähnliche Probleme auftreten. Eine Revision dieser Ostracoden ist deshalb unumgänglich.

Einen fast vollständigen Überblick über die Süßwasserostracoden Europas gibt neuerdings Löffler in der Illiesschen Limnofauna Europaea—Gustav Fischer Verlag, Stuttgart 1967. Aus der Artenliste ist zu ersehen, daß die Vertreter der Gattung *Candona* am zahlreichsten vorhanden sind—rd. 108 Spezies mit sicherer und weitere 35 mit unsicherer systemat. Stellung. Ein Wiedererkennen einer so großen Zahl von Arten ist natürlich mit beträchtlichen Schwierigkeiten verbunden, um so mehr, weil viele dieser Candonen sehr dürftig beschrieben sind. Außerdem wird die Determination noch dadurch erschwert, daß die vorliegenden Bestimmungsbücher entweder nicht fehlerfrei sind oder einen unklaren Überblick vermitteln. Das ist der Fall mit Bronstein (Fauna SSSR, 1947), der keine Gruppenverteilung der Candonen vorgenommen hat. Dies betrifft besonders die von ihm neu aufgestellten Arten.

In einer viel komplizierteren Lage befinden sich die Paläontologen, denen nur die Angaben über die Ostracodenschalen zu Vergleichszwecken zur Verfügung stehen. Von diesen sind aber die Umrisse meist nicht exakt abgebildet. Das gilt insbesondere für die Candonen der *compressa/rostrata*-Gruppe, weil sie alle äußerlich sehr ähnlich geformte Schalen haben. Noch weniger brauchbar sind aber die Schalen-Darstellungen der subterran lebenden Candonen; in der Tat sind diese nur als sehr vereinfachte Silhouetten entworfen.

Am schlimmsten steht es jedoch mit der Bestimmung der *Potamocypris*-Arten. Von den insgesamt 24 für Europa angeführten Formen, können sogar spezielisierte Zoologen nur fünf Arten zuverlässig bestimmen—und zwar dank der spezifischen Form ihrer Schalen:

P. variegata (Brady und Robertson) *P. producta* Sars

P. humilis Sars   *P. steueri* Klie

P. unicaudata Schäfer

Die übrigen 19 *Potamocypris*-Spezies lassen sich lediglich auf fünf (natürliche)

Gruppen verteilen, innerhalb derer die systematischen Zusammenhänge nicht geklärt sind (Tab. 1). Nur eine vergleichende Überprüfung sämtlicher Arten (nach dem Originalmaterial oder – fundort) innerhalb jeder einzelnen Gruppe kann die Verhältnisse erhellen. Anders ist jede Mühe vergebens.

TABELLE 1. Nicht eindeutig bestimmbare *Potamocypris*-Arten

1. Gruppe:	*P. smaragdina* (Vávra)	*P. almasyi* Daday
	P. longisetosa Bronstein	*P. dianae* M. Fox
2. Gruppe:	*P. similis* G. W. Müller	*P. tarnogradsky* Bronstein
3. Gruppe:	*P. zschokkei* (Kaufmann)	*P. compacta* Klie
	P. wolfi Brehm	*P. hambergi* Alm
	P. wolfi pyrenaica (Margalef)	
	P. hambergi rotundata Alm	
4. Gruppe:	*P. villosa* (Jurine)	*P. arcuata* Sars
	P. villosa crassipes (Masi)	*P. maculata* Alm
	P. vanoyei de Vos	
5.	*P. fulva* (Brady)	*P. pallida* Alm

Selber habe ich mich jahrelang mit der Systematik der Candonen, die über 40% aller Süßwasserostracoden Europas ausmachen befaßt, wobei ich der Morphologie der Genitalien beider Geschlechter eine große Aufmerksamkeit gewidmet habe. Dabei hat es sich gezeigt, daß der mittlere Fortsatz des männlichen Kopulationsorgans das am besten geeignete systematische Merkmal darstellt und—was besonders wichtig ist—es läßt sich sehr einfach herauspräparieren. Die übrigen Teile der männlichen und weiblichen Genitalien sind ebenfalls brauchbar, aber nicht so markant.

Bei sämtlichen *Candona*-Arten der *candida*-Gruppe habe ich elf verschiedene Bautypen von männlichen Kopulationsorganen festgestellt (Tab. 2). Es ist erwähnenswert, daß die mitteleuropäischen Arten sich als voneinander ganz gut abgesondert

TABELLE 2. Verteilung der Candonen der *candida*-Gruppe aufgrund des mittleren Fortsatzes am männlichen Kopulationsorgan

1. *neglecta*–Typus:	*C. neglecta* Sars *C. strumicae* Petkovski *C. bulgarica* n. sp. Sämtliche endemischen Arten im Ohridsee (außer *C. sketi* n. sp. und *C. ohrida* Holmes).
2. *angulata*–Typus:	*C. angulata* G. W. Müller
3. *fasciolata*–Typus:	*C. fasciolata* Petkovski *C. paionica* Petkovski *C. marginatoides* Petkovski *C. sketi* n. sp. im Ohridsee *C. bimucronata* Klie
4. *altoides*–Typus:	*C. altoides* Petkovski, *C. tuberculata* Lindner
5. *candida*–Typus:	*C. candida* (O. F. Müller) *C. ohrida* Holmes (= *C. cristatella* Klie)
6. *weltneri*–Typus:	*C. weltneri* Hartwig
7. *mülleri*–Typus:	*C. mülleri* Hartwig
8. *studeri*–Typus:	*C. studeri* Kaufmann
9. *improvisa*–Typus:	*C. improvisa* Ostermeyer
10. *natronophyla*–Typus:	*C. natronophila* n. sp.
11. *levanderi*–Typus:	*C. levanderi* Hirschmann

erweisen; jede Art hat einen eigenartig gestalteten mittleren Fortsatz des erwähnten Organs. Demgegenüber zeigt sich die entsprechende *Candona*-Gruppe vom Balkan in dieser Hinsicht mehr oder minder in Artenkreise aufgeteilt, d.h. alle Arten eines Kreises haben einander ähnlich gestaltete männliche Kopulationsorgane.

Die Vertreter der *fabaeformis*- und *acuminata*-Gruppe sind, genau wie die schon erwähnten Arten der *candida*-Gruppe aus Mitteleuropa, in Bezug auf das männliche Kopulationsorgan sehr deutlich voneinander getrennt. Auf dem Süd-Balkan kommt nur eine Abart von *C. fabaeformis* Fischer vor, während die *acuminata*-Angehörigen vollkommen fehlen.

Die Morphologie der Genitalien bestätigt also ziemlich vollständig die bisherige Gruppierung der Candonen, die auf der Beborstung des Mandibulartasters begründet wurde. Die neuen Merkmale erlauben aber viel tiefer gehende Erörterungen der phylogenetischen Stellung der einzelnen *Candona*-Arten zueinander. Von diesem Anhaltspunkt ausgehend bin ich geneigt, die Arten der Gruppen: *acuminata, candida* und *fabaeformis* in einem gemeinsamen Genus *Candona* zusammenzufassen. Die ehemaligen Gruppen sollen bestehen bleiben, jedoch nur innerhalb dieses Genus. Die übrigen Genera wären: *Pseudocandona* (*compressa* und *rostrata*-Gruppe), *Crypto-candona* und *Mixtocandona*. Die Genera *Metacandona* und *Nannocandona* benötigen eine Überprüfung, während *Typhlocypris* wegfallen müßte.

Bei der Untersuchung der Populationen von *C. neglecta* Sars vom Balkan fand ich die typische Art nur in den Gebirgsbiotopen. In den Niederungsgewässern wird sie durch vier Vikariaten vertreten, die in der äußeren Form der Schalen mit *C. neglecta* völlig übereinstimmen. Ein ganz anderes Bild bietet die Morphologie der Genitalien: vier vikariierenden Arten stehen sowohl unter sich als auch gegenüber *C. neglecta* sehr weit auseinander. Den männlichen Genitalien nach gehört nur ein Vikariat zum *neglecta*-Typus (*C. bulgarica* n. sp.), während die übrigen drei Arten zu drei anderen, verschiedenen Typen gehören (*C. fasciolata, C. altoides* und *C. natronophila* n. sp.) (Tab, 2). Ganz offensichtlich handelt es sich hier also um homöomorphe Spezies. Sie haben sich höchstwahrscheinlich unter ähnlichen Lebensbedingungen in den Kleingewässern herausgebildet.

Die Candonen des Ohridsees habe ich nur teilweise untersucht, trotzdem aber konnte ich sehr aufschlußreiche Beobachtungen machen. Es kommen nämlich im See über 15 bisher sicher klassifizierte Arten vor. Unter diesen gibt es je einen Vertreter der *compressa*-, *rostrata*- und *fabaeformis*-Gruppe. Der Rest von über einem Dutzend Arten gehört der *candida*-Gruppe an (Tab. 2). Das bedeutet eine erstaunliche Anhäufung von Angehörigen einer einzelnen *Candona*-Gruppe in einem See. Das ist fast die Hälfte aller bisher bekannten europäischen Arten der *candida*-Gruppe. Es ist weiterhin bemerkenswert, daß sie alle—mit einer einzigen Ausnahme (*C. sketi* n. sp.)—einen gleichartig gestalteten Fortsatz des männlichen Kopulations-organs besitzen. Sie bilden also einen Artenkreis von miteinander sehr nahe verwandten Arten des *neglecta*-Typus, wie es schon für die balkanischen Arten angedeutet wurde. Das ist eine Merkwürdigkeit der endemischen Arten des Ohridsees. Die zweite Besonderheit liegt darin, daß trotz der zweifellos engen Verwandtschaft alle diese Candonen voneinander sehr deutlich abweichende Schalen haben, die sogar schon äußerlich leicht unterscheidbar sind. Hier liegt also im Gegensatz zu der oben erwähnten Homöomorphie der Arten aus den Kleingewässern Divergenz vor. Obwohl die genauere räumliche Verteilung der Candonen im Ohridsee nicht bekannt ist, kann man annehmen, daß die Divergenz in der Schalenausbildung ebenso wie die Homö-omorphie ein Resultat der Anpassung ist, diesmal an verschiedene ökologische

Nischen des sehr mannigfaltig gestalteten Seebodens. Offenbar liegt hier ein gutes Beispiel für intralakustrische Speziation vor, die zu einer enormen Entfaltung einer einzelnen Ausgangsform (des *neglecta*-Typus) geführt hat. Weitere Beispiele über die intralakustrische Speziation im Ohridsee sind auch bei Isopoden, Amphipoden, Schnecken u. a. Tiergruppen festgestellt worden.

Die beiden oben behandelten Beispiele der Candonen aus der *candida*-Gruppe—einmal aus Kleingewässern und zum anderen aus dem Ohridsee—habe ich angeführt um zu zeigen, wie schwierig es ist, nur aufgrund eines einzelnen Merkmals, nämlich des Schalenumrisses, etwas über die phylogenetischen Zusammenhänge auszusagen. Beide Erscheinungen—die der Homöomorphie und der Divergenz—können denjenigen, der nur die Schalen behandelt, auf einen Irrweg führen.

Es gibt bestimmt weitere ähnliche und andere Probleme auch bei den übrigen Süßwasserostracoden Europas. Deshalb ist eine Revision dieser Tiergruppe unumgänglich. Die Revision soll nicht nur die Überprüfung der bestehenden Beschreibungen und Materialien umfassen, sondern sich auf neue weitergehende Freilanduntersuchungen ausdehnen. Nachdem die Arten rezent-zoologisch einwandfrei determiniert sind, müssen ihre Schalen auch sehr sorgfältig dargestellt werden, so daß sich auf diese Weise fossile und rezente Ostracoden umfassender vergleichen lassen.

DISCUSSION

GREKOFF: Sie haben gesagt, und natürlich haben Sie recht, dass man nicht nur nach Umriss der Schalen eine Art oder ein Genus bestimmen kann. Haben Sie in Ihren Arbeiten Beziehungen zwischen den inneren Organen und der Struktur von Schalen, insbesondere in der *Candona*-Gruppe, gefunden? Sie sagten, glaube ich, Sie hätten die Fein-Struktur der Schalen noch nicht untersucht; aber denken Sie dieses in Zukunft zu machen? Welches sind diese Beziehungen zwischen inneren Organen und der Schale?

PETKOVSKI: Ich habe bisher die Ostracoden rein zoologisch betrachtet; deshalb gibt es in meinen bisherigen Publikationen nicht viele Einzelheiten über den Schalenbau. Jetzt bereite ich aber eine Revision der Candonen vor und werde versuchen, die Arten zuerst nach ihren Weichteilen genau zu bestimmen, dann ihre Schalenmerkmale darzustellen. Dadurch wird sich feststellen lassen, ob es Unterschiede in der Morphologie der Schalen gibt und wie groß sie sind, insbesondere bei den Arten, die äußerlich gleich aussehen.

In manchen Fällen sind keine spürbaren Unterschiede zu erwarten. Man findet jedoch fast immer Schalenmerkmale, die für die einzelnen Arten spezifisch sind, so z.B. die ganz genaue Form, die Größe, verschiedene Profile der Schalen, des Gehäuses u.s.w.

Die von mir schon erwähnten Erscheinungen der Homöomorphie und der Divergenz, können die Paläontologen bei der Beurteilung der Verwandtschaft der Arten sehr leicht irreführen, da es in diesen Fällen keine Korrelation in der Ausbildung der Gliedmaßen und der Schale gibt. Ich werde versuchen, auch für solche Fälle irgendwelche Merkmale an den Schalen zu finden, die die Verwandtschaft richtig widerspiegeln.

OERTLI: Sie sagten in Ihrer Zusammenfassung, dass von den Schalen oft nur unexakte Umrisse abgebildet sind. Meinen Sie, wenn die Umrisse exakt abgebildet würden, dass dann die Schalen allein—dem Paläontologen—genügen könnten, um, wenigstens in vielen Fällen, die Arten zu bestimmen?

PETKOVSKI: Ich habe gesagt: Es sind nur unexakte Umrisse angegeben, das bedeutet aber nicht, daß die Bestimmung gelingen könnte, wenn diese exakt wären. Zur Bestimmung sind weitere Angaben nötig, wie z.B. Ausbildung des Schalenrandes, Lage des Innenrandes, Struktur der Schalenoberfläche und die übrigen, im Gespräch mit Herrn Dr. Grekoff bereits angeführten Charaktere.

NEALE: Welche Charakteristika des Weichkörpers sind Ihrer Meinung nach von der größten

Bedeutung für die Taxonomie (erste Antenne, vorletztes Glied des Putzfußes, Setae des Mandibulartasters sowie Geschlechtshöcker)?

PETKOVSKI: Die Klassifizierung der Candonen basiert auf der Beborstung des Mandibulartasters, was mit der Ausbildung der Schalen durchaus im Einklang steht. Weiter ist die Beschaffenheit des Putzfußes grundlegend für die Gruppen *Cryptocandona* und *mixta*. Ferner sind die Genitalien der Ostracoden besonders gut für die Taxonomie brauchbar, da sie vom Milieu am wenigsten beeinflußt worden sind. Beispielsweise bestätigt die Morphologie der Genitalien bie den Candonen nicht nur die Gruppenverteilung, sondern läßt weit tiefere Einblicke in der Phylogenie der Arten zu. Deshalb halte ich alle diese Merkmale des Weichkörpers für gut.

II. MORPHOLOGY AND FUNCTION

VARIATIONS PHÉNOTYPIQUES CHEZ UNE ESPÈCE TORTONIENNE DU GENRE *ELOFSONELLA* POKORNÝ

G. CARBONNEL
Université de Lyon, France

SUMMARY

This study underlines the importance of phenotypic variation in *Elofsonella amberii* sp. nov., a new Tortonian (Upper Miocene) species belonging to the Hemicytherinae. The horizon at which this was collected was a biocoenose of the oyster *Gryphaea gryphoides*. Three morphological patterns are described, namely the normal pattern with three dorsal ribs and a pitted surface, and the variants which have one or two dorsal ribs and in which the surface becomes smooth in 15% of the population.

The author considers that these changes are connected with a fresh-water influx into a marine environment and that it is necessary to be careful not to interpret ecological variations as stratigraphical evolution.

Cette étude s'inscrit dans le cadre des travaux antérieurs de V. Pokorný et P. Sandberg. Ils ont attiré l'un et l'autre l'attention sur le problème complexe des variations phénotypiques dans diverses familles d'Ostracodes.

Il est en effet très important, sinon indispensable, de pouvoir connaître l'ampleur des variations ornementales des Ostracodes et leur causalité. Ceci permettra d'éviter de superposer ces variations à celles résultant de l'évolution stratigraphique et de s'en servir comme telles.

LOCALISATION GÉOGRAPHIQUE ET STRATIGRAPHIQUE DU MATÉRIEL

Le matériel étudié provient des assises tortoniennes marines de la vallée du Rhône. Cette série, bien connue depuis la mise à jour récente de G. Demarcq (1962), se présente le plus souvent sous un faciès sableux à la partie supérieure, marneux (marnes bleues fossilifères) à la partie inférieure.

Cette succession stratigraphique, épaisse de 30 à 50 m dans le bassin de Valréas (Vaucluse), se termine fréquemment par une abondante accumulation en lentilles de valves de *Gryphaea gryphoïdes* (Schlot.) (= *Ostrea crassissima* Lamk.) en biocénose. C'est le cas au lieu dit 'la Savoyonne' (G. Demarcq, 1962) d'où provient le matériel.

Immédiatement au-dessus de ce banc à huîtres s'est déposée une série marno-sableuse et marneuse à intercalations ligniteuses du Tortonien lacustre.

C'est l'analyse morphologique d'une espèce d'ostracode contenu dans les sédiments remplissant les huîtres bivalves que je me suis efforcé de réaliser.

Cette microfaune est quantitativement abondante. Elle renferme au moins à ce jour une trentaine d'espèces.

De plus son état de conservation est particulièrement remarquable. Il m'a permis d'observer les variations de l'ornementation faisant l'objet de cette étude.

Parmi les publications relatives au Tortonien (celle ayant trait au stratotype de Tortona en Italie par Dieci et Russo (1964) ou celle de Ruggieri (1962) relative au Tortonien de l'Enna en Sicile), aucune espèce n'appartenant au genre *Elofsonella* Pokorný, n'a jamais été signalée. C'est la raison pour laquelle je suis obligé de

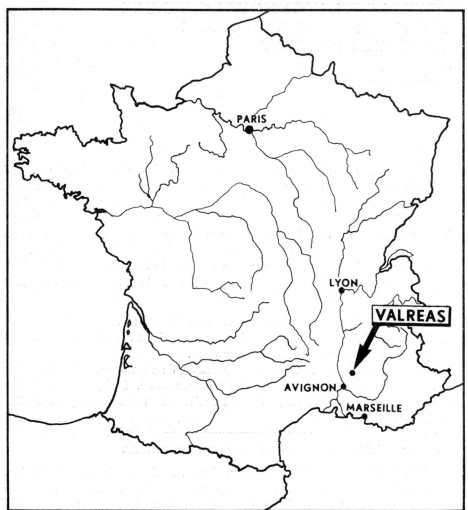

Fig. 1. Localisation géographique

proposer la création d'une espèce nouvelle. J'en donne la description ci-après puis j'envisagerai l'aspect morphologique et la causalité des variations phénotypiques qui l'atteignent.

Famille: *Cytheridae* Baird, 1850
Sous-Famille: *Hemicytherinae* Puri, 1953
Genre: *Elofsonella* Pokorný, 1955

Elofsonella amberii nov. sp.
Plate I, Figs 1–10

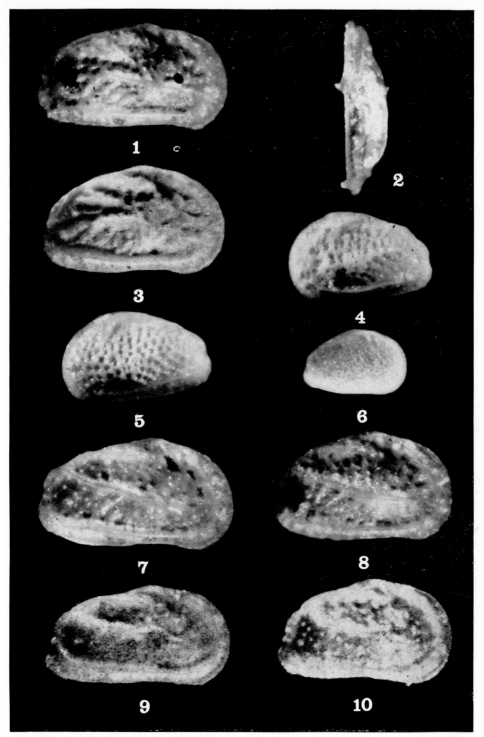

PLANCHE I. *ELOFSONELLA AMBERII ncv. sp.* (× 65 environ)

v.g. : valve gauche   v.d. : valve droite

Fig. 1. Holotype n° 135 008; ♂ v.d., type à 3 côtes, ponctué, 'normal'.
Fig. 2. idem, vue dorsale de la charnière.
Fig. 3. Paratypoïde n° 135 016, ♀ v.d. type 'normal'.
Fig. 4. Paratypoïde n° 135 010, stade larvaire v.g., type à 2 côtes, ponctué.
Fig. 5. Paratypoïde n° 135 011, stade larvaire, v.g., type à 1 côte, ponctué.
Fig. 6. Paratypoïde n° 135 017, stade larvaire.
Fig. 7. Paratypoïde n° 135 012, ♂ v.d., type à 2 côtes, à ponctuation réduites.
Fig. 8. Paratypoïde n° 135 013, ♀ v.d., type à 2 côtes, ponctué.
Fig. 9. Paratypcïde n° 135 014, ♂ v.d., type à 2 côtes, lisse.
Fig. 10. Paratypoïde n° 135 015, ♀ v.d., type à 2 côtes, lisse.

1965 *Elofsonella* sp. Carbonnel et Demarcq.

Nom: de Ambérieu-en-Bugey (Ain), localité où l'espèce a été observée pour la première fois.

Holotype: une valve droite ♂ no 135008 en dépôt au Département des Sciences de la Terre, Faculté des Sciences de Lyon (France).

Paratypoïde: 71 valves et carapaces adultes et stades larvaires no 135009 à 135018.

Locus typicus: commune de Visan (Vaucluse) près de la Ferme la Savoyonne, carte d'état-major au 1/20000 Valréas xxx–39, feuillet no 8; X: 808,35, Y: 238,96, z: 225.

Stratum typicum: Tortonien marin.

Autres localités: Ambérieu-en-Bugey (Ain), Tersanne (Drôme) découverte par G. Latreille, Lyon.

Diagnose

Un représentant appartenant au genre *Elofsonella* caractérisé par une exagération des caractères génériques de l'ornementation, et la présence de 2 côtes dorsales non jointives.

Description

de l'adulte ♂

Forme: bord dorsal: divisé en deux parties; l'une antérodorsale est inclinée vers l'avant, l'autre postérodorsale, la plus longue, est inclinée vers l'arrière; à leur jonction elles forment un angle de 160° juste à l'aplomb du tubercule oculaire.

bord ventral: concave à l'aplomb de l'angle dorsal, très faiblement convexe dans sa partie postérieure où il se raccorde avec le bord postérieur.

bord postérieur: angle postérodorsal arrondi mais net; il se poursuit verticalement sur 1/3 de la hauteur totale; un processus caudal faible mais toujours marqué, quelquefois crénelé, lui fait suite.

bord antérieur: largement arrondi; son point d'inflexion[1] se situe au tiers inférieur.

Ornementation: on note

—une côte antérieure parallèle au bord antérieur, s'étendant du tubercule oculaire au bord ventral où elle se poursuit par une côte ventrale parallèle au bord ventral; elle atteint le processus caudal.

—une côte médiane transversale oblique; elle naît à mi-hauteur de la carapace dans la zone postéromédiane, elle passe par le tubercule subcentral et atteint presque la côte antérieure, sans toutefois la rejoindre.

—une côte dorsale supérieure s'étendant de l'angle postérocardinal à la région oculaire.

—une côte dorsale inférieure parallèle à la côte médiane et plus courte que la côte dorsale.

Entre ces diverses côtes, la carapace est ornée de fossettes relativement grosses et alignées plus ou moins parallèlement à celles-ci.

Zone marginale: typique du genre avec repli interne, le rebord externe est denticulé antérieurement.

Canaux de pores marginaux: au nombre de 76 sur la zone marginale antérieure et droits, de 50 environ sur le reste de la zone marginale.

Canaux de pores normaux latéraux: du type simple.

[1] Le point d'inflexion est le point où la courbure du bord antérieur change de sens.

G

Charnière: typique du genre: hétérodonte. Elle comprend une dent antérieure dont l'extrémité distale est surbaissée, suivie par un sillon et une dent postérieure plus allongée.

La valve gauche est symétrique. La barre intermédiaire n'est pas crénelée, contrairement à la définition générique. La fossette antérieure est barrée inférieurement; la fossette postérieure s'ouvre librement sur l'intérieur.

Tubercule oculaire: bien développé à l'aplomb de l'angle dorsal. A l'intérieur lui correspond une dépression oculaire profonde.

Vestibule: peu visible sur l'holotype, antérieurement; il l'est plus chez d'autres individus.

Dimensions: longueur 0,725 mm
            hauteur 0,395 mm
            largeur 0,197 mm

Dimorphisme sexuel: il est marqué; le mâle est plus allongé et moins haut, la femelle plus courte et plus haute.

dimensions de ♀: longueur: 0,690 mm
                hauteur: 0,412 mm
                largeur: 0,182 mm

VARIANTS PHÉNOTYPIQUES

L'observation de l'ornementation de la carapace dans l'ensemble de la population, récoltée dans le sédiment de remplissage d'une huître, m'a conduit à noter les particularités suivantes. Elles se produisent à la fois sur les stades adultes ♂ et ♀ et sur les stades larvaires.

Ces variations se répartissent en trois *types morphologiques*.

Le stade adulte présente:

—un type à trois côtes transversales et ponctuations, présentes sur l'holotype (Pl. I, Fig. 1).

—un type à deux côtes transversales jointives et ponctuations; ces dernières peuvent disparaître partiellement ou totalement, aboutissant ainsi à un type à 2 côtes jointives, lisse (Pl. I, Figs 7–10).

Par contre je n'ai jamais pu observer un type morphologique à trois côtes et lisse. Quelque soit le cas envisagé la côte ventrale est toujours présente.

Au stade larvaire, l'examen de l'ornementation des valves révèle aussi des variations importantes. On distingue:

—un type à 2 côtes transversales et ponctué (Pl. I, Fig. 4).

—un type à 1 côte transversale; la carapace est entièrement ponctuée (Pl. I, Figs 5–6).

Dès le stade le plus jeune observé, la carapace présente toujours une ponctuation importante (Pl. I, Fig. 6).

La proportion de ces différents variants phénotypiques s'établit comme suit:

              type à 3 côtes   15%
              type à 2 côtes   57% dont 8 à 9% lisse
              type à 1 côte   28%

Les variants à une côte sont représentés seulement par des larves; ceux à trois côtes ne renferment que des adultes, au plus le stade larvaire $(n-1)$.

Parmi les variants à 2 côtes, on rencontre à la fois des adultes et des larves, mais seuls les adultes sont du type lisse.

Ces diverses observations replacées dans le cadre plus général de l'ornementation des ostracodes appellent certaines remarques.

On distingue, habituellement, une ornementation primaire et secondaire (Van Morkhoven, 1962, pp. 38–39). Selon cet auteur l'ornementation primaire concerne les traits fondamentaux, à savoir côtes, lobes et replis. L'ornementation secondaire concerne le plus souvent la réticulation entre les côtes primaires.

Chez *Elofsonella amberii*, il semble, en opposition avec ces idées, que l'ornementation primaire soit constituée par la réticulation. En effet tous les stades larvaires sont toujours ponctués.

L'ornementation secondaire serait composée des diverses côtes (dorsale inférieure et supérieure, médiane, ventrale). Elle serait acquise progressivement au cours de l'ontogenèse. Le type à 3 côtes correspond alors au stade adulte «normal».

En conséquence, dans cet échantillon, les variants phénotypiques présentent une réduction de l'ornementation secondaire chez les larves où pour, le même stade ontogénique, il existe à la fois le type normal à deux côtes et le type regressé à une côte (Pl. I, Figs 4–5).

Chez les adultes ♂ comme ♀, la réduction de l'ornementation porte non seulement sur le nombre de côtes, mais aussi sur l'ornementation primaire par disparition de la réticulation. On aboutit ainsi au type lisse à 2 côtes jointives.

ESSAI D'INTERPRÉTATION

Il faut rejeter comme causes éventuelles de ces modifications ornementales le dimorphisme sexuel. J'ai en effet montré plus haut que mâles et femelles sont également affectés par ces réductions.

A mon avis, ces variations s'inscrivent dans le cadre de celles qui affectent certaines espèces appartenant aux genres :
Cyprideis, Anomocytheridea, Haplocytheridea (formes européennes) et *Leptocythere* (P. Sandberg, 1965).

C'est néanmoins la première fois qu'on les observe chez les Hemicytherinae.

Cet essai d'interprétation, dans l'état actuel de nos connaissances, ne peut s'appuyer sur aucune recherche biochimique expérimentale relative aux modalités de la production de carbonate de calcium chez les ostracodes.

Pokorný (1965) et Neale (1965) soulignaient en effet combien nos connaissances étaient réduites sur l'ampleur et les modalités des variations phénotypiques de l'ornementation. Celles-ci, chez les espèces des genres cités plus haut, ont été attribuées à des arrivées d'eau douce dans un milieu marin. Sans nul doute, il en est de même ici.

La confirmation des influences d'eau douce nous est apportée à la fois par la microfaune d'ostracodes associée à *Elofsonella amberii* et par la position stratigraphique du prélèvement. Ce dernier est daté du sommet du Tortonien marin. Il lui succède une série lacustre ainsi que je l'ai indiqué précédemment. On peut alors vraisemblablement envisager de brèves arrivées d'eau douce en quantités limitées dès le sommet du Tortonien marin.

De plus la diminution de la salinité de ces eaux est attestée par la disparition partielle de l'ornementation chez une espèce appartenant au genre *Leptocythere* et associée à *Elofsonella amberii*.

Ainsi chez l'espèce étudiée, fondamentalement marine, la réponse à la diminution de la salinité se traduit, comme nous l'avons vu, par une réduction de l'ornementation. Celle-ci est inégale suivant les stades ontogéniques auxquels on s'adresse.

Le type à 3 côtes représenterait des individus parvenus au stade adulte lors de la venue d'eau douce. Elle aurait provoqué leur mort.

Le type lisse à 2 côtes jointives représenterait des individus adultes qui se seraient adaptés à une baisse de salinité des eaux lors d'un ou de plusieurs stades larvaires précédents.

Par ailleurs ces derniers s'adapteraient plus facilement puisque les modifications morphologiques semblent moindres. En effet ils conservent toujours leur ornementation primaire constituée par la réticulation fondamentale.

La salinité du biotope d'*Elofsonella amberii* semble donc supérieure à celle indiquée par les *Cyprideis* à tubercules phénotypiques. Dans la vallée du Rhône on les trouve d'ailleurs jusqu'à présent rarement associées.

Ainsi *Elofsonella amberii* serait par ses variants phénotypiques un indicateur fossile de salinité décroissante dans un milieu mixopolyhalin.

Je souhaite qu'une étude expérimentale des modifications morphologiques de la carapace, en corrélation avec les variations de salinité, soit réalisée chez *Elofsonella concinna* Jones. Il me semble en effet qu'elle est possible compte tenu de la limite inférieure de la salinité de cette espèce qui a pour valeur 16‰ (Elofson 1941). Cette étude confirmerait ou modifierait l'interprétation des faits exposés plus haut.

En conclusion, cette étude souligne la prudence avec laquelle le micropaléontologiste doit accepter l'existence de nouvelles espèces créées sur de faibles variations de l'ornementation. C'est ainsi qu'en 1965 M. el. A. A. Bassiouni créait une nouvelle sous-espèce *Elofsonella concinna neoconcinna* différenciée de *E. concinna concinna* par un réseau de ponctuations plus marqué. Il faudrait s'assurer qu'elle ne relève pas de variations analogues des conditions écologiques.

BIBLIOGRAPHIE

BASSIOUNI, M. EL A. A. 1965. Uber einige Ostracoden aus dem Interglazial von Esbjerg. *Meddr dansk geol. Foren.*, **15** (4), 507-18, Pl. 2.

CARBONNEL, G. AND DEMARCQ, G. 1965. Présence d'une faune d'Ostracodes marins dans les marnes et sables du Miocène supérieur de la région d'Ambérieu-en-Bugey (Ain). *C. r. hebd. Séanc. Acad. Sci., Paris*, **260**, 3116-19, 1 Table.

DEMARCQ, G. 1962. Etude stratigraphique du Miocène rhodanien, Thèse, Paris, no. **4723** et *Mém. Serv. Carte géol. dét. Fr.*, 1964, 400 pp., 53 Figs.

DIECI, G. AND RUSSO, A. 1964. Ostracodi tortoniani dell' Appennino settentrionale (Tortona, Montegibbio, Castelvetro). *Boll. Soc. paleont. ital.*, **3** (1), 38-88, Pls 9-17, text-fig. 6.

ELOFSON, O. 1941. Zur Kenntniss der marinen Ostracoden Schwedens, mit Berücksichtigung des Skagerraks. *Zool. Bidr. Upps.*, **19**, 215-534.

HARTMANN, G. 1965. The problem of polyphyletic characters in ostracods and its significance to ecology and systematics. *Pubbl. Staz. zool. Napoli* [1964], **33** (suppl), 32-44, Fig. 5.

KLIE, W. 1938. *Ostracoda, Muschelkrebse*. in F. Dahl (Edit.), **34** (3), 230 pp., 786 Figs.

MORKHOVEN, VAN F. P. C. M. 1962. *Post-Palaeozoic Ostracoda*. Elsevier edit., New York, *I*, 204 pp., 79 Figs, Table 1. *II*, 478 pp., 763 Figs.

NEALE, J. W. 1965. Some factors influencing the distribution of recent British Ostracoda. *Pubbl. Staz. zool. Napoli* [1964], **33** (suppl.), 247-307, Pl. 1, Fig. 11, Table 5.

OERTLI, H. J. 1960. Evolution d'une espèce d'*Echinocythereis* dans le Lutétien du Rio Isabena (Prov. Huesca, Espagne). *Revue Micropaléont.* **3** (3), 157-66, 3 Pls, 1 Fig.

POKORNÝ, V. 1955. Contribution to the morphology and taxionomy of the subfamily Hemicytherinae Puri. *Acta Univ. Carol. Geologica*, 25 pp., 19 text-figs.

——. 1964. The taxonomic delimitation of the superfamilies *Trachyleberidinae* and *Hemicytherinae* (Ostracoda, Crustacea). *Acta Univ. Carol., Geologica*, no. 3, 275-84.

——. 1965. Some paleoecological problems in Marine Ostracode Faunas, demonstrated on the Upper Cretaceous Ostracodes of Bohemia, Czechoslovakia. *Pubbl. Staz. zool. Napoli* [1964], **33** (suppl.), 462-79, Pls 1, 2, Tab. 1-3.

RUGGIERI, G. 1962. Gli Ostracodi marini del Tortoniano (Miocene-medio superiore) di Enna, nella Sicilia centrale. *Palaeontogr. ital.*, **56** (n. ser. 26), mem. no. 2, 1-68, Pls 1-17, Figs 1-15.

SANDBERG, P. 1965. Notes on some Tertiary and Recent brackish-water Ostracoda. *Pubbl. Staz. zool. Napoli*, [1964], **33** (suppl.), 496-514, Pls 1-3, Fig. 1.

TERQUEM, M. O. 1878. Les foraminifères et les Entomostracés Ostracodes du Pliocène supérieur de l'Ile de Rhodes. *Mém. Soc. géol. Fr.*, Paris, sér. 3, **1**, 81-128, Pls 10-14, Fig. 101.

WAGNER, C. W. 1957. *Sur les Ostracodes du Quaternaire récent des Pays-Bas et leur utilisation dans l'étude géologique des dépôts holocènes.* Mouton et Co, S-Gravenhage, édit. 259 pp., Pls 1-50, Figs 1-26.

——. 1965. Ostracods as environmental indicators in Recent and Holocene estuarine deposits in the Netherlands. *Pubbl. Staz. zool. Napoli* [1964], **33** (suppl.), 480-95, Figs 1-6.

DISCUSSION

KILENYI: You are supposed to have larger nodes developed in lesser salinities and several papers quote this without giving exact details on why this is considered to be salinity controlled. Is there any scientific evidence that there is a correlation between salinity and these phenotypic variations? In the Thames Estuary *Cyprideis torosa* shows an increase in nodosity in going from marine waters (34–35‰) down to a salinity of about 20‰. With lower salinities the nodes begin to disappear again. Do you think that this happens in your case and that the salinity does not have a straight-forward linear effect on phenotypic change?

CARBONNEL: Si on s'en tient aux travaux de Sandberg (Congrès de Naples), ceux-ci montrent que, lorsque la salinité diminue, l'ornamentation, composée de fossettes (chez les *Cyprideis*), tend à disparaître et la carapace devient ainsi presque lisse. Il semblerait que ce ne soit pas différent dans ce cas.

Il est maintenant indispensable d'étudier le contrôle de cette réduction de l'ornamentation. Est-elle due à une diminution de production du carbonate de calcium?

Seul un travail de biologie expérimentale permettrait d'étudier sur du matériel vivant si la fixation du carbonate de calcium est fonction de la salinité ou non et d'en préciser les limites. Ce n'est pas aux géologues d'apporter une réponse mais plutôt aux biologistes. Il est difficile actuellement d'en dire plus.

J'ajouterai qu'*Elofsonella concinna* a une limite de salinité inférieure voisine de 16‰ suivant le Dr Elofson. Il est donc possible de faire varier en laboratoire la salinité de cette espèce entre quelques millièmes et une valeur élevée à déterminer.

Sur les exemplaires observés au musée de Newcastle lors de ma visite durant ce symposium, j'ai pu remarquer que certains individus présentent une réduction analogue de l'ornamentation. Ceux-ci font partie de la collection Brady (exemplaires fossiles).

OERTLI: Dr Carbonnel has observed that when there is a fresh-water influence and salinity drops below 20‰ the ornamentation in his *Elofsonella* gets weaker and the reticulate pattern tends to disappear. I think that it is possible to imagine *Elofsonella* with normal reticulation living in an environment not far from the 20‰ threshold of your *torosa* and then when the milieu dropped below 20‰ the ornament disappearing.

CARBONNEL: Je voudrais ajouter que je suis tout à fait d'accord avec vous dans la mesure où des espèces appartenant au genre *Leptocythere* sont associées à *Elofsonella amberii* et présentent aussi une ornamentation réduite. J'ai cru comprendre que les zoologistes n'étaient pas d'accord sur la signification de la diminution de l'ornamentation chez les *Leptocythere*. Pour ma part je m'en tiens aux travaux déjà publiés et particulièrement celui de Sandberg où il montre que l'ornamentation diminue si la salinité baisse.

WHATLEY: I find the same thing in *Leptocythere* and I equate this with lack of calcium on reduction in salinity.

CARBONNEL: Il faudrait pour s'en assurer, utiliser du calcium radioactif et étudier son taux d'absorption par les valves en fonction de la salinité. C'est la seule façon d'effectuer un travail scientifique sur un problème de ce genre.

SYLVESTER-BRADLEY: I feel that there is a danger that we may generalize too much from the very few well documented examples. I should like to welcome this paper because it adds to the documentation. On the other hand I am quite convinced that there are other changes of ornament in other species of ostracods which, though they may look very similar, are controlled by quite other means. Dr Benson noticed such a variation in *Cythere lutea* where it was apparently controlled by geographic location. Geographic location may be correlated with a subsidiary temperature effect. We have in *C. lutea* a cline which may be a genocline or a phenocline and we do not know which. *Cytherissa lacustris* is another species which shows similar sorts of variation. Here again we get great variation in the development of tubercles and we do not know whether this is phenotypically or genotypically controlled. I believe that Morkhoven has suggested that this might be due to salinity control but I do not know of any evidence that points to this. Perhaps other people do.

BENSON: I think that there is a well-documented example of change of salinity not yet mentioned; not in a marine species, but in a fresh-water one. That is the problem that I believe both Klie and Sars ran into in studying *Limnocythere inopinata*. In this form, found in a north German river, there was a gradual transition in which you got increased nodosity with increasing salinity. This has been documented not only here but in a number of instances in Pleistocene material I have studied from Channing, Texas. This linkage between nodose and non-nodose forms and changes in the salinity gradients occurs numerous times. I am sure that there are other people here who have had this experience.

CARBONNEL: Pour les exemples de la coupe dont il s'agit les échantillons sont prélevés à la limite des deux formations. Si on franchit le niveau où se situent les huîtres la faune change. Elle est lacustre. On y trouve seulement des formes du genre *Candona*, *Paracandona*, et proche de *Eucypris*. En-dessous c'est différent, la salinité est normale avec une faune marine 'normale'.

ROME: Je crois que l'influence des variations de la salinité sur la sculpture des valves des ostracodes ne pourrait être démontrée que si l'on connaissait quelles sont les cellules qui sont responsables de la calcification. Les nombreuses coupes histologiques que j'ai faites n'ont pas permis de trouver de différence entre les cellules des tissues de l'épiderme. Toutes sont des cellules plates, réagissant de la même façon aux colorants.

ON THE QUESTION OF WHETHER THE PALAEOCOPE OSTRACODS WERE FILTER-FEEDERS

FRANCISZEK ADAMCZAK
University of Stockholm, Sweden

SUMMARY

The morphology of the free margin of the carapace of modern filter-feeding ostracods shows similarities to certain Palaeozoic forms of Palaeocopa. Analogous structures include the enlarged inner part of the ventral margin of valves, and the presence of two chambers in the domicilium. Comparative studies have been made on carapaces of Primitiopsidae, Beyrichiidae, Kloedenellidae and Cytherellidae (Cavellininae), and these show free margin structures similar to those of the Recent filter-feeding *Cyclasterope hendersoni* Brady.

INTRODUCTION

The free margin structure of the carapace of a primitiopsid ostracod (Palaeocopa) is the starting point of the present paper. This structure, approximately symmetrical as seen in transverse section, shows a fundamental similarity to the free margin of the carapace of the Recent *Cyclasterope hendersoni* Brady. The latter species is a perfect filter-feeding form and a similar mechanism of feeding may perhaps be imputed for the Palaeocopa.

The free margin structure of the palaeocope carapace is relatively constant in character. In general, it consists of a contact groove in one valve and a corresponding contact list in the opposite valve. The important feature, in such an arrangement of valves, is that when the carapace is closed the ventral part of the domicilium or shell-cavity is symmetrical and the valves do not overlap.

MECHANISM OF FEEDING IN MODERN OSTRACODS

It is Cannon (1926, 1933) who studied the feeding mechanism in modern ostracods. According to this author, there are two principal means of feeding: filter feeding and deposit feeding. The filter-feeding mechanism, which works on the principle of the suction pump, consists of four parts: (*i*) a filter situated underneath the adductor muscle, (*ii*) a means of producing a current of water through the filter (this is an important factor in feeding), (*iii*) a means of scraping the residue off the filter, and (*iv*) transporting the collected residue into the mouth.

In deposit-feeders the filter is absent. The current of water is produced by the vibratory plate of the epipodite of the first maxilla and a very small plate on the mandibular palp. The stream of water, and the particles of food sucked in with it, flows into the shell-cavity just before the hypostome itself and the projections of the free margin (Fig. 1). The projections, called 'bow-shaped projections' by Cannon and

93

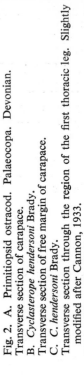

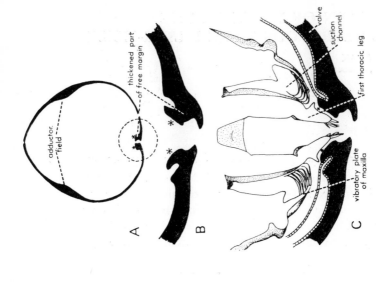

A

B

C

*

*

adductor field

thickened part of free margin

valve

suction channel

first thoracic leg

vibratory plate of maxilla

Fig. 2. A. Primitiopsid ostracod. Palaeocopa. Devonian. Transverse section of carapace.
B. *Cyclasterope hendersoni* Brady. Transverse section of free margin of carapace.
C. *C. hendersoni* Brady. Transverse section through the region of the first thoracic leg. Slightly modified after Cannon, 1933. (From Adamczak, 1968.)

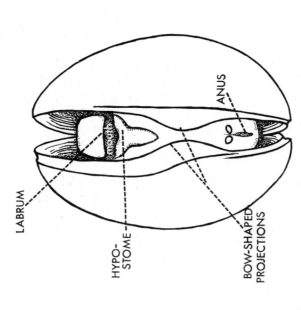

LABRUM

HYPO-STOME

BOW-SHAPED PROJECTIONS

ANUS

Fig. 1. *Cypridopsis vidua* (O. F. Müller). Podocopa. Recent. Ventral view of carapace. (From Cannon, 1926.)

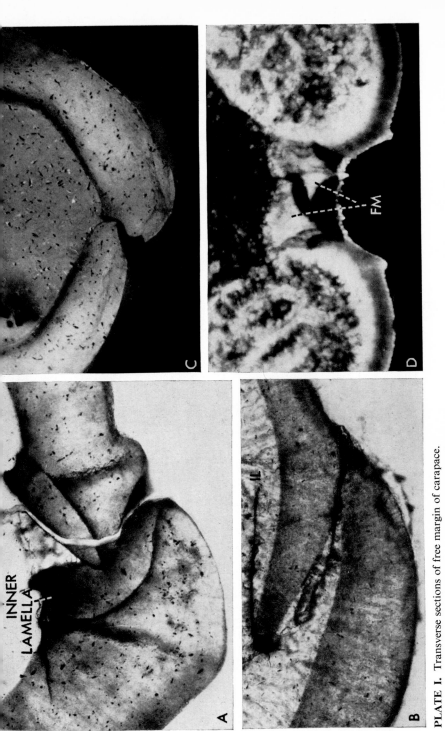

PLATE 1. Transverse sections of free margin of carapace.

A. *Bairdia cultrijugati* Krömmelbein. Podocopa. Devonian.
B. *Bairdiocypris* sp. Podocopa. Devonian.
C. *Cytherella* sp. Platycopa. Jurassic.
D. *Craspedobolbina clavata* (Kolmodin). Palaeocopa. Silurian. IL, inner lamella; FM, free margin.

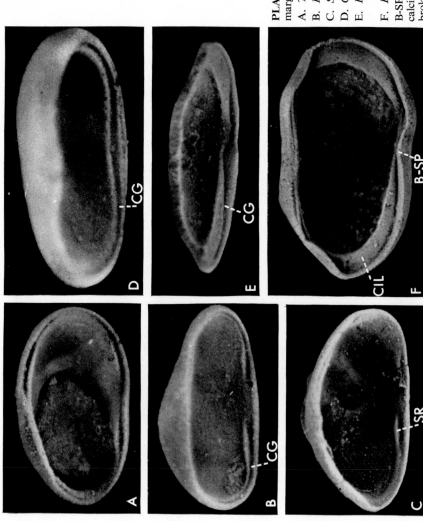

PLATE II. Interior view of larger valves showing free margin structures.

A. *Thlipsurella discreta* (Jones). Metacopa. Silurian.

B. *Kuresaaria gotlandica* Adamczak. Metacopa. Silurian.

C. *Silenis bassleri* (Sohn). Metacopa. Silurian.

D. *Cytherella ovata* (Roemer). Platycopa. Cretaceous.

E. *Bairdia cultrijugati* Krömmelbein. Podocopa. Devonian.

F. *Bairdia cultrijugati*.

B-SP, bow-shaped projections; CG, contact groove; CIL, calcified inner lamella; SR, stop-ridge. The figure E shows broken valve of *B. cultrijugati*, as seen from above.

'ventral lip' by Sohn (1960), are located on the ventral part of the valves. They are 'so situated that when the valves closed they overlap just behind the hypostome' (Cannon, 1926, p. 327).

FREE MARGIN STRUCTURE OF THE VALVES IN FILTER-FEEDING OSTRACODS

As far as the present writer is aware, the filter feeders are characterized by approximately symmetrical valves. The contact line of the carapace, as seen in ventral view, is always straight. A curved contact line in the ventral part of the carapace characterizes deposit-feeding forms.

The straight and symmetrically-arranged ventral margin of the valves in filter-feeding forms is important because of the function it fulfilled. In *C. hendersoni* it is thickened, and in both valves is provided with a kind of 'contact groove'. The thickened part of the valve is so shaped that it fits against the ventral part of the first thoracic leg and medially of it, so participating in the formation of the suction channel through which the water is drawn (Fig. 2). The edges of the free margin in this species are similar to those of the primitiopsid ostracod from the Middle Devonian rocks in the Holy Cross Mountains of Poland.

FREE MARGIN STRUCTURE OF THE VALVES IN PALAEOCOPE OSTRACODS

The free margin of the carapace of the primitiopsid form from the Devonian rocks is rather peculiar. It is characterized by a very thin (0·01 mm) carapace wall, distinct extra-domiciliar dimorphism and well-preserved adductor muscle scars. In a sectioned carapace it can be seen that the ventral part of the free margin is thickened and a kind of 'contact groove' occurs in each valve. This structure, very unusual in palaeocope ostracods, is strikingly similar to the free margin of the living myodocope—*C. hendersoni* (Fig. 2).

The free-margin structure of the valves in other palaeocope ostracods, although developed slightly differently, shows a situation in the ventral part of the carapace approximately similar to that of the primitiopsid form. It may be of interest to mention that the free margin structure of these representatives resembles that of the platycope ostracods (Plate I). The Recent representatives of the latter group feed in the same manner as the myodocope *C. hendersoni* (Cannon, 1933). The free margin structures of their carapace are the same as those of their Silurian progenitors. Moreover, these are also similar to those of the metacope ostracods.

FREE MARGIN STRUCTURE OF THE VALVES IN PODOCOPE AND METACOPE OSTRACODS

As seen in transverse section (Fig. 3), the ventral part of the domicilium of the carapace of the earliest known podocope ostracod—'*Bythocypris*' *ellipsiformis* Hessland—from the Lower Ordovician rocks in Sweden, is asymmetrical. The free margin of the valves is simple and there is no calcified inner lamella visible. The contact line of the carapace, as seen in ventral aspect (Hessland, 1949, Pl. IX, Fig. 8d), is curved and a kind of bow-shaped projection (ventral lip) may be observed on the left valve.

Analagous structures in some Silurian and Devonian pachydomellids and Devonian bairdiocypridids show important overlapping of the valves along the ventral part of the free margin and the presence of the ventral lip. Moreover, the Devonian representatives of these groups are provided with very thin but calcified inner lamellae (Plate IB; Fig. 3).

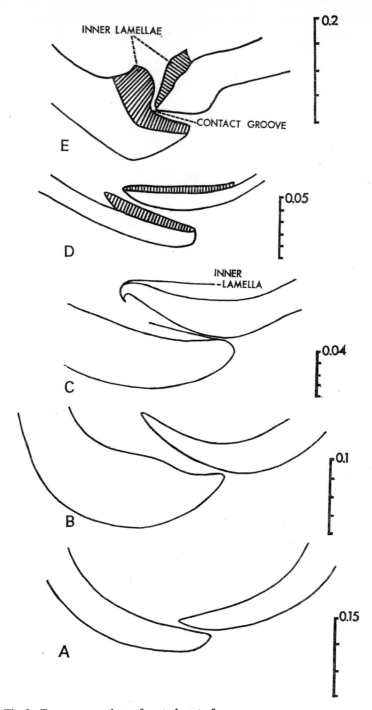

Fig. 3. Transverse sections of ventral part of carapace.
A. '*Bythocypris*' *ellipsiformis* Hessland. Podocopa. Ordovician.
B. Pachydomellid ostracod. Podocopa. Silurian.
C. *Bairdiocypris* sp. Podocopa. Devonian.
D. Bairdiid ostracod. Podocopa. Devonian.
E. *Bairdia cultrijugati* Krömmelbein. Podocopa. Devonian.

The free margin of the carapace of *Bairdia cultrijugati* Krömmelbein (from the Devonian), as seen in transverse section (Plate IA; Fig. 3), shows a well developed inner lamella and the contact groove. The contact groove in the latter form is impressed in the inner calcified lamella of the larger valve. From the point of view of homology, this is a fundamentally different arrangement of the free margin compared to that found in the palaeocope and platycope ostracods.

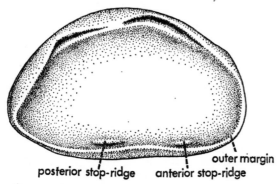

posterior stop-ridge   anterior stop-ridge   outer margin

Fig. 4. *Silenis bassleri* (Sohn). Metacopa. Silurian. Interior view of left valve. (From Adamczak, 1966.)

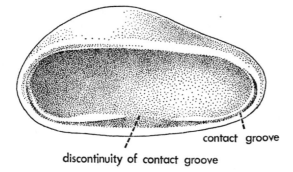

contact groove

discontinuity of contact groove

Fig. 5. *Kuresaaria gotlandica* Adamczak. Metacopa. Silurian. Interior view of left valve. (From Adamczak, 1967.)

The valves of the pachydomellids, bairdiocypridids and bairdiids are provided with the bow-shaped projections. In this they resemble the valves of the Recent deposit-feeder *Cypridopsis vidua* (O. F. Müller) (Cannon, 1926). In addition, it is also supposed that the Palaeozoic forms of podocope ostracods were presumably deposit-feeders. Furthermore, there are some bairdiocypridid-like forms from Ordovician and Silurian rocks which again differ. They lack calcified inner lamellae, but two ridges may be visible on the inner side of the larger valve (Plate IIC; Fig. 4). These ridges, from their assumed blocking function, have been named 'stop-ridges' (Adamczak, 1967). It is the writer's opinion (*op. cit.*) that these ridges led to the development of the contact groove. This opinion is partly supported by the fact that in several Silurian and Devonian metacope ostracods the contact groove of the larger valve is interrupted in the central part of the ventral margin (Plate IIB; Fig. 5). This interruption occurs in the same place as the discontinuity between the anterior and posterior stop-ridges in *Silenis bassleri* (Sohn). It may disappear and this process is gradual. It has already started in the Silurian (*Thlipsurella discreta* (Jones)) and it continues also in the Devonian (Plate IIA). An uninterrupted contact groove and symmetry of the ventral part of the domicilium is the result. Comparative studies of the free margin structure in palaeocope, platycope and metacope ostracods have shown that these elements are similar. This, however, does not necessarily mean that the metacopes are related to the two previous groups, but rather that the mechanism of feeding could lead to development of similar characters in different groups. The metacope ostracods dealt with here are particularly close to the early Podocopa.

REFERENCES

ADAMCZAK, F. 1966. Morphology of Two Silurian Metacope Ostracodes from Gotland. *Geol. För. Stockh. Förh.*, **88**, 462-75.
——. 1968. Palaeocopa and Platycopa (Ostracoda) from Middle Devonian rocks in the Holy Cross Mountains, Poland. *Stockh. Contr. Geol.,* **17**, 1-109.
CANNON, H. G. 1926. On the feeding mechanism of a freshwater ostracod *Pionocypris vidua* (O. F. Müller). *J. Linn. Soc.*, **36**, 325-75.
——. 1933. On the feeding mechanism of certain marine ostracods. *Trans. R. Soc. Edinb.*, **57**, 739-64.
HESSLAND, I. 1949. Investigation of the Lower Ordovician of the Siljan District, Sweden. *Bull. geol. Instn Univ. Upsala*, **33**, 97-408.
SOHN, I. G. 1960. Paleozoic species of *Bairdia* and related genera. Revision of some Paleozoic genera. *Prof. Pap. U.S. geol. Surv.*, **330-A**, 1-105.

DISCUSSION

HARDING: I would like to congratulate Dr Adamczak very much on this attempt at functional morphology of fossil material. His interpretation of the shell margin of fossil ostracods in the light of what is known of the function of the flanges and other structures of the shell in Recent forms is very stimulating.

SIGNIFICANCE OF CALCAREOUS NODULES IN MYODOCOPID OSTRACOD CARAPACES[1]

I. G. SOHN and L. S. KORNICKER
U.S. Geological Survey, Washington, D.C., and Smithsonian Institution, Washington, D.C.

SUMMARY

The integument of myodocopids is mineralogically different from those in the other ostracod orders (Podocopida, Platycopida, Palaeocopida, and Leperditicopida). The Myodocopida have amorphous calcium carbonate in the shell, the other orders have crystalline calcium carbonate. Spherical, hemispherical, discoidal and anastomosing calcareous nodules, ultramicroscopic to about 0·3 mm in size, form posthumously (less frequently *in vivo*) in the carapaces of myodocopid ostracods and also in the integuments of other crustacean groups. These artifacts have occasionally been erroneously used as criteria for classification of ostracod species. The following sequence of nodule formation is postulated: amorphous calcium carbonate—monohydrocalcite ($CaCO_3.H_2O$)—calcite. Some nodules resemble statoliths of Mysidacea, Foraminifera, brachiopod sclerites, holothurian plates, spherulites, and oolites. The nodules are a source of calcareous particles in sediments.

INTRODUCTION

Although living Myodocopida are abundant, their fossil representatives are scarce This paper describes artifacts in myodocopid shells that resemble microscopic organisms, it indicates some that were used in the classification of ostracod species; it explains the scarcity of fossil Myodocopida, and suggests that myodocopid ostracods and other Crustacea are a source of carbonate particles in marine sediments.

Criteria which distinguish Myodocopida from other ostracod orders are the number and type of appendages and organs, morphology of the carapace, and differences in the crystallographic habit of the calcified integument. In general, the Podocopida have crystallized calcium carbonate in their carapaces, while the carapaces of the Myodocopida have calcium carbonate that shows no birefringence in polarized light and is presumably amorphous or fine grained. This calcium carbonate is the source of nodules which form both in the live animal and after death.

In the fall of 1965 a valve of *Philomedes* (Crustacea, Ostracoda, Myodocopida), preserved in alcohol, was treated with Clorox (sodium hypochlorite) in order to remove muscle tissue for examination of the muscle scar pattern. Sodium hypochlorite does not have an adverse effect on podocopid valves. The myodocopid valve disintegrated in Clorox. This unexpected phenomenon prompted additional experimentation. Dry carapaces of the myodocopid *Cypridina hilgendorfii* Müller, 1890 (= *Vargula*), as well as specimens preserved in alcohol of *Aurila* sp. and '*Bairdia*' sp. (Podocopida),

[1] Publication authorized by the Director, U.S. Geological Survey.

99

were placed in Clorox (Sept. 29, 1965). After two days the myodocopids disintegrated, but the podocopids were and still are (Oct. 1966) undamaged. The myodocopid residue consisted of a variety of concretionary objects similar to those illustrated on Pl. I.

HISTORICAL REVIEW

Harding (1965) emphasized the fact that the ostracod carapace is similar in composition, formation and function to the cuticle of other Crustacea. The crustacean cuticle is discussed and illustrated by Dennell in Waterman (1960, p. 455, Fig. 2). Kelly (1901, p. 456) produced nodules by soaking crustacean integuments in water followed by maceration in Javelle water.

Dudich (1931, p. 113, Fig. 22) illustrated nodules that resemble those formed in myodocopid shells which he obtained by soaking a piece of cuticle of *Penaeus membranaceus* Risso (Decapoda) in distilled water. The animal had been preserved four years in 75% alcohol. Crystallization was complete after 24 days at 18° to 19°C, and no change was observed in the nodules after additional soaking for three months. He illustrated nodules in other groups of Crustacea. Calcium carbonate in Crustacea was studied by Kelly (1901), Schmidt (1924), Dudich (1929, 1931), Prenant (1927) and Richards (1951). Although calcareous nodules in the integument of the Crustacea have been recorded, their mode of formation was not elucidated, nor was their biological and geological significance recognized.

Müller (1894, pp. 93–95) described the mineralized portion of the ostracod shell as consisting of calcium carbonate and magnesium carbonate and he noted that it may be either amorphous, microgranular or prismatic. He stated (1894, p. 94) that nodules, which he correctly interpreted as post-mortem artifacts, were present in some valves of preserved Myodocopida. He also recorded that similar concretions are sometimes present in the shells of living Cypridinacea, and interpreted those concretions to be the result of starvation. In the German (1958, p. 73) and English (1965, p. 76) translations of Pokorný's textbook (1954), Müller is interpreted as attributing the concretions in living individuals to disease. Kesling (1951, p. 67) published the first X-ray diffraction study of the fresh-water podocopid *Cypridopsis vidua* (O. F. Müller, 1776). He determined the presence of calcite and found no indication of either aragonite or vaterite. Sohn (1958) published qualitative spectrographic analyses of podocopid ostracods and semiquantitative analyses of podocopid ostracods and the myodocopid *Vargula hilgendorfii*.

LABORATORY METHODS

Most of the experiments on the formation of nodules were made with specimens from about 30 cc of dry specimens of *Vargula hilgendorfii* (Müller, 1890) collected on September, 1954 at Zushi Beach, Kanagawa, Japan, by Dr Y. Haneda. The specimens were collected in order to demonstrate bioluminescence in Ostracoda (Johnson, 1955), consequently they were originally thoroughly sun-dried and bottled. Clear shells were used in the experiments. Specimens of Podocopida used as control material were from the collections of the U.S. Geological Survey and the U.S. National Museum.

Dry specimens of *V. hilgendorfii* were soaked at room temperature in tap water, distilled water, brackish water obtained from an aquarium containing *Limulus* maintained by J. M. Berdan, U.S. Geological Survey, artificial sea water obtained from M. A. Buzas, U.S. National Museum, and artificial sea water diluted with an equal amount of distilled water. Nodules developed in all these media. In Dec. 1965 five clear carapaces were placed in 70 % alcohol, and these are still free of concretions (Oct. 1966). Nodules were visible 17 hours after transfer to distilled water. The treated valves were disaggregated in Clorox in order to obtain individual concretions for study and illustration.

Selected wet specimens without nodules in the shells of non-luminous species in two cypridinid genera collected by Haneda in September 1952 at Hachijo Island, Japan, developed nodules in 24 hours of soaking in distilled water at room temperature (22–23 Sept. 1966). In this experiment artifacts were present in both adults and instars $\frac{2}{3}$ the size of adults.

In addition to the podocopid specimens that were discussed previously, a dry valve of *Chlamydotheca unispinosa* was placed in Clorox on 7 Dec. 1965, and that valve is still intact and unchanged (Oct. 1966). A dry valve of *Candona* sp. in Clorox remained unchanged for more than four weeks when observation was discontinued.

DESCRIPTION OF THE NODULES

The nodules developed in the laboratory are of four general types: spherical, hemispherical, discoidal, and anastomosing. They are solid bodies that vary in size from ultramicroscopic particles to 0·3 mm or larger in diameter. The anastomosing types are the largest. There is no correlation between the size and type of nodule and their position in the valves. Similar types form in different taxa. Because the nodules are artifacts, their size and shape have no classificatory or stratigraphic significance.

In the descriptions that follow, we will refer to the published record of similar things. The previously illustrated objects and our artifacts may not be genetically similar. For example, Enbysk and Linger (1966) described and illustrated statoliths (equilibrium receptors) of living Mysidacea (Crustacea, Malacostraca), and compared them with similarly shaped objects found in marine sediments. They stated that statoliths from formalin preserved specimens of *Neomysis rayii* are composed of calcium fluoride with an undetermined organic base. They noted, however, that a clump of 42 individual statoliths in organic material that stained dark red in rose bengal did not contain fluoride. This collection was from off British Columbia at 166 m and could well have been a fragment of the integument of a crustacean. A fragment of a treated myodocopid valve contained more than 50 individual discoidal nodules.

Spherical nodules

Minute perfect spheres up to approximately 0·1 mm in diameter are developed. Some resemble spherulites and oolites (Pl. I, Fig. 5), others are clumps that resemble Foraminifera (Pl. I, Figs 3a, b), or occur as chains (unfigured).

Hemispherical nodules

Hemispherical nodules have either flat or concave bases. The one illustrated (Pl. I Figs 1a–c) has a concave basal disc and an hemispherical dome. These are present as individuals, as rectilinearly joined chains (Pl. I, Figs 7a, b, 8a–c), and as clumps (Pl. I, Figs 4a–c). As yet we have not found descriptions of organic remains found in sediment that resemble this type. Dudich (1931, p. 114, Fig. 22) illustrated similar nodules formed in the cuticle of *Penaeus membranaceus* (Decapoda). Some of Dudich's artifacts resemble aragonitic tunicate spicules illustrated by Lowenstam (1963, Pl. 3, Fig. 7).

Discoidal nodules

These are rounded or sub-rounded, wafer-like, with either smooth surfaces or with sub-central, small pore-like areas. They form as single discs, or as strings and clumps of discs. When joined they may not be perfectly round, but each disc acts as a crystallographic unit when examined with polarized light. Some have one concave surface similar to the hemispherical type (Pl. I, Figs 2a–c, 6a–c), others are flat (Pl. I, Figs 9a–c). The flat ones resemble a Tertiary mysid statocyst [statolith] illustrated by Bandy and Kolpack (1963, p. 151, text-fig. 31B).

Structures described in abraded shells of *Cypridinella superciliosa* Jones, Kirkby and Brady (1874, p. 23, Pl. 5, Fig. 7d) (Myodocopida) from the 'Lower Limestone Group' (Lower Carboniferous) at Bathgate, Linlithgowshire, Scotland, may be discoidal nodules. Some nodules in *Cypridinodes* sp. collected in 1908 in the Philippine Islands and preserved in alcohol have pores and superficially resemble holothurian skeletal plates.

Anastomosing nodules

The illustrated nodules (Pl. I, Figs 10–12) are selected examples from many forms. Some appear to start developing as a discoidal concretion to which apophyses of varying width and length are added. The apophyses may divide, anastomose, or join discoidal platelets. Dudich (1931, p. 36, Fig. 2) illustrated similar nodules in the carapace of *Lophogaster typicus* (Mysidacea). Rioult (1966) illustrated a sclerite from a living *Terebratulina* (Brachiopoda) that resembles this type. He proposed the new paragenus *Eudesites* Rioult, 1966 for similar objects in sediments of Jurassic age.

Müller (1906, p. 23) described and illustrated blue-black radial pigment cells in *Codonocera* (Pl. 8, Figs 1, 9) and illustrated similar cells in the new species *Cypridina hesperida* (Pl. 3, Fig. 21) that resemble anastomosing nodules. We have observed similar pigment cells in the space between the outer and inner lamellae of *Codonocera* sp. from the Phillippine Islands. It seems unlikely that these chromatophores are related to nodule formation.

ELECTRON PROBE ANALYSIS

Dr K. Fredriksson, U.S. National Museum, allocated time to use the electron probe made available through NASA grant NsG 688. J. E. Merida, U.S. Geological Survey, prepared the polished surfaces of an untreated carapace and a treated carapace of *V. hilgendorfii*, and C. E. Fiori, U.S. National Museum, obtained beam scanning pictures for calcium, magnesium and phosphorus. Montages of selected areas on each of the two carapaces are illustrated (Pl. II, Figs 1–4). The areas that were analysed are shown on Pl. III (Figs 1, 2). The Ca is evenly distributed in the untreated carapace (Pl. II, Fig. 4) and concentrated within the nodules in the treated carapaces (Pl. II, Fig. 3). This indicates that there has been a rearrangement of the Ca due to soaking in water, and that the nodules form within the calcium-rich portion of the carapace. The Mg (Pl. II, Fig. 1) and P (Pl. II, Fig. 2) show the same phenomenon. It is interesting that P is more abundant along the inside and outside margins of the valve thickness in both treated and untreated (not illustrated) valves. The P in the nodules is concentrated from within the calcified middle zone of the shell.

Although P was not detected in the semi-quantitative spectrographic analysis of *V. hilgendorfii* (Sohn, 1958, p. 732, footnote 4), its presence was confirmed in electron probe analysis of additional specimens by E. J. Dwornik, U.S. Geological Survey. It

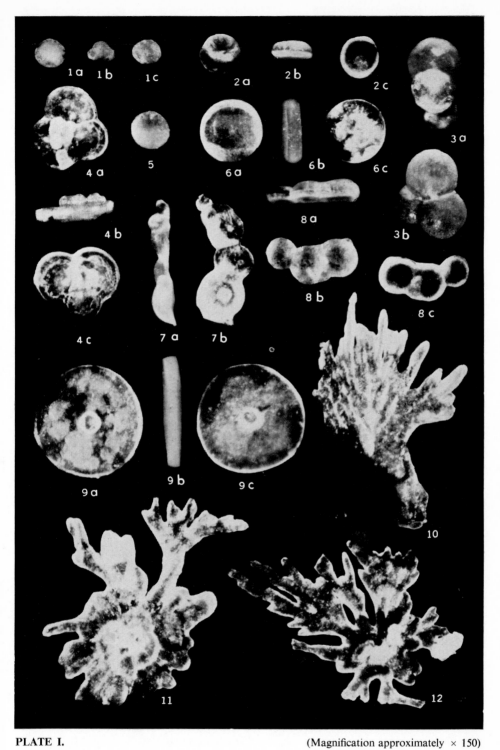

PLATE I.   (Magnification approximately × 150)

Figs 1-12. Nodules, artifacts formed in shells of *Vargula hilgendorfii* (Müller, 1890) by soaking in water and then macerating in Clorox.

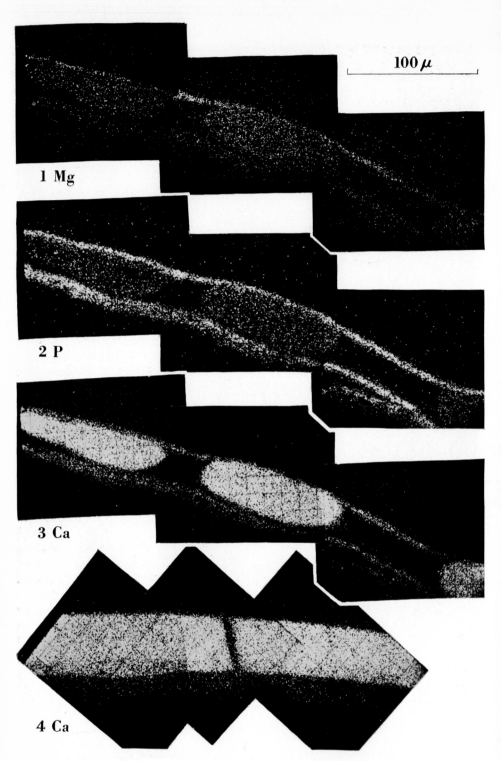

100 μ

1 Mg

2 P

3 Ca

4 Ca

PLATE II. (Magnification approximately × 350)

Fig. 1. Montage of electron probe photographs of a portion of a valve of *V. hilgendorfii* (Müller, 1890), after soaking in water to develop nodules, showing the distribution of Mg. The area on the valve is indicated on Pl. III, Fig. 1.

Fig. 2. Montage of the same area showing the distribution of P.

Fig. 3. Montage of the same area showing the distribution of Ca.

Fig. 4. Montage of electron probe photographs of a portion of an untreated valve of *Vargula hilgendorfii* (Müller, 1890) showing the distribution of Ca. The area on the valve is indicated on Pl. III, Fig. 2. The dark line is a fracture in the valve.

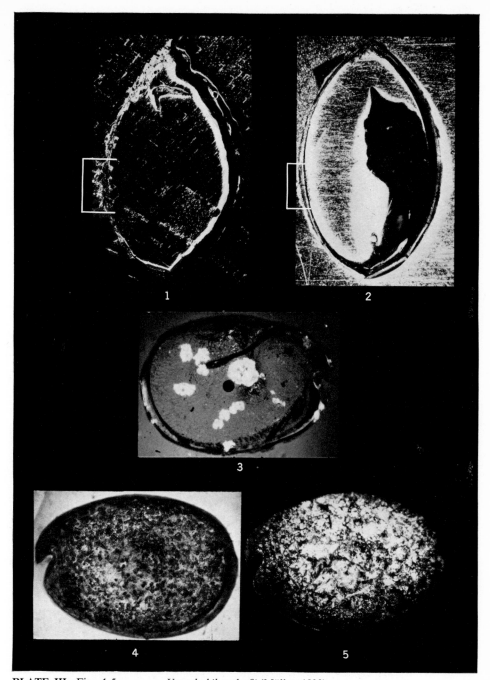

PLATE III. Figs. 1-5.   *Vargula hilgendorfii* (Müller, 1890)

Fig. 1. Polished surface showing the cross section (× *ca.* 30) of a carapace that was soaked in water to develop nodules. The area tested with the electron probe is shown by white lines.

Fig. 2. Polished surface showing the cross section (× *ca.* 30) of an untreated carapace. The area tested with the electron probe is shown by white lines.

Fig. 3. Right valve, exterior, in glycerin (× *ca.* 20) of a specimen in which a few nodules have formed, polarized light. U.S.N.M. 119844.

Figs 4, 5. Left valve, exterior, in glycerin, (× *ca.* 20) heated in water in order to develop nodules. 4. Transmitted light. 5. Polarized light. U.S.N.M. 119845.

should be emphasized that in order to obtain an adequate image for both P and Mg (Pl. II, Figs 1, 2) the area was scanned 10 times longer than for the Ca image. Consequently, the density of points on the photographs does not indicate the relative quantities of the respective elements.

MINERALOGY OF THE NODULES

Dr W. G. Melson, U.S. National Museum, advised in the petrographic study and was the first to suggest that the concretions may be a source of carbonate in marine sediments. Isolated nodules and those in treated valves are birefringent when viewed with polarized light; untreated valves are not birefringent. Examination with high magnification using oil immersion disclosed that some nodules have minute pores on their surface. Individual artifacts dissolve in dilute acid leaving a replica presumed to be the organic (protein?) matrix.

Eugene H. Roseboom, Jr., U.S. Geological Survey, examined several samples with the X-ray diffractometer. The results of the analyses are listed below:

1. Clear carapaces of *Vargula hilgendorfii* which had been sun-dried immediately after their capture gave no X-ray pattern, indicating a microcrystalline or amorphous phase, presumably a gel of calcium carbonate.

2. A second sample of sun-dried carapaces of *V. hilgendorfii* gave only lines of NaCl. After the NaCl was removed by washing in water and drying it briefly at 100°C, the sample gave no lines at all. The sample then was left in distilled water at room temperature for 66 hours, after which it gave a good pattern for monohydrocalcite ($CaCO_3.H_2O$).

3. Several carapaces of *V. hilgendorfii* from the same locality as the dry specimens, which had been preserved in alcohol since 1952 (U.S.N.M. 95646), were squeezed flat between two glass slides and X-rayed. These gave a poor quality but recognizable pattern of monohydrocalcite. The poor quality of the pattern is attributed to the method of mounting.

4. Additional specimens (U.S.N.M. 95646) were dried for 1 hour and then X-rayed. These gave a good pattern for monohydrocalcite.

5. Isolated nodules developed by soaking sun-dried valves of *V. hilgendorfii* in water gave an X-ray pattern indicating a mixture of monohydrocalcite and calcite. The proportions of each mineral are unknown but both phases showed strong peaks.

6. A sample of Recent '*Bairdia*' (Podocopida) carapaces gave a pattern of calcite only.

X-ray powder determination on duplicate samples of nodules were made by Mary E. Mrose, U.S. Geological Survey, and John S. White, Jr., U.S. National Museum. Both reported strong calcite lines and unidentifiable weak lines. It was noted that the formation of nodules is accelerated by heat. Valves of dry *V. hilgendorfii* were heated in water for 8 hours at 75°C. The nodules in these specimens had strong calcite and weak aragonite lines.

The following statement from the report of Dr Roseboom (14 Oct. 1966) suggests an explanation for the results of the above experiments: 'Brooks *et al.*, (1950) observed that at room temperature with time gels [of calcium carbonate] produced monohydrocalcite, calcite, or other phases and the monohydrocalcite in turn produced aragonite and the stable phase, calcite. Thus the phases observed in the ostracods [*V. hilgendorfii*] listed in order of decreasing metastability should be gel, monohydrocalcite, calcite (plus water). However, there is also a possibility that the shells were originally mono-

H

hydrocalcite and in the sun-dried sample, loss of water on drying produced a relatively dehydrated gel.'

The analytical methods used in this study (X-ray, petrographic microscope) do not give a positive test for the amorphous or gel phase of calcium carbonate. Therefore, its presence in carapaces containing crystalline phases of calcium carbonate cannot be excluded. For example, although specimens of *V. hilgendorfii* preserved in alcohol revealed by X-ray analysis the presence of monohydrocalcite, the possibility that the carapaces predominantly contain amorphous calcium carbonate is not excluded.

We conclude tentatively that the integuments of those Crustacea discussed in this paper contain calcium carbonate predominantly in the gel or amorphous phase. In nature the integument may contain smaller amounts of the more stable phases such as calcite or the intermediate monohydrocalcite. After death of the animal, the gel converts to nodules of calcite and monohydrocalcite. Presumably, nodules formed in the integument of living animals have a similar composition.

SOURCE OF CARBONATE IN NODULES

Dr Brian J. Skinner, U.S. Geological Survey, designed, performed, and interpreted the following experiments (written communication, Cristina C. Silber and Brian J. Skinner, 11 July 1966):

'De-ionized water was boiled for several hours to exhaust all its dissolved CO_2, in clean 500 cc flasks. These were then stoppered to avoid any CO_2 solution during cooling.

'When reduced to 25°C the flasks were arranged in groups of three, and into each flask a few carefully cleaned and handpicked ostracod [*V. hilgendorfii*, dry] shells were introduced. One flask remained stoppered, one sat open to the air with an air-water interface of approximately one square inch, and one flask had a 2 mm glass tube inserted to near the base and air was continuously bubbled through.

'Sets of flasks were stopped and the ostracods recovered and examined after 4, 8 and 16 days. The shells had partly crystallized in each case, and in no circumstance was any difference found between the reactions in the three flasks [in] a given group. The four day runs were less crystallized than the 8 and 16 day experiments, and a complete reaction was apparently reached by the eighth day.'

'The reasoning is this. If the Ca^{2+} of the calcite is in the shell and the CO_3^{2-} in the water, then having drastically different CO_2 contents in the water should produce very different crystallization rates, i.e. slowest or no action at all in the boiled and stoppered water, intermediate in the still, open water, rapid in the quickly saturated, bubbled water. There being no difference, apparently the water is not an essential CO_3 host. The crystallization reaction, therefore, apparently involves material already in the shell and is iso-chemical, the mechanism of the water being one of a catalyst.'

These experiments support our conclusion that the initial material is amorphous $CaCO_3$.

SCARCITY OF FOSSIL MYODOCOPIDA

Because many living Cypridinidae have firm carapaces and are abundant, they should be better represented in the geologic record. We suggest that the scarcity of fossil Myodocopida can be explained by the post-mortem formation of nodules in the myodocopid shell (Pl. III, Figs 3–5). After the nodules formed, the protein and chitin framework of the shell usually disintegrated; consequently, recognizable myodocopid fossils are rare. Bradley (1957, p. 658) established experimentally that decapod shells can be destroyed by bacteria in less than a year. Chitinoclastic bacteria were discussed by Seki (1965) who cited additional references.

RESEMBLANCE TO ORGANIC REMAINS

Certain organic remains that resemble the nodules were listed under the description of the artifacts. Although all the artifacts illustrated on Pl. I were developed in carapaces of only one myodocopid species (*V. hilgendorfii*), identical shapes form in other ostracod genera as well as in the integuments of other crustacean groups. Dudich (1931) illustrated nodules developed in the integuments of the following crustacean groups: Cladocera, Mysidacea, Isopoda, Amphipoda and Decapoda. Because similar calcareous particles can develop in different groups of animals, the assignment to a given taxon of isolated particles found in sediment should be based on corroborative evidence and not on size and shape alone.

USE IN CLASSIFICATION

Tressler (1949, p. 340) interpreted nodules in *Cycloasterope sphaerica* Tressler, 1949 as possible glands, and used their presence as one of the distinguishing features to differentiate his species from *C. americana* (Müller, 1890). Because nodules can form in shells of many myodocopid genera, their presence is not of classificatory significance.

The following list of ostracod taxa known to form nodules in their integument is preliminary. Additional observations may add significantly to this list.

Class Crustacea
Subclass Ostracoda
Order Myodocopida
Superfamily Cypridinacea
Family Cypridinidae

Codonocera stellifera (Claus, 1878), in Schmidt, 1924, p. 264.
Codonocera vanhöffeni (Müller, 1908), in Schmidt, 1924, p. 265, Fig. 136.
Crossophorus grimaldi Granata, in Granata and Caporiacco, 1946, p. 5, Pl. 1, Fig. 1.
Philomedes aspera (Müller, 1894) = *Euphilomedes*.
Pseudophilomedes ferulanus Kornicker, 1959.
Cypridina (*Vargula*) *mediterranea* (Costa, 1845), in Granata and Caporiacco, 1946,
    p. 6, Pl. 1, Fig. 2 = *Skogsbergia*.
Cypridinodes sp.
Vargula hilgendorfii (Müller, 1894).

Family Cypridinellidae

?*Cypridinella superciliosa* Jones, Kirkby and Brady, 1874, p. 22, Pl. 5, Figs 7a–d
    (Carboniferous).

Family Sarsiellidae

Sarsiella disparalis Darby, 1965.
Sarsiella greyi Darby, 1965, p. 38, Pl. 27, Fig. 9.
Sarsiella zostericola Cushman, 1906.

Family Cylindroleberidae

Cycloleberis biminiensis Kornicker, 1959, p. 243, Fig. 85E.

Cycloleberis sphaerica Tressler, 1949, p. 340, Fig. 13.
Cylindroleberis elliptica (Phillipi, ?1840), in Sars, 1887, p. 200, Pl. 4, Fig. 2.
Cylindroleberis norvegica (Sars, 1869), in De Vos, 1957, p. 6, Pl. 3, Fig. 1a.
Asterope muelleri Skogsberg, 1920, in De Vos, 1957, p. 8, Pl. 3, Fig. 2a = *Parasterope*.
Asteropina extrachelata Kornicker, 1959, p. 241, Fig. 87C, F = *Synasterope*.

SOURCE OF SEDIMENTARY CARBONATE

Lowenstam (1963) reviewed the available information on biogenetic minerals in sediments. His data (1963, Fig. 1) indicate that Arthropoda are equal in importance as secretory agents of calcium carbonate to Algae, Protozoa and Mollusca. Previously the presence of crustacean remains in sediment has been based on the occurrence of recognizable fossils or hard parts, such as ostracod shells or crab claws. Our data indicate that the Crustacea is also a source of sedimentary particles of varied size and shape. Although the myodocopid carapace is weakly calcified, the calcareous nodules formed in the carapaces of both benthonic and pelagic genera may contribute a significant volume of carbonate particles to marine sediments.

In addition to the Ostracoda, Dudich (1931) described and illustrated nodules in the integument of species in the following crustacean taxa: Cladocera, Mysidacea, Isopoda, Amphipoda and Decapoda. It is significant that the Cladocera are abundant in fresh water, suggesting that Crustacea may be a source of carbonate particles in fresh-water deposits as well as in marine deposits. Dr J. Laurens Barnard, U.S. National Museum, called our attention to nodules in the appendages of the amphipod *Bruzelia* sp., and we have noted similar nodules in the first and second antennae of the myodocopid ostracod *Euphilomedes aspera* (Müller, 1894). These nodules show typical calcite interference figures with polarized light, but we have not verified their mineralogy.

FOSSIL RECORD

Dr W. D. I. Rolfe, Hunterian Museum, Glasgow, Scotland, called our attention to a paper by Peach (1882) in which nodules are illustrated in a portion of the carapace of the Lower Carboniferous *Anthrapalaemon parki* Peach, 1882 = *Perimecturus parki*, (Peach) (Eumalacostraca) as well as in the Recent shrimp *Crangon vulgaris* (p. 80 Pl. 9, Figs 4g, h).

We have not observed and we do not know of any published record of structures in fossil ostracods that can be unquestionably identified as nodules. Structures in the shells of *Cypridinella superciliosa* Jones, Kirkby and Brady, 1874 are discussed under discoidal nodules. Palaeozoic ostracods with radial markings that superficially resemble anastomosing nodules have been described (Jones, Kirkby and Brady, 1874; Ruedemann, 1926; Ulrich, 1879). These markings are present both on the outside and inside surfaces of the valves and do not appear to be discrete bodies as are the artifacts in Recent Myodocopida.

ACKNOWLEDGEMENTS

We are grateful to our colleagues at the U.S. Geological Survey and the U.S. National Museum for support indicated in the body of the paper. Drs J. M. Berdan and K. M. Towe aided by numerous discussions and suggestions. Photographs are by J. A. Denson and D. H. Massie, U.S. Geological Survey, and the plates were assembled by Elinor Stromberg, U.S. Geological Survey, and Caroline Bartlett Gast, U.S. National Museum.

REFERENCES

BANDY, O. L. AND KOLPACK, R. L. 1963. Foraminiferal and sedimentological trends in the Tertiary section of Tecolote Tunnel, California. *Micropaleontology*, **9** (2), 117-70, 35 Figs.

BRADLEY, W. H. 1957. Physical and ecological factors of the Sagadahoc Bay tidal flat, Georgetown, Maine, Chap. 23 of Ladd, H. S., ed., Paleoecology. *Mem. geol. Soc. Am.*, **67**, 641-82, Pls 3-7, 10 Figs.

BROOKS, R., CLARK, L. M. AND THURSTON, E. F. 1950. Calcium carbonate and its hydrates. *Phil. Trans. R. Soc.*, *A*, **243**, 145-67, 1 Pl., 1 Fig.

CUSHMAN, J. A. 1906. Marine Ostracoda of Vineyard Sound and adjacent waters. *Proc. Boston Soc. nat. Hist.*, **32** (10), 359-85, Pls 27-38.

DARBY, D. G. 1965. Ecology and taxonomy of Ostracoda in the vicinity of Sapelo Island, Georgia. *Univ. Mich., Nat. Sci. Found. Project GB-26*, Rept. no. **2**, 76 pp., 33 Pls, 11 Figs.

DUDICH, E. 1929. Die Kalkeinlagerungen des Crustaceenpanzers in polarisiertem Licht. *Zool. Anz.*, **85** (9/10), 257-64, 9 Figs.

——. 1931. Systematische und biologische Untersuchungen über die Kalkeinlagerungen des Crustaceenpanzers in polarisiertem Licht: *Zoologica, Stuttg.*, **30**, 154 pp. 14 Pls, 27 Figs.

ENBYSK, B. J. AND LINGER, F. I. 1966. Mysid statoliths in shelf sediments off Northwest North America. *J. sedim. Petrol.*, **36** (3), 839-40, 1 Fig.

GRANATA, L. AND DI CAPORIACCO, L. 1946. Ostracodes marins receuillis pendant les croisières du Prince Albert Ier de Monaco. *Result. Camp. scient. Prince Albert I*, fasc. 107, 50 pp., 4 Pls.

HARDING, J. P. 1965. Crustacean cuticle with reference to the ostracod carapace. *Pubbl. Staz. zool. Napoli* [1964]. **33** (suppl.), 9-31, 28 Figs.

JOHNSON, F. H. 1955. The luminescence of biological systems. *Proc. Am. Ass. Adv. Sci.*, Conference on Luminescence, 1954, 452 pp.

JONES, T. R., KIRKBY, J. W. AND BRADY, G. S. 1874-1884. A monograph of the British fossil bivalved Entomostraca from the Carboniferous Formations. Pt. 1. The Cypridinidae and their allies. *Palaeontogr. Soc.*, 92 pp., 7 Pls.

KELLY, A. 1901. Beiträge zur mineralogischen Kenntnis der Kalkausscheidungen im Tierreich. *Jena. Z. Naturw.*, **35**, N.F., **28**, 429-94.

KESLING, R. V. 1951. The morphology of ostracod molt stages. *Illinois biol. Monogr.*, **21** (1-3), 324 pp., 96 Pls, 36 Figs, 5 charts.

KORNICKER, L. S. 1959. Ecology and taxonomy of Recent marine ostracodes in the Bimini area, Great Bahama Bank. *Publs Inst. mar. Sci. Univ. Tex.*, **5**, 194-300, 89 text Figs [1958].

LOWENSTAM, H. A. 1963. Biologic problems relating to the composition and diagenesis of sediments. In Donnelly, T. W. (ed.), *The earth sciences*. Univ. Chicago Press, Chicago, 137-95.

MÜLLER, G. W. 1890. Neue Cyprididen. *Zool. Jb.*, abt. f. Syst., **5**, 211-52, Pls 25-27.

——. 1894. Die Ostracoden des Golfes von Neapel und der angrenzenden Meeres-Abschnitte. *Fauna Flora Golf. Neapel*, Mongr. **21**, 404 pp., 40 Pls.

——. 1906. Die Ostracoden der Siboga-Expedition. *Siboga Exped.*, **30**, E. J. Brill, Leiden, 40 pp., 9 Pls.

PEACH, B. N. 1882. On some new crustaceans from the Lower Carboniferous rocks of Eskdale and Liddesdale. *Trans. R. Soc. Edinb.*, **30** (1), 73-90, Pls 7-10.

POKORNÝ, V. 1954. *Základy Zoologické Mikropaleontologie*. Ceskoslovenské Akademie Věd, Prague, 651 pp., 756 Figs.

——. 1958. *Grundzüge der zoologischen Mikropaläontologie*, **2**, Deutsch. Verlag der Wiss., Berlin, 453 pp., Figs 550-1077.

——. 1965. *Principles of zoological micropalaeontology*, **2**. Pergamon Press, London, 465 pp., Figs 550-1077.

PRENANT, M. 1927. Récherches sur la calcaire chez les êtres vivants. La stabilité du calcaire amorphe et le tégument des Crustacés. *Annls Physiol. Physicochim. biol.*, **4**, 818-43.

RICHARDS, A. G. 1951. *The integument of arthropods*. Univ. Minnesota Press, 411 pp., 65 Figs.

RIOULT, M. 1966. Présence de sclerites de brachiopodes dans l'Oxfordien de Normandie. *Bull. Soc. linn. Normandie*, (10) **6**, 71-76, 1 Pl.

RUEDEMANN, R. 1926. The Utica and Lorrain Formations of New York, Part 2, Systematic paleontology. No. 2, Mollusks, crustaceans and eurypterids. *Bull. N.Y. St. Mus.*, **272**, 227 pp., 28 Pls.

SARS, G. O. 1887. Nye bidrag til kindskaben om Middelhavets invertebratfauna. IV. Ostracoda mediterranea. *Arch. Math. Naturv.*, **12** (2-3), 173-324, 20 Pls.

SCHMIDT, W. J. 1924. *Die Bausteine des Tierkörpers in polarisiertem Licht*. Friedrich Cohen, Bonn, 528 pp., 250 Figs.

SEKI, H. 1965. Microbiological studies on the decomposition of chitin in marine environment. *Jap. J. Oceanogr.*, **21** (6), 139-47.

SOHN, I. G. 1958. Chemical constituents of ostracodes; some applications to paleontology and paleoecology. *J. Paleont.*, **32**, 730-6.

TRESSLER, W. L. 1949. Marine Ostracoda from T ortugas, Florida. *J. Wash. Acad. Sci.*, **39**, 335-43
25 Figs.
ULRICH, E. O. 1879. Descriptions of new genera and species of fossils from the Lower Silurian about Cincinnati. *J. Cincinn. Soc. nat. Hist.*, **2**, 8-30, Pl. 7.
Vos, A. P. C. DE. 1957. Liste annotée des ostracodes marins des environs de Roscoff. *Archs Zool. exp. gén.*, **95** (1), 1-74, 39 Pls.
WATERMAN, T. H. 1960. *The physiology of Crustacea*, **1**, Metabolism and growth. Academic Press, New York and London, 670 pp., illus.

DISCUSSION

PURI: I question your statement that you 'made' these objects. Professor Collier and I have been able to grow aragonite crystals in the laboratory and these crystals are biologically controlled. The size of the crystals associated with the micro-algae is a little under three microns. There are two types of crystals, aragonite and calcite, and the biologically controlled are aragonite. They have some features in common with the forms that you described, for instance these crystals will dissolve in fresh water.

SOHN: I am interested in your remarks, but it will be clear in the published text that they are inapplicable to these nodules.

BENSON: I was interested in what you said about magnesium since this is a very rare element in arthropods. According to Waterman's *Physiology of Crustacea*, it has an anaesthetic effect on the animal which thus tends to reject it in its metabolism. I think you have shown that the amount of magnesium present is extremely small. In fact that it only occurs in trace quantities. Am I right in thinking that you did not test for strontium which would confirm if aragonite was present, and do not most of your tests show simple calcite or varieties of calcite?

SOHN: As you know there is considerable pressure for time on available equipment. We analysed only for Ca, P and Mg with the electron probe. In fact we have had the specimens re-polished hoping that some day we may perhaps examine for additional elements. X-ray powder determination revealed calcite, monohydrocalcite and, in a specimen heated in water, weak aragonite lines.

SWAIN: I believe that these are most properly termed spherulites rather than concretions because of their radial structure. The process of their formation has been reproduced in the laboratory by diffusing sodium carbonate solution into a gelatine gel containing a small amount of ferrous chloride. The process was studied many years ago in trying to determine the origin of the spherulites in the clay ironstones of England underneath the coal beds.

SOHN: We did not use the term spherulite because we are not certain that all the nodules have radial structures.

ROBINSON: I am concerned as to the fate of ostracods hitherto allocated to the Myodocopa from the Tertiary, the Cretaceous and also the Palaeozoic. Does Dr Sohn imply that valves of all myodocopids were subject to such breakdown changes of shell into nodules or microconcretions on fossilization, or were there conditions which would allow their normal preservation?

Some Carboniferous forms, referred by Jones, Kirkby and Brady 1884, to *Polycope simplex* Jones and Kirkby (Pl. V, Figs I a-d) or *Bradycinetus rankinianus* Jones and Kirkby (Pl. II, Figs 22 a-c), were illustrated as having surfaces covered by a delicate meshwork reticulation with stronger projections at each intersection. Is it possible that this is a post-mortem change, and that the surface details are microconcretions upon a preserved shell, rather than a true valve ornament?

SOHN: The fact that rare Palaeozoic and younger myodocopid fossils have been found indicates that there must have been conditions which allowed their normal preservation. I have as yet not seen the structures illustrated in the above publication on specimens, consequently I am not certain whether they are primary or secondary in origin. I have, however, examined the specimens showing the 'radiate' structures mentioned on p. 88 of the above publication in '*Leperditia radiata* Ulrich', and I am certain that these structures are secondary in origin.

RELATIONSHIP BETWEEN THE FREE AND ATTACHED MARGINS OF THE MYODOCOPID OSTRACOD SHELL

LOUIS S. KORNICKER
Smithsonian Institution, Washington, D.C., U.S.A.

SUMMARY

Serial sections of Recent ostracods are used to relate the free and attached margins of the shell. The dorsal ligament is a continuation of the inner margin of the 'duplicature', and is not connected to the selvage as has been reported previously. In order to relate morphological features of the free and attached margins it is necessary to consider whether they are exterior or interior. Hinge structures may be exterior, interior, or both. Structures on the free margin such as the selvage, list and various ridges are all exterior. In the Myodocopa examined in this study, structures of the free margin are not homologous to hinge structures on the attached margin. The concept of distinguishing interior and exterior hinge structures might prove useful in classification.

INTRODUCTION

The present study grew out of an attempt to decipher details of the hinge structure of myodocopid ostracods through gross examination of closed, open, and detached valves. When the valves were separated, almost always the tearing of the ligament and inner membrane obscured the relationships of adjacent structures thus necessitating recourse to some other technique. The use of serial sections to eludicate these relationships was promoted by a paper by Dr J. P. Harding (1965), 'Crustacean cuticle with reference to the ostracod carapace', presented at the 1963 symposium 'Ostracods as ecological and palaeoecological indicators'. The excellent photographs of sections across the hinge and free margin of the myodocopid ostracods in Harding's paper made it seem quite feasible that the relationship between the margins could be readily determined by studying sections intermediate between the free and attached margins.

Essentially the same technique has been used previously by Fassbinder (1912) and Gauthier (1939) in studies of the relationship between the margins of the shells of podocopid ostracods. My conclusions do not agree with those of Fassbinder, who believed that the ligament is attached to the ends of the selvage. Gauthier's illustrations lead me to believe that his conclusion, although not stated, would be the same as mine —that the ligament begins and ends at the inner margin of the 'duplicature'.

As the study progressed I became interested in the morphology of the ostracod exoskeleton and the relationship between areas of rigid and flexible integument. It became apparent that in order to relate structures of the free and attached margins it would be useful to understand the arrangement and disposition of the laminae composing the exoskeleton. It also became apparent that it would be useful to have a point of reference in order to homologize structures of the free and attached margins. I believe that the methods used and concepts derived herein are of greater interest and possibly more significant than the final conclusions concerning the relationship between the structures of the free and attached margins of the valves, which was the original purpose of this study.

In the present paper the term *attached margin* is used for those portions of the shells connected dorsally by a ligament. The remaining margins are considered to be the *free margins*. The terminology is only slightly different from that of Pokorný (1965, p. 85): 'The two valves rest against each other along the free margin, but on the dorsal margin they are linked to each other by an elastic, chitinous strip, the ligament.'

I wish to thank Mrs Carolyn Bartlett Gast for final preparation of all figures from my camera lucida drawings, and for designing the cut-away drawings on Plate I. Photographs taken by the writer were mounted by Mr Ernest Giles Jr. under the supervision of Mrs Gast. I appreciate criticisms of the manuscript by Drs Horton H. Hobbs, Thomas E. Bowman, I. G. Sohn, and Vladimir Pokorný. I also wish to thank Dr Sohn for participation in many stimulating discussions which contributed immeasurably to the research. The microtome sections were prepared mostly by Mrs Vernetta Williams; a few were made by Mr Charles E. Cutress, Dr W. Duane Hope and myself.

Material

Species selected for detailed descriptions are *Pseudophilomedes ferulanus* Kornicker, 1959 (three specimens from the Atlantic shelf off North Carolina); *Codonocera* sp. (two specimens from the Philippines); *Cypridinodes* sp. (four specimens from the Philippines); *Scleroconcha* sp. (two specimens from the North Atlantic); *Azygocypridina* sp. (one specimen from Japan). More information concerning these species is given in the descriptive part of this paper. Unidentified specimens are representatives of undescribed species which will be described at a later date.

In addition to the above, the specimens of the following were sectioned and studied, but are not discussed in the present paper as the relationships between the free and attached margins were not basically different from those reported: Myodocopa: *Rutiderma dinochelata* Kornicker, 1959 (two dried specimens from Bimini, Bahamas); *Sarsiella zostericola* Cushman, 1906 (two females from Hadley Harbor, Mass.); *Vargula hilgendorfii* G. W. Müller, 1890 (two specimens from Japan); *Parasterope pollex* Kornicker, 1967 (one specimen from Hadley Harbor, Mass.); *Euphilomedes multichelata* Kornicker, 1959 (eleven specimens from Bimini, Bahamas); *Euphilomedes* sp. (two specimens from Bermuda). Podocopa: *Cypris pubera* O. F. Müller, 1776 (two specimens from Wisconsin); *Pontocypria* sp. (one specimen from Indian Ocean); *Mutilus* sp. (one specimen from Madagascar); *Haplocytheridea* sp. (one specimen from Florida); *Chlamydotheca* sp. (one specimen from Puerto Rico); *Cyprideis* sp. (one specimen from Woods Hole, Mass.). Platycopa: *Cytherella arostrata* Kornicker, 1963 (one dried specimen from Bimini, Bahamas).

Methods

All species except *Pseudophilomedes ferulanus* (its shells were soft and flexible) were decalcified with 'Suza Fixative', which has an acetic acid base, before sectioning. Attempts to section specimens having visible calcium carbonate without prior de-

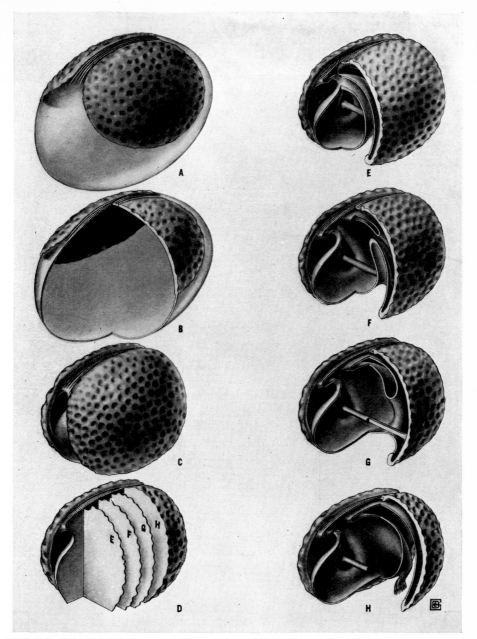

PLATE I. Schematic development of ostracod exoskeleton. A, hollow ellipsoid showing division into left and right shells, ligament, and vestment (anterior to left). B, same as Fig. A with quarter segment removed from anterior. C, same as Fig. A with vestment folded inward to form the body sheath. D, same as Fig. C with segment removed to show locations of cut-away illustrations in Figs. E-H. E-H, same as Fig. C except with segments removed to show complex evolutions of the vestment. (Adductor muscles illustrated to assist in orientation; all appendages except furca have been omitted.)

calcification were unsuccessful. Most specimens were embedded in paraffin; poly-ethylene glycol waxes were tried for a few, but were discontinued because of the effect of humidity differences on the hardness of the preparation. Sections were cut at 5–20 microns on a Spencer Model 829 microtome. Mallory's was used on all species. Some specimens were stained, for comparative purposes, with Chlorozol Black, and an Azan process reported by Hubschman (1962, p. 379). Differential staining of the exoskeleton cuticle was useful in tracing laminae from one part of the shell to another and also in interpreting relationships of parts of the exoskeleton. Camera lucida drawings of the dorsal margin were used to trace morphological features from the free to the attached margins. Photographs of the sections were made with a Polaroid camera.

Terminology

I found it necessary to coin the following terms for morphological features which had not previously been named (see Figs 1–2).

Anterior and *posterior junctures.* The anterior and posterior ends of the ligament. The junctures delimit the free and attached margins.

Taction margin. The margin of contact between the flexible integument and the shell. This margin is continuous around each shell.

Ligament line. The part of the taction margin of the shell between the anterior and posterior junctures.

Exterior surface of exoskeleton. The surface of the exoskeleton exposed to the external environment.

Interior surface of exoskeleton. The surface of the exoskeleton closest to the body cavity (haemocoele), separated from it only by the underlying hypodermis.

In addition to the above terms for previously unnamed features, I found it conveni-ent for clarity of description to give a new name to some features included under an older term (see Figs 1–2).

Vestment. The flexible membrane that forms the body sheath and lines each valve. The *vestment, shells* and *ligament* form the ostracod exoskeleton. The term vestment differs from the term 'inner lamella' (Moore, 1961, p. Q51) in excluding '. . . calcified marginal parts forming duplicature. . .'.

Infold. The infolded part of the shell margin. The term infold differs from the term 'duplicature' (Sylvester-Bradley, 1941, p. 8; Moore, 1961, p. Q51) in being defined as part of the shell and not part of the 'inner lamella'. This distinction is important because the infold along the attached margin is unrelated to the 'inner lamella'. The *infold* plus the *outer lamella* form the ostracod valve, or, as used herein, *shell.*

Axial line. This term is proposed as a substitute for 'line of concrescence' for shells whose margins are formed by folding.

A few additional terms are defined in later sections of this paper.

SUBDIVISIONS OF THE EXOSKELETON

The continuous nature of the ostracod exoskeleton is more easily visualized by con-sidering it as having developed from a hollow ellipsoid (Pl. I). The description of the arthropod exoskeleton by Maloeuf (1935; in Richards, 1951, p. 5) as '. . . funda-mentally an elongate, hollow, continuous ellipsoid modified by complex invaginations and evaginations' applies well to the ostracod. No evolutionary significance is intended in the developmental sequence illustrated in Plate I for, according to Calman (1909,

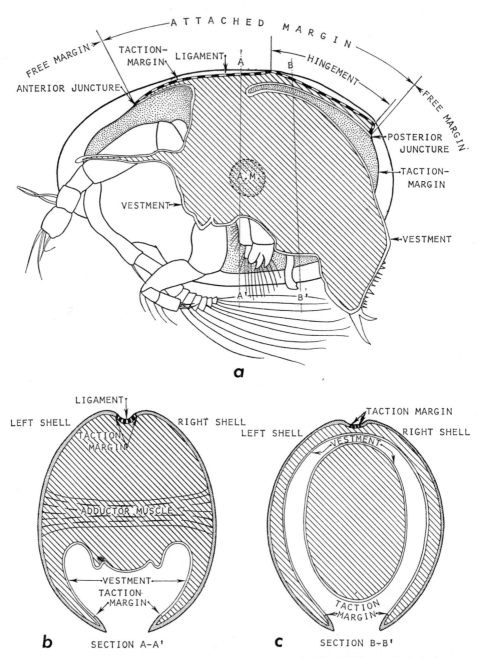

Fig. 1. Schematic diagrams of *Euphilomedes multichelata* labelled with morphological terms used in this paper (specimen length 0·93 mm.): (a) carapace and body with left half removed; (b) cross-section A-A' through adductor muscle; (c) cross-section B-B' posterior to adductor muscle. (Cross-hatched area represents body cavity; stippled area in fig. a is vestment surface; alternating black and white rectangles represents ligament, stippled area in Figs b-c is shell.)

p. 6), the carapace originates as a fold of the integument from the posterior margin of the cephalic region.

The continuity of the exoskeleton was stressed by Harding (1965, p. 28) who stated: 'It is important to realize that the shell is merely part of the integument which is a continuous sheet over all the body and limbs, and also lines both ends of the gut.' The continuous sheet is divided herein into four parts: left shell, right shell, vestment and ligament (see Fig. 1).

Shells

The *shells* are the mineralized or more rigid parts of the integumental fold forming the

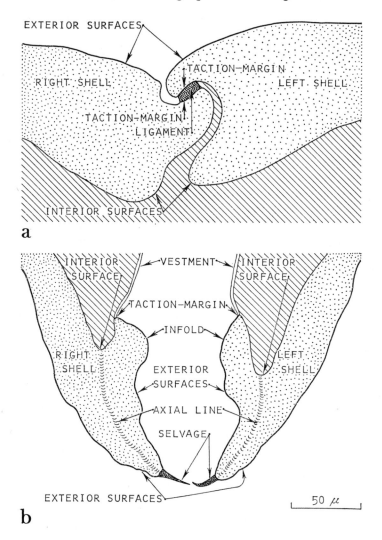

Fig. 2. Schematic diagrams of *Codonocera* sp. labelled with morphological terms used in this paper: (a) cross-section through hingement; (b) cross-section through ventral margins.

carapace. The term valve is avoided because of its different usage by zoologists and palaeontologists, but as used herein 'shells' are equivalent to 'valves' of the palaeontologist.

The shells of the myodocopids examined have on their outer surface a thin layer, generally unlaminated, that remains colourless, or stains light amber, over most of the shell. Beneath this layer is a thicker one consisting of two or three laminae that stain light to deep blue. This layer generally contains within it a network of fine reticulations. The filamentous structures forming the network stain brown. In this paper I have called the outer layer *epicuticle* and the blue layer *endocuticle*. The endocuticle generally has a middle lamina, the *middle layer*, that stains light blue. Between the *middle layer* and the epicuticle is a lamina of darker blue which I have called the *outer layer*. On the other side of the endocuticle is another dark blue layer, the *inner layer*.

Harding (1965, p. 16) interpreted the portion of the shell of *Cyclasterope* sp. that stains pale blue as having little sclerotization, and that part staining amber as being fully sclerotized. He considered red (not present on *Cyclasterope* sp.) as indicating sclerotization to a degree between that of the deep blue and amber layers. The pale blue part of the shell is probably hardened by calcification in the living animal. Harding (ibid., p. 6) observed the outer layer of the shell of *Cyclasterope* sp. to stain amber. On most of the specimens I examined the outer layer was unstained, but became amber in the area of the selvage, and amber to red in the area of the hingement. I believe that whether or not part of the outer layer remains unstained or becomes amber may depend in part on the staining technique or intensity of staining.

The thickness of epicuticle in the shell either remains constant as in *Azygocypridina* sp., or varies to form small nodes as in *Codonocera* sp. The outer surface of the epicuticle of *Scleroconcha* sp. contains small pits. The slender wedge-like selvage is formed of epicuticle but has a differently staining interior. In *Azygocypridina* sp. the selvage stains amber except for a grey centre; amber epicuticle separates the grey epicuticle from the underlying blue endocuticle. In *Scleroconcha* sp. the outer half of the selvage stains blue and the inner half stains amber. The selvages of other genera have additional slight differences in organization and reaction to staining. The list and associated marginal ridges stain amber and are formed by relatively thick sections of the epicuticle.

The outer layer of the endocuticle is poorly defined in *Codonocera* sp. and *Cypridinodes* sp. Where distinguishable it has a uniform thickness and is thinner and slightly darker than the middle layer. In *Pseudophilomedes ferulanus*, *Azygocypridina* sp., and *Scleroconcha* sp. it is well defined and stains a much deeper blue than the middle layer. The middle layer of the endocuticle is thicker than the outer and inner layers, forming 90% of the thickness in the shell of *Scleroconcha* sp. Difference in shell thickness is generally caused by variations in thickness of the middle layer. Large pits in the surface of the shell of *Cypridinodes* sp. are reflected by thin places in the middle layer. The prominent ridges of *Scleroconcha* sp. are the result of variations in the middle layer, and in part, of folding of the shell. Thickening of the shell along the attached margin to form hinge structures is generally caused by thickening of the middle layer, e.g. *Codonocera* sp. The inner layer of the endocuticle could be identified in all myodocopids examined, but is thin in some genera (e.g. *Scleroconcha* sp.) and thick in others (e.g. *Pseudophilomedes ferulanus*). It generally has a uniform thickness in each specimen, is laminated, and stains a deep-blue.

Vestment

As this paper is little concerned with the body sheath, the term vestment used herein usually refers to the flexible membrane lining the valves. The epicuticle and endocuticle are not visible as separate layers in the vestment, although in some specimens (e.g. *Azygocypridina* sp. *Scleroconcha* sp.) the vestment is laminated. At the margin of the vestment, where connected to the shell, a short segment of the vestment generally stains red (the remainder of the vestment is pale blue). In this segment the epicuticle and endocuticle are visible as separate layers.

Ligament

The ligament, which connects the shells dorsally, generally stains deep-blue with Mallory and is therefore easily distinguishable from the light blue shell and translucent vestment. It is also distinguishable from the vestment by its fibrous appearance. In the species studied, the shape of the ligament in cross-section varies considerably from stringy to wedge-shaped.

The epicuticle is continuous over the ligament where it stains amber or red, especially at the hingement. In most myodocopids examined, microlaminations in the endocuticle of the ligament could be followed into the shells on either side. *Codonocera* sp. has a wedge-shaped ligament which in some sections stains blue near its base and yellow and red toward the apex, dorsally. The yellow and red areas are seldom laminated and seem to have a denser structure than the blue areas. I am inclined to believe that the parts of the ligament that stain yellow and red are more fully sclerotized than the part that stains blue, but proof of this will require documentation by chemical analysis. It is possible that when the shells are closed the yellow and red part of the ligament acts as a fulcrum and the blue part is under compression. If this happens, the ligament of this species may assist in opening the shell when the ostracod relaxes its adductor muscles.

Relationship between shells, vestment and ligament

The entire margins of the shells are continuous with flexible integument—the ligament at the attached margin, and the vestment at the free margin (see Pl. IA). This margin of contact is the *taction margin*, and a recognition of its continuity is fundamental for the understanding of shell morphology. If this margin was not continuous the ostracod could not prevent sea water from freely mixing with fluids inside the body cavity.

The vestment and ligament meet at two places along the taction margin of each shell, at the *anterior juncture* located at the anterior end of the ligament, and at the *posterior juncture* at the posterior end. If the shells were removed the vestment and ligament would appear somewhat like a basket, with the ligament as its handle. The part of the taction margin of the shell between the anterior and posterior junctures may be termed the *ligament line*.

SUBDIVISIONS OF THE SHELL MARGINS

Along the anterior and posterior parts of the attached margin, the ligament is thin, broad and ribbon-like, and if it was of this nature along the whole of the attached margin, each shell could conceivably be moved a considerable distance from the other in any direction. However, along the middle part of the attached margin the ligament is thick and narrow, and acts like a strong leather hinge permitting the valves to open and close, but allowing little relative anterior-posterior movement of the shells. The shells themselves may be in contact, and the ligament visible in stained cross section only as a blue coloration between the shells. The term *hingement* is used herein for that part of the attached margin in which the shape or size of the ligament, or the configuration of the shell margins, can be interpreted as effectively restricting relative anterior-posterior movement of the shells. In the Myodocopa investigated, the hingement is generally more linear in lateral view than the remainder of the attached margin; it may have anterior and posterior elements, and is shorter than the attached margin. Groove and bar structures are present in the hingements of *Pseudophilomedes ferulanus* and *Codonocera* sp. As previously noted (p. 114) hinge structures are formed primarily by variations in thickness and shape of the middle layer of the endocuticle of the shell.

Harding (1965) observed that the free margin of the myodocopid shell folds inward. The folded over portion is termed herein the *infold*. Its edge is termed *inner margin*. The term inner margin has previously been used for the edge of the 'duplicature' (Moore, 1961, p. Q51) and the 'inner lamella' (Van Morkhoven, 1962, p. 60). Its usage herein differs in being applied to a structure in contact with the ligament as well as the vestment.

The infold is not limited to the free margin. Although generally reduced, it is also present along part of the attached margin in all myodocopids examined. The inner margin of the infold is connected to the vestment along the free margin and to the ligament along the attached margin. At the junctures, there is a smooth transition from vestment to ligament.

As indicated above, the *infold* plus the *outer lamella* are the major components of the ostracod shell; and structures on the infold such as selvage, list, and other ridges are part of the epicuticle layer of the shell.

In stained sections, the axis of the infold is often quite marked, being lighter in colour than the surrounding shell and having more distinct laminations. The axis serves as an aid in detecting the presence of an infold, and the *axial zone* might be considered the lateral boundary of the infold (Fig. 2b). The axial zone is equivalent to the 'zone of concrescence' except it originates by folding rather than by fusion. The term *axial line* is proposed as a substitute for 'line of concrescence' for shells the margins of which are formed by folding. The axial line seems to be confined to the endocuticle layer of the shell.

The reason for the distinct appearance of the axial zone is unknown. The differential effect of staining indicates that the composition of the shell in this zone differs from its surroundings. The continuous nature of the microlaminations passing through the zone and the surrounding shell is evidence of the absence of an adhesive strip. Where the microlaminations are absent or cannot be traced on to the surrounding shell, it is difficult to refute the possibility that the axial zone is a 'chitinous elastic strip' separating the 'lamellae'. However, the presence of the axial zone along sections of the attached margin where the vestment (inner lamella) is absent suggests that the infold is unrelated to the vestment (see Pls II I, IV G).

The infold varies in width. Behind the rostrum, below the sinus, and anterior to the caudal process, when present, it is usually wider than along the ventral margin of the shell. Along the attached margin, the infold progressively decreases in width from the anterior and posterior junctures, to the hingement, where it is extremely narrow or absent.

Zalányi (1929) classified ostracod free margins as monolamellar or bilamellar. He considered that both types are formed by fusing of the lamellae. In the monolamellar margin the fusion takes place at the ends of the lamellae without thickening the shell. In the bilamellar margin the facing sides of the lamellae are fused, resulting in the shell becoming thicker than the individual lamellae. The present study shows that thickening in the marginal area may be brought about simply by an increase in relative thickness of the individual microlaminations in the endocuticle of the fold area, and a fusion of lamellae is not required.

In some podocopid genera such as *Semicytherura* and *Cytheretta* (see Morkhoven, 1962, p. 70, Figs 62a, c) the interior side of an extremely wide infold is fused to the inner side of the outer lamella. Secondary fusion of this type was not observed in the Myodocopa studied.

RELATIONSHIP BETWEEN ATTACHED AND FREE MARGINS

The major subdivisions of the exoskeleton—the shells, the vestment, and the ligament —have an interior surface facing the hypodermis, and an exterior surface exposed to the external environment. Each surface is a continuum. Whether or not a part of the exoskeleton has an exterior and interior surface is a test of its relationship to the exoskeleton. For example, the selvage along the free margin has no interior surface—both of its sides are exterior (Fig. 2). Therefore, the selvage cannot be continuous with the ligament, which has both an interior and exterior surface, as erroneously concluded by Fassbinder (1912).

Previously on page 111, the taction margin was described as the margin of contact between the flexible integument (ligament and vestment) and each shell, and was shown to be continuous along the free and attached margins of each shell. If each shell is considered by itself, the taction margin is seen also to be a boundary separating interior and exterior surfaces of the shell margin (Figs 1, 2).

In order to relate morphological features on the free margin of a shell to features on the attached margin, it is necessary to consider whether they are exterior or interior. A fundamental rule that must apply is: *A structure on the exterior surface in one part of the shell cannot be identified with a structure on the interior surface in another part of the shell.* For example, a selvage on the exterior surface of the free margin cannot be considered to be homologous to an interior ridge along the attached margin.

Hinge structures may be exterior, interior, or both. The hingement of *Codonocera* sp. is an example of a species with both exterior and interior hinge elements (Fig. 2A). As may be seen in this figure, hinge structures are both above and below the ligament.

Structures on the infold such as the selvage, list and various ridges are all exterior structures. The flange of the outer lamella is also an exterior structure. As stated in the above rule, these features cannot be considered to be homologous to interior structures of the attached margin. Their relationship to the exterior structures of the attached margin probably varies considerably from species to species.

In the Myodocopa examined in this study, structures of the free margin are not homologous to hinge structures on the attached margin. A flange on the free margin of some species continues as a flange along the attached margin. The selvage present on the free margin of all myodocopids studied is considerably reduced along the dorsal margin, and is usually missing in the hingement area.

When the selvage is missing in the hingement area, it could be postulated that it has become part of the hinge structure. It is possible to discount this possibility in the Myodocopa investigated, because cross sections reveal that the hinge structures are formed by variations in thickness and shape of endocuticle, whereas, the selvage is formed of epicuticle.

In summary, exterior hinge structures of myodocopids do not seem related to exterior structures of the free margin. As the free margin has no interior structures, interior hinge structures must be considered to be independent. In all ostracods, identification of structures on the exterior surface of the attached margin with those of the free margin must be limited to exterior structures of the free margin.

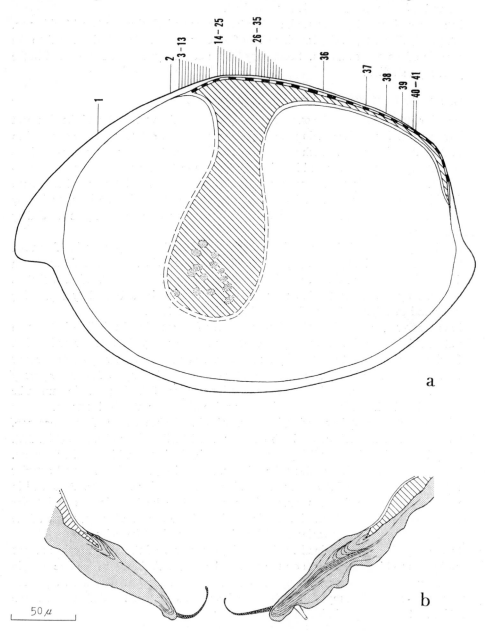

Fig. 3. Schematic diagrams of *Pseudophilomedes ferulanus* (specimen length 1·7 mm): (a) right shell with vestment lining from inside (numbers along dorsal margin show locations of cross-sections illustrated in Fig. 4); (b) cross-section through ventral margin, right shell on left. (The diagonal pattern represents body cavity; the pattern of alternating black and white rectangles in Fig. a represents ligament.)

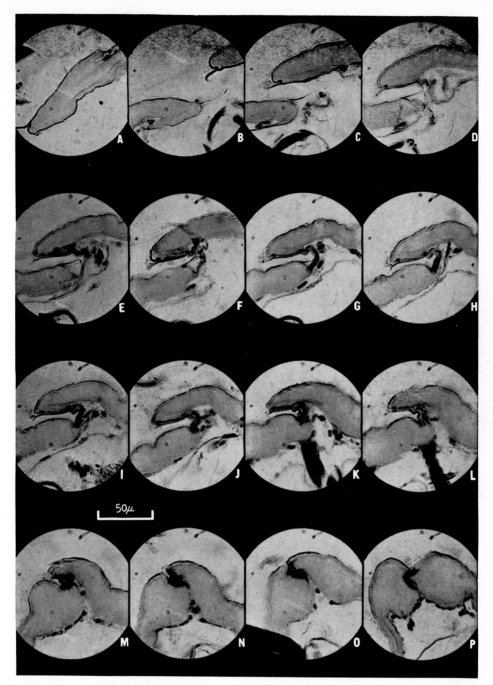

PLATE II. *Pseudophilomedes ferulanus.* A, cross-section of ventral margin of left shell; B-P, cross-sections along dorsal margin, right shell on left. (Figures B-P are equivalent to the following illustrations on Fig. 4, whose locations are indicated on Fig. 3: B-1; C-2; D-8; E-9; F-10; G-11; H-12; I-13; J-16; K-17; L-18; M-27; N-30; O-32; P-34.)

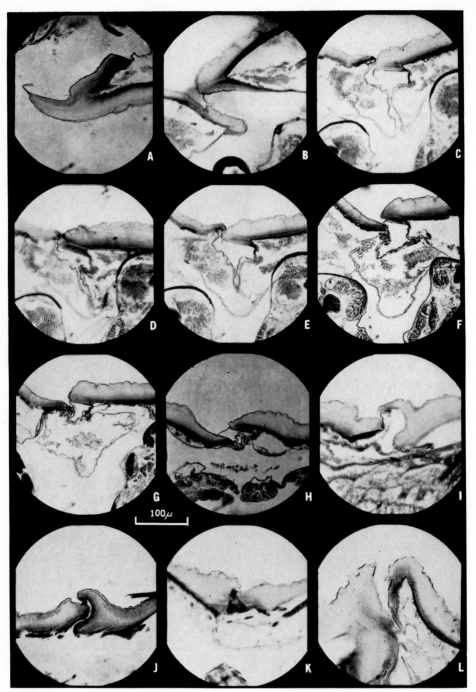

PLATE III. *Codonocera* sp. A, cross-sections of ventral margin of left shell; B-L, cross-sections along dorsal margin, right shell on left. (Figures B-L are equivalent to the following illustrations on Fig. 6, whose locations are indicated on Fig. 5: B-1; C-3; D-4; E-6; F-9; G-10; H-near 11; I-12; J-13; K-14; L-posterior to 19.)

The concept of distinguishing interior and exterior hinge structures might prove useful in classification, especially of podocopid ostracods in which hinge structures are of major importance. Two examples of podocopids that could be separated on this basis have been illustrated by Harding (1965, p. 21): *Notodromas* (ibid., Fig. 19) in which the shell is articulated above the ligament; and *Herpetocypris* (ibid., Fig. 20) in which the shell is articulated below the ligament.

In fossil ostracods that do not have living representatives, the position of hinge structures relative to the ligament might be difficult to determine. Knowledge that the ligament begins and ends at the inner margin of the infold might assist in determining the position of the ligament line.

DESCRIPTIONS

Pseudophilomedes ferulanus Kornicker, 1959
Figs 3, 4; Plate II.

Material

USNM	Locality data	Description
120508	From bottom sediments of Atlantic shelf, off North Carolina. Coll. in 1965	One ♀ in 111 10-micron sections on one slide.
120509	From bottom sediments of Atlantic shelf, off North Carolina. Coll. in 1965	One ♀ in 56 10-micron sections on two slides.
120510	From bottom sediments of Atlantic shelf, off North Carolina. Coll. in 1965	One ♀ in 166 10-micron sections on two slides

Morphology

The shells of *P. ferulanus* have an infold with a narrow vestibule and a selvage fringe (Fig. 3b; Pl. II A). The shell is laminated along the fold axis; laterally of the axis it is reticulate and layers are faint or not visible. The selvage fringe is in the plane of the fold axis.

The free margin along the anterodorsal part of each shell is similar to the ventral margin except for differences in width of the infold (Fig. 4.1; Pl. II B). Anterior to the juncture, the infold is very narrow, the vestibules are quite small, and the selvage fringe of the right shell is located laterally of the plane of the fold axis (Figs 4.2–3; Pl. II C).

Near the juncture, the vestment, at its point of contact with the inner margin of the left shell, appears as a 'bundle of deep blue fibres' (Fig. 4, 4). This is the beginning of the ligament. About 40 microns posterior to this point, the ligament connects the inner margins of the left and right shells, and the vestment forms an unbroken surface below the ligament (Fig. 4.11; Pl. II G); Figs 4.5–10, and Plate II D–F illustrate intermediate cross sections anterior to, and at the juncture, showing quite clearly that the ligament originates at the inner margin. At the anterior juncture the left shell has an infold but no vestibule, and the right shell is without an infold; both shells have a selvage (Figs 4.6, 7).

The ligament is shorter and thicker in each transverse section proceeding posteriorly along the dorsal margin (Figs 4. 12–23; Pl. II H–L). The selvage is progressively

I

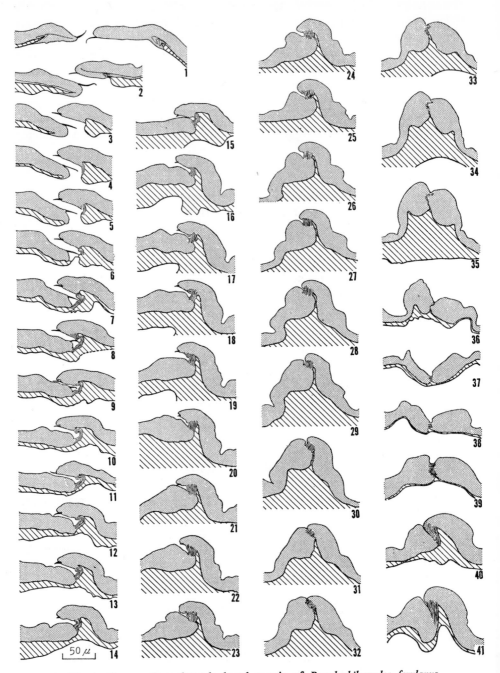

Fig. 4. Cross-sections through dorsal margin of *Pseudophilomedes ferulanus*, right shell on left. (Location of sections is indicated on Fig. 3. Cross-hatched area represents body cavity; selvage is black in figures.)

shorter on each shell and is no longer present on the right shell in the section illustrated by Fig. 4.10 and on the left shell in the section illustrated by Fig. 4.23.

Continuing posteriorly along the dorsal margin, the ligament in each transverse section is closer to the outer edge of each shell (Figs 4.24–32; Pl. II M–O). The shortening of the ligament posteriorly along the dorsal margin decreases the distance between the edges of each shell and they are in contact in the section illustrated by Fig. 4.32 and Pl. II O.

About 260 microns posterior to the anterior juncture, a ridge on the left shell fits into a groove on the right; this ridge-and-groove type of contact continues for a distance of about 200 microns along the dorsal margin (Figs 4.33–36; Pl. IIP). Along this part of the hingement, the ligament does not appear 'fibrous' and would be difficult to observe except for its having been stained deep-blue with Mallory. The ligament is horizontal in this area in contrast to its orientation in the anterior part, where it is vertical with the left shell overlapping the right.

Posterior to the ridge-and-groove structure, the dorsal margins of each shell are thin and connected by a narrow ligament (Figs 4.37, 38). This is followed along the margin by another short section having a ridge-and-groove type contact (Fig. 4.39). Posterior to that, the ligament again is prominent and vertical (Figs 4.40, 41). The posterior juncture was in poor condition on the specimen illustrated; however, on another specimen it was observed that the transition from attached to free margin is quite similar to that described above for the anterior juncture.

Cuticle

The shell has a thin epicuticle which remains unstained over most of the carapace. The selvage, which is structurally part of the epicuticle, has an inner layer of blue and outer layer of amber. The epicuticle continues across the ligament. A fairly thin outer layer of endocuticle stains blue. In some cross-sections of the ligament the part attributed to the outer layer stains red. The middle layer of endocuticle is almost colourless with a network of brown reticulations, and comprises about one half the shell thickness. The reticulations form laminae in some cross-sections. The inner layer of endocuticle stains pale blue and is less than one half the thickness of the shell. The parts of the ligament formed by the middle and inner layers of endocuticle stain deep blue. In some sections of the ligament, where the outer layer stains red, the middle layer stains amber.

Codonocera sp.
Figs 5, 6; Plates III, IV

Material

USNM	Locality data	Description
120511	San Miguel Harbor, Ticao Island, Philippines. Coll. 1908 by U.S. Bureau of Commercial Fisheries, Albatross Philippine Expedition	One ♀ with eggs in ovaries in 323 10-micron transverse sections on six slides.
120512	San Miguel Harbor, Ticao Island, Philippines. Coll. 1908 by U.S. Bureau of Commercial Fisheries, Albatross Philippine Expedition	One ♀ with eggs in ovaries in 641 5-micron transverse sections on seven slides.

Morphology

The free ventral margins of the left and right shells are similar; each has an infold with a distinct vestibule. The axis of the infold stains a lighter blue than the surrounding area and is marked by crenulations perpendicular to the axis. A selvage in the form of a lamellar prolongation is near the edge of each shell in the plane of its fold axis (Fig. 5b; Pl. III A).

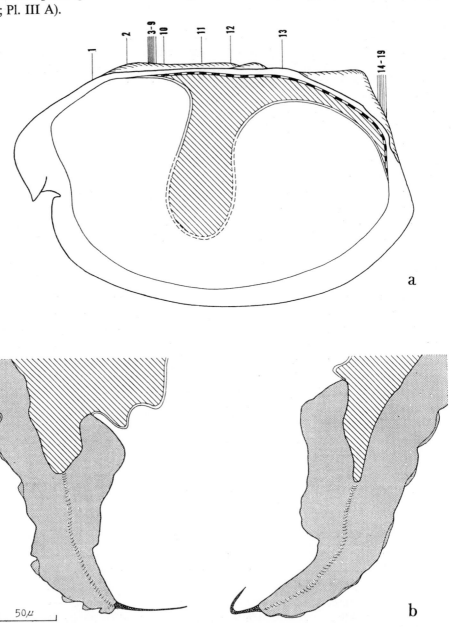

Fig. 5. Schematic diagrams of *Codonocera* sp. (specimen length 3·2 mm): (a) right shell with vestment lining from inside (numbers along dorsal margin show locations of cross-sections illustrated in Fig. 6; (b) cross-section through ventral margin, right shell on left. (The diagonal pattern represents body cavity; the pattern of alternating black and white rectangles in Fig. a represents ligament.)

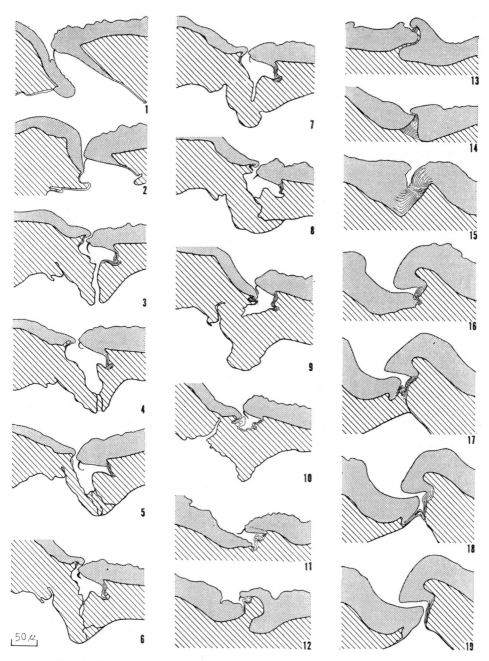

Fig. 6. Cross-sections through dorsal margin of *Codonocera* sp., right shell on left. (Location of sections is indicated on Fig. 5. Cross-hatched area represents body cavity; brick-like pattern in ligament indicates yellow or red, remainder of ligament is blue.)

123

On the anterodorsal part of the right shell, above the rostrum, the selvage is absent and the infold and vestibule small (Fig. 6.1; Pl. III B). Immediately anterior to the attached margins, the infolds are short on both shells, and the vestment is relatively thick and stains blue where attached to the infold of the left shell (Fig. 6.2). Anterior to the juncture, the vestments of each shell form flaps which are close together slightly below the infolds (Fig. 6.3; Pls. III C, IV C). The juncture is formed by the joining of the bottom of each flap as shown in Figs 6.4–5 and Pls III D, IV C–D. In Figs 6.3–7, which are in sequence and illustrate cross-sections only 5 microns thick, the juncture is observed only on the two middle sections (Figs 6.4–5). The sequence may also be followed on photographs equivalent to the text-figures as follows: Fig. 6.3 = Pls. III C, IV A–B; Fig. 6.4 = Pls III D, IV C; Fig 6.5 = Pl. IV D; Fig. 6.6 = Pls , III E.IV E Posterior to the juncture, the ligament and vestment are not connected to each other except for internal connective membranes (Figs 6.6–8; Pls III E, IV F). The ligament posterior to the juncture is blue and membrane-like with somewhat fibrous ends. The ligament is shorter in each transverse section proceeding posteriorly along the dorsal margin (Figs 6.10–12; Pls III F–H, IV G–H).

Posterior to the midpoint of the dorsal margin, a ridge and groove is present on the inner side of the left shell below the ligament attachment (Fig. 6.12; Pls III I–J, IV I–J) and the edge of the right shell fits into the groove. The ligament is attached at the edge of the right shell which has completely unfolded. A small infold, whose inner margin lies inside the groove, persists on the margin of the left shell. In this region the ligament is fairly short and is oriented vertically. Shortening of the ligament tends to pull the edges of the shells together (Fig. 6.14; Pl. IV K).

The dorsal margin of each shell is accuminate at the posterior end (Fig. 5a), and in that area the ligament is wedge-shaped in cross-section (Figs 6.15–16; Pls III K, IV L). The apex of the wedge stains yellow, orange, and red whereas the lower basal portion stains dark blue. The ligament continues along the posterior margin of the carapace and terminates at a juncture immediately above the caudal process (Figs 6.17–19; Pl. III L).

Cuticle

The shell has a thin layer of unstained epicuticle. The thickness of the epicuticle varies because of small ornamental nodes on the exterior surface. The epicuticle continues across the ligament, where in some cross-sections it is amber or red. The selvage is a structure of the epicuticle; it is wedge-shaped in cross-section and stains amber to orange along the free margin, becoming red along the attached margin. The epicuticle along the external surface of the infold stains amber.

The endocuticle is quite thick and predominantly middle layer, which stains light blue. A thin poorly defined outer layer stains a slightly deeper blue than the middle layer. A well defined inner layer is about one-fifth the thickness of the middle layer. In places, the inner part of the inner layer stains differently from the outer part, which stained medium blue in both specimens examined. The inner part stained deep blue in one specimen and bright red in the other. The red layer is fractured in places suggesting that it is more brittle, due to a higher degree of sclerotization, than the remainder of the decalcified cuticle. In cross-section and under oil immersion a network of fine reticulations is observed in the endocuticle, especially in the middle layer. The reticulations were not observed in the deep blue or bright red of the inner layer.

At the inner margin of the infold the endocuticle is as thin as the epicuticle and stains red. A short segment consisting of the layers of epicuticle and endocuticle lies between

the inner margin and the membraneous vestment, which stains pale blue and is un-
layered.

The bulk of the ligament is composed of endocuticle, and usually stains deep blue.
It cannot be differentiated into layers, but its colour is more similar to the outer and
inner layers than the middle layer. In some cross-sections microlaminations in the
ligament can be traced into the shells on either side.

In some cross-sections the outer part of the ligament stains red, the middle part
yellow, and the inner blue, indicating a progressive increase in sclerotization from the
internal to external surfaces of the ligament. This suggests that the inner part of the
ligament might be under compression when the valves are closed, and the ligament
might assist in opening the shells when the adductor muscle is relaxed. In some cross-
sections blue cuticle and amber cuticle form streaks perpendicular to the ligament
surfaces.

Cypridinodes sp.
Figs 7, 8; Plate V

Material

USNM	Locality data	Description
120513	San Miguel Harbor, Ticao Island, Philippines. Coll. 1908 by U.S. Bureau of Commercial Fisheries, Albatross Philippine Expedition	One ♀ with eggs in ovaries in 226 10-micron transverse sections on five slides.
120514	San Miguel Harbor, Ticao Island, Philippines. Coll. 1908 by U.S. Bureau of Commercial Fisheries, Albatross Philippine Expedition	One ♀ with eggs in ovaries in 147 10-micron longitudinal sections on three slides.
120515	San Miguel Harbor, Ticao Island, Philippines. Coll. 1908 by U.S. Bureau of Commercial Fisheries, Albatross Philippine Expedition	One ♀ with eggs in ovaries 272 transverse sections on five slides.
120516	San Miguel Harbor, Ticao Island, Philippines. Coll. 1908 by U.S. Bureau of Commercial Fisheries, Albatross Philippine Expedition	One ♀ with eggs in ovaries in 313 sections on six slides.

Morphology

Each shell has a short infold with a narrow vestibule and a selvage fringe (Fig. 7b;
Pls V A, B). In the anterodorsal part of the shell the infold is longer than on the ventral
margin, the vestibule correspondingly larger, and the selvage is lacking on the right
shell (Fig. 8.1; Pl. V C). Immediately in front of the juncture, the right shell is without
an infold, and the vestment is relatively thick where it connects with each shell (Fig.
8.2; Pls V D–F). Fig. 8.3 and Pl. V G show the ligament connecting the shells; the
section shown in Pl. V G is 30 microns from the section shown in Pl. V F, where the
shells are unattached. In each section proceeding posteriorly along the dorsal margin,
the ligament is thicker and shorter, and near the posterior part of the carapace, is
visible only as a blue coloration of the shells (Figs 8.4–11; Pls V H–L).

Cuticle

The shell has a thin unstained epicuticle. Its external surface is uneven, caused by variation in thickness of the epicuticle. The selvage which is structurally part of the epicuticle, stains amber, except for a small pale blue part situated laterally and proximally. A thin amber epicuticle layer lies between the endocuticle and the blue part of the selvage. The epicuticle on the external surface of the infold is relatively thick and stains amber near the selvage. Between the selvage and the ligament the epicuticle stains red. The epicuticle continues across the ligament.

An outer layer of endocuticle is indicated by faint deepening of the blue of the

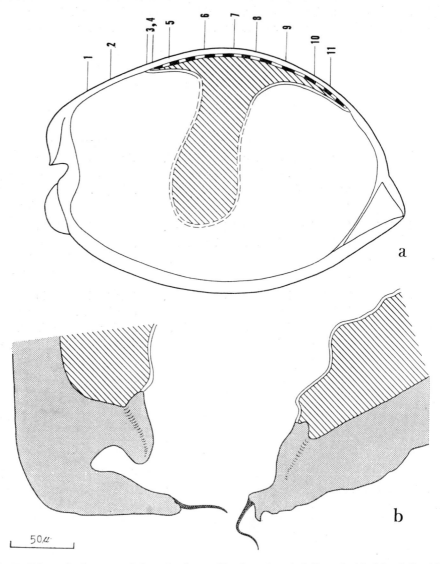

Fig. 7. Schematic diagrams of *Cypridinodes* sp. (Specimen length 2.64 mm): (a) right shell with vestment lining from inside (numbers along dorsal margin show locations of cross-sections illustrated in Fig. 8); (b) cross-section through ventral margin, left shell on left. (The diagonal pattern represents body cavity; the pattern of alternating black and white rectangles in Fig. a represents ligament.)

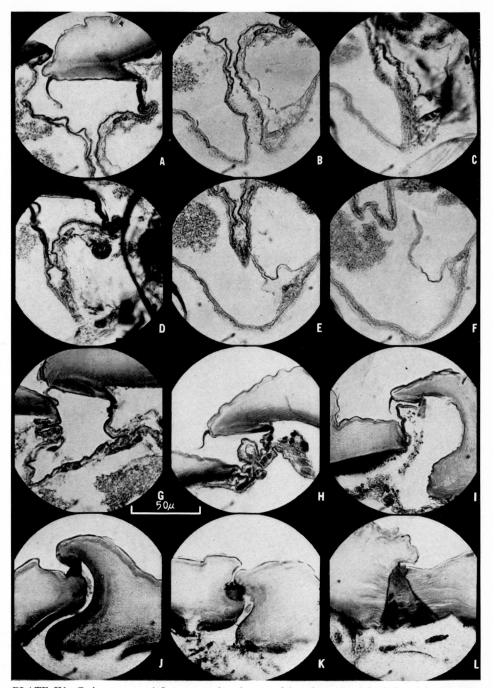

PLATE IV. *Codonocera* sp. A-L, cross-sections in area of dorsal margin, right shell on left. (Figures are equivalent to the following illustrations on Fig. 6, whose locations are indicated on Fig. 5: A & B, 3; C-4; D-5; F-8; E-7; G-9; H-11; I-12; J-13; K-between 13 & 14; L-14.)

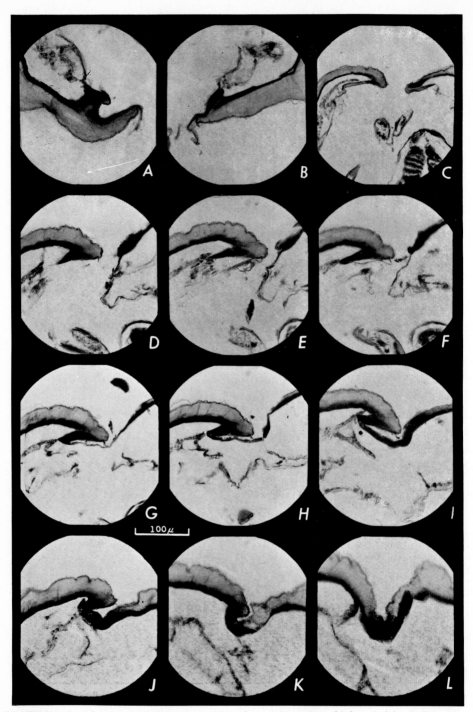

PLATE V. *Cypridinodes* sp. A-B, cross-sections of ventral margins of left and right shells, respectively; C-L. cross-sections along dorsal margin, left shell on left. (Figs C-L are equivalent to following illustrations on Fig. 8, whose locations are indicated on Fig. 7: C-1; D to G-between 2 & 3; H-3; I-4; J-7; K-9; L-between 9 & 10.)

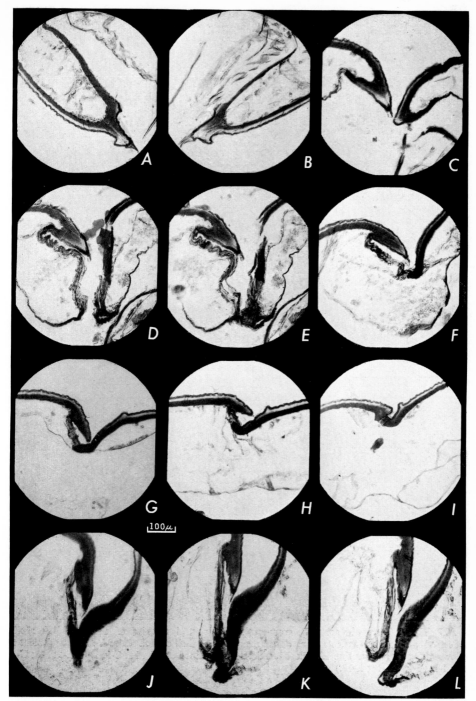

PLATE VI. *Azygocypridina* sp. A-B, cross-sections of ventral margins of right and left shells, respectively; C-L, cross-sections along dorsal margins, right shell on left. (Figs C-L are equivalent to following illustrations on Fig. 10, whose locations are indicated on Fig. 9: C-1; D-2; E-3; F-8; G-9; H-10; I-11; J-13; K-14; L-15.)

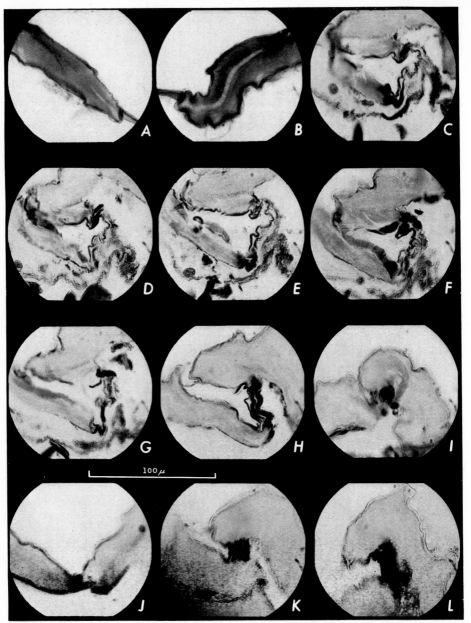

PLATE VII. *Scleroconcha* sp. A-B, cross-sections of ventral margins of right and left shells, respectively; C-L, cross-sections along dorsal margin, right shell on left. (Figs C-L are equivalent to following illustrations on Fig. 12, whose locations are indicated on Fig. 11: C-3; D-4; E-5; F-6; G-7; H-9; I-11; J-13; K-16; L-17.)

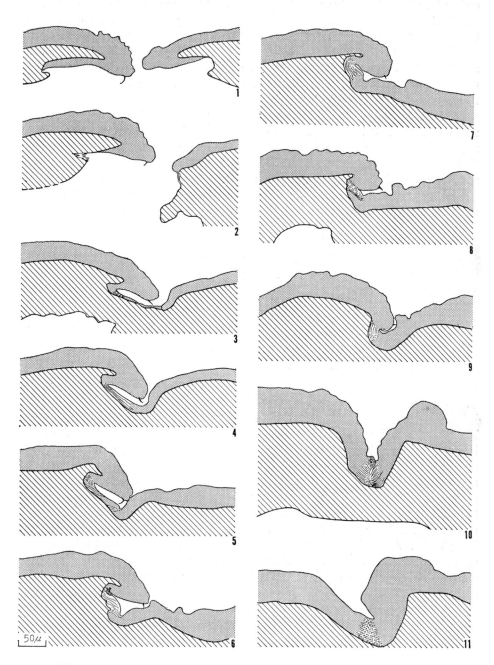

Fig. 8. Cross-sections through dorsal margin of *Cypridinodes* sp., left shell on left. (Location of sections is indicated on Fig. 7. Cross-hatched area represents body cavity; brick-like pattern and dashes in ligament indicate yellow or red, remainder of ligament is blue.)

endocuticle immediately beneath the epicuticle in some parts of the shell, especially in marginal areas. The middle layer of endocuticle stains pale blue and comprises about three-quarters of the shell thickness. It contains a network of fine reticulate structures, which in some parts of the shell form vague laminae. Large pits in the shell surface are depressions in the middle layer. The prominent list of the left shell margin is formed by thickening of the middle layer. The fold axis of the list is more prominent than that in the shell margin bearing the selvage.

The inner layer of endocuticle stains deep blue and forms about one-quarter of the shell's thickness. It is laminated but does not contain reticulate structures. In the areas of the list and near the ligament, the whole endocuticle is uniformly a deep blue similar to the blue of the inner layer. The ligament stains dark blue, but in some cross-sections part of the ligament is amber or red. In some cross-sections of the hingement the outer part of the ligament stains red and the inner blue. In cross-section the ligament has a twisted appearance caused by a diagonal pattern of colour variations.

Azygocypridina sp.
Figs. 9, 10; Plate VI

Material

USNM	Locality data	Description
120519	South coast of Hokkaido, Japan. Station 5036, Coll. 1906. U.S. Fisheries Steamer Albatross	One ♀ with eggs in brood chamber in 624 14-micron transverse sections on 33 slides.

Morphology

The ventral margins of the left and right shells are similar in that each contains a wide infold enclosing a deep vestibule and a selvage consisting of a lamellar prolongation along the outer edge of the shell (Fig. 9b; Pl. VI A–B). A selvage is not present on the anterodorsal margin of the right shell. The infolds on the anterodorsal shell margins are not as wide as along the ventral margins (Fig. 10.1; Pl. VI C). Immediately in front of the anterior juncture, the vestment at its attachment to the left shell is thick and crinkly, and the infold is narrow. The right shell is yellow near its infold (Fig. 10.2; VI D). At the juncture the thickened vestment on the left shell joins the right shell near the yellow area (Figs. 10.3–5; Pl. VI E). Immediately following the juncture the unfolded right shell is joined to the inner margin of the infold by the ligament (Figs 10.6–7). Figs 10.3–7 are illustrations of consecutive sections along the dorsal margin and represent a total thickness of about 70 microns. The ligament is shorter and the shells closer together in each section proceeding posteriorly along the dorsal margin to a point near the middle of the margin (Figs 10.8–11; Pl. VI G–I). From the middle of the dorsal margin to a point near the posterior juncture, the ligament is short and the shells almost in contact (Figs 10.11–12). In front of the posterior juncture the ligament is long and connects the edge of the right shell with the infold of the left (Fig. 10.13; Pl. VI J). The posterior juncture is similar to the anterior juncture (Fig. 10.14; Pl. VI K). Behind the posterior juncture the shells are not connected by a ligament and the vestment is attached to the infold of each shell (Fig. 10.15; Pl. VI L). A selvage is continuous around the exterior surface of the margin of the left shell, but is absent along the attached margin of the right shell.

Cuticle

Azygocypridina sp. has a thin amber epicuticle over most of the shell. It is continuous across the ligament where it stains orange to red. The wedge-shaped selvage is structurally part of the epicuticle and stains deep amber. The epicuticle lines the exterior surface of the infold and becomes thicker at the inner margin. The list is structurally part of the epicuticle and is formed by its thickening.

The inner layer of endocuticle stains dark blue and represents about one-half the thickness of the shell. The middle layer of endocuticle stains pale blue and is slightly thicker than the outer layer. The outer layer stains dark blue and is well defined.

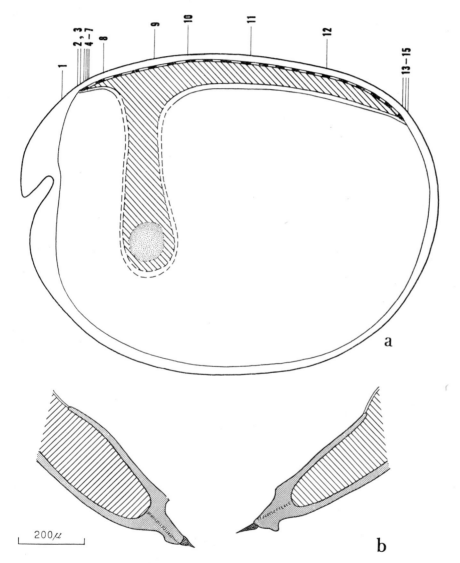

Fig. 9. Schematic diagrams of *Azygocypridina* sp. (Specimen length 2·2 mm): (a) right shell with vestment lining from inside (numbers along dorsal margin show locations of cross-sections illustrated in Fig. 10; (b) cross-section through ventral margin, right valve on left. (The diagonal pattern represents body cavity; the pattern of alternating black and white rectangles in Fig. a represents ligament.)

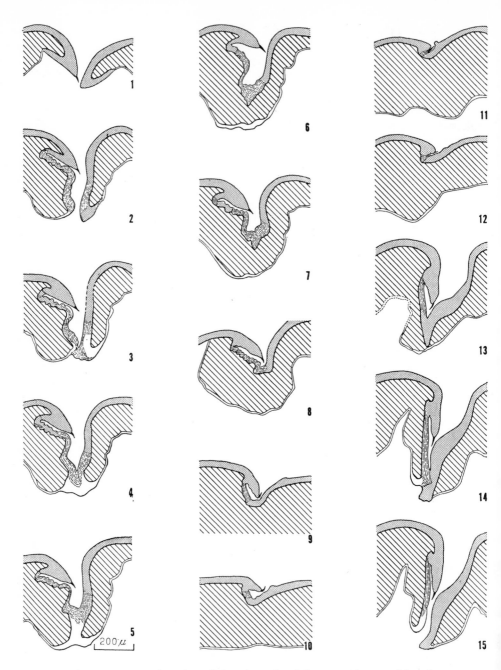

Fig. 10. Cross-sections through dorsal margin of *Azygocypridina* sp., right valve on left. (Location of sections is indicated on Fig. 9. Cross-hatched area represents body cavity; dashed pattern in ligament indicates yellow, remainder of the ligament is blue.)

130

In some cross-sections of the ligament, the outer layer of endocuticle stains dark blue, and the inner layer amber. Microlaminations in the middle layer of the ligament can be traced into the middle layer of endocuticle of the shell on either side. Microlaminations are not visible in the outer or inner layers of the endocuticle of the ligament.

At the inner margin of the infold the endocuticle is almost as thin as the epicuticle, and, with the epicuticle, forms a short segment that stains red and is continuous with the vestment. The vestment is fairly thick and has white and blue laminations.

Scleroconcha sp.
Figs 11, 12; Plate VII

Material

USNM	Locality data	Description
120517	Station V-17-75, Lamont Geological Observatory. North Atlantic.	One ♀ with eggs in brood chamber in 473-6-micron sections on five slides.
120518	Station V-17-75, Lamont, Geological Observatory. North Atlantic.	One ♀ with eggs in brood chamber. Distorted uncalcified sections on one slide.

Morphology

The ventral margin of each shell has an infold with a small vestibule; a long selvage fringe is located just within the plane of the fold axis (Fig. 11b; Pl. VII A–B). The free margin in the anterodorsal part of the shell, above the rostrum, is similar to the ventral margin except for the selvage, which is shorter on both shells, and thin membrane-like on the right shell (Fig. 12.1). Immediately in front of the juncture on the left shell, both the infold and selvage are relatively short; on the right shell the infold and selvage are absent (Fig. 12.2). Four 6-micron sections posterior to that shown in Fig. 12.2 are distorted and therefore have not been illustrated. The anterior juncture is shown in the four serial sections in Figs 12.3–6, and Pl. VII C–F. These sections show the ligament originating at the inner margin of each shell. The ligament is shorter in each transverse section proceeding posteriorly along the dorsal margin to about the highest point of the carapace (Figs 12.7–11; Pl. VII G–I). Anterior to the high point the ligament is oriented more or less vertically, and the dorsal margin of the left shell overlaps the right. From a point posterior to the high point of the carapace to near the posterior juncture, the ligament is short, stout and horizontal, and the left shell does not overlap the right (Figs 12.12–15; Pl. VII J). Anterior to the posterior juncture the ligament is vertical and the dorsal margin of the left shell overlaps the right (Figs 12.16–17; Pl. VII K–L). Posterior to the juncture the vestment is again attached to the infold at the free margin (Fig. 12.18).

Cuticle

Scleroconcha sp. has a thin epicuticle that is not stained by Mallory over most of the shell. It continues across the ligament where it stains light amber to red. The external surface of the epicuticle has minute punctations. The wedge-shaped selvage is structurally part of the epicuticle and stains amber except for a small blue part. Ridges on the infold's exterior surface are relatively thick places in the epicuticle, and stain amber.

The outer layer of endocuticle is distinct, but only about twice the thickness of the epicuticle. The light blue middle layer of endocuticle forms about 90% of the shell's thickness. It contains a network of faint reticulations and is divided in places into an inner and outer part. Ridges on the carapace generally reflect thickening of the middle

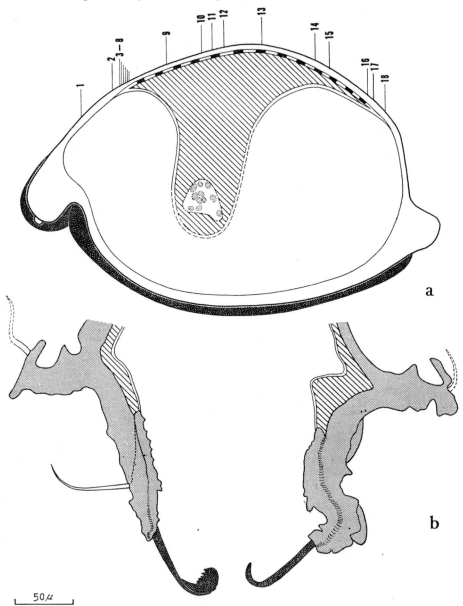

Fig. 11. Schematic diagrams of *Scleroconcha* sp. (Specimen length 2·2 mm): (a) right shell with vestment lining from inside (numbers along dorsal margin show locations of cross-sections illustrated in Fig. 12); (b) cross-section through ventral margin, right shell on left. (The diagonal pattern represents body cavity; the pattern of alternating black and white rectangles in Fig. a represents ligament.)

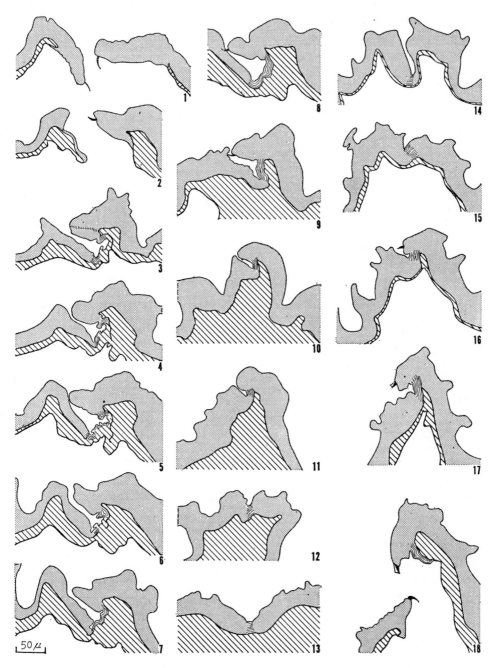

Fig. 12. Cross-sections through dorsal margin of *Scleroconcha* sp., right shell on left. (Location of sections is indicated on Fig. 11. Cross-hatched area represents body cavity.)

133

layer, but the major ridges are, in part, folds in the shell, and reflected by indentations of the interior surface of the shell. A very thin layer, about the thickness of the epicuticle, is visible in parts of the shell and probably represents the inner layer of endocuticle.

The cuticle of the ligament stains deep blue and is laminated. In some cross-sections the outer layer of endocuticle can be traced across the ligament and into the shells on either side. In the hingement area the outer part of the ligament stains red in some cross-sections. The vestment stains a deeper blue than the endocuticle, and laminations are visible in places.

REFERENCES

BOWMAN, T. E. AND KORNICKER, L. S. 1967. Two new crustaceans: The parasitic copepod *Sphaeronel-lopsis monothrix* (Choniostomatidae) and its myodocopid ostracod host *Parasterope pollex* (Cylindroleberidae) from the southern New England coast. *Proc. U.S. natn. Mus.*, **123**, No. 3613, 28 pp., 7 Figs, 2 Pls.

CALMAN, W. T. 1909. *Treatise on Zoology, Part VII, Appendiculata, 3rd Fascicle, Crustacea*, pp. 56-70, ed. R. Lankester, A. and C. Black, London. 346 pp., 194 Figs.

CUSHMAN, J. A. 1906. Marine Ostracoda of Vineyard Sound and adjacent waters. *Proc. Boston Soc. nat. Hist.*, **32** (10), 359-85, Pls 27-38.

FASSBINDER, K. 1912. Beiträge zur Kenntnis der Süsswasserostracoden. *Zool. Jb.*, Jena, **32**, 533-76, 61 Figs.

GAUTHIER, H. 1939. Sur la Structure de la Coquille Chez Quelques *Cypidopsides* à Furca Réduite et Sur la Validité du Genre *Cyprilla* (Ostracodes). *Bull. Soc. zool. Fr.*, **64**, 203-28.

HARDING, J. P. 1965. Crustacean cuticle with reference to the ostracod carapace. 9-31, 28 Figs. In H. Puri (ed.) Ostracods as Ecological and Palaeoecological Indicators, *Pubbl. Staz. zool. Napoli* [1964]. **33** (suppl.), 612 pp.

HUBSCHMAN, J. H. 1962. A simplified Azan process well suited for crustacean tissue. *Stain Technol.*, **37** (6), 379-80.

KORNICKER, L. S. 1959. Ecology and taxonomy of Recent Marine ostracods in the Bimini area, Great Bahama Bank. *Publs Inst. mar. Sci. Univ. Tex.*, 1958, **5**, 194-300, 89 Figs.

——. 1963. Ecology and classification of Bahamian Cytherellidae (Ostracoda). *Micropaleontology*, **9** (1), 61-70, 44 Figs.

MALOEUF, N. S. R. 1935. The role of muscular contraction in the production of configurations in the insect skeleton. *J. Morph.*, **58**, 41-86.

MOORE, R. C. 1961. Glossary of Morphological terms applied to Ostracoda. Pp. Q47-Q56, Pt. 3. In Arthropoda of Part Q in Moore, *Treatise on invertebrate paleontology*. University of Kansas Press, Lawrence, Kansas. xxiii+442 pp., 334 Figs.

MORKHOVEN, F. P. C. M. VAN. 1962. *Post-Palaeozoic Ostracoda*, vol. 1. Elsevier Publishing Company, Amsterdam, London, New York. vii+204 pp., 79 Figs.

MÜLLER, G. W. 1890. Neue Cyprididen. *Zool. Jb.*, Abt. f. syst., **5**, 211-52, Pls 25-27.

MÜLLER, O. F. 1776. *Zoologiae danicae prodromus, seu animalium Daniae et Norvegiae indigenarum characters, nomina, et synonyma imprimus popularium*. 8 vo. Havinae. i-xxxii, 1-282 pp.

POKORNÝ, V. 1965. *Principles of zoological micropalaeontology*, **2**, 465 pp., 1077 Figs. [Translated by K. A. Allen, edited by John W. Neale.] Pergamon Press, Oxford, England.

RICHARDS, A. G. 1951. *The integument of Arthropods*. University of Minnesota Press, Minneapolis, Minnesota. 410 pp. 65 Figs.

SCOTT, H. W. 1961. Shell morphology of Ostracoda. Q21-Q37, Pt. 3. In Arthropoda of Part Q in Moore, *Treatise on invertebrate paleontology*. The University of Kansas Press, Lawrence, Kansas. xxiii+442 pp., 334 Figs.

SYLVESTER-BRADLEY, P. C. 1941. The shell structure of the Ostracoda and its application to their palaeontological investigation. *Ann. Mag. nat. Hist.* (11), **8**, 1-33, 18 Figs.

ZALÁNYI, B. 1929. Morph-systematische Studien über fossile Muschelkrebse. *Geologica hung. Ser. Palaeontologica*, Fasc. **5**, 1-152, 35 Figs. 4 Pls. Budapest.

DISCUSSION

BENSON: I think it is worth noticing that the ligament line can probably be traced in podocopids such as *Quasibuntonia*. It seems that in podocopid amphidont hinges the selvage, or what has been called the selvage, continues dorsally past the juncture (using Dr Kornicker's terms) where the attached and taction margins meet. It then goes over the top of the terminal elements of the hinge so that the median element in the amphidont hinge which we referred to in the past, is in fact quite distinct and separate from the terminal elements as a part of this marginal structure. I think that in future we can learn a great deal about the development of the parts of the terminal elements such as the stepped nature of the anterior tooth in the right valve which may, in fact, be two stages in its development. The inner part of the step is perhaps a remnant of the penultimate or incomplete stage of hinge development, and the under part, which then makes the major or most prominent part of the tooth, may come from the interior and would be encompassed within the vestment.

BATE: Just one comment on the ligament. In the podocopids, I feel that the term ligament is not strictly accurate; personally I do not think that it is homologous with the mollusc ligament. Work on *Pontocypris* and *Propontocypris* suggests that the elastic, chitinous part in the hinge area is simply an elastic part of the whole chitinous envelope which completely covers the calcareous carapace. In the hinge area this may perhaps have some elastic properties, but to call it a ligament, as if it was an isolated bar as in molluscs, is misleading. I am talking in terms of podocopids, not myodocopids about which I have no first hand knowledge.

KORNICKER: I do not believe that the molluscan specialist has a stronger claim than we do over the term ligament. I think that it is a good name for the fibrous structure connecting the dorsal margins of the valves. The ligament is continuous with the carapace and does not cover it.

K

RECHERCHES SUR LA MORPHOLOGIE DE L'ORGANE COPULATEUR MÂLE CHEZ QUELQUES OSTRACODES DU GENRE *CANDONA* BAIRD (FAM. CYPRIDIDAE BAIRD)

DAN L. DANIELOPOL
Institut de Spéologie 'E. G. Racovitza', Bucarest, Roumanie

'Especially the structure of the penis appears to be significant. Indeed, just as in several other groups of Arthropods this organ appears to have been the seat of the initial morphological changes leading to speciation. Unfortunately, the morphological interpretation of the structural complexities of this organ is still uncertain. To carry out the homologies of its different joints will probably prove the most fascinating and fruitful morphological problem that the Ostracod group has to offer.'

T. SKOGSBERG, 1928

SUMMARY

The male copulatory organ of *Candona neglecta* Sars (*sensu* G. W. Müller, 1900) is described in detail. The different parts are compared with similar structures found in *Candona fasciolata* Petkovski, *Candona weltneri* Hartwig, *Candona* aff. *C. compressa* (Koch)-Brady, *Candona levanderi* Hirschmann, *Candona fabaeformis* Fischer, representing the *candida-neglecta*, *rostrata-compressa* and *fabaeformis-acuminata* groups. This is the first time that a systematic study of this organ in the Cyprididae has been undertaken.

The male copulatory organ in the *Candona* examined consists of two penes, fused together in a basal part. Each penis in its turn consists of a number of pieces forming a 'gaine pénienne'. These pieces are: a part of the vas deferens, a strongly chitinized piece called the labyrinth, the muscular copulatory tube, the sleeve with the copulatory pouch, a median chitinous piece or 'M' piece, and a distal piece or 'D' piece. The different pieces are joined together by chitinous strips.

In order to discuss the functional significance of the pieces, the copulatory organ of *Candona* aff. *C. compressa* (Koch)-Brady is described in the erected position. Finally the possibility of using a study of the copulatory organ to clarify the taxonomy of the genus *Candona* in particular, and the Cyprididae in general, is envisaged.

RÉSUMÉ

L'auteur décrit en détail l'organe copulateur mâle chez *Candona neglecta* Sars (*sensu* G. W. Müller 1900). Les différentes pièces sont comparées avec les formations similaires appartenant à *Candona fasciolata* Petkovski, *Candona weltneri* Hartwig, *Candona* aff. *C. compressa* (Koch)-Brady, *Candona levanderi* Hirschmann, *Candona fabaeformis* Fischer, représentant les groupes *candida-neglecta*, *rostrata-compressa* et *fabaeformis-acuminata*.

C'est pour la première fois qu'on entreprend une étude systématique sur la morphologie de cet organe chez les ostracodes de la famille Cyprididae Baird.

L'organe copulateur mâle, chez les *Candona* éxaminés, est formé par deux corps péniens, soudés entre eux dans la partie basale. Chaque corps comporte à son tour un complexe de pièces, maintenues dans une *gaine pénienne*.

Ces pièces sont: une partie du *vaisseau déférent*, une pièce fortement chitinisée nommée *labyrinthe*, le *tube copulateur* musculeux, le *manchon* avec la *bourse copulatrice*, une pièce chitineuse médiane ou *pièce M*, une pièce distale ou *pièce D*.

Les différentes pièces sont liées entre elles par des bandes chitineuses.

Pour discuter la valeur fonctionnelle des pièces, l'auteur décrit l'organe copulateur de *Candona* aff. *C. compressa* (Koch)-Brady, en érection.

Dans la dernière partie, on envisage les possibilités d'utiliser l'étude de l'organe copulateur pour la systématique du genre *Candona*, en particulier, et pour les Cyprididae en général.

L'appareil génital mâle chez les Cyprididae possède plusieurs particularités interessantes: premièrement les testicules tubuleux sont situés dans la cavité de la coquille, chez la majorité des ostracodes étant placés à l'intérieur du corps; secondement il y a une formation chitineuse contractile, l'organe de Zenker, qui assure l'évacuation des spermatozoïdes géants; enfin l'organe copulateur, situé à l'extérieur du corps est fortement chitinisé et possède une structure complexe qui jusqu'à présent est très peu connue.

Parmi les ostracodes du sous-ordre Podocopa, seul l'organe copulateur des Cythéridae est mieux connu, grâce aux recherches de Hirschmann, 1912; Klie, 1937; Hoff, 1944; Hart, 1962, 1967. Chez les Cyprididae dont le type d'organisation est différent des Cytheridae, la majorité des investigations ont porté spécialement sur l'aspect général de la formation en question, les pièces internes avec leurs correlations intimes et leurs valeurs phylogénétiques étant négligées (voir G. W. Müller, 1900; Sars, 1922-1928; Schäfer, 1937; Klie, 1938; etc.). Un progrès dans la connaissance de l'organe copulateur a été marqué dernièrement par les recherches de Petkovski 1959–66 sur la morphologie de la *pièce médiane*, chez les *Candona* du groupe *candida-neglecta*. Vu les données mentionnées la nécessité et l'intérêt d'une étude approfondie et détaillée de l'organe copulateur des Cyprididae, n'est plus à discuter.

Le but initial de cette communication était de passer en revue la morphologie de l'organe copulateur chez les principaux groupes de Cyprididae. Hélas, nous nous sommes vite aperçus que les différents types d'organes copulateurs, de structure très diverses, représentent à l'état actuel de nos connaissance des problèmes qui doivent être résolus séparément. C'est pourquoi nous avons opté pour l'étude d'un seul groupe, à notre avis plus homogène du point de vue phylétique; les ostracodes du genre *Candona* Baird (*sensu* G. W. Müller, 1912), en sont un excellent matériel. En ce qui suit nous présenterons la morphologie détaillée de l'organe copulateur chez l'espèce *Candona neglecta* Sars (*sensu* G. W. Müller, 1900), qui sera comparé avec les formations similaires de *C. fasciolata* Petkovski,[1] *C. weltneri* Hartwig, *C.* aff. *C. compressa* (Koch)-Brady, *C. lobipes* Hartwig, *C. fabaeformis* Fischer et *C. levanderi* Hirschmann,[2] représentant des groupes *candida-neglecta*, *rostrata-compressa* et *acuminata-fabaeformis*.[3]

---

[1] Nous adressons nos vifs remerciements à M le Dr T. Petkovski, qui a bien voulu vérifier la détermination du matériel de *C. neglecta* Sars et *C. fasciolata* Petkovski utilisée pour ce travail.

[2] Nous remercions Mlle F. Caraion pour l'amabilité de nous avoir cédé quelques mâles de *C. levanderi* Hirschmann.

[3] Les descriptions détaillées de l'organe copulateur, des espèces que nous avons cité ici seront publiées séparément.

Pour la précision et la facilité des descriptions nous proposerons ici un système d'orientation et de nomenclature des pièces. L'étude comparative entreprise sur les sept espèces citées plus haut nous permettra de démontrer la valeur systématique des différentes pièces. Pour comprendre la valeur fonctionnelle de chaque pièce, nous avons cru utile de décrire l'organe copulateur en érection, chez *Candona* aff. *C. compressa* (Koch)-Brady. Dans la dernière partie de ce travail nous ferons quelques

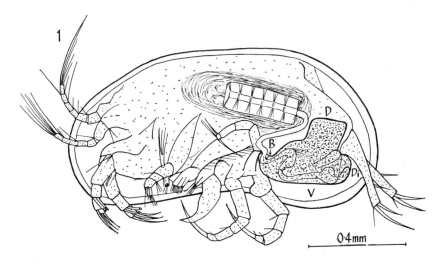

Fig. 1. *Candona neglecta* Sars – mâle, vue latérale.

Liste des abréviations dans les figures
L – latérale; M – médiale; B – antérieur; D – dorsal; Di-posterieur; V – ventral; a – face latérale; b – face médiale; c – bande chitineuse; C_1 – C_3 – bandes chitineuses de la face latérale; C_4 – C_6 bandes chitineuses de la face médiale; d – *labyrinthe* (d_1 – rameau ascendant, d_2 – rameau transversal, d_3 – rameau descendant, d_4 – rameau terminal); e – *bourse copulatrice*; f – *manchon*; g – *pièce M* (gb – région basale, gd – région distale) h – *pièce D*; i – *tube déférent*; k – capuchon; 1 – *tube copulateur*; r.d.m. – rainure distale médiane; p.m. – pli lamellaire médiale; pl – pli lamellaire latérale; d – δ – tubercules distales.

remarques sur les perspectives des études morphologiques de l'organe copulateur chez les *Candona* en particulier, et chez les Cyprididae en général.

MÉTHODE D'ÉTUDE, ORIENTATION ET NOMENCLATURE

L'organe copulateur peut être facilement détaché du corps de l'animal et examiné sur toutes ses faces. Jusqu'à présent la majorité des chercheurs avaient étudié l'organe copulateur seulement sur des préparats totaux où il est impossible de surprendre tous les détails. C'est grâce à des dissections très fines que nous avons séparé les deux *corps péniens* et en continuant la dissection sur l'un d'entre eux nous avons réussi à isoler et à étudier directement les différentes pièces, spécialement celles internes qui sont très peu connues.

Pour l'orientation des pièces nous avons tenu compte des avis émis par le grand zoologiste roumain E. G. Racovitza (1923), à propos de ses études sur les isopodes: «Il faut adopter un système d'orientation spécial pour chaque groupe ayant une orientation naturelle semblable et pour que ces systèmes soient «permanents» il ne faut point qu'ils soient basés sur des conceptions théoriques, mais sur l'orientation naturelle de la majorité des espèces du groupe envisagé» (p. 77).

Nous considérons donc que le *corps pénien* des *Candona* possède une face latérale et une face médiale, une région basale ou antérieure, par laquelle on fait l'insertion au corps de l'animal, une région distale ou postérieure, enfin une face ventrale et une face dorsale. Chaque pièce possède à son tour un côté externe, un côté interne. Les *corps péniens* soudés entre eux dans la région basale des faces médiales forment le *complexe copulateur*.

Chez les *Candona* chaque *corps pénien* est formé à son tour par un *système de pièces internes* maintenu dans une *gaine pénienne*. Les *pièces internes* sont reliées entre elles ainsi qu'à la *gaine pénienne* par des *bandes chitineuses*. Nous avons identifié les suivantes *pièces internes*: une partie du *canal déférent*; le *tube copulateur*, probablement musculeux; *le labyrinthe*, une formation fortement chitinisée; *le manchon*, qui recouvre le *tube copulateur*; la *bourse copulatrice*, un prolongement distal du *manchon*; *la pièce M*, ou *pièce chitineuse médiane*; *la pièce D*, ou *pièce chitineuse distale*. Enfin la *gaine pénienne* possède une face latérale et une face médiale dont la forme dès bords dorsales et distales est fort diverse, pouvant former des lobes proéminents. La portion basale du *corps pénien* par laquelle on fait l'insertion au corps de l'animal pourait être nommée *région de connexion* (Fig. 3).

Nous tenons à remarquer que certains noms que nous proposons ici ont été déjà employés par d'autres auteurs. C'est ainsi que les noms de *tube copulateur* et *labyrinthe* ont été proposés par Hirschmann (1912)[1]; le nom de *pièce médiane* est l'équivalent de l'«inner mittelforzats» de Müller (1900) et Petkovski (1959), et pourrait être homologuer avec la pièce des Entocytherinae «*clasping apparatus*» Hoff (1944). Le terme de *région de connexion* est équivalent au terme «*connecting piece*» utilisé par Hoff (1944) (p. 330).

LA MORPHOLOGIE DE L'ORGANE COPULATEUR DE *Candona neglecta* SARS

Chez *Candona neglecta* Sars l'organe copulateur, massif, est formé par deux corps péniens applatis sur leur axes latéro-médiales (Figs. 1, 2). Chaque *corps pénien* possède une *gaine pénienne* et un *système de pièces internes*.

La *gaine pénienne* (Figs 2–5) présente une face latérale (Figs 3 et 5) qui recouvre toutes les pièces internes, excepté la région distale de la *pièce D*. Un lobe carré présentant des rainures chitineuses fait sailli du côté dorsal. Sur le côté interne de la face latérale de la *gaine pénienne* il y a une bande chitineuse oblique (Fig. 3 C_2) qui fait la liaison entre celle-ci et la *pièce M*. Dans la région de contact l'extrémité distale de la bande chitineuse forme avec la face latérale un pli lamellaire (Fig. 3). Des bandes chitineuses transversales (Figs 3 et 5, C_1 et C_3) renforcent la face latérale de la *gaine* en la reliant au *labyrinthe*. La face médiale de la première étant plus large que son opposé latérale a dans sa position basale le bord postérieur ainsi que le bord dorsal largement arrondi. Du côté interne de la face médiale il y a trois bandes de chitines (Figs 4 C_4, C_5) qui font la liaison entre la *gaine pénienne* et le *labyrinthe*. La même face est traversée par un pli lamellaire externe en diagonale (Fig. 4, pm.). Dans la région de connexion on peut observer aussi des bandes chitineuses de renforcement (Fig. 2, C_6 et Fig. 4, C_6). La face latérale et la face médiale s'unissent sur le bord dorsal dans le tiers basal.

[1] Le terme de *labyrinthe* a chez l'auteur cité un sens plus large: 'In Aufsatz finden wir ein äusserst kompliziertes System von Leitungsröhren (vielleicht Rinnen), aus starkem und schwachem Chitin, Hebeln, Hacken und Leisten, die den Ductus ejaculatorius, bis zum Genitalöffnung leiten . . . Diesem Aggregate von Chitingebilden, das auch in den einfechsten Fällen ein chaotiches Bild darstellt, gebe ich, den Namen des Penislabyrinthes' (Lab) (p. 31).

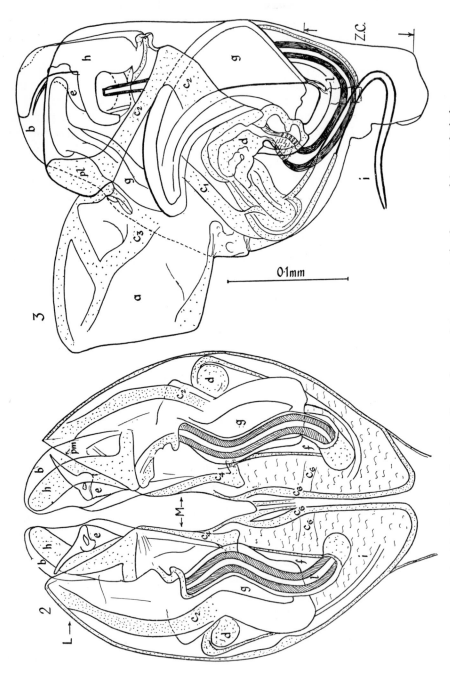

Figs 2-3. *Candona neglecta* Sars. 2. *complexe copulateur, vue ventrale.* 3. *corps pénien, vue latérale.*

0.1mm

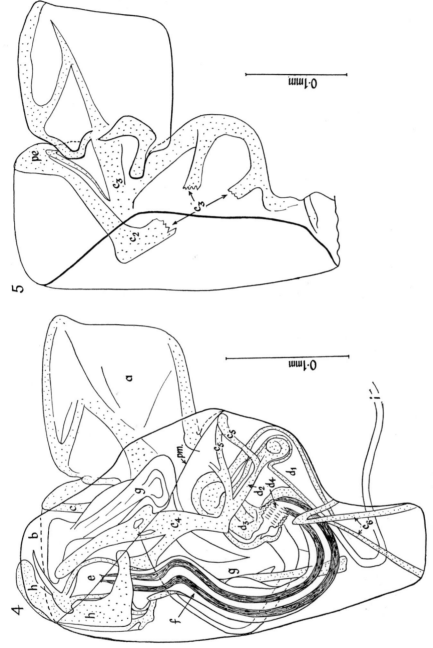

Figs 4-5. *Candona neglecta* Sars – *corps pénien*. 4. vue médiale. 5. la face latérale de la *gaine pénienne*, vue latérale (détail).

141

Les *pièces internes* sont représentées par des formations tubuleuses, le *canal déférent*, le *labyrinthe*, le *tube copulateur*, le *manchon* avec la *bourse copulatrice*, et des formations lamellaires, la *pièce M* et la *pièce D*.

Le *canal déférent* (Figs 2–4), est un tube spiralé mince placé dans la région dorsale et antérieure du *corps pénien*. Dans la portion distale du *canal déférent* on fait le passage vers le *labyrinthe*.

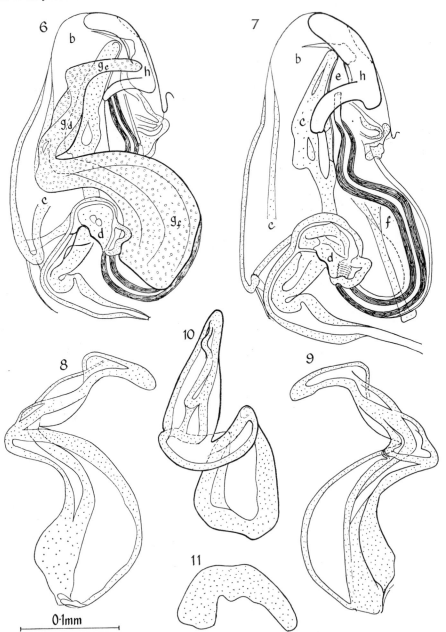

0·1mm

Figs 6-11. *Candona neglecta* Sars – *corps pénien*. 6. la face médiale de la *gaine pénienne* et les pièces internes, vue latérale (détail). 7. la face médiale de la *gaine pénienne* et les *pièces internes*, moins la *pièce M*, vue latérale (détail). 8. la *pièce M*; vue latérale. 9. vue médiale. 10. vue dorsale. 11. vue basale (détail).

Le *labyrinthe* (Figs 3, 4 et 6, 7 et 13) est une formation massive, chitineuse située dans la région antérieure de la *gaine pénienne*. A l'intérieur du *labyrinthe* il y a un canal qui est en continuation de celui *déférent* et qui passe par quatre régions que nous dénommons le *rameau ascendant*, le *rameau transversal*, le *rameau descendant* et le *rameau terminal* (Fig. 13, d_1-d_4). Ce dernier possède une cavité carrée avec les parois pliées. A l'extérieur des bandes chitineuses relient le *labyrinthe* à la *gaine pénienne* et à la *pièce M*.

Le *tube copulateur* (Figs 2–7, 13, 29), placé dans la région ventrale de la *gaine pénienne* est relié par son extrémité basale au *rameau final* du *labyrinthe*, tandis que l'extrémité distale repose dans la *bourse copulatrice*. Les parois du *tube copulateur* sont renforcées par un tissu, probablement musculaire (Fig. 29).

Le *manchon* (Figs 2, 4, 7, 13), une formation membraneuse qui recouvre le tube copulateur, est visible surtout dans la région ventrale. Une bande chitineuse longitudinale renforce la paroi dorsale de cette formation. Dans le tiers distal du *manchon* la paroi ventrale va se plier en continuant avec celle de la *bourse copulatrice*.

La *bourse copulatrice* (Figs 12–13) plus élargie dans sa région basale présente un bord ventral concave et un bord dorsal convexe. Des rainures chitineuses viennent renforcer cette formation. A l'extrémité distale de la *bourse* il y a l'*orifice génital* par où sort le *tube copulateur* au moment de la copulation. Cette région fortement chitinisée nous la nommerons *capuchon*. Elle possède une épine dont le rôle probable est de fixer la bourse au niveau de l'orifice génital femelle.

La *pièce chitineuse médiane*, ou *pièce M* (Figs 2, 3, 6, 8–11), est placée au centre de la *gaine pénienne* étant fixée par plusieurs bandes chitineuses, l'une transversale reliant la portion basale de la *pièce M* au *labyrinthe*, l'autre oblique reliant la région médiane, d'une part au *labyrinthe*, d'autre part à la *gaine pénienne*. En examinant la *pièce M* en vue latérale, après avoir enlevé la face latérale de la *gaine pénienne* (Fig. 6) on distingue une portion basale antérieure, dilatée, et dont le bord antéro-ventral est largement courbé, une portion distale, postérieure, mince, formée par deux rameaux l'un dorsal, l'autre postérieur et perpendiculaire sur le premier. En vue latérale à l'angle extérieur de ces deux rameaux il y a une proéminence (plus évidente en vue dorsale). Des rainures chitineuses sillonnent la surface de cette pièce. Nous avons figuré la *pièce M* séparément, en position latérale (Fig. 8), médiale (Fig. 9) et dorsale (Fig. 10) pour pouvoir nous faire une idée plus exacte des différents détails morphologiques. En vue dorsale, le rameau dorsal est plus large dans sa portion proximale. La portion basale de la *pièce M* possède une concavité (Fig. 11) dans laquelle peuvent s'encastrer une partie du *labyrinthe*. La région qui relie la portion basale à la portion distale et en vue dorsale[1] plus développée du côté latérale. Dans la portion distale de la *pièce M* il y a une rainure chitineuse médiane qui continue dans la portion basale sur le bord ventral. Le bord dorsal de la pièce est entièrement chitinisé. Dans le tiers basal la rainure chitineuse dorsale s'unit avec une rainure chitineuse médiane. Des rainures chitineuses peuvent être observées aussi sur la face médiane.

La *pièce chitineuse distale*, ou *pièce D* (Figs 7, 13, 14) est placée dans la région postéro-ventrale du corps pénien, en recouvrant la bourse copulatrice, en partie, du côté latéral. La base de la *pièce D* a la forme d'un anneau qui embrasse la *bourse copulatrice*. Du côté ventral l'anneau continue avec une lamelle largement arquée.

---

[1] Nous avons examiné et figuré le rameau dorsal de la *pièce M* des *Candona* du groupe *candida-neglecta* dans un plan parallèle avec le plan du support.

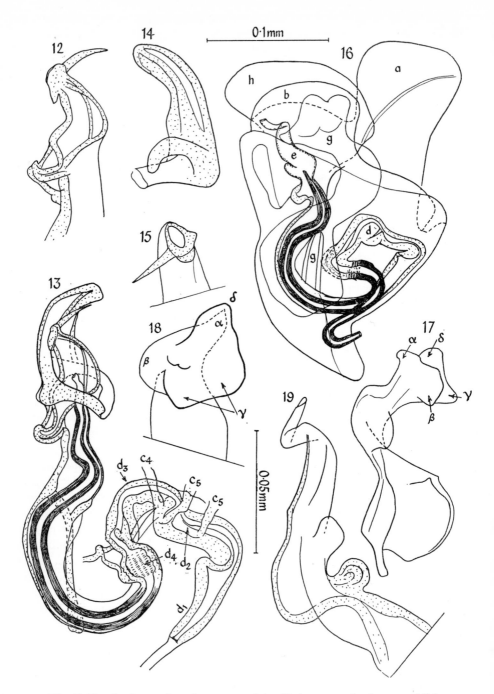

Figs 12-19. *Candona neglecta* Sars – *corps pénien*. 12. *bourse copulatrice*, vue médiale (détail). 13. *pièces internes*, vue médiale (détail). 14. *pièce D*, vue latérale. 15. *bourse copulatrice* extrémité distale (détail). 16-19. *Candona lobipes* Hartwig. 16. *corps pénien*, vue médiale. 17. *pièce M*, vue médiale. 18. *pièce M*, vue apicale (détail). 19. *bourse copulatrice*, vue latérale.

SUR LE VALEUR SYSTÉMATIQUE DE L'ORGANE COPULATEUR CHEZ
  QUELQUES ESPÈCES DU GENRE *Candona* BAIRD

Après avoir passé en revue les principaux détails de l'organe copulateur de *Candona neglecta* Sars, pour nous rendre compte de la valeur systématique des différentes pièces nous présenterons quelques particularités morphologiques des formations similaires appartenant à *Candona fasciolata* Petkovski, *C. weltneri* Hartwig, *C.* aff. *C. compressa* (Koch)-Brady, *C. lobipes* Hartwig, *C. levanderi* Hirschmann et *C. fabaeformis* Fischer, représentant des principales ligniées de diversification du genre *Candona* Baird (gr. *candida-neglecta, rostrata-compressa* et *fabaeformis-acuminata*).

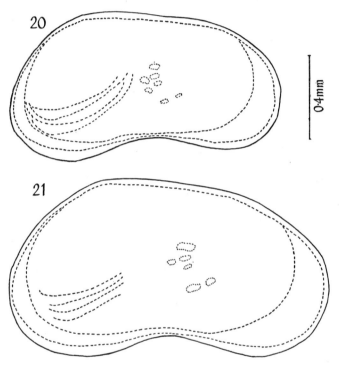

Figs 20-21. *Candona neglecta* Sars – mâle, valve droite, vue latérale externe. 21. *Candona fasciolata* Petkovski – mâle, valve droite, vue latérale externe.

Une analyse comparative de la gaine pénienne des sept espèces citées montre les différences suivantes: la face latérale a la même étendue que la face médiale chez *C. neglecta* (Fig. 3), *C. fasciolata* Petkovski (Fig. 22) et *C. weltneri* Hartwig, plus grande chez *C. lobipes* Hartwig (Fig. 16) et *C.* aff. *C. compressa* (Koch)-Brady (Fig. 33), ou plus petite chez *C. fabaeformis* Fischer (Fig. 43) et *C. levanderi* Hirschmann (Fig. 44).

Le lobe de la face latérale a des formes très variables étant carré chez *C. neglecta* Sars (Fig. 5) et *C. fasciolata* Petkovski ou divisé chez *C. levanderi* Hirschmann (Fig. 44); sa position peut être dorsale—groupe *candida-neglecta* et *fabaeformis-acuminata* —ou postérieure—groupe *rostrata-compressa*. La face médiale possède un seul pli chez toutes les espèces examinées, en plus chez *C. fabaeformis* Fischer il y a un deuxième dans la région basale (Fig. 43). La région distale de la face médiane peut être largement arrondie, comme chez *C. neglecta* Sars (Fig. 3,4) et *C. fasciolata* Petkovski (Fig. 22), elle peut être anguleuse comme chez *C. fabaeformis* Fischer (Fig.

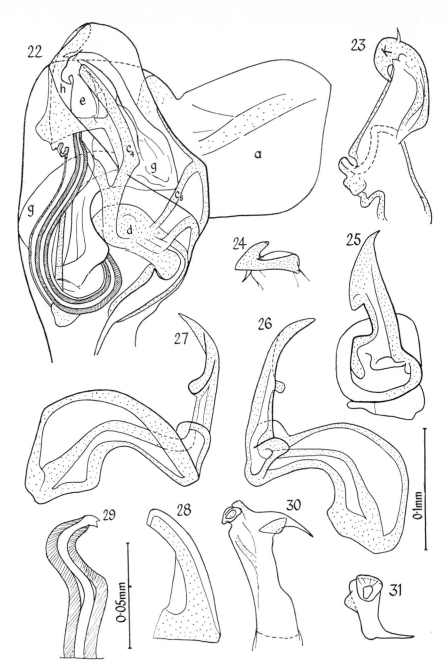

Figs 22-31. 22-28. *Candona fasciolata* Petkovski – *corps pénien*. 22. vue médiale
23. *bourse copulatrice*, vue médiale (détail). 24. *bourse copulatrice*, vue apicale (détail).
25. *pièce M*, vue dorsale (détail). 26. *pièce M*, vue latérale. 27. *pièce M*, vue médiale.
28. *pièce D*, vue latérale (détail). 29. *Candona neglecta* Sars – *tube copulateur* (détail).
30-31. *Candona weltneri* Hartwig – *corps pénien*. 30. *bourse copulatrice*, vue latérale. 31.
bourse copulatrice, vue apicale (détail).

43); des formes particulières existent chez *C. aff. C. compressa* (Koch)-Brady (Fig. 33) et *C. lobipes* Hartwig (Fig. 16). La face médiale et la face latérale s'unissent chez les *Candona* du groupe *candida-neglecta* (Figs 3, 22) et *fabaeformis-acuminata* (Figs 43, 44) sur le tiers de la région dorsale, tandis que chez les *Candona* du groupe *rostrata-compressa* elles sont unies sur la moitié de cette région (Figs 16, 32).

Parmi les *pièces internes*, les *pièces chitineuses M et D* sont les plus remarquables. La *pièce M* chez *C. neglecta* Sars possède un rameau postérieur perpendiculaire sur le rameau dorsal, tandis que chez *C. fasciolata* Petkovski (Figs 25, 27) la même formation est très réduite ayant une forme lancéolaire et recourbée vers le côté latéral et ventral. Chez *C. aff. C. compressa* (Koch)-Brady (Figs 35, 37) et *C. lobipes* Hartwig (Figs 17, 18) sur la face médiale dans la région centrale de la *pièce M* il y a une tubérosité, très évidente en vue dorsale (Fig. 18). Chez *C. aff. C. compressa* (Koch)-Brady l'extrémité apicale de la même pièce est pourvue d'un repli lamellaire (Figs 36, 37); chez *C. lobipes* Hartwig dans la même région il y a quatre tubérosités placées en croix (Figs 17, 18). La *pièce D* est fortement développée chez les *Candona aff. C. compressa* (Koch)-Brady (Fig. 34) et *C. lobipes* Hartwig (Fig. 46); elle est réduite chez les *Candona* des groupes *candida-neglecta* et *fabaeformis-acuminata* (Figs 14, 28, 43, 44). Chez *C. neglecta* Sars et *C. fasciolata* Petkovski il y a un anneau chitineux sur lequel est fixée une lamelle légèrement courbée; chez *C. weltneri* la lamelle est à peine ébauchée, tandis que chez *C. lobipes* Hartwig (Fig. 46) la lamelle est fortement développée suivant l'axe transversal, à l'encontre de ce qui se passe chez *C. aff. C. compressa* (Koch)-Brady (Fig. 34) où le développement maximal a suivi l'axe longitudinal. Il est intéressant de remarquer que chez *Heterocypris barbara* Gauthier et Brehm et *Herpetocypris* sp. deux espèces éloignées du point de vue phylétique, des *Candona*, la *pièce M* et *la pièce D* sont absentes.

La *bourse copulatrice* peut avoir non seulement des formes très diverses mais aussi des formations chitineuses fort variables: chez *C. neglecta* Sars (Figs 12, 15) la *bourse* est largement courbée sur le bord dorsal et possède un *capuchon* pourvu d'une forte épine; chez *C. fasciolata* Petkovski (Figs 23, 24) le bord dorsal est plus allongé, le *capuchon* est recourbé du côté ventral et l'épine est très réduite; *C. weltneri* Hartwig possède près de l'orifice génital une languette placée de façon perpendiculaire sur l'axe longitudinale et une épine chitineuse (Figs 30, 31); *C. aff. C. compressa* (Koch)-Brady et *C. lobipes* Hartwig ont des bourses avec un pli très évident dans la région basale (Figs 19, 40, 41). La région distale de la bourse est tordue et s'amincie; près de l'orifice génital il y a de faibles formations chitineuses en opposition avec les *Candona* du groupe *candida-neglecta*. Chez *C. levanderi* Hirschmann (Fig. 45) la bourse est très allongée et il semble que l'extrémité distale est transformée en crochet; l'orifice génital est sous-apical.

La forme du *labyrinthe* du *manchon* et du *tube copulateur* sont à peu près semblables chez tous les *Candona* que nous avons examinés. Nous devons remarquér toutefois quelques particularités dans le système de chitinisation du *labyrinthe* des *Candona* appartenant aux groupes *rostrata-compressa* et *fabaeformis-acuminata*. Chez le premier groupe le *rameau ascendant* est faiblement chitinisé tandis que chez le second la même formation est recouverte d'un épais manchon.

En examinant quelques exemplaires de *Hungarocypris madaradzi* Örley, une espèce éloignée, du point de vue phylétique, des *Candona*, nous avons constaté que la forme et la position du *labyrinthe*, du *manchon* et du *tube copulateur*, sont fort différentes.

Enfin, nous tenons à remarquer que parmi les espèces étudiées *Candona fasciolata* Petkovski est du point de vue phylétique très proche de *C. neglecta* Sars et bien que la

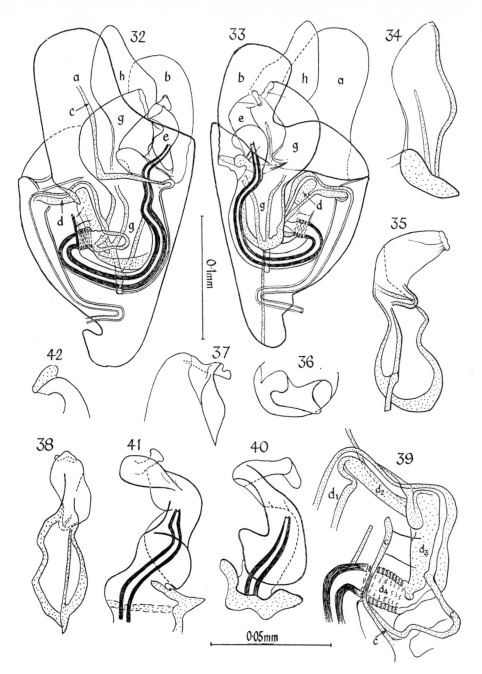

Figs 32-42. *Candona aff. C. compressa* (Koch) – Brady – *corps pénien*. 32. vue latérale. 33. vue médiale. 34. *pièce D*, vue latérale. 35. *pièce M*, vue médiale. 36. *pièce M*, vue apicale (détail). 37. *pièce M*, vue latérale. 38. *pièce M*, vue dorsale. 39. *labyrinthe*, vue latérale. 40. *bourse copulatrice*, vue ventrale. 41. *bourse copulatrice*, vue médiale. 42. *bourse copulatrice,* vue apicale (détail).

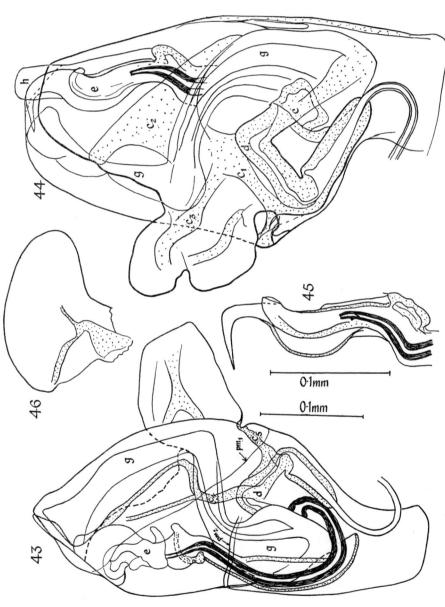

Figs 43-46. 43. *Candona fabaeformis* Fischer – corps pénien droit, vue médiale. 44-45. *Candona levanderi* Hirschmann. 44. corps pénien, vue latérale. 45. bourse copulatrice, vue médiale. 46. *Candona lobipes* Hartwig. corps pénien, pièce D, vue latérale (détail).

0·1mm

0·1mm

morphologie des valves (Figs 20, 21) et les appendices sont à peu près semblables, les différences morphologiques de l'organe copulateur sont si évidentes (Figs 2-19 et 21-26) qu'elles ne laissent point de doute sur leur validité.

L'ORGANE COPULATEUR EN ÉRECTION CHEZ *Candona* aff. *C. compressa* (KOCH)-BRADY

Pour comprendre la valeur fonctionnelle des différentes pièces que nous avons décrites ci-dessus, il nous a paru très utile de présenter aussi la position des pièces chez l'organe copulateur en érection de *C.* aff. *C. compressa* (Koch)-Brady.

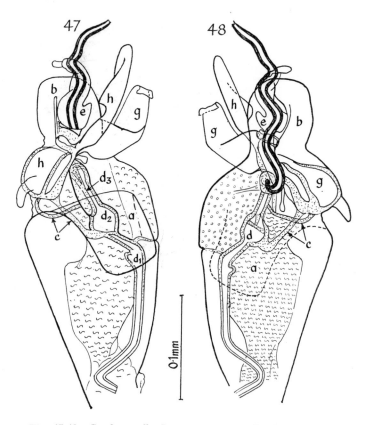

Figs 47-48. *Candona aff. C. compressa* (Koch) – Brady. *Corps pénien* droit en érection. 47. vue latérale. 48. vue médiale.

Nous devons remarquer que l'érection des *corps péniens* se fait indépendamment, et c'est pourquoi nous avons trouvé plusieurs exemplaires dont seulement un des *corps péniens* était en phase d'érection. En examinant un *corps péniens* en érection on observe: le *rameau transversal* et le *rameau descendant* du *labyrinthe* projetés en avant vers l'extrémité distale (Figs 47, 48). Le *labyrinthe* fait une torsion de 90° ses parois internes, sur le parcours du *rameau transversal* formant une cavité ovale.

Grâce aux bandes chitineuses qui lient le *labyrinthe* aux faces latérales médiales de la *gaine pénienne*, celle-ci va s'élargir permettant le déplacement vers l'avant des autres pièces. Le lobe latéral de la *gaine* va se plier vers l'extérieur et le pli de la face médiale va disparaître.

La *pièce M* liée au *labyrinthe* par une ou deux bandes chitineuses, occupera une position oblique et en avant.

Les modifications de la position du *labyrinthe* entraineront aussi un changement de la position du *tube copulateur*, du *manchon*, de la *bourse copulatrice* et de la *pièce D*. Au moment où l'érection atteint son plein, toutes les pièces internes sont projetées dans la région distale de la *gaine pénienne* et une partie du *tube copulateur* va sortir par l'*orifice génital* de la *bourse copulatrice*.

La *pièce D* et la *pièce M* viennent probablement en contact avec le lobe génital femelle. Elles pourraient servir ou bien à la préhension de ce lobe, ou bien avoir un rôle sensoriel. L'extrémité distale de la *bourse copulatrice* avec ses crochets se fixe probablement au niveau de l'orifice génital femelle. Le *tube copulateur* doit pénétrer à l'intérieur de l'organe génital femelle. Il est difficile à expliquer la fonction du *labyrinthe*: servirait-il au stockage éphémère des spermatozoïdes ou au contraire serait-il une sorte de pompe pour la propagation en avant des spermatozoïdes? Enfin, il reste encore un problème à éclaircir: quel est le mécanisme qui projette vers l'avant le *labyrinthe* ainsi que toutes les autres pièces? Chez les Cythéridae le mécanisme d'érection est facile à expliquer grâce à la structure très évidente de la *région basale* avec ses bandes musculaires et ses piliers chitineux. Chez les *Candona* examinés nous n'avons pas pu mettre en évidence des formations musculaires similaires à celles des Cythéridae. Tout ce que nous avons observé est un tissu plus dense dans la région médiane de la *gaine pénienne*, qui vient envelopper la région basale des pièces internes. L'érection se ferait-elle grâce à la turgescence de ce tissu, qui dans la phase normale occupe l'espace existant entre les *pièces internes* et la *gaine copulatrice*? Des recherches supplémentaires tenteront de vérifier cette hypothèse.

CONCLUSIONS

(*a*) L'organe copulateur chez les sept espèces de *Candona* Baird (*sensu* G. W. Müller, 1912) étudiées—représentants des principaux groupes d'espèces—est constitué par deux *corps péniens*, soudés entre eux, chacun ayant un complexe de *pièces internes* —une partie du *vaisseau déférent*, le *labyrinthe*, le *tube copulateur*, le *manchon* avec la *bourse copulatrice*, la *pièce M*, et la *pièce D*—maintenu dans une *gaine pénienne* par l'intermède de bandes chitineuses. Les *pièces internes* possèdent des détails difficilement visibles à cause de l'opacité de la *gaine pénienne*. C'est pourquoi la dissection du *corps pénien*, l'isolement et l'examination attentives de chaque formation est souhaitable.

La compréhension exacte de l'organisation de l'appareil copulateur permet pour la première fois d'homologuer les pièces chez différentes espèces. L'analyse comparative chez les *Candona* étudiées démontre la grande diversité de formes revêtues par toutes les structures et spécialement par la *gaine pénienne*, la *pièce M*, la *pièce D* et la *bourse copulatrice*.

(*b*) L'étude minutieuse de chaque pièce du *corps pénien* corroborée avec les renseignements morphologiques fournis par les valves et les appendices représente des éléments qui pourront avancer la systématique assez embrouillée des Cyprididae.

Les différences morphologiques très évidentes existantes entre l'organe copulateur —voir la *pièce M*—de *C. fasciolata* Petkovski et *C. neglecta* Sars, deux espèces dont les valves et les appendices sont à peu près similaires, démontrent que chez certaines Cyprididae la différenciation spécifique a opéré en principale sur la première formation. L'examination de l'organe copulateur des *Candona* nous a permis d'entrevoir également-

L

ment la possibilité de l'existence de caractéristiques à valeur superspécifique, comme la forme et la position des lobes de la *gaine pénienne* de la *pièce D*. De même, en comparant les formations similaires appartenant à des Cyprididae éloignées du point de vue phylétique—voir *Candona, Hungarocypris, Heterocypris*, nous avons constaté l'existence de la différenciation supergénérique, comme la forme et la position du *manchon* du *tube copulateur*, du *labyrinthe* ainsi que la présence ou l'absence de certaines *pièces internes*.

Donc à notre avis toutes les formations de l'organe copulateur peuvent être utilisées avec succès dans la taxonomie des ostracodes Cyprididae.

(*c*) En présentant l'organe copulateur en érection chez *Candona* aff. *C. compressa* (Koch)-Brady, nous avons tenté d'entrevoir la valeur fonctionnelle des différentes pièces. C'est ainsi que la *pièce M* et la *pièce D* pourraient avoir un rôle préhensile ou sensoriel, la *gaine pénienne* et le *manchon* ont certainement un rôle protecteur, la *bourse copulatrice* avec son extrémité distale fortement chitinisée servirait à la fixation sur l'orifice génital femelle, et enfin le *vaisseau déférent*, le *labyrinthe* et le *tube copulateur*, conduisent les spermatozoïdes.

(*d*) Des recherches futures devront être faites sur la morphologie de certaines pièces, comme le *manchon*, la face médiale de la *gaine pénienne*, ainsi que sur la position des bandes et rainures chitineuses. Il serait nécessaire aussi de définir exactement les structures morphologiques qui produisent l'érection du *corps pénien*.

BIBLIOGRAPHIE

HART, C. W. JR. 1962. A revision of the Family Entocytheridae. *Proc. Acad. nat. Sci. Philad.*, **114** (3), 121-47, 18 Figs.
——. AND HART, D. G. 1967. The entocytherid ostracods of Australia. *Proc. Acad. nat. Sci. Philad.*, **119** (1), 1-51, 95 Figs.
HIRSCHMANN, N. 1912. Beitrag zur Kenntnis der Ostrakodenfauna des finnischen Meerbusens. *Acta Soc. Fauna Flora fenn.*, **36** (2), 1-15, 47 Figs.
HOFF, C. C. 1944. New American species of the Ostracod Genus Entocythere. *Am. Midl. Nat.*, **32** (2), 327-57, 33 Figs.
KLIE, W. 1938. Krebstiere oder Crustacea III, Ostracoda, Muschelkrebse. *Tierwelt Dtl.*, **34**, 1-230, 786 Figs.
——. 1943. Die mänliche Kopulationsorgane einiger *Loxoconcha* Arten aus der Adria. *Arch. Hydrobiol*, **40**, 71-78.
MÜLLER, G. W. 1900. Deutschlands Süsswasserostracoden. *Zoologica, Stuttg.*, **12**, 1-112 (Apud Klie, 1938).
PETKOVSKI, T. 1959. Süsswasser-Ostracoden aus Jugoslavien, VI. *Acta Mus. maced. Sci. nat.*, **6** (3), 53-75.
——. 1960. Zur Kenntnis der Crustaceen des Prespasees. *Fragm. balcan.*, **3** (15), 117-31, 39 Figs.
——. 1961. Zur Kenntnis der Crustaceen des Skadar (Scutari) Sees, *Acta Mus. maced. Sci. nat.*, **8** (2), 29-52, 36 Figs.
——. 1966. Ostracoden aus einigen Quellen der Slovakei. *Acta Mus. maced. Sci. nat.*, **10** (4), 91-107, 38 Figs.
RACOVITZA, G. E. 1923. Notes sur les Isopodes. Orientation de l'isopode et essais de nomenclature pratique des bords et faces de son corps et de ses appendices. *Archs Zool. exp. gén.*, **61** (4), 75-122.
SARS, G. O. 1922-28. *An account of the Crustacea of Norway. Vol. IX. Ostracoda*, 1-277, 119 Pls. Bergen, Bergen Museum.
SCHÄFER, H. W. 1937. *Candona levanderi* Hirschmann ein für Deutschland neuer Muschelkrebs. *Zool. Anz.*, **119**, 211-17. Fig. 14
SKOGSBERG, T. 1928. Studies on marine Ostracods. Part II. External morphology of the genus *Cythereis* with descriptions of twenty-one new species. *Occ. Pap. Calif. Acad. Sci.*, No. 15 155 pp., Pls 1-VI (apud Klie, W., 1943).

DISCUSSION

FERGUSON: In 1942 Hoff grouped the *Candona* of Illinois principally on the basis of the shape of the female genital lobe. Is there any mention in Danielopol's paper of the correlation between the structure of the male genital organs and the genital lobe of the female?

GUILLAUME (for DANIELOPOL): There is nothing in the paper about the female, only the male.

McGREGOR: For the past 100 years most authors have illustrated their own versions of the haemopenes of various species of the genus *Candona*. Dr Danielopol's excellent paper assists in understanding the anatomy and certain aspects of the functional morphology of this organ. Dr R. V. Kesling and I have been investigating the same problem and concur in most aspects with the findings of Dr Danielopol. Although the diagrams show considerable detail in sclerotization, I should stress that the haemopene is a rather simple structure when viewed from the standpoint of functional morphology.

McKENZIE: Were any special dissection techniques used?

GUILLAUME (for DANIELOPOL): Dr Danielopol says in his paper: 'It is not difficult to separate the body of the animal and to examine all its aspects. By careful and minute dissection the two penes have been separated and continuing the dissection on one of these the different parts have been successfully separated and studied, especially the unknown internal parts.'

ROME: L'étude du Dr Danielopol est très remarquable, tant par sa précision que par son illustration. La fine structure du pénis qu'il met en lumière apporte au point de vue morphologique des données nouvelles. Mais je ferai une réserve au sujet de l'emploi de ces détails morphologiques en systématique. Ils sont en effet très difficiles à mettre en évidence; il faut une rare habilité pour faire une dissection aussi minutieuse. Peut-il être utile de faire dépendre la systématique d'éléments aussi peu accessibles?

McGREGOR: In reference to Dr McKenzie's question concerning dissection techniques, one may induce erection of the haemopene in *Candona* by partially anaesthetizing the animals. The haemopenes are often erected posteriorly before they are rotated 180° and may be maintained in this position by closing the valves to prevent retraction into the carapace. One can then observe the internal anatomy of the haemopene and study its functional morphology.

McKENZIE: I understood that in Dr Danielopol's work several constituent parts of the penes were dissected out. While it is indeed simple to dissect off the penes from the body I have experienced difficulty in dissecting out the parts of each penis.

McGREGOR: When the haemopene is erected, the difficulty of dissection is decreased considerably.

D. L. DANIELOPOL: Réponse aux discussions qui on suivis la communication 'Recherches sur la morphologie de l'organe copulateur mâle chez quelques ostracodes du genre *Candona* Baird.'

Bucarest, le 11 septembre 1967

Chers collègues,

Ne pouvant participer personnellement à cet intéressant Symposium, ma contribution a été communiquée grâce à l'obligeance et amabilité de M. le Dr J. Neale et Mlle M-Cl. Guillaume. Je leur en remercie infiniment.

Bien que je regrette de ne pas avoir pu répondre directement aux remarques faites par MM. Dr E. Fergusson, M. D. Mc.Gregor, Dr McKenzie et Dr D. R. Rome, je dois dire que les répliques données par Mlle Guillaume, à ma place sont exactement ce que j'aurai répondu.

A la question posée par M. D. R. Rome je dirai que la dissection est souhaitable car il y a les pièces internes qui possèdent des détails difficilement a apercevoir à cause de la gaine pénienne. En ce qui concerne les descriptions morphologiques trop détaillées de l'organe copulateur bien que longues et fastidieuses leur utilité et importance est indiscutable car, comme chez d'autres groupes d'arthropodes la formation en question est le siège de nombreux caractères paléogénétiques. Or à l'avenir, grâce aux données fournies par cet organe nous arriverons probablement à mieux comprendre l'évolution des Cyprididae.

Dan L. Danielopol

THE FUNCTIONAL MORPHOLOGY OF ENTOCYTHERID OSTRACOD COPULATORY APPENDAGES, WITH A DISCUSSION OF POSSIBLE HOMOLOGUES IN OTHER OSTRACODS[1]

C. W. HART, Jr., and DABNEY G. HART
Academy of Natural Sciences of Philadelphia, U.S.A.

SUMMARY

In studying entocytherid ostracods it has become increasingly evident that their thin, almost completely transparent shells are of little taxonomic value. Aside from the anterior and posterior spines on a few North American species, size, and scattered setae, the shells bear little in the way of ornamentation—and the shells of even the Australian species are virtually indistinguishable from those of most of the North American species.

Study of the entocytherid ostracods has been concentrated, therefore, on the appendages, which have proved to have many valuable characters. A summary is presented of the terminology that has developed dealing with the copulatory apparatus of male entocytherids, and a number of new terms are proposed to designate previously undescribed parts of the apparatus.

In addition, the homologies existing between widely divergent genera of entocytherids are discussed, and it is shown how this same terminology can be applied to the genital apparatus of at least one species of free-living marine ostracods

It has become increasingly evident that the thin, almost completely transparent shells of entocytherid ostracods are of little taxonomic value, and because of this illustrations of shells are usually limited to mere outline drawings. Aside from the anterior and posterior spines on a few North American species (*Uncinocythere caudata* Kosloff, 1955; *U. thektura* Hart, 1965a; *U. cassiensis* Hart, 1965a; and *Dactylocythere amphiakis* Hart and Hart, 1966), size, and scattered setae, the shells bear little in the way of ornamentation—and the shells of European, Australian, and North American species are usually indistinguishable from one another. Furthermore, muscle scars are so ill-defined among entocytherids as to render them useless in the systematics of the group.

Study of the entocytherid ostracods has, therefore, been concentrated on the appendages, which have proved to have many valuable characters. Over the years a rather extensive terminology dealing with the copulatory apparatus (= copulatory complex) of the entocytherids has been developed by workers on that group, and this terminology is here reviewed and several new terms proposed which, hopefully, will aid in the exposition of possible homologies.

[1] This work was done with the aid of National Science Foundation grant GB-4197 to the senior author.

It is also hoped that this paper will serve as an aid to workers on related ostracods—allowing the various parts of the copulatory structures to be compared and discussed, rather than merely figured as they so often are. As a tentative gesture in that direction, we here discuss possible homologies existing between the copulatory apparatus of entocytherids and that of a hemicytherid—comparisons which we hope can be similarly made with other ostracods.

While essentially the same copulatory apparatus appears to be present in all entocytherids, for the sake of clarity the large and heavily sclerotized apparatus of the entocytherid genus *Geocythere* has been used for Figs 5–8.

THE COPULATORY APPARATUS OF ENTOCYTHERID OSTRACODS

The copulatory apparatus of male entocytherids is a conspicuous, paired organ situated in the posterior shell region, and is of great importance as a diagnostic feature in the group. While it is conventional to illustrate only one half of this apparatus, it should be understood that each part exists in duplicate. Fig. 1 shows a composite copulatory apparatus, including all parts presently recognized (Roman type), together with several proposed names (Italic type) for proximal structures which have seldom been figured or discussed.

For the sake of a uniform terminology, we propose that the anterior and posterior directions indicated in Fig. 1 be used to designate portions of the peniferum and its associated fingers and clasping apparatus, although those directions will be 'true' directions only when the apparatus is in its copulatory position (see discussion below, and also Fig. 8). It should be remembered that these directions apply only to the peniferum, fingers, and clasping apparatus; not to the *zygum, sterinx, tropis,* and *pastinum.*

Terminology

The peniferum is that portion of the copulatory apparatus which bears the penis (usually in its distal portion) and its proximal portion forms a hinge on which the entire apparatus may be turned through a 180° arc about the zygum (see below). Although the overall plan of the peniferum is similar in all species, variations in proportions and modifications in details are numerous and bizarre—affording features to delineate species, genera, and subfamilies. The proximal portion of the peniferum also serves as a place of origin of a complex system of muscles, but until more is known of the homologies of these muscles we feel it advisable not to illustrate them.

Marshall (1903: 133-5, 140-1) figured and briefly discussed the copulatory apparatus of *Entocythere cambaria,* but, following Zenker's earlier work (1854: 47), apparently misinterpreted the junction of the zygum (see below)—believing it to be a half of the so-called 'basal piece'. While Marshall understood that 'all parts (of the copulatory apparatus) are paired', it seems probable from his illustrations that he believed the zygum to be paired also, while, as far as we can determine, it is a single structure in entocytherids. (It may be interpreted otherwise in hemicytherids, however.) What we now call the peniferum was considered by Marshall to be the other half of the 'basal piece'.

The peniferum as a discrete structure was discussed in some detail by Hoff (1944: 328–9), who apparently introduced the terms *penis, horizontal ramus, vertical ramus, internal border, external border,* and *teeth* at that time. However, Hoff followed Marshall's terminology, calling the more conspicuous structure the 'base', and

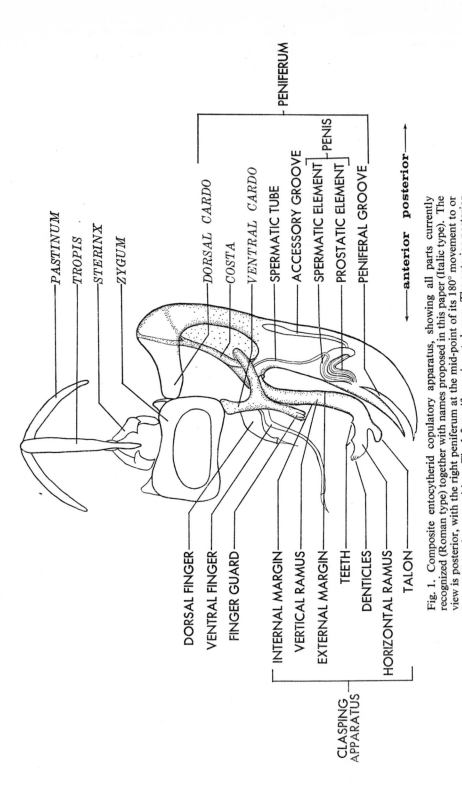

PASTINUM

TROPIS

STERINX

ZYGUM

DORSAL CARDO

COSTA

VENTRAL CARDO

SPERMATIC TUBE

ACCESSORY GROOVE

SPERMATIC ELEMENT ⎤
                   ⎥ PENIS
PROSTATIC ELEMENT ⎦

PENIFERAL GROOVE

PENIFERUM

DORSAL FINGER

VENTRAL FINGER

FINGER GUARD

INTERNAL MARGIN

VERTICAL RAMUS

EXTERNAL MARGIN

TEETH

DENTICLES

HORIZONTAL RAMUS

TALON

CLASPING APPARATUS

◄─── anterior posterior ───►

Fig. 1. Composite entocytherid copulatory apparatus, showing all parts currently recognized (Roman type) together with names proposed in this paper (Italic type). The view is posterior, with the right peniferum at the mid-point of its 180° movement to or from the copulatory position. The left peniferum is not shown. The anterior-posterior directions refer only to reference positions with regard to the peniferum.

156

referring to 'distal' and 'proximal portions of the base' when discussing what we now call the peniferum—a shorter term introduced by Hobbs and Walton (1961: 379) to encompass the distal portion of the base, but which, by usage, has already come to include both Hoff's distal and proximal portion of the base.

So that homologies can be more clearly discussed, we here introduce the terms *dorsal cardo* and *ventral cardo* (Latin *cardo* = hinge) for those parts of the peniferum (Hoff's 'proximal portion of the base') which serve as hinges by which it articulates with the zygum. The term *costa* (Latin *costa* = rib) may be used for any thickened portion of the peniferum which appears to perform a strengthening function.

In North American species of the subfamily Entocytherinae, the penis, which is a distinct sclerotized organ, may be variously situated on the peniferum, either in a sub-horizontal position and only slightly curved (as in *Uncinocythere ambophora* Walton and Hobbs, 1959: 115), or being curved to a greater degree—sometimes almost approaching a full circle (as in *Geocythere gyralea* Hart, 1965b: 257)—and it appears to be attached in its proximal portion to a muscle extending dorsally toward the dorsal-most portion of the peniferum. It is probable that this muscle is used to pivot the

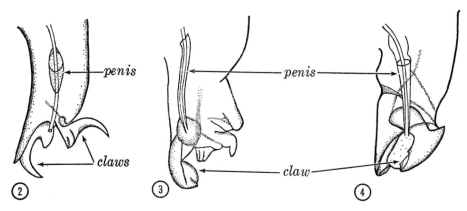

Fig. 2. Terminal portion of peniferum of the recently described Indian ento-cytherid (*Microsyssitria indica*) commensal on marine isopod *Sphaeroma terebrans*, showing two terminal claws.
Fig. 3. Terminal portion of peniferum of Australian entocytherid (*Chelocythere kalganensis*) commensal on fresh-water crayfish *Cherax* sp., showing anterior movable claw.
Fig. 4. Terminal portion of peniferum of Australian entocytherid (*Lichnocythere tubrabucca*) commensal on fresh-water crayfish *Euastacus hystericosus*.

penis at the time of copulation. Depending on the species, the penis is situated any-where along the distal one-third of the peniferum. The penis of Australian ento-cytherids (Hart and Hart, 1967) appears to be invariably vertically situated (Figs 3, 4) —exhibiting differences in length relative to the entire peniferum and varying from straight to slightly sinuous. The form and arrangement of the penis is one of the most important diagnostic features of the Australian subfamily Notocytherinae.

Similarly, the penis of a newly described marine entocytherid from India (Fig. 2)—although only weakly developed—shows a vertical orientation. Because it is not associated with the sclerotized structure mentioned above encasing the terminal portion of the spermatic tube, we do not believe the penis of either the Australian or the Indian entocytherids to be homologous with the penis of North American ento-cytherids. We believe that the penis of North American entocytherids is homologous with the movable anterior peniferal claw found in the Indian and Australian species mentioned above, and that in North American species this claw has become integrally associated with the terminal portion of the spermatic tube (which lies adjacent to it in the Indian and Australian species) and has migrated proximally to varying levels on the peniferum.

The term penis was first used by Hoff (1944: 329) for the sclerotized structure encasing the terminal portion of the spermatic tube, and later Hobbs III (1965: 161) noted that in the North American genus *Plectocythere* the penis was actually divisible into two separate elements—the dorsal one of which he designated the *spermatic element*; the ventral one, the *prostatic element*. Later Hobbs and Hart (1966) pointed out that the penis in the genera *Ascetocythere*, *Cymocythere*, and *Phymocythere*, is also in two parts.

The *accessory groove*, first illustrated by Hobbs (1955: 327) in the genus *Dactylocythere*, may serve some reproductive function, but at the present this is unclear.

Hobbs (1966: 8) proposed the term *peniferal groove* for a terminal bifurcation present in some species of the genus *Dactylocythere*. This bifurcation also appears to be present in some species of the genera *Uncinocythere* and *Ankylocythere*.

While the terminal portion of the peniferum in all known North American entocytherids apparently bears no movable structures (with the exception of the subterminal penis), certain Australian entocytherids (*Chelocythere* and *Lichnocythere*) appear to bear at least one movable claw-like structure, and a recently discovered and newly described entocytherid commensal on wood-boring isopods (Hart, Nair, and Hart, 1967) bears two apposed terminally-located claws (Figs 2–4).

The term *finger guard* was proposed by Hobbs (1955: 329) for that extension of the ventral cardo alongside the dorsal and ventral fingers. This structure is present in the North American genera *Dactylocythere*, *Sagittocythere*, *Phymocythere*, *Harpagocythere*, *Cymocythere*, and *Ascetocythere*—and a possible homologue is found in the Australian genus *Riekocythere*.

The spermatic tube of the entocytherids forms a proximal loop at approximately the level of the ventral cardo, and appears to be somehow interconnected with the distal portion of that cardo (Fig. 7). A bifurcation of this tube, at varying locations in the loop, leads proximally in the direction of the zygum, but we have been unable to trace this for any considerable distance beyond the place where it leaves the loop. Distally the spermatic tube terminates in the sclerified penis mentioned above. It seems probable that the dorsal and ventral fingers are somehow interconnected with the spermatic tube, but we have been unable to resolve the connection. Indeed Marshall (1903: 134) asserted that one of the fingers—possibly the ventral one—contained the sperm duct, and Rioja (1940: 605, Figs 8, 9) illustrated the loop of the spermatic duct as terminating in the dorsal finger. This interpretation probably arose from Marshall relying too heavily on the earlier work of Zenker, who described the copulatory apparatus of distantly related ostracods in which the copulatory duct (= spermatic duct) is not an integral part of the peniferum, but a separate entity whose homologies we do not understand at present (see spermatic tube of hemicytherid ostracod, Fig. 9). Subsequently Hoff (1944: 329) showed that there was no duct in either of the two fingers and postulated that the fingers are probably tactile organs and that they may also assist in directing or holding certain structures in position during copulation.

The term *clasping apparatus* (or clasping piece), used by Marshall (1903: 133) was probably taken from Zenker's *Apparat zum Festhalten* (1854: 47). This refers to the heavily sclerotized, apparently movable appendage articulating with the mid-portion of the peniferum in a socket near the ventral cardo and in the vicinity of the loop of the spermatic tube. It, too, probably serves as a tactile organ, and because of its characteristic shapes—often used in separating species and genera—is probably associated with some key-lock mechanism. The apparatus may be either straight or, more often, L- or C-shaped—and the *horizontal ramus* may bear *teeth* on its internal margin, *denticles* on its tip, and/or one or more *talons* on its external margin. This usage of 'teeth', 'denticles', and 'talons' to designate the excrescences on the parts of the horizontal ramus, has in spite of its apparent redundancy, proved to be useful in discussing the horizontal ramus.

Fig. 5. Posterior portion of entocytherid ostracod showing copulatory apparatus held up inside shell in non-copulatory position. Note vertical position of pastinum.

The term *zygum* (Latin *zygum* = yoke) is here proposed as a name for what Zenker (1854) apparently called the 'Basis', Marshall (1903: 140) called 'half of the basal piece', Rioja (1941: 184) called the 'pieza intermediaria', and Hoff (1944: 329) called the 'connecting piece'. The zygum is a single chitinous piece, suspended in the posterior shell region by a system of chitinous rods (see three dimensional drawings, Figs 6, 8, to visualize its position). It is probably homologous with the fused first podomeres of adjacent walking legs—a supposition supported by its relationship to the *tropis* and the *pastinum* (see below), and by the presence of setae on the zygum of the hemicytherid ostracod shown in Fig. 9a. These setae, single in the dorsal portion and double in the ventral, may be homologous with the single seta in the posteroventral portion of the first podomere of the walking legs and the double setae in the mid-dorsal portion of the first podomere of the first walking leg—although, admittedly, considerable torsion would have had to take place to make this true.

While the zygum figured in most of the entocytherid illustrations in this paper is the comparatively simple, heavily sclerotized form found in the genus *Geocythere*, between genera it is tremendously variable. Similar heavy zygums are found in the genus *Donnaldsoncythere*, but the part seems to be so variable that it could, were it easier to see, probably be useful as a specific character.

While the zygum of the genus *Geocythere* (Figs 5–8) is heavy and shows no median bars, that of *Entocythere heterodonta* (= *Ankylocythere heterodonta*) figured by Rioja (1941: 183, Fig. 1) and called the 'pieza intermediaria', shows traces of median bars.

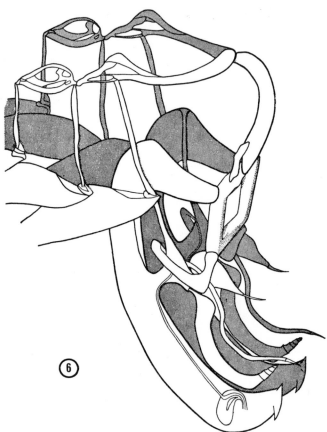

Fig. 6. Three-dimensional drawing of copulatory apparatus and adjacent legs, with copulatory apparatus lowered so that distal portion extends beyond shell. Note horizontal position of pastinum.

The *tropis* (Latin *tropis* = keel), a heavy chitinous rod connecting the zygum to the pastinum, is probably formed by the fusion of the posterior vertical internal skeletal rods that support the legs. When viewed laterally the tropis may appear as a single rod in some species (Figs 5–8), or as a double bow-shaped rod in others (Figs 9–12). When viewed from a caudal aspect it usually appears to be a single rod, but in ostracods other than entocytherids (see Skogsberg 1928: 155, Fig. 1, for instance) it sometimes appears as two unfused rods.

The paired *sterinx* (Greek *sterinx* = support) form connecting or buttressing pieces lateral to the tropis, and probably serve to stabilize the zygum in a horizontal position, while at the same time allowing some freedom of anterior-posterior angular motion, as indicated in Fig. 8.

The *pastinum* (Latin *pastinum* = fork) appears to be homologous with the horizontal chitinous skeletal rods connecting the vertical rods supporting the legs. When

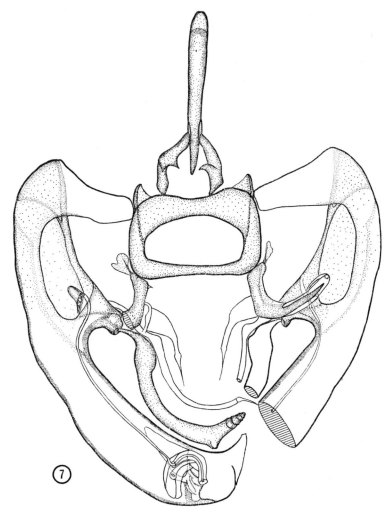

Fig. 7. Posterior view of copulatory apparatus, with peniferums mid-way in 180° swing into copulatory position. Distal portion of left ventral cardo removed to show socket into which clasping apparatus fits; distal portion of right peniferum removed for clarity.

viewed from a caudal aspect it can be seen to be fork-shaped—anteriorly connected with the lateral skeletal apodemes and posteriorly fused to form a structure from which the tropis (and hence the entire copulatory complex) hangs. This is most easily understood by referring to Figs 6 and 8.

Function

The copulatory organs of the entocytherids are complex, but possibly more easily understood than those of some other ostracods. Indeed some of the parts appear to be

simpler than homologous parts in other ostracods, and we believe that a better understanding of the function and homologies can be gained by beginning our comparisons
with entocytherids and working, perhaps backwards, into other groups. Admittedly
this is probably due to our ignorance of other ostracod genitalia, but for lack of
another beginning we will start with the entocytherids.

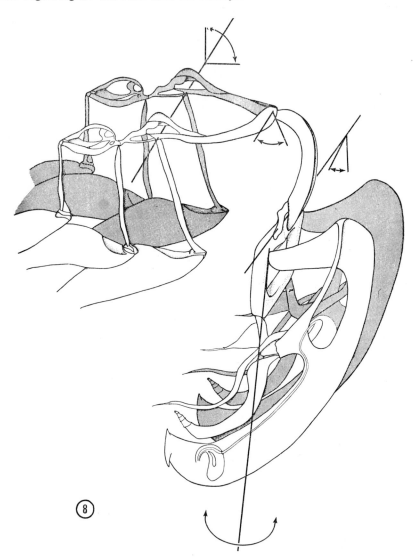

Fig. 8. Three-dimensional drawing of copulatory apparatus and adjacent legs,
with copulatory apparatus lowered and peniferum swung through 180° into
copulatory position. Arrows show angular motion through which various parts
move.

In simple terms, the copulatory apparatus of entocytherids apparently serves to
raise, lower, and position its integral organs of sperm transfer so that they are able to
introduce sperm into the seminal receptacle of the female.

When the ostracod is not concerned with copulation, the apparatus is held high
within the shell and usually somewhat between the bases of the last pair of legs (Fig. 5).

To accomplish this the pastinum may be raised to almost a vertical position, thus holding the tropis and its associated apparatus in that position.

At the time of copulation the pastinum may be lowered into an almost horizontal position (Fig. 6), thus lowering the attached apparatus so that its distal portion extends beyond the shell.

Then, before copulation can take place, the tropis probably swings slightly caudal, allowing the proximal portion of the peniferums to clear the legs—at which time the two peniferums swing 180° laterally to each side on the zygum (Figs 7, 8), thus reversing their non-copulatory position and making it possible for the organs of sperm transfer to face anteriorly.

The pivotal points utilized in these manoeuvres are shown in Fig. 8, and the probable angular motions are shown by appropriate arrows.

PROBABLE HOMOLOGIES EXISTING BETWEEN ENTOCYTHERIDS AND HEMICYTHERIDS

Fig. 9 shows the copulatory apparatus of a hemicytherid ostracod. With some apparent variations in the distal portion of the peniferum, its functions and parts seem to be essentially similar to those found in entocytherids.

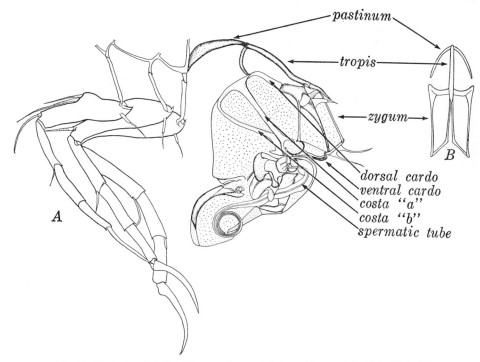

Fig. 9. A. Lateral view of legs and copulatory apparatus of a hemicytherid ostracod, showing parts homologous with those of previously discussed entocytherid ostracods. B. Posterior view of pastinum, tropis, and zygum of same ostracod.

The pastinum and tropis shown in Fig. 9 compare well with those of entocytherids, but the paired sterinx appears to be absent in the hemicytherid.

The zygum (Figs 9a, b) is rather delicate in its construction when compared with that of *Geocythere* (Fig. 7), and consists of four vertical bars (two median and two lateral) connected dorsally and ventrally. While at first glance this does not appear to be similar to the zygum of the entocytherid *Geocythere*, the median pair of bars appear to be homologous with the traces of median bars illustrated by Rioja (1941: 183, Fig. 1) for *Entocythere heterodonta* (= *Ankylocythere heterodonta*).

The peniferum articulates with the zygum just as it does among the entocytherids: a dorsal cardo and a ventral cardo forming hinges at the dorsolateral and ventrolateral portions of the zygum.

Costa 'b' in Fig. 9 appears to be homologous with the costa designated in Fig. 1, while costa 'a' is probably homologous with the thickened marginal portion of the peniferum adjacent to the costa shown in Fig. 1.

We do not believe any portion of the spermatic tube shown in Fig. 9 to be homologous with the entocytherid penis—and, indeed, if any portion of the hemicytherid peniferum shown in Fig. 9 is homologous with it, it is probably the thickened portion shown underneath the coiled terminal portion of the spermatic tube.

As the peniferum shown in Fig. 9 is not in the copulatory position recognized in the entocytherids, the heavy falcate structure anterior (posterior, according to our proposed terminology) to the coiled spermatic tube may be a homologue of the posterior claw of the Indian and Australian entocytherids shown in Figs 2 and 4. This may be borne out by a study of the peniferal muscles which we plan to undertake.

Function

As shown in Figs 10 and 11, the copulatory apparatus of this hemicytherid appears to function much as does that of entocytherids, as far as lowering the apparatus from between the posterior legs is concerned. However, Fig. 12 is conjectural, representing a hypothetical peniferal position equivalent to that shown in Fig. 8 for the entocytherids. It seems probable that, because the spermatic tube of the hemicytherids appears to be loosely coiled and not an integral part of the distal portion of the peniferum (contrary to the condition found among entocytherids), the sperm may be introduced into the female by an uncoiling of the spermatic tube from the peniferal position shown in Fig. 11.

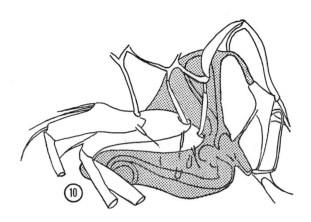

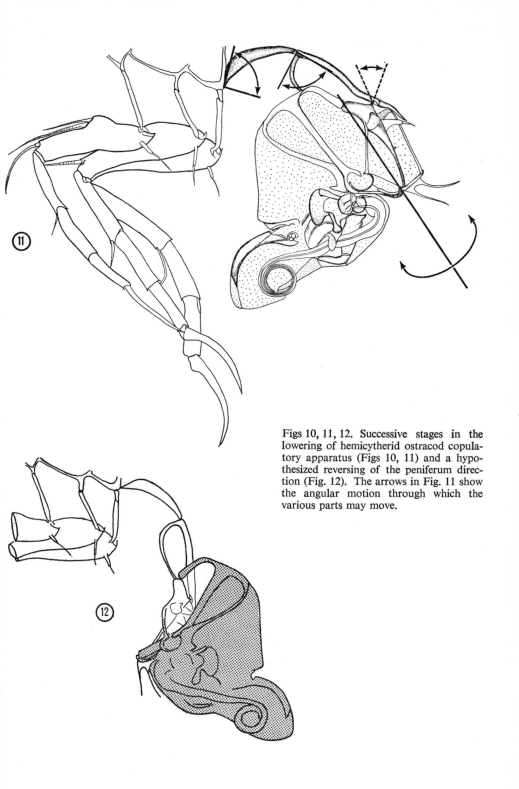

Figs 10, 11, 12. Successive stages in the lowering of hemicytherid ostracod copulatory apparatus (Figs 10, 11) and a hypothesized reversing of the peniferum direction (Fig. 12). The arrows in Fig. 11 show the angular motion through which the various parts may move.

REFERENCES

HART, C. W., JR. 1965a. Three new entocytherid ostracods from the western United States with new locality data for two previously described western entocytherids. *Crustaceana*, **8** (2), 190-6.
——. 1965b. New entocytherid ostracods and distribution records for five midwestern states. *Trans. Am. microsc. Soc.*, **84** (2), 255-9.
—— AND HART, D. G. 1966. Four new entocytherid ostracods from Kentucky, with notes on the troglobitic *Sagittocythere barri*. *Notul. Nat.*, **388**, 10 pp.
—— AND HART D. G. 1967. The entocytherid ostracods of Australia. *Proc. Acad. nat. Sci. Philad.*, **119** (1), 1-51.
——., NAIR, N. BALAKRISHNAN AND HART, D. G. 1967. An entocytherid ostracod commensal on the wood-boring marine isopod, *Sphaeroma terebrans*, from Kerala, India. *Notul. Nat.*, **409**, 11 pp.
HOBBS, H. H., JR. 1955. Ostracods of the genus *Entocythere* from the New River system in North Carolina, Virginia, and West Virginia. *Trans. Am. microsc. Soc.*, **74** (4), 324-33, Figs. 1-10.
——. 1966. A new genus and three new species of ostracods with a key to the genus *Dactylocythere* (Ostracoda: Entocytheridae). *Proc. U.S. natn. Mus.*, **122** (3587), 1-10.
—— AND HART, C. W., JR. 1966. On the entocytherid ostracod genera *Ascetocythere*, *Plectocythere*, *Phymocythere* (gen. nov.), and *Cymocythere*, with descriptions of new species. *Proc. Acad. nat. Sci. Philad.*, **118** (2), 35-61.
HOBBS, H. H., JR., AND WALTON, M. 1961. Additional new ostracods from the Hiwassee drainage system in Georgia, North Carolina, and Tennessee. *Trans. Am. microsc. Soc.*, **80** (4), 379-84, 8 Figs.
HOBBS, H. H., III. 1965. Two new genera and species of the ostracod family Entocytheridae, with a key to the genera. *Proc. biol. Soc. Wash.*, **78** (19), 159-64.
HOFF, C. 1944. New American species of the ostracod genus *Entocythere*. *Am. Midl. Nat.*, **32** (2), 327-57, Figs 1-33.
KOZLOFF, E. N. 1955. Two new species of *Entocythere* (Ostracoda, Cytheridae) commensal on *Pacifastacus gambelii* (Girard). *Am. Midl. Nat.*, **53** (1), 156-61, Figs 1-24.
MARSHALL, W. S. 1903. *Entocythere cambaria* n.g. n.s., a parasitic ostracod. *Trans. Wis. Acad. Sci. Arts Lett.*, **14** (1), 117-44, Pls X-XIII, Figs 1-30.
RIOJA, E. 1940. Estudios carcinologicos. V. Morfologia de un ostracodo epizoario observado sobre *Cambarus* (*Cambarellus*) *montezumae* Sauss. de Mexico, *Entocythere heterodonta* n.sp. y descripcion de algunos de sus estados lavarios. *An. Inst. Biol. Univ. Méx.*, **11** (2), 593-609, Figs 1-6.
——. 1941. Estudios carcinologicos. VI. Estudio morfologico del esquelto interno de apodemas quitinoso de *Entocythere heterodonta* Rioja (Crust. Ostracodos). *An. Inst. Biol. Univ. Méx.*, **12** (1), 177-91, Figs 1-9.
SKOGSBERG, T. 1928. Studies on marine ostracods. Part II. External morphology of the genus *Cythereis* with descriptions of twenty-one new species. *Occ. Pap. Calif. Acad. Sci.*, XV, 155 pp.
WALTON, M. AND HOBBS, H. H., JR. 1959. Two new eyeless ostracods of the genus *Entocythere* from Florida. *Q. Jl Fla Acad. Sci.*, **22** (2), 114-20, Figs 1-20.
ZENKER, W. 1854. Monographie der ostracoden. *Arch. Naturgesch.*, 87 pp.

DISCUSSION

MCGREGOR: Is the copulatory position similar to that reported for free living ostracods?
HART: Quite honestly, I am not familiar with how free living ostracods do it. In entocytherids the male opens his valves and positions himself over the dorsal portion of the female; the female fits snugly up between the valves of the male—thus allowing at least one member of the pair (the female) to hold on to the host during copulation. The copulatory organs are lowered into position, reverse themselves in direction, and somehow are introduced between the posterior portion of the female valves. The clasping apparatus may play a part in separating the female valves.
FRYER: Does the size of the sperms in these entocytherids in any way resemble that of the giant sperms of some other fresh-water ostracods?
HART: Some years ago I saw what I thought was the sperm of an entocytherid, but it disappeared into the mounting medium which was apparently of the same refractive index. I could never find it again. As I remember, it appeared to be shaped somewhat like a brittle-star, but it could also have been an artifact—so I would advise you not to take this as an acceptable description of it. It did not appear to be particularly large in relation to the ostracod.
BATE: Could you give an outline of the ontogeny of these interesting species? Is the life

cycle entirely commensal? Do they actually lay the eggs on the crayfish or its gills or do the young in fact have a free swimming stage?

HART: In 1942 Hoff (*American Midland Naturalist*, **27** (1), 64) reported having found free living entocytherids in crayfish burrows. He also quoted Klie, who in 1931 (*Archives de Zoologie Expérimentale et Générale*, **71** (3), 333-44) described an entocytherid from the waters of an Indiana cave. I think this latter reference is in error, and that the ostracods were actually from either *Orconectes p. pellucidus* or *Cambarus tenebrosus*, which were reported from the same cave by the same expedition by Fage in 1931 (*Archives de Zoologie Expérimentale et Générale*, **71** (3), 361-74). They do, however, apparently lay their eggs on the host. From India we have recently received 10 isopods on which were 224 commensal ostracods, ranging from eggs up through seven moult stages. In this Indian material the eggs averaged 84 microns in length, and are oval. The first two instars have *anlagen* of the first pair of walking legs, and the first instar appears to be attached to the host by a gelatinous thread extending to the remains of the egg mass. The slightly larger second instar still has the *anlagen* of the first pair of legs, and may be attached to the host by a sticky substance (pure conjecture). In the third instar there are two legs showing the typical entocytherid claw used for holding on to the host; the next instar has four legs; the next, six.

BENSON: Rosalie Maddocks of the Smithsonian Institution has recently found a species of *Pontocypria* commensal on the tube feet of a starfish. To my knowledge this is one of the first commensal marine forms that has been described. The second antennae are modified forming a hook apparatus by which they attach to the tube feet of the starfish. Seen from the bottom the starfish appears to have lines of these commensals orientated in military fashion along the rows of tube feet.

KAESLER: Are these entocytherids specific to various parts of the body so that the differences could be phenetic rather than indicating separate species?

HART: No, they are true species—although it has been shown that some species of crayfish will have one species of ostracod living predominantly among the mouth parts, another in the thorax region, and still another on the abdomen. I think that the enitocytherid ostracods perform a cleaning function, serve as a sort of live toothbrush, ridding the interstices of the host's body of detritus.

KORNICKER: I know that in this group it is difficult to identify the females, but since the female genitalia are perhaps a more conservative feature, would it not be useful to examine them with a view to connecting this group up with other ostracod groups?

HART: About the only ones that have any obvious genitalia to speak of belong to the genus *Dactylocythere*. In this genus there is a rather conspicuous apparatus in the posterior portion of the female shell. For years it was called a 'ruffled skirt'. For want of a more dignified name we now call it the *amiculum*. I should be interested to know if this can be homologized with some device in other ostracod groups.

SYLVESTER-BRADLEY: I congratulate the author on his beautiful diagrams. Skogsberg always hoped that the study of the copulatory apparatus would elucidate what we now call the Trachyleberididae. I do think that once we understand the functions of its parts it may prove of great systematic significance.

M

MORPHOLOGIE DE L'ATTACHE DE LA FURCA CHEZ LES CYPRIDIDAE ET SON UTILISATION EN SYSTÉMATIQUE

DOM REMACLE ROME, O.S.B.
University of Louvain, Belgium

SUMMARY

E. Triebel pointed out that the use of the furcal attachment had been neglected in systematics. In further publications he used this element to define several genera. The author has studied the furcal attachment in a large number of genera and species. After describing its morphology in *Herpetocypris reptans*, descriptions and figures are given of the furcal attachments in the different subfamilies of the Cyprididae. It will be seen that this allows the easy comparison of genera with one another and their clear distinction. This anatomical detail does not supercede the other details which are used for diagnoses, but it states the differences precisely.

RÉSUMÉ

E. Triebel a montré qu'on avait négligé en systématique l'utilité de l'attache de la furca. Dans la suite de ses travaux cet élément lui a servi à définir plusieurs genres. Nous basant sur cette opinion, nous avons étudié ce détail d'organisation dans un grand nombre de genres et d'espèces. Après une description de la morphologie de l'attache en prenant pour exemple celle de *Herpetocypris reptans*, nous donnons la description et la figuration des attaches dans différentes sous-familles des Cyprididae. On verra qu'elles permettent aisément de rapprocher des genres les uns des autres ou de les séparer nettement. Il va de soi que ce détail anatomique n'exclut pas les autres détails qui servent aux diagnoses, mais il les précise.

INTRODUCTION

E. Triebel (*Genotypus und Schalen-Merkmale der Ostracoden Gattung Stenocypris*, p. 7, note 2. Senckenbergiana, Bd. 34. 1953) faisait remarquer qu'en systématique on n'avait pas attaché assez d'importance à l'attache des furca (*Furkal-Stützen* ou *Chitinstützen der Furka*), alors que dans bien des cas elle offre des caractères très utiles. Il donnait comme exemple la distinction facile entre les genres *Eucypris* et *Cypricercus*, grâce à la forme de l'attache de la furca. Surtout que l'absence de mâle parmi les espèces européennes de ce dernier genre ne permet pas d'observer dans la partie antérieure des valves l'enroulement en spirale des testicules, qui constitue le caractère principal du genre de G. O. Sars.

Au cours de ces dernières années, en étudiant des Ostracodes de provenances très différentes, nous avons pu constater combien cette remarque était judicieuse. Il faut ajouter que la facilité de mettre cette attache en évidence au cours de la dissection en rend l'utilisation fort aisée.

Le matériel dont nous disposons nous a permis de décrire et de figurer l'attache d'un assez grand nombre de genres et d'espèces de la famille des Cyprididae qui ont des furca bien développées. Plusieurs espèces rares nous ont été fournies par feu le Professeur Fox.

Rappelons, qu'avec les secondes antennes, les furca contribuent à la marche de ces Ostracodes. Aprés leur extension vers l'avant, les secondes antennes se fixent par leurs griffes dans le sédiment, de même que les furca après leur flexion vers l'avant. Les secondes antennes par leur flexion vers l'arrière et les furca par leur extension vers l'arrière entraînent l'Ostracode vers l'avant. Ces derniers nouvements nécessitent le plus de puissance. Dans la propulsion vers l'avant les deux furca agissent simultanément.

Sauf lorsque les furca sont soudées, comme chez *Macrocypris*, elles peuvent cependant agir indépendemment l'une de l'autre. On l'observe lorsque l'Ostracode couché latéralement tente de reprendre sa position normale.

L'articulation qui permet ces mouvements se fait entre la furca et les bandelettes de chitine qui constituent l'attache de la furca. Celle-ci sert aussi à l'insertion des muscles moteurs.

MORPHOLOGIE

Les deux attaches s'étendent obliquement de chaque côté de l'arrière du corps, jusqu'aux environs du début de l'intestin postérieur. L'attache est formée de plusieurs côtes superposées, mince aux bords elle s'épaissit vers le milieu.

Nous distinguons dans une attache: l'extrémité proximale, la partie médiane et l'extrémité articulaire.

Nous prenons pour exemple l'attache de *Herpetocypris reptans* (Fig. 1).

Extrémité proximale

Cette extrémité a l'aspect d'une fourche, dont une branche se dirige vers la région dorsale du corps: la branche dorsale, *d*. L'autre, tout en restant dans la région moyenne du corps a tendance à s'infléchir vers la région ventrale: la branche ventrale, *v*. Les deux branches font entre elles un angle aigu. Toutes deux sont issues d'une côte centrale. La branche dorsale est courbée en S, la branche ventrale est droite et se termine en une petite palette.

Partie médiane

La partie médiane est étroite. Elle présente une faible courbure à convexité dorsale. Ses bords sont minces, elle s'épaissit vers le milieu. L'extrémité distale de la partie médiane porte une expansion triangulaire (expansion ventrale) qui rejoint sa symétrique au bord ventral. Cette expansion existe sous d'autres formes dans beaucoup de genres.

Extrémité articulaire

Vue latéralement l'extrémité articulaire se présente sous la forme d'un prolongement acéré qui se compose de deux éléments superposés, l'un d'eux est pourvu dorsalement d'une petite masse globuleuse. Une coupe axiale légèrement oblique montre que ces deux éléments sont nettement séparés, l'un surmonté de la petite masse globuleuse, l'autre de forme plus simple.

Mécanisme articulaire

L'extrémité articulaire s'engage dans une cavité triangulaire allongée du bord proximal antérieur de la furca. Du fond de cette cavité se projette vers l'avant une languette étroite qui s'insinue entre les deux éléments de l'extrémité articulaire.

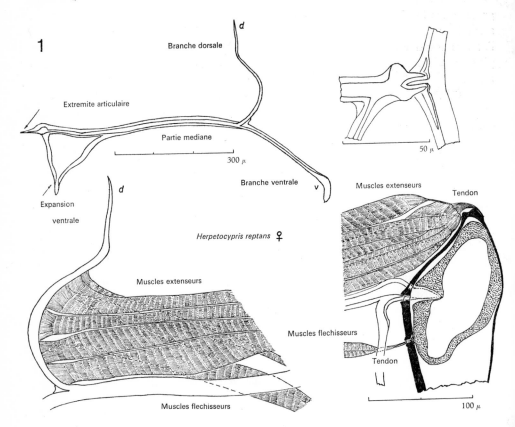

Fig. 1. La Morphologie de l'attache de la furca de *Herpetocypris reptans*.

Le bord proximal antérieur de la furca est divisé inégalement par le point où se situe l'articulation: le segment proximal est plus long que le segment distal. Il en résulte que le mouvement d'extension de la furca aura une plus grande amplitude que le mouvement de flexion. L'extension exagérée peut en effet amener la furca dans le prolongement du corps.

Musculature

Les muscles qui régissent les mouvements des furcas sont insérés le long de la concavité de la branche dorsale de l'attache. Les muscles extenseurs forment un puissant faisceau, ils se rejoignent distalement sur un tendon qui s'insère en arrière de l'extrémité proximale postérieure de la furca. Le faisceau des muscles fléchisseurs (en traits interrompus), qui ne se compose que de deux muscles, passe sous le faisceau des muscles extenseurs, sous la partie médiane de l'attache et s'insère par un long tendon au dessous de l'articulation.

Lorsque les muscles de gauche et de droite agissent simultanément les deux furca sont entraînées dans le même mouvement; agissant indépendemment, les muscles entraînent les furca dans des sens différents.

L'attache est indépendante de l'ensemble des bandelettes de chitine du corps. L'opposition aux mouvements des muscles moteurs de la furca est réalisée par des muscles insérés distalement au bord antérieur des branches dorsale et ventrale de la partie proximale; ils se dirigent vers la partie antérieure du corps, mais nous n'avons pas pu nous rendre compte de l'endroit où ces muscles s'insèrent proximalement.

CARACTÈRES DE L'ATTACHE UTILES EN SYSTÉMATIQUE

La partie de l'attache la plus importante en systématique est l'extrémité proximale. C'est du reste la plus facile à observer. Elle peut ne présenter qu'une fourche très simple dont les branches font des angles plus ou moins ouverts, ou présenter des dispositions plus compliquées: les bandelettes de chitine qui forment l'extrémité proximale s'écartent puis se rejoignent pour donner des œillets d'allure et de position différentes. La forme et la longueur des branches, voire l'absence de la branche ventrale sont des caractères importants.

La partie médiane présente peu de caractères: sa longueur et sa largeur, sa courbure, de fines expansions dorsales ou ventrales le long d'un de ses bords peuvent entrer en ligne de compte. L'extrémité de cette partie porte plus ou moins distalement, l'expansion ventrale qui rejoint sa symétrique au bord ventral du corps: la forme en est caractéristique et constante dans toute une famille, ou dans un genre seulement.

L'exiguité de l'extrémité articulaire la rend difficilement observable. Toutefois sa forme générale, plus ou moins aiguë, son inclinaison, ventrale ou dorsale, peuvent être caractéristiques.

FORMES DE L'ATTACHE DE LA FURCA

CANDONINAE

Chez les Candoninae l'angle que forment les deux branches de l'extrémité distale est plus ou moins obtus. La branche dorsale est courte et légèrement courbée. La branche ventrale est très longue, toujours terminée par un large anneau bordé d'épaisses bandelettes de chitine. La partie médiane est souvent courte et porte vers son extrémité distale chez les mâles et chez les femelles l'expansion ventrale qui rejoint sa symétrique au bord ventral du corps, elle est souvent longue et courbée. L'extrémité articulaire est acérée inclinée tantôt ventralement, tantôt dorsalement, ou dans le prolongement de la partie médiane.

Candona candida (O. F. Müller) ♀ (Fig. 2). Les deux branches de l'extrémité proximale font un angle légèrement obtus. La branche dorsale, courte et courbée, est dirigée vers l'arrière. La longue branche ventrale un peu courbée, se termine par un anneau irrégulier de la base duquel se détache une fine bandelette acérée. La partie médiane décrit une courbe à concavité dorsale, l'expansion ventrale est située vers les 2/3 de la longueur, elle est longue, large, courbée en S et son extrémité distale est tournée vers l'avant. L'extrémité articulaire est acérée et inclinée ventralement.

Candona neglecta G. O. Sars ♀ (Fig. 2). Les deux branches de l'extrémité proximale font un angle presque droit. La branche dorsale est peu courbée, elle se divise à son extrémité distale en deux parties d'inégale largeur et dirigées en sens opposé. La branche ventrale est droite et se termine par un anneau régulier. La partie médiane décrit une courbe à concavité dorsale, l'expansion ventrale est située vers les 3/4 de la longueur, elle est longue et large, peu courbée vers l'avant et se termine en pointe acérée. L'extrémité articulaire est longue, acérée et dans le prolongement de la partie médiane.

Candona neglecta G. O. Sars ♂ (Fig. 2). Les deux branches de l'extrémité proximale

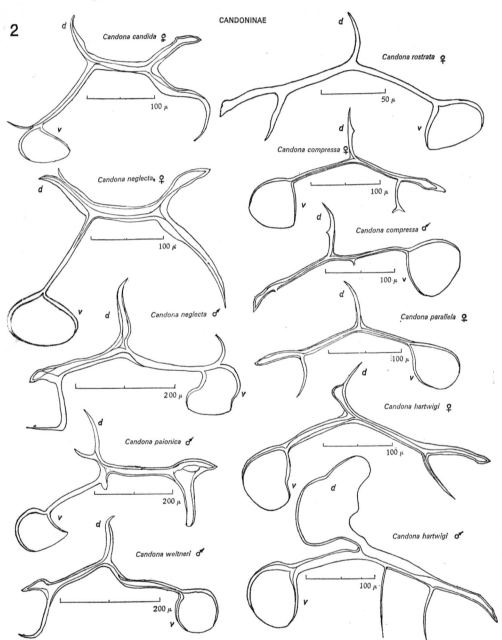

Fig. 2. L'attache de la furca chez la sous-famille Candoninae.

font un angle presque droit. La branche dorsale est légèrement courbée, son extrémité distale est tournée vers l'avant. La branche ventrale est très légèrement courbée, elle se termine par un anneau très irrégulier, de la partie dorsale duquel se détache une bandelette de chitine. La partie médiane est droite, l'expansion ventrale est située à son extrémité distale, elle est étroite, droite, puis se plie à angle droite vers l'arrière. L'extrémité articulaire est peu allongée, peu acérée et dans le prolongement de la partie médiane.

Candona paionica Petkovski ♂ (Fig. 2). Les deux branches de l'extrémité proximale font un angle presque droit. La branche dorsale est légèrement courbée en S, son extrémité distale est tournée vers l'avant et porte à proximité de sa base une bandelette de chitine courbée. La branche ventrale est courte et peu courbée, elle se termine par un anneau irrégulier, à sa jonction avec la partie médiane elle porte un court mammelon. La partie médiane est courte et droite, l'expansion ventrale est située vers l'extrémité distale, elle est formée de deux bandelettes de chitine séparées par une plaquette transparente. L'extrémité articulaire est longue et acérée, elle présente en son milieu une ouverture allongée, elle est dans le prolongement de la partie médiane.

Candona weltneri Hartwig ♂ (Fig. 2). Les deux branches de l'extrémité proximale font un angle obtus. La branche dorsale est courte légèrement courbée en S, son extrémité distale est tournée vers l'avant. La branche ventrale est sinueuse et se termine par un anneau régulier. La partie médiane est courte et droite, l'expansion ventrale est située à l'extrémité distale, elle est étroite, inclinée et courbée vers l'arrière. L'extrémité articulaire est longue et acérée, elle est dans le prolongement de la partie médiane.

Candona rostrata Brady et Robertson ♀ (Fig. 2). Les deux branches de l'extrémité proximale font un angle obtus. La branche dorsale est courte et régulièrement courbée, à concavité postérieure. La branche ventrale est presque droite et se termine par un anneau plus large que long. La partie médiane est presque droite, l'expansion ventrale est située aux 2/3 de la longueur, elle est étroite, droite et inclinée vers l'arrière. L'extrémité articulaire est courte, acérée et dans le prolongement de la partie médiane.

Candona hartwigi G. W. Müller ♀ (Fig. 2). Les deux branches de l'extrémité proximale font un angle aigu. La branche dorsale est courte et courbée vers l'arrière, une des bandelettes de chitine qui la forment s'écarte proximalement jusque vers le milieu en décrivant un arc irrégulier. La branche ventrale est très légèrement courbée et se termine par un anneau allongé dont un des arcs est irrégulier. La partie médiane est droite, l'expansion ventrale est située aux 2/3 de la longueur, elle est mince, presque droite et fortement inclinée vers l'arrière. L'extrémité articulaire est courte, acérée et dans le prolongement de la partie médiane.

Candona hartwigi G. W. Müller ♂ (Fig. 2). Les deux branches de l'extrémité proximale sont très peu distantes. La branche dorsale est très longue, mince et courbée très irrégulièrement; vers l'arrière elle porte une très large plaquette chitineuse transparente qui se prolonge le long de la partie médiane. La branche ventrale est courte et se termine par un anneau allongé. La partie médiane est droite, à son bord ventral elle porte deux fines expansions. L'extrémité articulaire est acérée et inclinée ventralement.

Candona compressa (Koch) ♀ (Fig. 2). Les deux branches de l'extrémité proximale font un angle droit. La branche dorsale est droite, son bord postérieur est formé de deux courbures. La branche ventrale est longue et mince, elle est terminée par un anneau dont un des arcs est presque droit. La partie médiane est légèrement courbée, l'expansion ventrale est située vers les 5/7 de la longueur, elle est peu inclinée vers l'arrière et se termine en s'élargissant. L'extrémité articulaire est longue, acérée et dans le prolongement de la partie médiane.

Candona compressa (Koch) ♂ (Fig. 2). Les deux branches de l'extrémité proximale font un angle presque droit. La branche dorsale est un peu courbée vers l'arrière, son bord postérieur est formé de deux courbures. La branche ventrale est longue, légèrement sinueuse et se termine par un anneau régulier, elle porte à son bord ventral une mince expansion (brisée). La partie médiane est droite, l'expansion ventrale ne montre que son origine. L'extrémité articulaire est longue, acérée et dans le prolongement de la partie médiane.

Candona parallela G. W. Müller ♀ (Fig. 2). Les deux branches de l'extrémité proximale font un angle très obtus. La branche dorsale est courbée vers l'arrière. La branche ventrale est droite et se termine par un anneau allongé. La partie médiane est droite, l'expansion ventrale est située vers les 3/4 de la longueur, elle est mince et régulièrement courbée vers l'arrière. L'extrémité articulaire est longue, acérée et dans le prolongement de la partie médiane.

Candona zucki Hartwig ♀ (Fig. 3). Les deux branches de l'extrémité proximale font

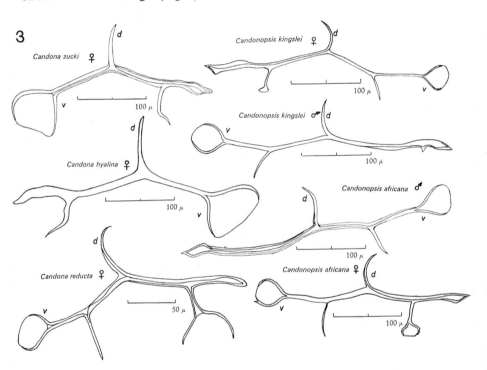

Fig. 3. L'attache de la furca chez la sous-famille Candoninae.

un angle obtus. La branche dorsale est peu courbée vers l'arrière. La branche ventrale est droite et se termine par un anneau dont un des arcs est presque droit. La partie médiane est très légèrement courbée, l'expansion ventrale est située aux 5/7 de la longueur, elle est mince et courbée brusquement vers l'arrière. L'extrémité articulaire est longue, acérée et dans le prolongement de la partie médiane.

Candona hyalina Brady et Robertson ♀ (Fig. 3). Les deux branches de l'extrémité proximale font un angle obtus. La branche dorsale est longue, fort légèrement courbée vers l'avant. La branche ventrale est droite et se termine par un anneau régulier, plus large que long. La partie médiane est très légèrement courbée, l'expansion ventrale est située aux 4/5 de la longueur et régulièrement courbée vers l'arrière. L'extrémité articulaire est acérée et un peu inclinée ventralement.

Candona reducta Alm ♀ (Fig. 3). Les deux branches de l'extrémité proximale font un angle droit. La branche dorsale est longue, très fortement courbée vers l'arrière. La branche ventrale est légèrement courbée et se termine par un anneau très petit, son bord postérieur porte une longue expansion droite. La partie médiane est légèrement courbée, l'expansion ventrale est située aux 7/10 de la longueur, elle a une base large et se divise distalement en deux branches courbées. L'extrémité articulaire est allongée, acérée et dans le prolongement de la partie médiane.

Candonopsis kingsleii (Brady et Robertson) ♀ (Fig. 3). Les deux branches de l'extrémité proximale font un angle légèrement obtus. La branche dorsale est courte et fortement courbée vers l'arrière. La branche ventrale est presque dans le prolongement de la partie médiane, elle se compose de deux segments droits séparés à l'endroit d'où se détache de son bord postérieur une courte expansion, elle se termine par un anneau très petit. La partie médiane est droite et porte un renflement distal à son bord dorsal, l'expansion ventrale est située aux 3/5 de la longueur, elle est courte, droite et s'évase distalement. L'extrémité articulaire est acérée et inclinée dorsalement.

Candonopsis kingsleii (Brady et Robertson) ♂ (Fig. 3). Les deux branches de l'extrémité proximale font un angle aigu. La branche dorsale est courte et courbée vers l'arriére. La branche ventrale est presque dans le prolongement de la partie médiane, elle se compose de deux segments droits séparés à l'endroit d'où se détache de son bord postérieur une courte expansion, elle se termine par un anneau très petit. La partie médiane est droite et porte ventralement à son extrémité distale un petit prolongement anguleux. L'extrémité articulaire est acérée et dans le prolongement de la partie médiane.

Candonopsis africana Klie ♀ (Fig. 3). Les deux branches de l'extrémité proximale font un angle presque droit. La branche dorsale est courte et fortement courbée vers l'arrière. La branche ventrale est presque dans le prolongement de la partie médiane, elle est courbée et se termine par un anneau très petit; vers son milieu se détache au bord postérieur une expansion très étroite, légèrement courbée. La partie médiane est courbée, son expansion ventrale est située vers les 2/3 de la longueur: elle est courte et droite et s'évase distalement. L'extrémité articulaire est acérée et inclinée dorsalement.

Candonopsis africana Klie ♂ (Fig. 3). Les deux branches de l'extrémité proximale font un angle presque droit. La branche dorsale est courte et fortement courbée vers l'arrière. La branche ventrale est presque dans le prolongement de la partie médiane, elle est courbée et se termine par un anneau très petit; vers son milieu se détache au bord postérieur une expansion étroite. La partie médiane est fort courbée et ne porte pas d'expansion ventrale. L'extrémité articulaire est acérée et dans le prolongement de la partie médiane.

CYCLOCYPRIDINAE

L'extrémité proximale des Cyclocypridinae est caractérisée par une très longue branche dorsale fortement courbée vers l'arrière, et souvent anguleuse vers le milieu de sa longueur. La branche ventrale est large de courbure nette ou presque nulle. L'extrémité articulaire diffère beaucoup d'un genre à l'autre.

Cyclocypris ovum (Jurine) ♂ (Fig. 4). Les deux branches de l'extrémité proximale font un angle presque droit. La branche dorsale est longue, très courbées vers l'arrière, anguleuse au milieu de sa longueur. La branche ventrale est longue, droite et très large. La partie médiane est courte, s'élargit distalement et porte à son extrémité distale l'expansion ventrale (brisée). L'extrémité articulaire est acérée et dans le prolongement de la partie médiane.

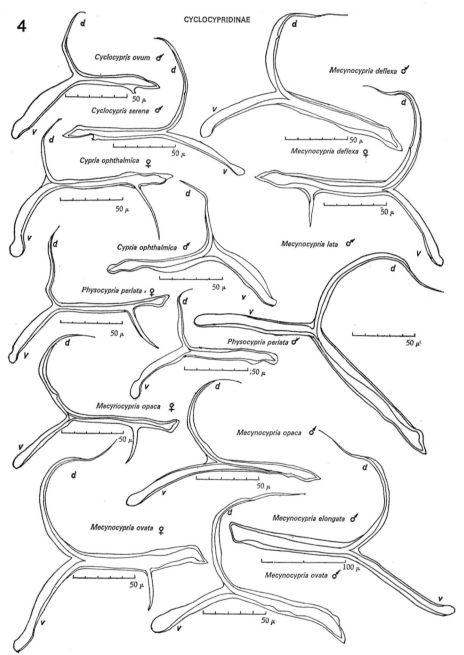

Fig. 4. L'attache de la furca chez la sous-famille Cyclocypridinae.

Cyclocypris serena (Koch) ♂ (Fig. 4). Les deux branches de l'extrémité proximale font un angle aigu. La branche dorsale est longue, fortement courbée vers l'arrière, l'angle qu'elle forme en son milieu est très effacé. La branche ventrale est longue terminée par une courte plaquette. La partie médiane est droite et porte à son extrémité distale l'expansion ventrale (brisée). L'extrémité articulaire est acérée et dans le prolongement de la partie médiane.

Cypria ophthalmica (Jurine) ♀ (Fig. 4). Les deux branches de l'extrémité proximale font un angle très obtus. La branche dorsal est longue, fort courbée vers l'arrière, anguleuse aux 2/3 de sa longueur. La branche ventrale est longue, large et courbée, elle se termine par une plaquette arrondie. La partie médiane est presque droite, à son extrémité distale elle porte l'expansion ventrale, longue et peu courbée vers l'avant. L'extrémité articulaire est peu acérée et dans le prolongement de la partie médiane.

Cypria ophthalmica (Jurine) ♂ (Fig. 4). Les deux branches de l'extrémité proximale font un angle très obtus. La branche dorsale est longue, fort courbée et anguleuse aux 3/4 de sa longueur. La branche ventrale est très large et presque droite. La partie médiane est légèrement concave au bord dorsal, très convexe au bord ventral. L'extrémité articulaire est dans le prolongement de la partie médiane; un des deux éléments qui la composent est arrondi, l'autre est acéré.

Physocypria perlata D. R. Rome ♀ (Fig. 4). Les deux branches de l'extrémité proximale font un angle très obtus. La branche dorsale est longue et sinueuse. La branche ventrale est large, presque droite et se termine par une plaquette arrondie. La partie médiane est droite, vers les 4/5 de sa longueur elle porte l'expansion ventrale, assez large et courbée vers l'arrière. L'extrémité articulaire est acérée et légèrement inclinée dorsalement.

Physocypria perlata D. R. Rome ♂ (Fig. 4). Les deux branches de l'extrémité proximale font un angle obtus. La branche dorsale est longue et légèrement sinueuse. La branche ventrale est large, courbée et se termine par une plaquette allongée. La partie médiane est droite. L'extrémité articulaire est dans le prolongement de la partie médiane, un des deux éléments qui la composent est arrondi, l'autre est acéré.

Mecynocypria opaca (G. O. Sars) ♀ (Fig. 4). Les deux branches de l'extrémité proximale font un angle droit. La branche dorsale est très longue, très fortement courbée et présente un angle effacé vers le milieu de sa longueur. La branche ventrale est droite, peu allongée et assez large. La partie médiane est droite, vers les 2/3 de sa longueur, elle porte l'expansion ventrale, large et courbée vers l'avant. L'extrémité articulaire est peu acérée et légèrement inclinée dorsalement.

Mecynocypria opaca (G. O. Sars) ♂ (Fig. 4). Les deux branches de l'extrémité proximale font un angle aigu. La branche dorsale est très longue, très fortement courbée et présente un angle effacé à 1/3 de sa longueur. La branche ventrale est longue et légèrement courbée. La partie médiane est droite. L'extrémité articulaire est acérée et dans le prolongement de la partie médiane.

Mecynocypria deflexa (G. O. Sars) ♀ (Fig. 4). Les deux branches de l'extrémité proximale font un angle aigu. La branche dorsale est très longue, régulièrement courbée sur les 3/4 de sa longueur, puis se plie brusquement; elle porte un petit ergot au milieu de la courbure. La branche ventrale est longue et large. La partie médiane est droite et très large et porte vers les 4/5 de sa longueur l'expansion ventrale, courte, droite et large. L'extrémité articulaire est acéré et dans le prolongement de la partie médiane.

Mecynocypria deflexa (G. O. Sars) ♂ (Fig. 4). Les deux branches de l'extrémité proximale font un angle aigu. La branche dorsale est très longue, régulièrement courbée sur un peu plus de la moitié de sa longueur, puis se plie brusquement, elle porte un petit ergot au dessus du milieu de la courbure. La branche ventrale est longue et large. La partie médiane est droite et large. L'extrémité articulaire est acérée et légèrement inclinée ventralement.

Mecynocypria ovata D. R. Rome ♀ (Fig. 4). Les deux branches de l'extrémité proximale font un angle aigu. La branche dorsale est très longue et très fortement courbée, à son extrémité distale la courbure est inversée. La branche ventrale est longue, large, légèrement courbée en S. La partie médiane est droite et très large et porte vers les 2/3 de sa longueur l'expansion ventrale, courte, large et droite. L'extrémité articulaire est courte, acérée et légèrement inclinée ventralement.

Mecynocypria ovata D. R. Rome ♂ (Fig. 4). Les deux branches de l'extrémité proximale font un angle presque droit. La branche dorsale est très longue, fortement courbée, elle se plie vers la moitié de sa longueur. La branche ventrale est courte, droite et très large. La partie médiane est droite et très large. L'extrémité articulaire est courte et dans le prolongement de la partie médiane.

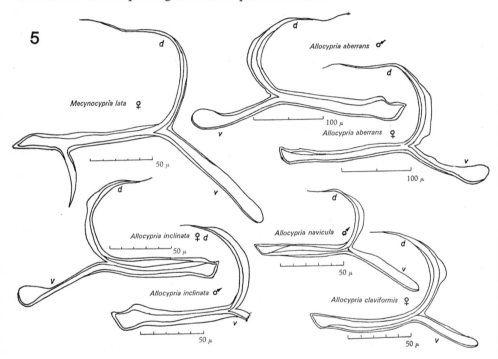

Fig. 5. L'attache de la furca chez la sous-famille Cyclocypridinae.

Mecynocypria lata D. R. Rome ♀ (Fig. 5). Les deux branches de l'extrémité proximale font un angle légèrement obtus. La branche dorsale est très longue, fortement et presque uniformément courbée. La branche ventrale est longue et peu large. La partie médiane est droite et très large et porte vers les 4/5 de sa longueur l'expansion ventrale large et légèrement courbée vers l'arrière. L'extrémité articulaire est courte et dans le prolongement de la partie médiane.

Mecynocypria lata D. R. Rome ♂ (Fig. 4). Les deux branches de l'extrémité proximale font un angle légèrement obtus. La branche dorsale est très longue régulièrement et fortement courbée, à son extrémité distale la courbure est inversée. La branche ventrale est longue, droite et large. La partie médiane est droite et très large. L'extrémité articulaire est courte, acérée et dans le prolongement de la partie médiane.

Mecynocypria elongata D. R. Rome ♂ (Fig. 4). Les deux branches de l'extrémité proximale font un angle aigu. La branche dorsale est très longue, fortement et presque uniformément courbée. La branche ventrale est longue et peu large. La partie médiane est droite et très large. L'extrémité articulaire est courte, son extrémité distale est dirigée dorsalement.

Allocypria inclinata D. R. Rome ♀ (Fig. 5). Les deux branches de l'extrémité proximale font un angle aigu. La branche dorsale est longue, fortement courbée, au delà de la moitié de sa longueur son bord antérieur porte un court renflement. La branche ventrale est longue, droite, étroite et se termine par une large plaquette. La partie médiane est droite et très large. L'extrémité articulaire est courte, se termine brusquement et son extrémité peu acérée est dirigée dorsalement.

Allocypria inclinata D. R. Rome ♂ (Fig. 5). (La branche ventrale était brisée.) La branche dorsale est longue et fortement courbée, ne présente pas de renflement à son bord antérieur. La partie médiane est droite et large. L'extrémité articulaire est courte, se termine brusquement et son extrémité acérée est dirigée dorsalement.

Allocypria claviformis (G. O. Sars) ♀ (Fig. 5). Les deux branches de l'extrémité proximale font un angle droit. La branche dorsale est longue, fortement courbée, vers la moitié de sa longueur son bord antérieur porte un fort renflement. La branche ventrale est courte et large. La partie médiane est large et courbée, sa concavité dorsale se raccorde sans interruption avec le concavité de la branche dorsale. L'extrémité articulaire est courte, son extrémité acérée est dirigée dorsalement.

Allocypria navicula D. R. Rome ♂ (Fig. 5). Les deux branches de l'extrémité proximale font un angle droit. La branche dorsale est longue et fortement courbée. La branche ventrale est droite et large. La partie médiane est droite et large. L'extrémité articulaire est courte, se termine assez brusquement et son extrémité acérée est dirigée dorsalement.

Allocypria aberrans D. R. Rome ♀ (Fig. 5). Les deux branches de l'extrémité proximale font un angle presque droit. La branche dorsale est très longue, fortement courbée et son bord antérieur est droit sur une longue partie. La branche ventrale est courte, large, légèrement courbée et se termine par une large plaquette. La partie médiane est droite et large. L'extrémité articulaire est courte, se termine brusquement et son extrémité acérée est dirigée dorsalement.

Allocypria aberrans D. R. Rome ♂ (Fig. 5). Les deux branches de l'extrémité proximale font un angle aigu. La branche dorsale est très longue, fortement courbée et son bord antérieur est droit sur une courte partie. La branche ventrale est courte, large, légèrement courbée et se termine par une large plaquette. La partie médiane est droite et très large. L'extrémité articulaire est courte, se termine brusquement et son extrémité acérée est dirigée dorsalement.

CYPRIDINAE

Cypris bispinosa H. Lucas ♀ (Fig. 6). Les deux branches de l'extrémité proximale font un angle très aigu. La branche dorsale est courte et légèrement courbée vers l'arrière. La branche ventrale est longue, courbée et son extrémité distale est acérée. La partie médiane est longue fortement courbée à concavité dorsale. L'extrémité articulaire est acérée et dans le prolongement de la partie médiane.

Chlamydotheca incisa (Claus) ♀ (Fig. 6). Les deux branches de l'extrémité proximale font un angle aigu. La branche dorsale est longue et fortement courbée vers l'arrière. La branche ventrale est courte et légèrement courbée. La partie médiane est longue et fortement courbée à concavité dorsale, elle porte à l'extrémité distale l'expansion ventrale, mince et courbée vers l'arrière. L'extrémité articulaire est acérée et dans le prolongement de la partie médiane.

Eucypris virens (Jurine) ♀ (Fig. 6). Les deux branches de l'extrémité proximale font un angle obtus. La branche dorsale est courte et peu courbée. La branche ventrale est très longue et courbée. La partie médiane est longue et presque droite. L'extrémité articulaire est courte, dans le prolongement de la partie médiane, son extrémité acérée est dirigée dorsalement.

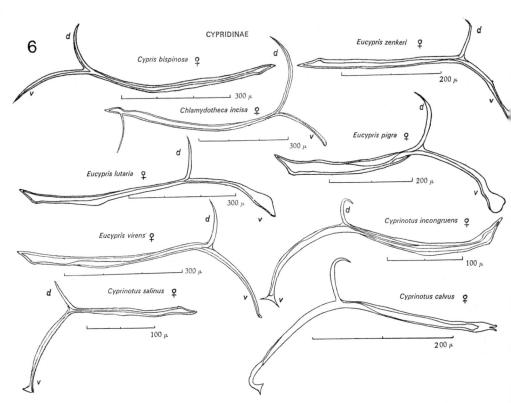

Fig. 6. L'attache de la furca chez la sous-famille Cypridinae.

Eucypris zenkeri (Chyzer) ♀ (Fig. 6). Les deux branches de l'extrémité proximale font un angle obtus. La branche dorsale est courte et droite sur presque toute sa longueur et se plie brusquement vers son extrémité distale. La branche ventrale est longue, peu courbée et se termine par une plaquette allongée. La partie médiane est longue et droite. L'extrémité articulaire est allongée, acérée et dans le prolongement de la partie médiane.

Eucypris lutaria (Koch) ♀ (Fig. 6). Les deux branches de l'extrémité proximale font un angle droit. La branche dorsale est courte et droite. La branche ventrale est longue, courbée et se termine par une plaquette longue et large. La partie médiane est longue, droite à son bord dorsal, un peu convexe à son bord ventral. L'extrémité articulaire est courte et dans le prolongement de la partie médiane, son extrémité acérée est dirigée dorsalement.

Eucypris pigra (Fischer) ♀ (Fig. 6). Les deux branches de l'extrémité proximale font un angle aigu. La branche dorsale est assez longue et fortement courbée. La branche ventrale est très longue, courbée et se termine par une plaquette longue et large. La partie médiane est large, très légèrement courbée, à concavité dorsale. L'extrémité articulaire est longue, acérée et nettement dirigée dorsalement.

Cyprinotus incongruens (Ramdohr) ♀ (Fig. 6). Les deux branches de l'extrémité proximale font un angle aigu. La branche dorsale est très petite et très brusquement courbée à son extrémité distale. La branche ventrale est longue, large et très courbée, elle se termine en s'évasant. La partie médiane est longue, large et courbée, à concavité dorsale. L'extrémité articulaire est longue, acérée et fortement inclinée dorsalement.

Cyprinotus calvus D. R. Rome ♀ (Fig. 6). Les deux branches de l'extrémité proximale font un angle presque droit. La branche dorsale est courte, fortement courbée à son extrémité distale. La branche ventrale est très longue, peu courbée, sauf à son extrémité distale où sa courbure est très forte, avant de se terminer en s'évasant. La partie médiane est longue, large et droite. L'extrémité articulaire est courte, dans le prolongement de la partie médiane; les deux éléments qui la composent semblent ne pas se superposer.

Cyprinotus salinus (Brady) ♀ (Fig. 6). Les deux branches de l'extrémité proximale font un angle légèrement obtus. La branche dorsale est courte et droite. La branche ventrale est très longue et peu courbée, elle se termine en s'évasant. La partie médiane est longue et droite. L'extrémité articulaire est acérée et dans le prolongement de la partie médiane.

Cypricercus fuscatus (Jurine) ♀ (Fig. 7). Les branches de l'extrémité proximale font un angle très obtus. La branche dorsale est courte et courbée. Les bandelettes de chitine qui la forment s'écartent pour former l'œillet signalé par E. Triebel, qui permet de distinguer aisément *Eucypris* de *Cypricercus*. Cet œillet se trouve dans la concavité de la courbure de la branche dorsale (soit donc à sa partie postérieure). C'est à son bord que s'insèrent les muscles moteurs de la furca. L'extrémité distale de la branche est nettement divisée. La branche ventrale est longue, peu large, peu courbée et se termine par une large plaquette de forme irrégulière. La partie médiane est longue, peu large, peu courbée à concavité dorsale, elle porte à son extrémité distale une très courte expansion ventrale courbée. L'extrémité articulaire est courte et dans le prolongement de la partie médiane. Un de ses éléments est arrondi, l'autre est acéré.

Cypricercus affinis (Fischer) ♀ (Fig. 7). Les branches de l'extrémité proximale font un angle très obtus. La branche dorsale est assez longue et fort courbée. Les bandelettes de chitine qui la forment s'écartent pour former l'œillet qui se trouve dans la concavité de la courbure (soit donc à sa partie postérieure). C'est à son bord que s'insèrent les muscles moteurs de la furca. L'extrémité distale de la branche dorsale ne semble pas divisée. La branche ventrale est longue, étroite, sinueuse et se termine par une très courte plaquette. La partie médiane est très longue étroite et courbée, à concavité dorsale, elle porte à son extrémité distale une courte expansion ventrale presque droite, dirigée vers l'avant. L'extrémité articulaire est très courte, très étroite (au point qu'on ne distingue pas ses deux éléments) et ne semble pas acérée.

Strandesia. Dans ce genre l'extrémité proximale est très différente de celle des autres genres. On y retrouve bien la branche dorsale et la branche ventrale, mais au lieu de présenter une fourche dont les branches font un angle très net, elles sont rejointes par une bandelette de chitine courbée qui avec la branche dorsale forme un œillet situé antérieurement (au contraire, chez *Cypricercus* l'œillet est situé dans la courbure de la branche dorsale, soit donc postérieurement). Plus proximalement encore les deux branches sont rejointes par un filet de chitine tangent ou séparé de l'œillet.

Strandesia mulargiae G. Anichini ♀ (Fig 7) La branche dorsale est longue et mince, légèrement sinueuse d'abord, puis se courbe très fort. L'œillet est large. La branche ventrale est longue, droite et étroite, elle s'évase à son extrémité distale pour entourer une large plaquette. Le filet de chitine qui rejoint les deux branches est tangent à l'œillet sur une très courte longueur. La partie médiane est longue, large et courbée, à concavité dorsale; elle porte à son extrémité distale une expansion ventrale, courte, mince et courbée. L'extrémité articulaire est longue, acérée et dans le prolongement de la courbure de la partie médiane.

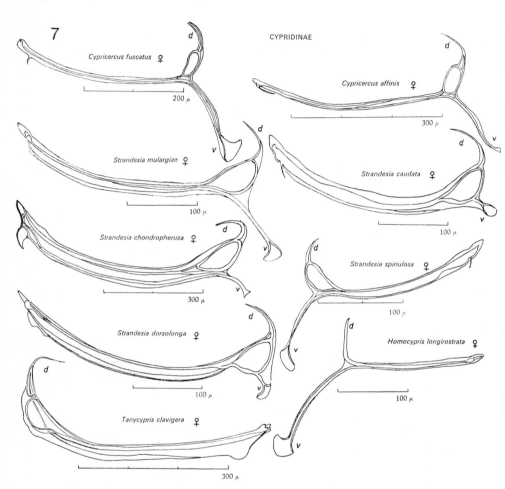

Fig. 7. L'attache de la furca chez la sous-famille Cypridinae.

Strandesia chondropherusa D. R. Rome ♀ (Fig. 7). La branche dorsale est longue, large, presque droite d'abord, puis se courbe très fort en restant large. L'œillet est allongé. La branche ventrale est courte, courbée et étroite, elle s'évase distalement. Le filet de chitine qui rejoint les deux branches est tangent à l'œillet sur près de toute sa longueur. La partie médiane est longue, très large et fortement courbée, à concavité dorsale; elle porte à son extrémité distale une expansion ventrale qui a la forme d'un onglet. L'extrémité articulaire est allongée et dans le prolongement de la courbure de la partie médiane.

Strandesia dorsolonga D. R. Rome ♀ (Fig. 7). La branche dorsale est longue, étroite, droite d'abord, puis se courbe très fortement en s'amincissant. L'œillet est large. La branche ventrale est très courte, elle s'évase distalement pour entourer une plaquette allongée transversalement. Le filet de chitine qui rejoint les deux branches est séparé d'elles et de l'œillet sur toute sa longueur. La partie médiane est longue, large et fort courbée, à concavité dorsale; elle porte à son extrémité distale une plaquette allongée. L'extrémité articulaire est longue, acérée et dans le prolongement de la courbure de la partie médiane.

Strandesia caudata Klie ♀ (Fig. 7). La branche dorsale est longue, large et presque droite d'abord, puis se courbe très fortement en s'amincissant. L'œillet est allongé. La branche ventrale est très courte et s'évase distalement pour entourer une plaquette allongée. Le filet de chitine qui rejoint les deux branches leur est parallèle, ainsi qu'à la courbure antérieure de l'œillet, et s'en éloigne fort peu. La partie médiane est longue, très large et courbée, à concavité dorsale; elle porte à son extrémité distale une expansion ventrale courte, mince et courbée. L'extrémité articulaire est large, peu acérée et dans le prolongement de la courbure de la partie médiane.

Strandesia spinulosa Akatova ♀ (Fig. 7). La branche dorsale est longue, étroite, légèrement sinueuse d'abord puis courbée. L'œillet est allongé et irrégulier. La branche ventrale est longue, étroite et très légèrement courbée, elle s'évase fortement à son extrémité distale pour entourer une plaquette allongée obliquement. Le filet de chitine qui rejoint les deux branches suit tout leur contour à une courte distance. La partie médiane est longue, large et courbée, à concavité dorsale; elle porte à son extrémité distale une expansion ventrale très courte, très mince et courbée. L'extrémité articulaire est large, peu acérée et dans le prolongement de la courbure de la partie médiane.

Tanycypris clavigera (G. W. Müller) ♀ (Fig. 7). L'extrémité proximale ne possède pas de branche ventrale. La branche dorsale est fortement courbée, elle est suivie par une fenêtre arrondie, formée par l'entrecroisement des bandelettes de chitine issues de la partie médiane. Celle-ci est très large, et courbée, à concavité dorsale; elle porte ventralement à son extrémité distale une plaquette arrondie. L'extrémité articulaire est formée de deux éléments globuleux.

Homocypris longirostrata D. R. Rome ♀ (Fig. 7). Les branches de l'extrémité proximale font un angle très obtus. La branche dorsale est courte, droite et étroite. La branche ventrale est longue, étroite, courbée et s'évase distalement pour se terminer par une large plaquette. La partie médiane est peu allongée et étroite. L'extrémité articulaire est courte, peu acérée, légèrement dirigée dorsalement.

ISOCYPRIDINAE

Isocypris quadrisetosa D. R. Rome ♀ (Fig. 8). Les branches de l'extrémité proximale font un angle droit. La branche dorsale est très longue, courbée en S et assez large. La branche ventrale est très courte, droite et terminée par une plaquette triangulaire. La partie médiane est droite et large, elle porte à son extrémité distale l'expansion ventrale, très mince et droite, précédée par une longue et large plaquette transparente. L'extrémité articulaire est courte, acérée et dans le prolongement de la partie médiane.

Amphibolocypris exigua D. R. Rome ♂ (Fig. 8). Les branches de l'extrémité proximale font un angle obtus. La branche dorsale est longue, droite et se courbe en s'élargissant à son extrémité distale. La branche ventrale est très longue, sinueuse et étroite, elle se termine par une plaquette très large, bordée distalement par un renforcement de chitine La partie médiane est droite et étroite, elle porte à son extrémité distale l'expansion ventrale dirigée vers l'avant, qui se divise en deux filaments acérés. L'extrémité articulaire est courte, peu acérée et dans le prolongement de la partie médiane.

N

DOLEROCYPRIDINAE

Dolerocypris fasciata (O. F. Müller) ♀ (Fig. 8). Les branches de l'extrémité proximale font un angle obtus. La branche dorsale est très longue, large et peu courbée d'abord, puis se plie brusquement vers les 2/3 de sa longueur et devient droite et très mince. La branche ventrale est courte, droite, étroite et se termine par une très courte plaquette. La partie médiane est très large et courbée, à concavité dorsale; son bord dorsal présente vers son milieu un renflement. L'extrémité articulaire est courte, dans le prolongement de la partie médiane; un de ses éléments est acéré, l'autre, plus court et globuleux.

Dolerocypris sinensis G. O. Sars ♀ (Fig. 8). Les branches de l'extrémité proximale font un angle légèrement aigu. La branche dorsale est très longue, large d'abord et

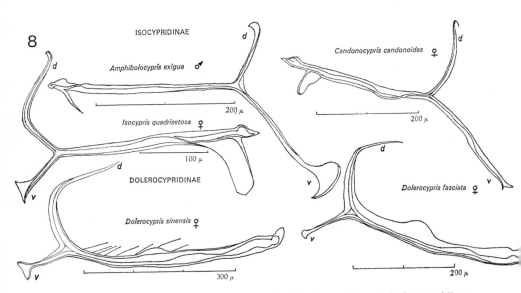

Fig. 8. L'attache de la furca chez les sous-familles Isocypridinae et Dolerocypridinae.

courbée, puis devient droite en s'amincissant. La branche ventrale est très courte, droite, assez large, et s'évase distalement pour entourer une plaquette allongée transversalement. La partie médiane est large, peu courbée, à concavité dorsale, son bord dorsale présente un étroit renflement de part et d'autre duquel s'insère un faisceau de muscles extenseurs de la furca (indiqué par des traits). L'extrémité articulaire est courte et dans le prolongement de la partie médiane; un de ses éléments est très peu acéré, l'autre est globuleux.

Candonocypris candonoides G. O. Sars ♀ (Fig. 8). Les deux branches de la partie proximale font un angle obtus. La branche dorsale est longue, étroite, peu courbée et se termine par un petit crochet. La branche ventrale est très longue, droite, étroite et se termine par une très petite plaquette. La partie médiane est droite, une des bandelettes de chitine qui la forme fait saillie à son bord ventral; elle porte à son extrémité distale l'expansion ventrale, courte, très mince et peu courbée, précédée d'une plaquette transparente. L'extrémité articulaire est courte acérée et dans le prolongement de la partie médiane.

HERPETOCYPRIDINAE

Tous les genres de cette sous-famille ont comme particularité de l'attache de la furca : l'expansion triangulaire transparente bordée de bandelettes de chitine. Les deux expansions, gauche et droite, se rejoignent au bord ventral du corps.

Herpetocypris reptans (Baird) ♀ (Fig. 1). Les deux branches de l'extrémité proximale font un angle obtus. La branche dorsale est longue, étroite et courbée en S. La branche ventrale est longue, étroite, droite et se termine par une longue plaquette. La partie médiane est longue, étroite, légèrement courbée à convexité dorsale; elle porte à son extrémité distale l'expansion triangulaire transparente, bordée de bandelettes de chitine, l'une droite, l'autre sinueuse, qui ne se rejoignent pas. L'extrémité articulaire est courte, acérée et dans le prolongement de la partie médiane.

Herpetocypris lenta D. R. Rome ♀ (Fig. 9). Les branches de l'extrémité dorsale font un angle aigu. La branche dorsale est très longue, étroite et très fortement courbée en S allongée. La branche ventrale est courte, très légèrement courbée et se termine par une très petite plaquette. La partie médiane est longue, large et droite; elle porte à son extrémité distale l'expansion ventrale transparente bordée de bandelettes chitineuses, l'une presque droite, l'autre concave, qui ne se rejoignent pas. L'extrémité articulaire est courte, acérée et dans le prolongement de la partie médiane.

Herpetocypris romei G. Anichini ♀ (Fig. 9). Les branches de l'extrémité proximale font un angle presque droit. La branche dorsale est longue, étroite, fortement courbée dans un seul sens et se termine par une plaquette allongée. La branche ventrale est longue, droite, assez large, s'évase à son extrémité pour entourer une plaquette allongée transversalement. La partie médiane est longue, large, et légèrement courbée, à concavité ventrale; elle porte à son extrémité distale l'expansion triangulaire transparente, bordée de bandelettes de chitine, l'une droite, l'autre concave qui se rejoignent en formant un court bec tourné vers l'avant. L'extrémité articulaire est courte, acérée et dans le prolongement de la partie médiane.

Herpetocypris flumendosae G. Anichini ♀ (Fig. 9). Les branches de l'extrémité proximale font un angle très obtus. La branche dorsale est longue, large et fortement courbée dans un seul sens, elle se termine par une plaquette allongée. La branche ventrale est longue, étroite, très légèrement courbée et s'évase à son extrémité pour se terminer par une petite plaquette La partie médiane est longue, étroite, très légèrement courbée, à concavité ventrale; elle porte à son extrémité distale l'expansion triangulaire transparente bordée de bandelettes de chitine, l'une droite et très oblique, l'autre courbée, qui se rejoignent pour former un très petit bec tourné vers l'avant. L'extrémité articulaire est très petite, acérée et dans le prolongement de la partie médiane.

Herpetocypris chevreuxi (G. O. Sars) ♀ (Fig. 9). Les branches de l'extrémité proximale font un angle droit. La branche dorsale est très longue, étroite et fortement courbée en S. La branche ventrale est courte, droite et se termine par une très petite plaquette. La partie médiane est longue, droite et large; son bord dorsal porte une courte bandelette de chitine courbée où s'insère un des muscles extenseurs de la furca; elle porte à son extrémité distale l'expansion ventrale triangulaire transparente bordée de bandelettes de chitine, l'une presque droite, l'autre concave, qui ne se rejoignent pas. L'extrémité articulaire est longue et très légèrement dirigée ventralement.

Herpetocypris agilis D. R. Rome ♀ (Fig. 9). Les branches de l'extrémité proximale font un angle aigu. La branche dorsale est longue, étroite et fortement courbée en S. La branche ventrale est très longue, large et nettement courbée. La partie médiane est longue, large et peu courbée; son bord dorsal porte une très courte bandelette de chitine courbée où s'insère un des muscles extenseurs de la furca; elle porte à son extrémité distale l'expansion ventrale triangulaire transparente, bordée de bandelettes de chitine, toutes deux concaves, qui ne se rejoignent pas. L'extrémité articulaire est longue, large, acérée et son extrémité est légèrement dirigée ventralement.

Stenocypris anisoacantha D. R. Rome ♀ (Fig. 9). Les branches de l'extrémité

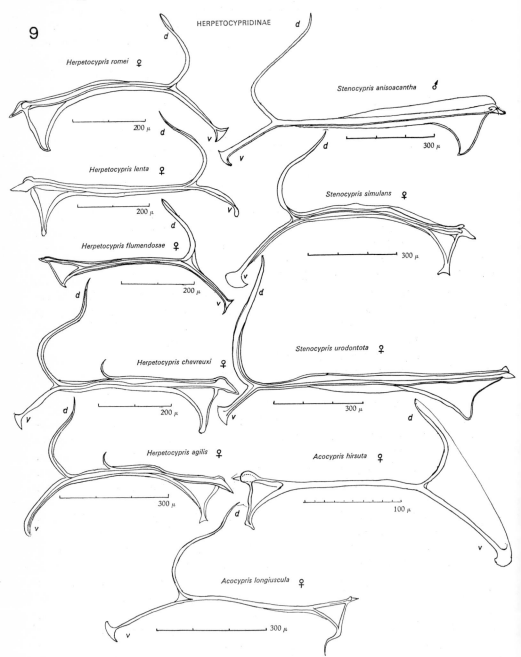

Fig. 9. L'attache de la furca chez la sous-famille Herpetocypridinae.

proximale font un angle presque droit. La branche dorsale est très longue, étroite et courbée en S. La branche ventrale est courte assez large et s'évase à son extrémité pour entourer une courte plaquette transverse. La partie médiane est longue, droite et large; elle porte à son extrémité distale l'expansion ventrale triangulaire transparente bordée de minces bandelettes de chitine, l'une concave, l'autre sinueuse qui se rejoignent pour former un bec. L'extrémité articulaire est très courte, et dans le prolongement de la partie médiane; un de ses éléments est acéré, l'autre globuleux.

Stenocypris urodontota D. R. Rome ♀ (Fig. 9). Les branches de l'extrémité proximale font un angle obtus. La branche dorsale est longue, large et légèrement courbée en S. La branche ventrale est courte, large et se termine par une plaquette transverse. La partie médiane est longue, droite et large: elle porte à son extrémité distale l'expansion ventrale triangulaire transparente, sa base est très longue, elle est bordée de bandelettes de chitine d'épaisseur inégale, l'une presque droite, l'autre concave, qui se rejoignent en formant une courbe. L'extrémité articulaire est courte et dans le prolongement de la partie médiane, un de ses éléments est acéré, l'autre globuleux.

Stenocypris simulans D. R. Rome ♀ (Fig. 9). Les branches de l'extrémité proximale font un angle obtus. La branche dorsale est longue, étroite et courbée en S. La branche ventrale est longue, légèrement courbée et assez large, elle se termine par une large plaquette. La partie médiane est longue, large et courbée, à convexité dorsale; elle porte à son extrémité distale l'expansion ventrale triangulaire transparente, bordée de larges bandelettes de chitine concaves qui se rejoignent en une surface terminée par un bord concave. L'extrémité articulaire est longue, l'un de ses éléments est acéré, l'autre est globuleux et échancré.

Acocypris hirsuta D. R. Rome ♀ (Fig. 9). Les branches de l'extrémité proximale font un angle aigu. La branche dorsale est longue, étroite et courbée en S. La branche ventrale est très longue, étroite et se termine par une large plaquette. Les deux branches sont jointes par un mince filet de chitine. La partie médiane est longue, droite et étroite; elle porte à son extrémité distale l'expansion ventrale triangulaire transparente, bordée de larges bandelettes de chitine, l'une droite, l'autre concave qui se rejoignent en une surface terminée obliquement en pointe. L'extrémité articulaire est courte et dans le prolongement de la partie médiane, un de ses éléments est acéré, l'autre fortement globuleux.

Acocypris longiuscula D. R. Rome ♀ (Fig. 9). Les branches de l'extrémité proximale font un angle presque droit. La branche dorsale est très longue, étroite et sinueuse. La branche ventrale est longue, très étroite et se termine par une plaquette allongée transversalement. La partie médiane est longue, étroite et un peu sinueuse; elle porte à son extrémité distale l'expansion triangulaire transparente bordée de bandelettes de chitine qui se rejoignent pour former un long prolongement courbé. L'extrémité articulaire est très courte et dans le prolongement de la partie médiane, un de ses éléments est très acéré, l'autre globuleux.

Ilyodromus olivaceus (Brady et Norman) ♀ (Fig. 10). Les branches de l'extrémité proximale font un angle droit. La branche dorsale est longue, large et courbée en S. La branche ventrale est longue, large et courbée et se divise à son extrémité. La partie médiane est longue, large et sinueuse; elle porte à son bord dorsal une courte bandelette de chitine courbée où s'insère un muscle extenseur de la furca; à son extrémité distale elle porte l'expansion ventrale triangulaire bordée de larges bandelettes de chitine presque droites qui se rejoignent. L'extrémité articulaire est courte, acérée et dans le prolongement de la partie médiane.

Ilyodromus fontinalis (Wolf) ♀ (Fig. 10). Les branches de l'extrémité proximale font un angle aigu. La branche dorsale est assez longue, étroite, courbée brusquement, son bord postérieur porte un petit crochet. La branche ventrale est longue, large, droite puis courbée et acérée à son extrémité distale. La partie médiane est droite et large; à proximité de la branche dorsale elle porte une courte bandelette de chitine courbée où s'insère un muscle extenseur de la furca; cette bandelette est jointe au petit crochet de la branche dorsale par un mince filet de chitine et forme ainsi avec lui un anneau irrégulier. L'expansion ventrale triangulaire transparente de l'extrémité distale est bordée d'étroites bandelettes de chitine qui ne se rejoignent pas. L'extrémité articulaire est courte, acérée et dans le prolongement de la partie médiane.

Ilyodromus fontinalis (Wolf) ♂ (Fig. 10). Les branches de l'extrémité proximale font

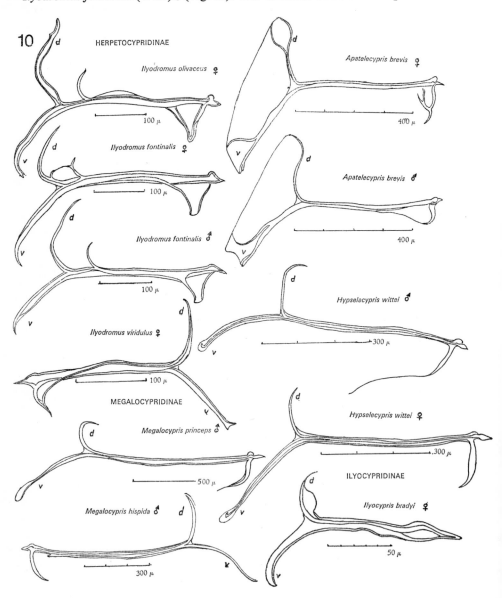

Fig. 10. L'attache de la furca chez les sous-familles Herpetocypridinae, Megalocypridinae et Ilyocypridinae.

un angle presque droit. La branche dorsale est longue, étroite et irrégulièrement courbée. La branche ventrale est longue, large, droite puis courbée et acérée à son extrémité distale. La partie médiane est droite et large; elle porte à son bord dorsal une mince bandelette de chitine courbée où s'insère un muscle extenseur de la furca; elle porte à son extrémité distale l'expansion ventrale triangulaire transparente bordée d'étroite bandelettes de chitine qui ne se rejoignent pas. L'extrémité articulaire est longue, acérée et dans le prolongement de la partie médiane.

Ilyodromus viridulus (Brady) ♀ (Fig. 10). Les branches de l'extrémité proximale font un angle obtus. La branche dorsale est longue, étroite et peu courbée. La branche ventrale est longue, étroite, droite et se termine par une courte plaquette. La partie médiane est droite et large; à son extrémité distale l'expansion ventrale triangulaire transparente est bordée de bandelettes de chitine fort courbées, l'une convexe, l'autre concave, qui se rejoignent en un prolongement sinueux très mince.

MEGALOCYPRIDINAE

Dans cette sous-famille l'attache de la furca est caractérisée par une extrémité proximale en fourche souvent irrégulière, et par une expansion ventrale courbée vers l'avant.

Megalocypris princeps G. O. Sars ♂ (Fig. 10). Les branches de l'extrémité proximale font un angle légèrement aigu. La branche dorsale est courte, étroite et fortement courbée. La branche ventrale est longue, étroite, presque droite et terminée par une plaquette allongée. La partie médiane est étroite, très légèrement courbée, à concavité dorsale; à son extrémité distale l'expansion ventrale est large, irrégulièrement courbée vers l'avant et s'amincit distalement. L'extrémité articulaire est courte et dans le prolongement de la partie médiane; un de ses éléments est acéré, l'autre globuleux.

Megalocypris hispida G. O. Sars ♂ (Fig. 10). Les branches de l'extrémité proximale font un angle presque droit. La branche dorsale est longue, étroite et peu courbée. La branche ventrale est longue, très étroite, légèrement courbée en S. La partie médiane est longue, étroite et courbée, à concavité dorsale; à son extrémité distale l'expansion ventrale est étroite et régulièrement courbée vers l'avant. L'extrémité articulaire est très courte, acérée et dans le prolongement de la partie proximale.

Apatelecypris brevis (G. O. Sars) ♀ (Fig. 10). Les branches de l'extrémité proximale font un angle très obtus. La branche dorsale est longue, très mince et se divise vers son extrémité en deux parties convexes, pour entourer une plaquette arrondie transparente. La branche ventrale est très longue, droite et se divise vers son extrémité en deux parties, l'une droite, l'autre courbée, pour entourer une plaquette de contour irrégulier. Les deux branches sont jointes par un mince filet de chitine. La partie médiane est longiue, étroite et légèrement sinueuse; à son extrémité distale l'expansion ventrale est étrote, sinueuse, ses bandelettes de chitine se ramifient. L'extrémité articulaire est courte et dans le prolongement de la partie médiane; un de ses éléments est très acéré, l'autre globuleux.

Apatalecypris brevis (G. O. Sars) ♂ (Fig. 10). Les branches de l'extrémité proximale font un angle obtus. La branche dorsale est longue, très mince et courbée en S. La branche ventrale est longue, large et se divise vers son milieu en deux parties, l'une droite, l'autre courbée, pour entourer une plaquette de contour très irrégulier. Les deux branches sont jointes par un mince filet de chitine. La partie médiane est longue, peu large, légèrement sinueuse; à son extrémité distale l'expansion ventrale est large, courbée en S vers l'avant et se termine par un mince filet qui rejoint la partie médiane. L'extrémité articulaire est très courte et dans le prolongement de la partie médiane; un de ses éléments est acéré, l'autre globuleux.

Hypselecypris wittei D. R. Rome ♀ (Fig. 10). Les branches de l'extrémité proximale font un angle obtus. La branche dorsale est courte, étroite et courbée. La branche ventrale est longue, large, droite et se termine par un crochet au milieu d'une plaquette allongée. La partie médiane est longue, droite et large; à son extrémité distale l'expansion ventrale est large, presque droite sauf à l'extrémité acérée, courbée vers l'avant. L'extrémité articulaire est longue et dans le prolongement de la partie médiane; un de ses éléments est acéré, l'autre globuleux.

Hypselecypris wittei D. R. Rome ♂ (Fig. 10). Les branches de l'extrémité proximale font un angle droit. La branche dorsale est droite d'abord, puis se courbe brusquement vers l'arrière. La branche ventrale est longue, peu large, droite et se termine par un petit crochet au milieu d'une plaquette allongée. La partie médiane est longue, droite et étroite; à son extrémité distale l'expansion ventrale est peu large, droite et dirigée vers l'avant; elle se termine par un filet de chitine qui borde toute la partie ventrale du corps. L'extrémité articulaire est courte et dans le prolongement de la partie médiane, un de ses éléments est acéré l'autre globuleux.

ILYOCYPRIDINAE

Ilyocypris bradyi G. O. Sars ♀ (Fig. 10). Les deux branches de l'extrémité proximale font un angle obtus. La branche dorsale est courte et fortement courbée, elle est pourvue le long de son bord concave d'une plaquette transparente. La branche ventrale est longue, large et courbée. La partie médiane est droite. L'extrémité articulaire est très longue, fortement courbée au bord dorsal et très acérée; elle montre en son milieu une ouverture allongée.

NOTODROMATINAE

Dans l'angle que forment les branches de l'extrémité proximale se trouvent des bandelettes de chitine isolées ou formant un réseau.

Notodromas monacha (O. F. Müller) ♀ (Fig. 11). Les branches de l'extrémité proximale font un angle obtus. La branche dorsale est longue, large et courbée en S vers l'avant; une bandelette de chitine, issue de la partie médiane la rejoint en s'incurvant et forme avec la branche un anneau postérieur; le long de son bord antérieur sont issues des bandelettes qui forment un anneau. La branche ventrale est très longue, très étroite; de sont bord antérieur sont issues des bandelettes qui forment un réseau; un de ses éléments est triangulaire et se prolonge en une bandelette mince, parallèle à la branche ventrale; vers son extrémité s'en détache vers l'avant une bandelette courte et fort courbée, jointe à la branche dorsale par un mince filet de chitine. La partie médiane est longue, large et droite; elle porte à son extrémité distale une très mince expansion ventrale sinueuse, inclinée vers l'avant. L'extrémité articulaire est longue, acérée et dirigée dorsalement.

Notodromas monacha (O. F. Müller) ♂ (Fig. 11). Les branches de l'extrémité proximale font un angle obtus. La branche dorsale est courte et se divise dorsalement pour former un anneau. La branche ventrale est très longue, très mince et sinueuse; à son bord antérieur un petit crochet s'en détache, il est rejoint par un mince filet de chitine à une bandelette de chitine sinueuse issue de la branche dorsale, pour former un anneau irrégulier. La partie médiane est longue, étroite et droite; elle porte à son extrémité distale une courte et large expansion ventrale en forme d'onglet. L'extrémité articulaire est longue, acérée et dirigée dorsalement.

Notodromas persica var. dalmatina Petkovski ♀ (Fig. 11). Les branches de l'extrémité proximale font un angle aigu. La branche dorsale est courte, large et courbée, de son bord antérieur se détachent des bandelettes de chitine qui forment un anneau ir-régulier. La branche ventrale est très longue, mince et presque droite. La partie médiane est longue, peu large et courbée, à concavité dorsale; vers les 3/5 de sa longueur elle porte une courte expansion ventrale, légèrement courbée vers l'avant, bordée de minces bandelettes de chitine. L'extrémité articulaire est longue, acérée et dirigée dorsalement.

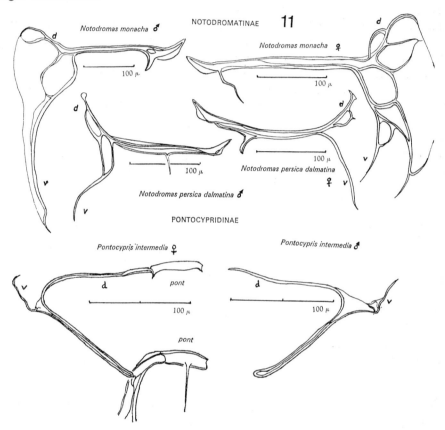

Fig. 11. L'attache de la furca chez les sous-familles Notodromatinae et Ponto-cypridinae.

Notodromas persica var. dalmatina Petkovski ♂ (Fig. 11). Les branches de l'extrémité proximale font un angle obtus. La branche dorsale est longue, mince, peu courbée et se termine par une petite plaquette. La branche ventrale est très longue, mince et courbée en S. Les deux bords antérieurs des branches se détachent des bandelettes de chitine qui forment deux anneaux allongés. La partie médiane est longue, large et presque droite; elle porte vers les 3/5 de sa longueur une expansion ventrale droite, courte et très mince. L'extrémité articulaire est longue, acérée et dirigée dorsalement.

Il aurait été intéressant de poursuivre notre revue des attaches de la furca en abordant les genres marins des Cyprididae. Le manque de matériel frais nous en a empêché.

Toutefois nous avons disposé de *Pontocypris intermedia*. Il nous a semblé utile de signaler certaines particularités, que nous avons retrouvées chez *Pontocypris dispar* et chez *Pontocypris setosa*, sans pouvoir les représenter.

L'extrémité proximale de la furca se termine en une crosse qui s'unit à sa symétrique par un pont de chitine souple et transparent situé au bord ventral du corps, au milieu duquel se trouve un poil impair. L'articulation entre la furca et son attache est située à l'origine de la crosse dans une petite cavité arrondie. La longue branche dorsale de l'extrémité proximale est, elle aussi, unie à sa symétrique par un pont de chitine souple et transparent situé au bord dorsal du corps. Comme chez les autres Cyprididae les muscles extenseurs de la furca sont insérés proximalement à la branche dorsale, et distalement à l'extrémité de la crosse; les muscle fléchisseurs, à cette même branche et sous la cavité articulaire.

Pontocypris intermedia Brady ♀ (Fig. 11). L'extrémité proximale est formée d'une branche dorsale très longue, droite et étroite unie à sa symétrique par le pont de chitine; et d'une branche ventrale très réduite, composée d'une mince bandelette de chitine sinueuse, jointe à la branche dorsale par une étroite plaquette transparente. La branche dorsale fait avec la partie médiane un angle aigu dont le sommet est largement arrondi. La partie médiane est longue, droite et étroite. L'extrémité articulaire est très courte, dans le prolongement de la partie médiane; un de ses éléments est acérée, l'autre est globuleux.

Pontocypris intermedia Brady ♂ (Fig. 11). L'attache à la même forme que celle de la femelle, mais elle est plus petite.

Nous remercions vivement Monsieur P. Piette, assistant à l'Institut de Zoologie, qui s'est chargé du travail fastidieux de la préparation des planches, et Mademoiselle Delrue qui a fait la dactylographie.

DISCUSSION

KORNICKER: Were any special techniques used in preparing the material?

ROME: Il est très facile de faire la dissection de l'attache de la furca des Cyprididae. Même quand il y a une vingtaine d'années je n'avais pas l'intention de séparer les attaches, dans des préparations de Cyprididae de cette époque je les ai retrouvées entières. En disséquant les furcae on obtient presque toujours les attaches. Elles sont généralement très résistantes et faciles à observer au cours de la dissection, car elles mesurent environ le tiers de la longuer du corps de l'ostracode.

PETKOVSKI: Diesem ausgezeichneten Vortrag kann ich nichts hinzufügen, nur will ich bemerken, daß *Dolerocypris pellucida* Klie ein *Tanycypris* ist. Ich habe meine Exemplare von *D. pellucida* aus den Reisfeldern Mazedoniens mit Ihren *Tanycypris clavigera* (G. W. Müller) aus Südafrika verglichen und konnte feststellen, daß diese zwei Arten identisch sind. Die Hinterrandborste der Furka ist typischerweise zerspalten, nicht einfach. Dr Hartmann hat an der chilenischen Küste auch eine *Dolerocypris* in einer submarinen Süßwasserquelle gefunden, die er als eine neue Spezies beschrieben hat. Sie ist auch eine *Tanycypris*.

ROME: Le Dr Triebel a défini le genre *Tanycypris* en se basant sur l'attache de la furca, qui sépare nettement ce genre de *Dolerocypris*.

PETKOVSKI: Ja, das weiß ich schon. Ich wurde aber erst durch Ihre Abbildungen veranlaßt, meine *D. pellucida* zu überprüfen, da Sie so viel Einzelheiten geben, die einen vollständigen Vergleich ermöglichen. Wenn auch *D. pellucida* nicht vollkommen identisch mit *Tanycypris clavigera* ist, so gehört sie jedenfalls diesem Genus an. Ich glaube, beide Arten sind doch identisch.

HARDING: Do you find much variation from one individual to another?

ROME: Je n'ai pas remarqué de différences individuelles dans la forme de l'attache. Par exemple des individus d'une même espèce d'*Herpetocypris*, recueillis en des endroits très distants en Sardaigne, avaient une attache exactement la même.

HARDING: To me the great value of Dom Rome's work on the furcal attachments is that they provide characters that enable one easily to distinguish females of closely related species of genera like *Candona*.

Rome: Yes, indeed.

McGregor: Are the movements of the furca related in any way to the erection of the haemopenes in *Candona* or other genera? When the haemopenes are rotated and everted the furca is occasionally in motion. Are there any muscle attachments between the furca and haemopenes which assist in rotation of the latter?

Rome: Les muscles agissent uniquement sur la furca, et je n'ai pas vu de muscles qui pourraient s'attacher au pénis.

The operation of the copulatory apparatus is quite separate from that of the furca. There is sometimes a ventral appendage called 'l'expansion ventrale', a joint to support the part of the venter round the female organs.

Delorme: Est ce que vous avez un exemple de l'attache de la furca de l'espèce *'lyocypris bradyi*?

Rome: Oui.

Delorme: Ou *Ilyocypris gibba*?

Rome: Je n'avais pas d'exemplaires de cette espèce.

THE REPRODUCTIVE POTENTIAL, LIFE HISTORY AND PARASITISM OF THE FRESHWATER OSTRACOD *DARWINULA STEVENSONI* (BRADY AND ROBERTSON[1])

DON L. McGREGOR[2]
Michigan State University, Michigan, U.S.A.

SUMMARY

This study represents one portion of a more general investigation of the taxonomy and selected aspects of the ecology of the ostracod fauna of Gull Lake, Michigan, initiated in 1964.

The widespread distribution, unique morphological characteristics, geologic history, parthenogenetic reproduction, retention of young through embryological development to the first instar, and high incidence of parasitism in *Darwinula stevensoni* suggested the opportunity to investigate several ecological aspects of the type species of the genus *Darwinula*, of potential interest to both zoologists and palaeontologists.

In Gull Lake, *D. stevensoni* reached a maximum density at a depth of about 6 metres throughout the year. The population density decreased markedly between 6 and 9 metres. No specimens were found at depths greater than 12 metres. More than 95% of the adults and juveniles present in sediment cores were in the upper five centimetres of sediment.

The reproductive period of this species in Gull Lake begins in May and is effectively completed by October. The number of young per individual increased from a maximum of 3 in May to 15 in August, 1966. Although the reproductive potential of a given individual is strongly correlated with temperature and may vary with depth, most adults produce only one brood of young per year. The mean reproductive potential of *D. stevensoni* at depths of 3 and 6 metres is about 11 young per individual; that at 9 metres depth is about 13.

Eggs are released and develop within the carapace of the parent. Juveniles produced by first and second reproductive season adults are released during the summer and early autumn and over-winter as juveniles. These young mature and reproduce the following summer. Surviving members of this age class over-winter as adults during their second winter and reproduce again the following summer. Most second reproductive season adults die following release of young and are replaced by over-wintering juveniles entering their first reproductive season. Thus, an almost complete turnover of the adult portion of the population occurs each year at depths of 6 metres or less. The percentage of second year adults that survive and produce young in the third reproductive season is not known.

[1] This paper represents a contribution from the Department of Zoology, and No. 158 from the W. K. Kellogg Biological Station of Michigan State University.
[2] Now at: Department of Zoology, University of Georgia, Athens, Georgia, 30601, U.S.A.

194

Darwinula stevensoni harbours an unidentified parasitic rotifer which is presumed to be strictly ectoparasitic. Parasitism varies with water depth, size and age of individuals, number of young, and geographically. Only 2·1% of 1,963 juveniles were infected whereas 26% of 2,169 adults were parasitized. The greatest incidence of infection occurred at depths of 6 metres. Older individuals within the same year class were more heavily parasitized than the recently moulted adults.

Temperature appears to be the most important environmental factor which regulates the general distribution pattern and density of *D. stevensoni* in Gull Lake. It is hoped that laboratory experiments currently in progress will provide additional information regarding the effects of temperature, nutrition, and parasitism on the reproductive potential, growth and development of this species.

RÉSUMÉ

Cette étude ne représente qu'une partie d'un travail plus général sur la taxonomie et certains aspects choisis de l'écologie de la faune des ostracodes du Lac Gull, Michigan, commencé en 1964.

La vaste distribution, les caractéristiques morphologiques particulières, l'histoire géologique, la reproduction parthénogénétique, la conservation des jeunes pendant le développement embryonnaire jusqu'au premier stade, et la grande fréquence du parasitisme de *Darwinula stevensoni* étaient autant de raisons d'étudier quelques aspects écologiques de l'espèce type du genre *Darwinula*, qui pourront intéresser aussi bien les zoologistes que les paléontologistes.

Dans le Lac Gull *D. stevensoni* atteignait toute l'année un maximum de densité à une profondeur d'environ 6 mètres. La densité de population diminuait nettement entre 6 et 9 mètres. Aucun spécimen n'a été trouvé à des profondeurs supérieures à 12 mètres. Plus de 95% des adultes et des juvéniles présents dans les échantillons de sédiment se trouvaient dans les 5 premiers centimètres.

La période de reproduction de cette espèce dans le Lac Gull commence en Mai et se termine en Octobre. Le nombre de jeunes par individu est passé d'un maximum de 3 en Mai à 15 en Août 1966. Bien que la capacité de reproduction d'un individu soit très nettement en relation avec la température et qu'elle puisse varier avec la profondeur, la plupart des adultes ne produisent qu'une ponte par an. La capacité moyenne de reproduction de *D. stevensoni* est d'environ 11 jeunes par individu à des profondeurs de 3 à 6 mètres; à 9 mètres elle est à peu près de 13.

Les œufs sont pondus à l'intérieur de la carapace du parent et ils s'y développent. Les juvéniles issus des adultes de la première et de la seconde saisons de reproduction sont libérés pendant l'été et le début de l'automne, et passent l'hiver à l'état juvénile. Ces jeunes deviennent mûrs et se reproduisent l'été suivant. Les membres survivants de cette classe d'âge hivernent à nouveau et se reproduisent encore l'été suivant. La plupart des adultes de la deuxième période de reproduction meurent après la libération des jeunes et sont remplacés par les juvéniles qui ont passé l'hiver et qui entrent dans leur première saison de reproduction. Ainsi, un renouvellement presque complet de la partie adulte de la population se produit chaque année à des profondeurs de 6 mètres ou moins. On ne connaît pas le pourcentage des adultes de deuxième année qui survivent et produisent des jeunes à la troisième saison de reproduction.

Darwinula stevensoni héberge un rotifère parasitique non identifié qui est supposé être strictement ectoparasite. Le parasitisme varie avec la profondeur de l'eau, la taille et l'âge des individus, le nombre des jeunes, et géographiquement. Seulement 2,1%

des 1963 juvéniles étaient infestés, tandis que 26% des 2169 adultes étaient parasités. La plus grande fréquence d'infestation se rencontrait à une profondeur de 6 mètres. Les individus plus âgés à l'intérieur d'une même classe annuelle étaient plus fortement parasités que les adultes ayant mué récemment.

La température semble être le facteur externe le plus important qui règle le modèle général de distribution et la densité de *D. stevensoni* dans le Lac Gull. Des expériences au laboratoire, actuellement en cours, et très prometteuses, apporteront des renseignements supplémentaires concernant les effets de la température, de la nutrition et du parasitisme sur la capacité de reproduction, la croissance et le développement de cette espèce.

INTRODUCTION

In recent years, increased attention has been focused on the trophic structure, population dynamics and energetics of many groups of benthic organisms in diverse types of aquatic ecosystems. Yet, when one peruses the literature relating to studies of this nature on the Ostracoda, the paucity of information is indeed striking. The small size of most fresh-water ostracods, difficulty in making specific determinations, and the general disregard of this group by benthic ecologists may account for the limited number of ecological studies on the fresh-water Ostracoda.

This study represents a preliminary attempt to establish a basic framework for identifying and formulating more detailed and sophisticated approaches to the study of some of the aforementioned characteristics of ostracod populations. This investigation is limited in scope and is based almost entirely upon field studies. However, future work, integrating laboratory and field observation and experimentation, on *Darwinula stevensoni* and other species of fresh-water ostracods is currently being planned. More quantitative and experimental approaches are not only desirable, but are essential in elucidating the importance of benthic ostracods in the population, community, and trophic interactions of ecological significance in aquatic ecosystems.

This study represents one portion of a more general investigation of the taxonomy and certain aspects of the ecology of the ostracod fauna in Gull Lake, Michigan, initiated during the summer of 1964. Most data presented herein were collected from August 1965, to August 1966.

Previous Work

Darwinula stevensoni and other species of this genus are unique among the fresh-water Ostracoda because they are the only fresh-water representatives which retain the young through embryological development to the first instar. Members of this genus are characterized by weakly calcified, unornamented valves, with muscle scars arranged in a rosette pattern differing from that found in all other extant or fossil Ostracoda. The furca is rudimentary, and the morphologically similar second and third pairs of thoracic appendages are dissimilar to the first pair. Additional morphologic and taxonomic characteristics of the genus *Darwinula* may be found in Swain (1961), Howe (1962), Van Morkhoven (1963), and Hartmann (1965).

Brady (1870) was the first to report on *Darwinula stevensoni* (Brady and Robertson). He figured the carapace and mentioned that the species would be named and described later in collaboration with David Robertson. *Polycheles stevensoni* was described (Brady and Robertson, 1870) later the same year, but was changed (Brady and Robertson, 1872) to *Darwinella stevensoni* due to preoccupation of the former name.

Descriptions based upon dried specimens in the 1872 publication were later emended (Brady and Robertson, 1874). *Darwinella*, however, was also preoccupied and the name *Darwinula* was erected in 1885 (Brady and Robertson). In 1889, Brady and Norman presented another description and designated *Darwinula stevensoni* as the genotype.

Subsequent descriptions of *D. stevensoni* or its synonyms have been given by various workers including Vávra (1909), Müller (1912), Alm (1916), Sars (1922–28), Turner (1895), Furtos (1933), Hoff (1942), Staplin (1963). and Swain and Gilby (1965), Pinto and Sanguinetti (1958) presented a redescription of the genotype of *Darwinula* in order to clarify some of the nomenclatural problems in the literature.

Distribution

Darwinula stevensoni is one of the most widely distributed species of fresh-water ostracods. Some of the distribution records for *D. stevensoni* are summarized below:

United States
Ohio	Furtos (1933)
Illinois	Hoff (1942)
Michigan	Tressler (1947), Moore (1939), McGregor (this paper)
Massachusetts	Furtos (1935)
Georgia	Turner (1895)
Tennessee	Cole (1966)
Kentucky	Cole (1966)
Mississippi	McGregor (this paper)
Virginia	Nichols and Ellison (1967)
Texas	Cole (1966)

Mexico	Furtos (1936)
Nicaragua	Swain and Gilby (1965), Hartmann (1959, from Swain and Gilby, 1965)
Great Britain	Brady (1870), Brady and Robertson (1872), Brady and Norman (1889)
France	Sars (1922–28)
Sweden	Sars (1922–28)
Norway	Sars (1922–28)
Switzerland	Sars (1922–28)
The Netherlands	de Vos (1954)
Italy	Fox (1965, 1966)
Poland	Sywula (1965)
Yugoslavia	Klie (1941), Petkovski (1960, 1961), Stanković (1960)
Turkey	Hartmann (1964)
U.S.S.R.	Bronstein (1947), Hartmann (1964)
Iran	Hartmann (1964)
Sumatra	Klie (1933)
Java	Klie (1933)
Africa	Hartmann (1964), Klie (1935, 1939)

The type locality of *Darwinula stevensoni* is the East Anglian Fen District in England (Brady and Norman, 1889). Brady (1870) first reported *D. stevensoni* from the Ouse and Scheldt rivers of England and Holland, respectively. Additional distribution records for the East Anglian Fen District (Brady and Robertson, 1870, 1872) were summarized by Brady and Norman (1889). Other areas inhabited by *D. stevensoni* include: Lakes Ohrid, Prespa, Dojran (Yugoslavia); Maggiore, Mergozzo, Varese, Comabbio, Monate (Italy); Ijssel (The Netherlands); Van (Norway); Pereslavskoye (U.S.S.R.); Ngebel, Bedali (Java); Toba (Sumatra); Lothing and many others (Great Britain); Nicaragua, Managua (Nicaragua); Erie, Douglas, Gull (United States). Several river habitats include, the Nene, Cam, Ouse, Deben (Great Britain), Warta (Poland) and Rappahannock (United States).

The seasonal and spatial distribution of *D. stevensoni* has received limited study. Alm (1916, p. 225) stated that this species should be found throughout the entire year on the basis of its 'viviparous' reproduction.

Pinto and Purper (1965) illustrated the seasonal occurrence of *Darwinula* in a recent study, but did not specify the species. Moore (1939) collected *D. stevensoni* at depths of from 4–9 metres in Douglas Lake, Michigan. Klie (1933) reported specimens from depths of approximately 4 metres and Cole (1966) found them at depths of about 0·3 metre. I have collected adults and late instar juveniles at depths ranging from a few millimetres, near Saline, Michigan, to 12 metres in Gull Lake.

Life histories of several species of fresh-water ostracods have been studied by Ferguson (1944, 1958) and Hoff (1943). Kenk (1949), Winkler (1960), Ferguson (1957), and Cole (1953) reviewed the distribution of numerous living and fossil Ostracoda of Michigan, but made no mention of *Darwinula stevensoni*.

The geologic range of the genus *Darwinula* extends from the Upper Carboniferous (Pennsylvanian) to Recent according to Swain (1961) and Van Morkhoven (1963). Swain (1961) also reported a questionable record from the Ordovician Period. Several authors, notably, Brady *et al.*, (1874), Wagner (1957), and Staplin (1963), have recorded *D. stevensoni* from Quaternary deposits.

Other extant species of *Darwinula* have been reported from România (Danielopol and Vespremeanu, 1964), Brazil (Pinto and Kotzian, 1961), Africa (Klie, 1935, 1939; Menzel, 1916), Java (Menzel, 1923; Klie, 1933), Patagonia (Daday, 1902), Uruguay (Klie, 1935b), Holland (Delachaux, 1928), and Rennell Island, Pacific Ocean (Harding, 1962). Sohn (1965), Pinto and Purper (1965), and Sandberg (1965) reported *Darwinula*, respectively, from Lake Tiberias, Israel, Brazil, and Mexico, but did not mention the species encountered. Fossil species have been described, for example, from the Miocene (Méhes, 1908), and Jurassic (Bate, 1967).

Reproduction and Development

Darwinula stevensoni is presumably a parthenogenetic species. Brady and Robertson (1870) briefly described and figured the copulatory apparatus of a male specimen, and later restated the description (1874). Brady and Norman (1889) incorporated the above description and figure in their monograph on the marine and fresh-water Ostracoda of the North Atlantic and North-western Europe. Male characteristics other than the copulatory apparatus were not described in these papers nor have any subsequent investigators reported finding male *D. stevensoni*.

I have isolated and reared numerous late instar individuals through at least one reproductive cycle and on all occasions only females were produced. In addition, no males have been encountered in more than 10,000 specimens collected from Gull Lake and other localities in the present study. The validity of the record by Brady and Robertson (1870) must be considered highly suspect until adequate evidence is provided to prove the existence of males in this species.

Kesling (1951, 1953, 1961a,b), Van Morkhoven (1962) and Sandberg (1964) have reviewed some of the early studies on embryological development, egg laying, and instar growth and development of fresh-water and marine ostracods. Sandberg (1964) commented on the advantage of brood care in passive dispersal and establishment of new populations by parthenogenetic species. Pokorný (1952) reviewed several examples of possible parthenogenetic reproduction in fossil ostracods.

The number of eggs or nauplii produced by different ostracod species may vary from two to sixty, with some young maturing before the last eggs or larvae are released (Van Morkhoven, 1962). Kesling (1951) reported a maximum of 53 young produced by one individual of the parthenogenetic ostracod *Cypridopsis vidua* (O. F. Müller). The eggs were released over a 35-day period in laboratory culture. Kesling (1951) further noted that development from the egg stage to sexually mature individuals lasted approximately one month at room temperature. Sandberg (1964) found a maximum of 11 eggs and nauplii, with a range between 6 and 11, in the 'brood chamber' of *Cyprideis castus* Benson. Elofson (1941) and Weygoldt (1960) reported a maximum of 42 and 30 eggs, plus nauplii, respectively, in *Cyprideis torosa* (Jones).

Kesling (1953) noted some of the causes for variation in form within an instar, such as diet, effects of temperature, sexual dimorphism, parasitism and individual variation. He also stated (1953, p. 101) that:

> The number of instars in the ontogeny of an ostracod appears to be constant for a genus. Insofar as is known, it is constant for certain families, but relatively few species have been studied and there may be exceptions.

Sandberg (1964) reviewed several systems of instar designation used by different authors and some of the problems encountered in their use when working with living and fossil species. In the present discussion the first instar is number one and later instars are numbered consecutively.

The ontogenetic development of *Darwinula stevensoni* has been discussed by Scheerer-Ostermeyer (1940). Some of the apparent differences in the ontogenetic development of *D. stevensoni* from that reported in other fresh-water ostracods may be due, as Kesling noted (1951, p. 94), to different interpretations of appendages. Because

TABLE 1. Ontogenetic development and relative size range of instars of *Darwinula stevensoni*. Modified from Scheerer–Ostermeyer (1940, p. 366) in the style of Kesling (1951, pp. 94-95)

Instar		Length (microns)	Height (microns)
1 Al An Md Mx — — —	(Fc)		
2 Al An Md Mx (L1) — —	(Fc)	160–175	90–100
3 Al An Md Mx (L1) — —	(Fc)	223(R)–217(L)	126(R)
4 Al An Md Mx (L1) — —	(Fc)	250(R)	133(R)
5 Al An Md Mx L1 (L2) —	(Fc)	290(R)–270(L)	130(R)
6 Al An Md Mx L1 L2 (L3)	(Fc)	350(R)–320(L)	168(R)
7 Al An Md Mx L1 L2 L3	(Fc)	440(R)–431(L)	245(R)
8 Al An Md Mx L1 L2 L3 (Rp)	(Fc)	518(R)–508(L)	250(R)
9 Al An Md Mx L1 L2 L3 Rp	—	700	

Al = Antennule
An = Antenna
Md = Mandible
( ) = Anlagen, incomplete
(R) = Right valve
(L) = Left valve

Mx = Maxilla
L1 = First thoracic leg
L2 = Second thoracic leg
L3 = Third thoracic leg
Fc = Furca
Rp = Reproductive organs

O

the principle concern here is the total number and relative size of the instar stages in this species, the sequence of ontogenetic development and relative size described by Scheerer-Ostermeyer (1940, p. 366) is presented in Table 1. I have followed the style of Kesling (1951, pp. 94–95) in portraying the developmental process of *D. stevensoni* described by Scheerer-Ostermeyer (1940). Some modification was necessary owing to a different interpretation of appendages.

Parasitism of ostracods by several groups of invertebrates has been reviewed recently by Neale (1965). Ostracods have been reported to be parasitized by acantho-cephalans (Hoff, 1942; Ward, 1940; Baer, 1952, p. 115), isopods, (Sars 1899, Stephensen 1938), cestodes (Scott, 1891; Rome, 1947, in Kesling, 1953), and tremadotes (Macy and Demott, 1957). Commensalism or suspected, but unproven, parasitism, by ostracods has also been reported by de Vos and Stock (1956), Elofson (1941), and Harding (1966).

STUDY AREA

Gull Lake (Fig. 1) is located in the south-western corner of the Lower Peninsula of Michigan, in Kalamazoo and Barry counties, between latitudes 42°20′–42°30′ and longitudes 85°20′–85°30′. The lake basin, of glacial ice block origin, was formed during retreat of the Wisconsin glacier which had covered Barry and Kalamazoo counties (Martin, 1957).

The lake drainage basin is small and most water comes from springs and a few small streams. Gull Lake drains into the Kalamazoo River which, in turn, empties into Lake Michigan. The lake covers an area of 821·5 hectares and has a maximum depth of 33·5 metres (Taube and Bacon, 1952). Major physiographic features, morainic high-lands and dissected glacial outwash plains, in the area surrounding Gull Lake reflect the impact of the recession of the Wisconsin glacier (Deutsch *et al.*, 1960). The regional soils are the well-drained, Grey-Brown Podzolic or Brunizem Great Soil Groups described by Whiteside *et al.* (1963).

Preliminary studies on nutrient cycling and primary productivity levels of Gull Lake have been initiated only recently and no detailed chemical or physical measurements have been published. Yet, the lake may be classified generally as oligotrophic, on the basis of its physiographic and chemical-physical properties, when compared with other lakes nearby.

Gull Lake is well oxygenated at all depths throughout the year except in the deepest portions of the hypolymnion during the latter phase of summer and autum stratifica-tion. Alkalinity, measured as mg/l $CaCO_3$, generally ranges between 140–160, and the pH, between 7·5–8·5. The lake stratifies thermally by late May or early June and begins the autumn turnover sometime in November. By mid-January the lake usually freezes over and remains ice-covered through most of March and, occasionally, early April. Secchi disc readings reach a maximum of about 10 metres and calcium carbonate deposits cover most of the littoral areas. Steep slopes on all sides of the lake reflect its glacial origin and a majority of the surface area is more than nine metres in depth.

METHODS

An Ekman dredge, free-fall corer with removable plastic inner-liner, modified Glemacher frozen core sampler (Gleason and Ohlmacher, 1965), and a modified aquatic sweep net were used for sample collection. Eighty-seven Ekman, 15 core, and 22 net samples from Gull Lake, and 7 net and hand samples from other areas were

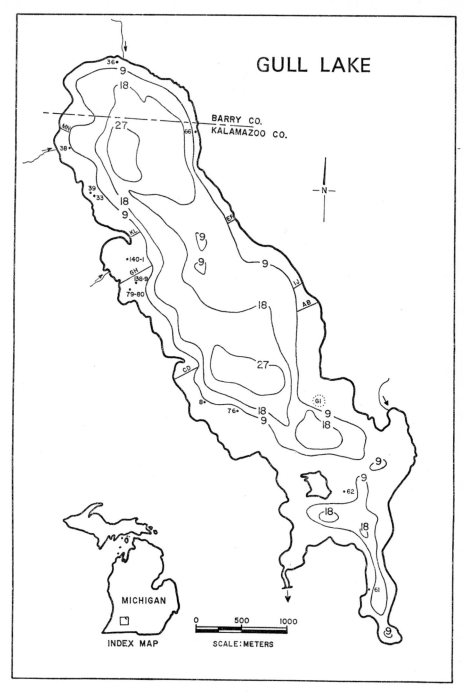

Fig. 1. Map of Gull Lake showing transect and individual sample localities, and general lake morphometry. Contour interval is nine metres.

201

examined. The modified aquatic net was used in areas suitable for wading or for manipulation from a boat. Ekman samples generally were collected from depths of three metres or greater. Transect and individual sample localities are shown in Fig. 1.

All Ekman and net samples were sieved in the field using the modified aquatic net. This net accommodated a second, finer mesh, detachable nylon net for retention of ostracods. Samples were examined in the laboratory immediately following collecting or stored temporarily at 4°C. No preservatives were added as separation and examination of living animals proved to be more satisfactory.

Samples were hand sorted and all non-swimming ostracods visible to the unaided eye were separated from the sediment by employing the water surface tension technique. The sediment is drained and exposed to air; the non-wettable ostracod valves are held by the water surface tension when the sediment is again immersed. On several occasions, a sample was first hand sorted and then sorted again using other techniques, including treatment with weak acids and bases, differential heating, saturated sugar and salt solutions, wet sieving through graded sieve series, behavioural responses to light quality and quantity, elutriation, and microscopic examination of sediment. In no instance were specimens of *Darwinula stevensoni* found in excess of three per cent of the number previously removed by hand sorting.

Most of the above sorting techniques induced the adults to release their young or eggs and were, therefore, unsatisfactory. Adult and juvenile *D. stevensoni* were sorted further, counted, separated into size categories, anaesthetized with Tricaine Methanesulfonate (MS 222) and dissected under a dissecting microscope. The number of young within the carapace, parasites, and several other observations were recorded for each individual. Most specimens were preserved for future study.

Sediment cores obtained using the free-fall corer were removed from a plastic innerliner, 4·3 cm in diameter. This corer was allowed to free-fall from approximately 3 metres above the sediment-water interface. The liner was marked for separation of the top 5 cm of sediment from the underlying 10 cm sediment interval. Retrieval of cores was dependent upon a clay plug to hold the sediment in the liner. After removing the liner, water was siphoned off to a point about 4 cm above the substrate-water interface and the core was allowed to slide slowly out the coring end. Sediment below the 15 cm interval was discarded and that between the 5 cm and 15 cm interval was collected in a labelled container. The upper 5 cm of sediment, and the water above, were collected in a second container. Both samples were immediately frozen in dry ice and were later thawed and sieved in the laboratory.

The modified Glemacher frozen core sampler employs a mixture of dry ice and acetone in the freezing compartment for *in situ* freezing of the sediment. The dry ice and acetone, separated prior to penetration of the sediment, are released by a lever and spring tripping mechanism when a messenger is lowered. Complete freezing of the 3 cm diameter sediment core occurs within three minutes following messenger release (McGregor, unpublished). The frozen cores, ranging in length from 18–23 cm, were sectioned at 5 cm intervals and sieved in the laboratory.

RESULTS

A total of 6,274 adult and juvenile *Darwinula stevensoni* from Gull Lake and nearby localities were studied. The number of specimens examined for parasites or young is summarized on the next page.

Although a larger number of replicated samples would have been desirable, the time involved in collecting samples, sieving, ostracod separation, and dissection precluded

	Total Examined	Number Adults	Number Juveniles	Reproductive Potential	Parasites Adult	Juvenile
Gull Lake	5993	2906	3087	2590	2169	1879
Other Areas	281	68	213	53	53	84

analyses of a larger sample series at this time. Approximately 50,000 ostracods representing genera in Gull Lake other than *Darwinula* are also being investigated.

Twenty sediment cores of more than 20 cm in length were collected using the free-fall corer at depths of 3, 6, 9, 12, and 15 metres along Transect AB on 5 June, 1966. Three cores at 3 metres, four cores at 6 metres and one at 9 metres contained *D. stevensoni*. Seven sediment cores were collected *in situ*, employing the modified Glemacher frozen core sampler, at a depth of 5 metres along Transect MN on 15 October, 1965.

Only one adult of 12 juvenile and 10 adult *D. stevensoni* present in cores obtained using the free-fall corer was found at depths below 5 cm. This specimen, however, was in one of two cores which previously had been labelled for possible contamination of the 5–15 cm sediment interval. Thirteen adults and twenty-seven juveniles were present in the frozen core samples, but only one juvenile was found below the 5 cm interval. These data indicate that this species normally burrows in sediment at depths less than 5 cm and may be sampled adequately with the Ekman dredge or other bottom dredges suitable for use in loosely consolidated sediments.

The secondary nylon monofilament screen cloth used with the modified aquatic net had a mesh opening of 202 microns. Consequently, only those instars with length and height dimensions greater than 202 microns were retained. Good agreement with expected instar development was observed in all juvenile samples. Generally, only the seventh, eighth and adult instars were present; earlier instar individuals always comprised less than one per cent of the sample. Measurements of a limited number of eighth and ninth instar animals were somewhat greater than those reported by Scheerer-Ostermeyer (1940), but were in general agreement. Further discussion of juvenile-adult percentages will be in reference to the seventh and eighth juvenile instars and the ninth or adult instar.

Darwinula stevensoni was abundant in the littoral areas of Gull Lake and reached a maximum density at a depth of about 6 metres throughout the year. The population density decreased markedly between 6 and 9 metres and only a few individuals were found at depths greater than 9 metres. Although an unequal number of samples was collected from different depths the majority of specimens were from depths of 3 and 6 metres.

Variation in number of individuals sampled at a given depth and date, or at different times during one month, probably represented a reasonable degree of sample bias. Much of the variability, however, reflected the geographic variation in population density and the number of late instar juveniles that had reached the size necessary for retention by the collecting process discussed above.

The percentages of adult and juvenile ostracods in replicated and non-replicated samples from Gull Lake at depths of <3, 3, 6, 9, and 12 metres are presented in Tables 2 and 3. At <3, 3, and 6 metres the percentage of adults was maximal during the summer months and was followed by a sharp decline in the early autumn. The percentage of adults continued to decrease through the autumn and winter seasons and generally reached a minimum by early or late spring depending upon the depth from which samples were collected. More than 90 per cent of the total number of ostracods col-

lected from all depths, in August of 1965–66, were adults. The adult component of
most samples collected at depths of <3, 3, and 6 metres during late March, 1966
ranged between about 25–30 per cent. By late May and early June the percentage of
adults in the population again began to increase (Tables 2 and 3).

TABLE 2. Percentage of Adult and Juvenile *Darwinula stevensoni* from samples
collected at depths of <3 and 3 metres in Gull Lake

Date	Sample No.	Transect	N	% Adult	% Juvenile
		<3 metres			
1965					
7–9	33		42	100·0	0·0
7–19	36		4	100·0	0·0
8–2	39		8	100·0	0·0
8–2	40	KL	49	98·0	2·0
8–3	44	MN	2	50·0	50·0
8–3	45	KL	43	95·4	4·6
9–14	61		13	100·0	0·0
9–14	62		8	37·5	62·5
9–14	60, 64	GI	99	77·8	22·2
9–22	65	GI	86	62·8	37·2
9–27	66		22	54·5	45·5
10–4	70, 71	GI	164	57·3	42·7
10–13	72, 73	GI	64	60·9	39·1
10–18	78	GI	105	60·0	40·0
10–18	79, 80		98	55·1	44·9
11–1	81	GI	75	49·3	50·7
12–2	102	GI	46	58·7	41·3
1966					
1–4	109	GI	18	50·0	50·0
1–11	111	GI	24	58·3	41·7
2–11	117	AB	1	00·0	100·0
2–21	124	AB	6	33·3	66·7
2–24	129	AB	1	00·0	100·0
3–21	134, 135	AB	3	33·3	66·7
3–26	138, 139		67	25·4	74·6
3–26	140, 141		36	30·6	69·4
5–31	172	GI	59	74·6	25·4
6–15	186	GI	9	66·7	33·3
7–2	207	AB	35	97·1	2·9
		3 metres			
1965					
8–3	41	MN	61	100·0	0·0
8–3	46	KL	38	97·4	2·6
8–5	50, 53	AB	17	100·0	0·0
11–4	82	AB	137	35·0	65·0
11–4	92	EF	184	47·3	52·7
11–4	97	GH	365	31·8	68·2
11–4	87	CD	167	62·3	37·7
1966					
2–24	131, 132	AB	241	23·2	76·8
3–30	142	AB	159	24·5	75·5
4–29	156	AB	50	18·0	82·0
5–19	164	AB	36	27·8	72·2
6–1	174, 180	AB	89	29·2	70·8
6–15	187, 188	AB	121	24·0	76·0
7–1	197, 198	AB	2	100·0	0·0
7–15	212, 213	AB	152	88·8	11·2
8–1	223, 224	AB	29	89·7	10·3

The seasonal change in the adult-juvenile ratio may be expressed more clearly, for example, using data from samples collected at depths of <3 metres. The combined numbers of adult *D. stevensoni* collected each month were divided by the total monthly sample to obtain the following adult percentages. In August of 1965, adults comprised 96 per cent of 102 collected; in September, 70 per cent ($N = 228$); October, 61 per cent ($N = 262$); 1966 January, 55 per cent ($N = 42$); March, 27 per cent ($N = 106$); May, 75 per cent ($N = 59$), and July 97 per cent ($N = 135$).

TABLE 3. Percentage of Adult and Juvenile *Darwinula stevensoni* from samples collected at depths of 6, 9 and 12 metres in Gull Lake

Date	Sample No.	Transect	N	% Adult	% Juvenile
		6 metres			
1965					
*7–27	8		1	100·0	0·0
8–2	38		33	100·0	0·0
8–3	47	KL	35	97·1	2·9
8–5	51–54	AB	53	87·5	12·5
10–13	76		28	46·4	53·6
11–4	83	AB	123	55·3	44·7
11–4	93	EF	63	49·2	50·8
11–4	98	GH	98	52·0	48·0
12–8	105	EF	511	12·9	87·1
1966					
2–21	125	IJ	82	24·4	75·6
3–30	148, 149	IJ	248	28·2	71·8
4–29	157	AB	54	38·9	61·1
5–19	165	AB	42	56·8	43·2
6–1	175, 181	AB	79	74·7	25·3
6–15	189, 190	AB	189	47·1	52·9
7–1	109, 200	AB	338	47·6	52·4
7–15	214, 215	AB	177	76·8	23·2
8–1	225, 226	AB	185	76·8	23·2
		9 metres			
1965					
8–3	48	KL	61	96·7	3·3
8–5	55, 56	38AB	7	100·0	0·0
11–4	84	AB	6	100·0	0·0
11–4	94	EF	24	87·5	12·5
11–4	89	CD	7	85·7	14·3
12–8	106	EF	93	24·7	75·3
1966					
2–5	114	IJ	4	75·0	25·0
3–30	144	AB	1	100·0	0·0
5–19	166	AB	7	100·0	0·0
6–1	182	AB	9	100·0	0·0
6–15	192	AB	3	66·7	33·3
7–15	216, 217	AB	10	100·0	0·0
8–1	227, 228	AB	14	100·0	0·0
		12 metres			
1965					
11–4	100	GH	1	100·0	0·0
12–8	107	EF	1	100·0	0·0
2–21	126	IJ	1	0·0	100·0
* 1964					

Many of the over-wintering juveniles had moulted to the ninth or adult instar by mid-May, 1966, and contributed to the high percentage of adults reported at depths of <3 metres. Development and moulting proceeded less rapidly at greater depths owing, in part, to the lower temperatures. Consequently, the adult component at 3 metres (Table 2) was low until the eighth instar juveniles completed the final moult in early June. At 6 metres the percentage of adults remained somewhat higher during the winter months as a smaller fraction of the over-wintering juveniles had attained the size of those juveniles at shallower depths. The high adult percentages at 6 metres during May and early June (Table 3) probably represented an over-estimation in light of the reproductive potential data discussed later.

The small number of specimens collected from depths of 9 and 12 metres made comparison of adult-juvenile percentages difficult (Table 3). This portion of the total population also is considered later in the discussion of reproductive potential.

Estimates of ostracod density per square metre presented in Table 4 are useful in predicting the contribution of young by adults and assist in understanding the life cycle of *Darwinula stevensoni* in Gull Lake. Only data from replicated samples collected along Transect AB, during the summer of 1966, are considered (Refer to Tables 2 and 3 and Fig. 1).

The percentages of adults and juveniles collected on 15 June–1 July and 15 July–1 August at 6 metres depth, were almost identical (Table 3). Yet, when one considers the density estimates on these dates, changes in the size structure of the population become more evident. The number of adults and juveniles per square metre was almost double, on 1 July, the corresponding estimates for 15 June (Table 4).

TABLE 4. Estimates of the mean number of adult and juvenile *Darwinula stevensoni* per square metre at depths of 3, 6 and 9 metres along Transect AB during the period from 1 June–1 August 1966 (A = adult; J = juvenile)

1 June		15 June		1 July		15 July		1 August	
A	J	A	J	A	J	A	J	A	J
				3 metres					
560	1356	624	1756	43	—	2906	366	560	65
				6 metres					
1270	430	1916	2152	3466	3811	2928	883	3057	926
				9 metres					
194	—	43	22	—	—	215	—	301	—

The small number of replicated samples may account for some of the variation in density. I believe the major difference in density is due, however, to the number of newly moulted over-wintering juveniles that were contributing to the adult and late instar classes by 1 July. This explanation also is tenable for estimates on 1 July and 1 August at 6 metres. The abrupt decrease in juvenile density is believed to result from the rapid contribution of juveniles to the adult class by mid-July and early August. The slight decrease in adult density from 1 July to 15 July may reflect a replacement of second reproductive season adults by first reproductive season adults during the latter part of the summer.

Study of the reproductive cycle, discussed below, indicated that at least two adult age groups were represented in the summer collections. The adult age structure exhibited marked changes in 15 July and 1 August samples and was comprised primarily of first reproductive season adults. The juvenile density, on 1 June, at 6 metres, was much lower than expected and is believed to reflect considerable sampling bias.

The variable nature of the substrate and the small number of ostracods at depths of 3 and 9 metres, respectively, renders density estimates less reliable in terms of absolute numbers. Yet, high juvenile densities in June and an abrupt reversal of adult-juvenile densities later in the summer correspond to the general trend noted at 6 metres. The decrease in density at 9 metres compared with that at 6 metres is obvious.

The reproductive potential and cycles of fresh-water ostracods have received limited study. Observations of the number of eggs or young in one or a few specimens of *D. stevensoni* are scattered throughout the literature. However, I know of no study on the reproductive potential of *D. stevensoni* over a single reproductive cycle.

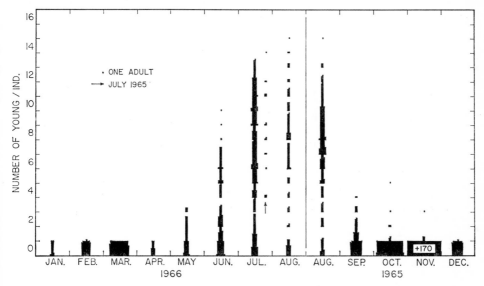

Fig. 2. Annual reproductive cycle and reproductive potential of *Darwinula stevensoni* in Gull Lake. The total number of specimens examined each month, from left to right, is: Jan. (23), Feb. (79), Mar. (175), Apr. (30), May (83), Jun. (217), Jul. (475 and 20), Aug. (181), Aug. (324), Sep. (152), Oct. (252), Nov. (477), Dec. (99).

Some difficulty was encountered in studying the reproductive potential of *Darwinula stevensoni* from field samples. When the population reproductive cycle is initiated, successive addition and development of eggs within the carapace and ultimate release of first instar young proceeds at varying rates. The difficulty arises in distinguishing adults exhibiting the maximum production of young from those which are adding or losing young.

The thin, translucent carapace of *D. stevensoni* permits direct observation of the developing young during most of the reproductive season. However, eggs and young become so closely packed prior to release of the first individuals that accurate counts cannot be made without dissection of each animal. In addition, parasites attached to the hypodermis in the area in which the young are carried are easily confused with the developing eggs. Consequently, each adult ostracod was anaesthetized and dissected alive, in order to obtain accurate counts of the number of eggs or young and parasites.

The annual reproductive cycle of *D. stevensoni* and number of eggs and young in the carapace of each dissected female are presented in Fig. 2. These data represent combined monthly counts for 2587 individuals collected from Gull Lake. One specimen from 1964, two animals which died before young were counted, and several young released during observation were not included in Fig. 2.

In Gull Lake the reproductive period of D. stevensoni begins in May and is effectively completed by October. Only 8 adults examined in late fall and early winter were found with young or eggs. The number of developing young per individual increased from a maximum of 3 in May to 15 in August. The sharp decrease in the number of young per individual from August to September reflects widely spaced sampling periods. Had samples been taken more frequently the shape of a curve drawn for Fig. 2 would probably approximate a normal curve.

Information derived from study of the reproductive cycle and reproductive potential assists in understanding the juvenile-adult percentages and density data presented earlier. On the other hand, the generalized reproductive potential and reproductive cycle presented in Fig. 2 do not reveal certain variables which are important in estimating future population densities.

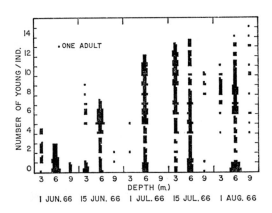

Fig. 3. Differential rates of egg deposition and development of young along Transect AB depth gradient during the summer of 1966. Number of animals represented by histograms at each depth, on consecutive dates, may be found in Tables 2 and 3.

The differential rates at which eggs were released into the carapace of adults at depths of 3, 6 and 9 metres during the summer of 1966, are shown in Fig. 3. These histograms show the retarded development and addition of young at increasing depths. At a depth of 3 metres the number of eggs and young per individual progressed from a maximum of 4, on 1 June, to 13 on 15 July. By 15 July, a large number of the 3 and 6 metre adults had begun to release their young whereas those adults collected at depths of 9 metres had no first instar young in the carapace. Although samples from depths of 3 and 9 metres were small or absent on certain dates, the maximum number of young increased by 5 individuals about every 15 days at depths of 3 and 6 metres. The corresponding increase at 9 metres depth occurred over a period of about one month.

The presence of both adults and juveniles through the winter and spring of 1966 suggested the possibility of determining the contribution of young by at least two age classes of adults during the following reproductive season.

Over-wintering adults from the summer of 1965 were observed to begin yolk deposition as early as November 1965. These individuals were suspected to begin releasing eggs into the carapace during the following reproductive period prior to the moulting and production of young by over-wintering juveniles. Thus, one would have a means of separating first and second year adult age classes early in the reproductive season. Once the juveniles matured and began producing young, recognition of this age class would be dependent on some other criteria.

The relative maximal developmental state of the young carried by each adult was perhaps the most useful single criterion for distinguishing first and second reproductive season adults. The number of parasites within the carapace proved to be an additional aid in recognizing first and second reproductive season individuals, at least during the early part of the summer. Recently moulted adults generally had few parasites, if any, whereas many of the older ostracods were found with three or more parasites. Another criterion which aided was the pigmentation of the hepatopancreas. Although pigmentation varied with depth and was measured only qualitatively, recently moulted individuals usually exhibited much less pigment than older animals. Detailed records 'were kept on the qualitative changes in pigmentation of both adults and juveniles but are not considered in the present study.

Adults examined for parasites and young from 1 June through 1 August 1966 were grouped and placed into three major catagories. Group 1 consisted of adults with at least one young, with movable appendages, in the first instar stage. Group 2 included adults with developing embryos having no moving appendages, and Group 3 was comprised of adults with no eggs or young free within the carapace.

The number of first and second reproductive season adults comprising Groups 1, 2, and 3 are presented in Table 5. On 1 June 1966, all adults in Groups 2 and 3 at depths of 3 and 6 metres were considered second reproductive season adults, based upon yolk deposition, developmental state of young, degree of parasitism and pigmentation of the hepatopancreas. Ostracods in Group 3 at 9 metres depth were also second reproductive season, or possibly older, adults. By 15 June more than half of the individuals from 3 metre samples consisted of recently moulted ostracods; greater than 90 per cent on 15 July and 100 per cent of the animals on 1 August were considered first reproductive season adults.

The change in percentage dominance of first and second reproductive season adults from 6 metre samples, occurred approximately 15 days later than the change at 3 metres depth. Less than 10 per cent of the ostracods collected on 15 July and 1 August were second reproductive season adults.

These data indicated an almost complete turnover of the adult portion of the population, at depths of 6 metres or less, each year. Juveniles produced by first and second reproductive season adults are released during the summer and early fall and over-winter as juveniles. The following summer these young mature and reproduce. Surviving members of this age class over-winter as adults during their second winter and reproduce for a second time the following summer. The data presented in Table 5 indicate that most second reproductive season adults die following release of young and are replaced by maturing offspring produced during their first reproductive season. Juveniles produced by the second reproductive season adults of the previous year also contribute to the replacement of the two-year-old adults. Most adults of both age classes produce only one brood of young per year.

Over-wintering juveniles mature during mid-July and early August at depths of 9 metres. Over-wintering adults did not begin releasing their young until early August. This study was terminated in August of 1966; consequently, no data are available concerning the potential contribution of young by adults in their first reproductive season. The development and release of young by second reproductive season adults suggests that the contribution of young by late maturing first reproductive season adults may be limited or entirely lacking. Several of the early maturing adults had from 4–6 eggs within the carapace by 1 August. Second reproductive season adults at 9 metres appear to have a slightly higher reproductive potential than animals at

TABLE 5. Age structure and reproductive potential of adult *Darwinula stevensoni* in Groups 1, 2 and 3. Samples collected along Transect AB at depths of 3, 6 and 9 metres during the summer of 1966. Group 1 includes only those adults having at least one first instar young with movable appendages; Group 2 includes adults with young which have no movable appendages; Group 3 includes adults with no young or eggs free in the carapace. Percentages with totals more or less than 100 per cent result from rounding to nearest tenth

Date	Group	N	Per-centage	$\overline{X}$ No. young/ adult	Range No. young	First Reprod. season Adults	Second Reprod. season Adults
				3 metres			
	1	0	0·0	0·0	—	0	0
1 June	2	24	96·0	2·3	1–4	0	24
	3	1	4·0	0·0	—	0	1
	1	0	0·0	0·0	—	0	0
15 June	2	14	48·3	6·3	1–10	1	13
	3	15	51·7	0·0	—	15	0
	1	0	0·0	0·0	—	0	0
1 July	2	2	100·0	3·5	2-5	0	2
	3	0	0·0	0·0	—	0	0
	1	92	68·7	8·0	3–13	84	8
15 July	2	36	26·9	4·4	1–7	32	4
	3	6	4·5	0·0	—	6	0
	1	25	96·2	8·2	4–11	25	0
1 Aug.	2	1	3·9	7·0	7	1	0
	3	0	0·0	0·0	—	0	0
				6 metres			
	1	0	0·0	0·0	—	0	0
1 June	2	35	59·3	1·4	1–2	0	35
	3	24	40·7	0·0	—	0	24
	1	0	0·0	0·0	—	0	0
15 June	2	76	87·4	4·4	1–7	0	76
	3	11	12·6	0·0	—	6	5
	1	73	45·3	9·0	4–12	0	73
1 July	2	65	40·4	4·4	1–9	64	1
	3	23	14·3	0·0	—	23	0
	1	68	50·0	8·4	1–13	65	3
15 July	2	49	36·0	3·1	1–7	48	1
	3	19	14·0	0·0	—	16	3
	1	89	63·1	7·9	2–14	82	7
1 Aug.	2	18	12·8	4·6	1–8	17	1
	3	34	24·1	0·0	—	33	1
				9 metres			
	1	0	0·0	0·0	—	0	0
1 June	2	0	0·0	0·0	—	0	0
	3	9	100·0	0·0	—	0	9
	1	0	0·0	0·0	—	0	0
15 June	2	2	100·0	1·5	1–2	0	2
	3	0	0·0	0·0	—	0	0
	1	0	0·0	0·0	—	0	0
15 July	2	9	90·0	8·2	1–10	1	8
	3	1	10·0	0·0	—	1	0
	1	10	71·4	11·7	8–15	0	10
1 Aug.	2	4	28·6	5·3	4-6	4	0
	3	0	0·0	0·0	—	0	0

shallower depths. At 9 metres, egg development is much slower than that of animals living under higher temperature regimes at shallower depths. The generally smaller and more uniform size of eggs and the retarded development of first instar young apparently provides space for a greater number of young than is possible for adults at lesser depths.

The mean number of young per adult shown for Group 1 on successive dates (Table 5) represents an underestimation of the reproductive potential of ostracods from all three depths. The means of 8·0 and 8·2 young per individual on 15 July and 1 August at 3 metres depth represent, at best, only minimal estimates. The mean number of young per individual would have been higher had sampling intervals been more closely spaced.

Six metre samples on 1 July show a mean maximum of 9·0 young per individual. When adults with three or more parasites and those which had obviously begun releasing young are excluded, the reproductive potential is 10·1 young per individual ($N = 45$). At 9 metres the mean maximum of 11·7 young per individual for second reproductive season adults represents the highest and possibly most accurate reproductive potential estimate.

Ostracods that had not produced the maximum number of young, those with maximal production, and those which had released some portion of the brood were included in calculating the mean number of young per individual from the Group 1 category. A more realistic estimate of the mean reproductive potential of *D. stevensoni* would be about 11 young per individual at 3 and 6 metres, and 13 young per individual at 9 metres. The latter estimates are based upon observations of the probable maximal development of young in several hundred adults from depths of 3, 6 and 9 metres.

TABLE 6. Mean number of first and second reproductive season adults per square metre calculated from data presented in Table 5. Density estimates for second reproductive season adults are given on second line; first line gives estimates for first reproductive season adults

1 June	15 June	1 July	15 July	1 August
		3 metres		
0	344	0	2497	560
560*	280	43	258	0
		6 metres		
0	129	1873	2777	2842
1270	1744	1593	151	194
		9 metres		
0	0	0	43	86
194	43	0	172	215

* One adult added to total adults examined ($N = 26$).

A second density estimate is made possible by differentiating adult age classes. The estimated number of first and second reproductive season adults, per square metre along Transect AB, during the summer of 1966, is presented in Table 6. Density estimates presented in Table 4 were calculated from the mean number of adults and juveniles in replicated Ekman dredge samples. First and second reproductive season adults were not considered separately in the latter estimates.

The estimated number of second reproductive season adults per square metre at

6 metres depth was 1270, 1744, 1595, 151, and 194, respectively, on the dates shown in Table 6. These estimates differ somewhat from those presented in Table 4, because the number of ostracods dissected was often less than the total collected. Yet the point here is to show the decreased variability in second reproductive season adult estimates for June and early July compared with total adult estimates in Table 4, and the rapid decline of second reproductive season adults following release of young. The first estimate (Table 4) gives a clearer representation of the size structure of the population; however, the second estimate assists in understanding the turnover in age structure of the adult segment of the total *D. stevensoni* population.

Darwinula stevensoni harbours an unidentified parasitic rotifer on which work is currently in progress. The parasite is almost always found in the posterior region of the carapace where the ostracod eggs are carried, and is presumed to be strictly ectoparasitic.

More than 4,000 adult and juvenile ostracods were dissected and examined for parasites. Only 41 or 2·1 per cent of 1,963 juveniles were infected. The infected animals were generally in the penultimate instar and the greatest incidence of infection occurred in a relatively shallow, sheltered bay area on the west side of Gull Lake (sample localities 138, 140–1).

Parasitism varies with water depth, size and age of individuals, number of young, and geographically. Variation in the percentage of parasitized adults collected along four transects in Gull Lake on the same date is shown in Table 7. Here, the number of parasitized individuals varies both geographically and with depth.

TABLE 7. Geographic and depth variation in the number of parasitized adult *Darwinula stevensoni* collected along four transects on 4 November 1965

Transect	No. Adults	% Parasitized	$\bar{X}$ Parasites/ Infected Ind.	Range No. Parasites
		3 metres		
AB	47	31·9	1·7	0–3
EF	87	23·0	1·7	0–5
GH	99	6·1	1·7	0–3
CD	103	25·2	1·8	0–5
		6 metres		
AB	21	42·9	2·4	0–4
EF	9	0·0	0·0	—
GH	51	0·0	0·0	—
CD	—	—	—	—
		9 metres		
		No Parasites ($N = 24$)		
		12 metres		
		No Parasites ($N = 1$)		

Approximately 26 per cent or 562 specimens of a total 2,169 adults examined had at least one parasite within the carapace. A total of 1,400 parasites were found in infected ostracods and the maximum number per individual ostracod was nine. More than 200 ostracods had three or more parasites within the carapace.

The seasonal, geographic, and depth variation in parasitized adults from Gull Lake is presented in Table 8. Although the incidence of infection varied geographically at a given depth, the percentages of parasitized individuals generally were highest at depths of 6 metres. About 40 per cent of the ostracods were infected at 6 metres; less than 10 per cent of the ostracods, in most <3 metre samples, were parasitized. Only 3 of 105 ostracods from depths of 9 metres were infected.

TABLE 8. Seasonal, geographic, and depth variation in the number of parasitized adult *Darwinula stevensoni* collected along Transect AB, Station GI, and other localities in Gull Lake

Date	Locality	No. Adults	% Infected	$\bar{X}$ No. Parasites/ Infected Ind.	Range No. Parasites
		<3 metres			
21 Feb. 1966	AB	2	0·0	—	—
21 Mar. 1966		1	0·0	—	—
2 July 1966		34	0·0	—	—
22 Sept. 1965	GI	54	11·1	3·7	0–8
4 Oct. 1965		94	0·0	0·0	—
13 Oct. 1965		39	10·3	2·0	0–4
18 Oct. 1965		63	6·4	1·3	0–2
1 Nov. 1965		35	2·9	2·0	0–1
2 Dec. 1965		27	3·7	1·0	0–1
4 Jan. 1966		9	0·0	0·0	—
11 Jan. 1966		14	7·1	3·0	0–3
31 May 1966		41	4·9	1·5	0–3
15 June 1966		4	0·0	0·0	—
14 Sept. 1965	62	3	0·0	—	—
27 Sept. 1965	66	12	0·0	—	—
18 Oct. 1965	79–80	43	48·8	2·3	0–6
26 Mar. 1966	138	17	47·1	2·3	0–5
26 Mar. 1966	140–1	11	36·4	1·3	0–2
		3 metres			
4 Nov. 1965	AB	47	31·9	1·7	0–3
24 Feb. 1966		56	42·9	3·0	0–7
30 Mar. 1966		25	64·0	2·8	0–8
29 Apr. 1966		9	0·0	—	—
19 May 1966		10	40·0	3·8	0–7
1 June 1966		25	32·0	2·5	0–5
15 June 1966		29	13·8	5·0	0–6
1 July 1966		2	0·0	—	—
15 July 1966		134	22·4	2·3	0–5
1 Aug. 1966		26	36·9	2·0	0–4
		6 metres			
4 Nov. 1965	AB	21	42·9	2·4	0–4
21 Feb. 1966		18	55·6	3·4	0–7
30 Mar. 1966		50	30·0	1·7	0–3
29 Apr. 1966		21	61·9	3·2	0–7
19 May 1966		25	44·0	2·1	0–6
1 June 1966		59	45·8	2·1	0–8
15 June 1966		87	43·7	2·4	0–5
1 July 1966		161	36·7	3·2	0–9
15 July 1966		136	38·2	2·6	0–9
1 Aug. 1966		141	53·2	2·6	0–7
13 Oct. 1965	76	13	0·0	—	—
8 Dec. 1965	EF	48	33·3	2·4	0–7
30 Mar. 1966	IJ	70	31·4	3·0	0–7

The low incidence of infection in juveniles compared with that in adults illustrates the variation in parasitism in relation to the size structure of the population. Parasitism of mature and immature ostracods produced during the same reproductive season was much greater in recently matured individuals, even though the age of both groups was essentially the same.

Parasitism also varies with the age structure of the population. The percentages of infected ostracods comprising adult age classes discussed above (Table 5) are presented in Table 9. These data are presented in the manner of construction used for Table 5; Groups 1, 2 and 3 were explained in the latter table and may be compared with the two right-hand columns for age class composition (Table 5) not repeated in Table 9.

TABLE 9. Parasites of *Darwinula stevensoni* in Groups 1, 2 and 3, collected at depths of 3 and 6 metres along Transect AB, during the summer of 1966. Explanation of Group designations given in Table 5

Date	Group	N	% Parasitized	Range Parasites	Adults with $\bar{X}$ Para./Ind.	Parasitized $\bar{X}$ young/Ind.
			3 metres			
	1	0	0·0	—	0·0	0·0
1 June	2	24	33·0	0–5	2·5	2·1
	3	1	0·0	—	0·0	0·0
	1	0	0·0	—	0·0	0·0
15 June	2	14	28·6	0–6	5·0	5·8
	3	15	0·0	—	0·0	0·0
	1	0	0·0	—	0·0	0·0
1 July	2	2	0·0	—	0·0	0·0
	3	0	0·0	—	0·0	0·0
	1	92	20·7	0–5	2·1	8·5
15 July	2	36	30·6	0–5	2·7	5·0
	3	6	0·0	—	0·0	0·0
	1	25	28·0	0–4	2·0	7·7
1 Aug.	2	1	0·0	—	0·0	0·0
	3	0	0·0	—	0·0	0·0
			6 metres			
	1	0	0·0	—	0·0	0·0
1 June	2	35	45·7	0–3	1·6	1·3
	3	24	45·8	0–8	3·0	0·0
	1	0	0·0	—	0·0	0·0
15 June	2	76	42·1	0–5	2·3	3·8
	3	11	54·6	0–5	3·2	0·0
	1	73	53·4	0–9	3·9	9·7
1 July	2	65	26·2	0–4	1·9	4·7
	3	23	8·7	0–1	1·7	0·0
	1	68	50·0	0–9	2·8	7·1
15 July	2	49	28·6	0–5	1·9	2·9
	3	19	21·1	0–6	3·5	0·0
	1	89	69·7	0–7	2·7	7·2
1 Aug.	2	18	44·4	0–5	2·6	4·3
	3	34	14·7	0–4	2·0	0·0

Only second reproductive season adults were parasitized at 3 metres on 1 June and 15 June. More than one-half of the individuals were first reproductive season adults in the 15 June samples, yet none of these ostracods was infected. Data from 15 July to 1 August, at 3 metres, indicate that first reproductive season adults were parasitized soon after reaching maturity. Older individuals within the same year class (Groups 1 and 2) were more heavily parasitized than the more recently moulted adults (Group 3). Mid-July and August samples, at 6 metres, also indicated progressively increased infection of older individuals. First reproductive season adults comprised more than 90 per cent of the latter samples. The high incidence of parasitized adults in Groups 2 and 3 during June, at 6 metres, reflects the fact that over 95 per cent of the animals were second reproductive season adults. On 1 July, Group 1 included only second reproductive season adults whereas Groups 2 and 3 were almost exclusively first reproductive season adults; the percentage of infected Group 1 adults was twice that of Group 2.

The mean number of young per individual in parasitized adults (Table 9) compared to the mean number of young per individual in the total sample (Table 5) does not appear to reflect a reduction of the reproductive potential in infected animals. However, no comparison with the reproductive potential of non-parasitized adults is given. Reduction of reproductive potential resulting from parasitism is currently being investigated experimentally under controlled laboratory conditions and will be reported later (McGregor and Esch, in preparation).

DISCUSSION

Reference to *D. stevensoni* as a representative of the fresh-water Ostracoda has been emphasized in this study. This species, however, has been reported to occur in waters of varying salinity (Brady and Robertson, 1872; de Vos, 1954; Neale, 1965; Hagerman, 1967). Neale (1965) listed a maximum salinity range of about 0·8‰ but Hagerman (1967) recently reported this species in water with a salinity of 2–3‰.

Swain (1961) and Van Morkhoven (1963) recorded the genus *Darwinula* in both fresh and mixohaline water. Sohn (1951) listed the inferred salinity habitats of two fossil species of *Darwinula* and reported (in Wagner, 1965) culturing an unidentified species of *Darwinula* from Lake Tiberias in varying salinity solutions of up to 20 per cent of normal marine water. Sandberg (1965) found mixed assemblages of ostracod valves, including *Darwinula*, in the Laguna de Tamiahua, Veracruz, Mexico, and suggested the presence of fresh-water faunas at times of high outflow from the lagoon. Harding (1962) described a new species of *Darwinula* from Lake Te-Nggano, Rennell Island, Pacific Ocean; this lake was reported to have a salinity of 4·56‰.

Numerous authors, e.g. Kornicker (1958, 1961, 1963), Kornicker and Wise (1960), Benson (1959, 1966), Tressler (1957), Sandberg (1964, 1965), McHardy and Bary (1965), and Elofson (1941), have reviewed work or have investigated various aspects of the ecology of benthonic and planktonic Ostracoda. The somewhat restricted scope of the present study has made direct comparison with the findings of other authors difficult.

Moore (1939) studied the vertical distribution of *Cypria* spp. and *Candona* spp. in the sediments of Douglas Lake, Michigan, and found, respectively, 97 and 66 per cent of the specimens in the upper four centimetre stratum. Some *Candona* spp. were found about 17 cm below the sediment surface. Cole (1953) reported finding approximately 93 per cent of macrobenthic organisms between the substrate surface and the

P

14 cm sediment interval in Douglas Lake; only the general depth distributions of several ostracod species were considered. Greater than 95 per cent of the juvenile and adult *Darwinula stevensoni* in sediment cores from Gull Lake were present in the upper five centimetres of sediment.

Bronstein (1947) and Sandberg (1964) reviewed some of the density estimates of the ostracod *Cyprideis torosa* in various areas of the Soviet Union. Values ranging from about 1,000–50,000 individuals per square metre were common. The maximum estimate was 670,900 per square metre on which Sandberg (1964, pp. 19–20) commented: 'Even though the extreme values given by some of those workers may be misleading, it is evident that the ostracods are quite numerous in the bottom sediments of brackish-water bodies.'

Kozhov (1963) reported estimates of more than 10,000 individuals per square metre for *Candona unguiculata* Bronstein and *Cytherissa elongata* Bronstein in littoral areas of Lake Baikal, U.S.S.R. On the other hand, Tressler (1957a) reported an average of only 57 individuals per square metre with a range from 4 to 152 per square metre in the Great Slave Lake, Canada.

The above densities, excepting Tressler (1957a), represent estimates based upon the combined number of adults and juveniles of a given species. The highest combined density estimate for adult and late instar juveniles of *D. stevensoni* from Gull Lake was 7,277 per square metre, on 1 July, 1966. Had all instar stages been sampled, the latter estimate, undoubtedly, would have been much higher. Considerably higher densities of *D. stevensoni* have been found in small lakes in the area surrounding Gull Lake but cannot be considered in detail at this time. This species occurs in a variety of habitats in south-western Michigan, including alkaline bogs, small spring fed streams, marl lakes and occasionally on the upper surfaces of *Sphagnum* in acid bog lakes during times of high water levels.

A detailed description of the annual temperature cycle of Gull Lake has not been published and is not the intent of this paper. However, a few general temperature characteristics of the lake are reviewed below in relation to the general depth and density distribution of *D. stevensoni*. The water temperature at depths of 3, 6, 9 and 12 metres increases from about 1·9°C in early February to a maximum of approximately 27°C by early July. In February, water temperatures are almost identical at the above depths and generally range between 1·8°–2·0°C, but temperature maxima at 3, 6, 9 and 12 metres, respectively, are: about 27°C, early July; 24°C, mid-July; 22°C, mid-August; and 15°C, early September.

The general time lag at which temperature maxima are reached at the above depths corresponds closely to a similar time lag in development and release of eggs and first or second instar young at these respective depths during the reproductive season of *Darwinula stevensoni*.

Second reproductive season adults began releasing young by mid- and late June whereas first reproductive season adults did not begin until early July at 3 metres depth. Second and first reproductive season adults at 6 metres were releasing young, respectively, by 1 July and 15 July (Table 5). Maximal reproductive potentials and initial release of first instar individuals at 9 metres were observed for second reproductive season adults on 1 August. The release of first, or occasionally second, instar juveniles by second reproductive season adults generally begins about 15 days before that in first reproductive season adults; release of young by the latter begins during approximately the same periods of the month that maximum temperatures are reached at the depths stated above.

Temperature may be one of the more, if not most, important environmental factors which regulate the general distribution pattern and density of *D. stevensoni* in Gull Lake. I believe the low temperatures and consequent shortened growing seasons at depths greater than 9 metres may account for the lower density. This species may be entirely unable to replace itself at these depths due to the time necessary for production of a single brood of young. Other factors such as substratum, food availability, predation, et cetera, may be important geographically in regulating the densities and patterns of distribution within the more general limits influenced by temperature.

It is hoped that laboratory experiments currently in progress will provide additional information regarding the effects of temperature, nutrition and parasitism on the reproductive potential, growth and development of *D. stevensoni*. The effect of parasitism on reproductive potential, and life table information can best be obtained under controlled laboratory conditions and it is hoped will strengthen field observations discussed herein. Greater emphasis on quantification and replication of future field and laboratory experiments will perhaps aid in better understanding the selected ecological aspects of ostracod populations stated briefly in the introductory remarks.

ACKNOWLEDGEMENTS

Acknowledgement is made to a number of individuals and institutions which have contributed in varying capacities to this study. My wife, Jerry, deserves special thanks for assistance in many aspects of this work. Drs J. Whitfield Gibbons, Donald C. McNaught, Messrs Richard Burbidge, Peter Rich, John Hesse, and Arthur Wiest assisted in the field sampling. Dr J. W. Gibbons collected *Darwinula stevensoni* from ponds near Corinth, Mississippi. Mr Harold Allen and Mrs Dolores Johnson assisted, respectively, in the preparation of figures and tables. I thank the staff of the Zoology Department and the W. K. Kellogg Biological Station of Michigan State University for providing financial support and facilities. Gratitude is expressed to members of my guidance committee and, particularly, to Dr T. Wayne Porter, chairman, for advice and support. Dr Robert G. Wetzel kindly assisted in various aspects of this work including translation of several papers. Appreciation is expressed to Dr Marie-Claude Guillaume, University of Paris, for providing a French translation of the summary and to Drs T. W. Porter and R. G. Wetzel for their valuable comments on the manuscript. Most of the work was supported by a National Institutes of Health predoctoral fellowship, 1–Fl–GM–22, 970–01 and 2–Fl–GM–22, 970–02, from the Institute of General Medical Sciences to which I owe my sincere thanks.

REFERENCES

ALM, G. 1916. Monographie der schwedischen Süsswasser-Ostracoden nebst systematischen Besprechungen der Tribus Podocopa. *Zool. Bidr. Upps.*, **4**, 1-247.
BAER, J. G. 1952. *Ecology of animal parasites*. Univ. Illinois Press, Urbana, Illinois, 224 pp.
BATE, R. H. 1967. The Bathonian Upper Estuarine Series of eastern England. I: Ostracoda. *Bull. Br. Mus. nat. Hist., A. Geology*, **14** (2), 21-66.
BENSON, R. H. 1959. Ecology of Recent ostracodes of the Todos Santos Bay region, Baja California, Mexico. *Paleont. Contr. Univ. Kans.*, Arthropoda, Art. **1**, 1-80.
——. 1966. Recent marine Podocopid Ostracodes. *Oceanogr. mar. Biol. Ann. Rev.*, **4**, 213-32.
BRADY, G. S. 1870. The microscopic fauna of the English Fen District. *Nature, Lond.*, **1**, 483-4.
—— AND ROBERTSON, D. 1870. The Ostracoda and Foraminifera of tidal rivers. With an analysis and descriptions of the Foraminifera, by H. B. Brady. *Ann. Mag. nat. Hist.*, (4) **6**, 1-33.
—— AND ROBERTSON, D. 1872. Contributions to the study of the Entomostraca. VI. On the distribution of the British Ostracoda. *Ann. Mag. nat. Hist.*, (4) **9**, 48-63.

—— AND ROBERTSON, D. 1874. XV. Contributions to the study of the Entomostraca. IX. On Ostracoda taken amongst the Scilly Islands, and on the anatomy of *Darwinella stevensoni*. *Ann. Mag. nat. Hist.*, (4) **13**, 114-19.

—— AND ROBERTSON, D. 1885. Genus *Darwinula*. In JONES, T. R. 1885. On the Ostracoda of the Purbeck Formation; with notes on Wealden species. *Q. Jl geol. Soc. Lond.*, **41**, 346.

——, CROSSKEY, H. W. AND ROBERTSON, D. 1874. A monograph of the post-Tertiary Entomostraca of Scotland including species from England and Ireland. *Palaeontogr. Soc.*, **28**, 1-232.

—— AND NORMAN, A. M. 1889. A monograph of the marine and freshwater Ostracoda of the North Atlantic and of northwestern Europe. Section I. Podocopa. *Scient. Trans. R. Dubl. Soc.*, Ser. 2, **4**, 63-270.

BRONSTEIN, Z. S. 1947. Ostracoda presnii vod. *Fauna S.S.S.R. Rakoobraznie*, **2** (1). *Akad. Nauk. S.S.S.R., N.S.* **31**, 1-339 (Russian with English summary).

COLE, G. A. 1953. Notes on the vertical distribution of organisms in the profundal sediments of Douglas Lake, Michigan. *Am. Midl. Nat.*, **49** (1), 252-6.

COLE, M. E. 1966. Four genera of ostracods from Tennessee (*Darwinula, Limnocythere, Ilyocypris*, and *Scottia*). *J. Tenn. Acad. Sci.*, **41** (4), 35-146.

DADAY, E. 1902. Microskopische Süsswassertiere aus Patagonien. *Természetr. Füz.*, **25**, 298.

DANIELOPOL, D. L. AND VESPREMEANU, E. E. 1964. The presence of ostracods on floating fen soil in Rumania. *Fragm. balcan.*, **5** (7), 29-36.

DELACHAUX, T. 1928. Faune invertébrée d'eau douce des hauts plateau de Pérou (région de Huancavelica, département de Junin) récoltée en 1915 par feu Ernest Godet (calanides, ostracodes, rotateurs nouveaux). *Bull. Soc. neuchâtel. Sci. nat.*, N.S. **1**, 45-77.

DEUTSCH, M., VANLIER, K. E. AND GIROUX, P. R. 1960. Ground-water hydrology and glacial geology of the Kalamazoo area, Michigan. *Rept. Mich. geol. Surv.*, **23**, 1-122.

ELOFSON, O. 1941. Zur Kenntnis der marinen Ostracoden Schwedens, mit besonderer Berücksichtigung des Skagerraks. *Zool. Bidr. Upps.*, **19**, 215-534.

FERGUSON, E., JR. 1944. Studies on the seasonal life-history of three species of fresh-water Ostracoda. *Am. Midl. Nat.*, **32** (3), 713-27.

——. 1957. Ostracoda (Crustacea) from the northern Lower Peninsula of Michigan. *Trans. Am. microsc. Soc.*, **76** (2), 212-18.

——. 1958. Seasonal life history studies of two species of fresh-water ostracods. *Anat. Rec.*, **131** (3), 549-50.

FOX, H. M. 1965. The ostracods of the Lago Maggiore. *Memorie Ist. ital. Idrobiol.*, **19**, 81-89.

——. 1966. Ostracods from the environs of Pallanza. *Memorie Ist. ital. Idrobiol.*, **20**, 25-39.

FURTOS, N. C. 1933. The Ostracoda of Ohio. *Bull. Ohio biol. Surv.*, **29**, 411-524.

——. 1935. Fresh-water Ostracoda from Massachusetts. *J. Wash. Acad. Sci.*, **25**, 530-44.

——. 1936. On the ostracods from the cenotes of Yucatan and vicinity. *Publs Carnegie Instn.*, **457**, 98-115.

GLEASON, G. R. AND OHLMACHER, F. J. 1965. A core sampler for *In situ* freezing of benthic sediments. *Am. Soc. Limnol. Oceanog. mar. tech. Sci.*, **1**, 737-41.

HAGERMAN, L. 1967. Ostracods of the Tvärminne area, Gulf of Finland. *Commentat. biol.*, **30** (2), 1-12.

HARDING, J. P. 1962. *Mungava munda* and four other new species of ostracod crustaceans from fish stomachs. *The Natural History of Rennell Island, British Solomon Islands*, **4**, 51-62.

——. 1966. Myodocopan ostracods from the gills and nostrils of fishes. In *Some Contemporary Studies in Marine Science* (Barnes, H. ed.). George Allen and Unwin Ltd., London. pp. 369-74.

HARTMANN, G. 1959. Beitrag zur Kenntnis des Nicaragua—Sees unter besonderer Berücksichtigung seiner Ostracoden. *Zool. Anz.*, **162**, 270-94.

——. 1964. Asiatische Ostracoden. Systematische und zoogeographische Untersuchungen. In *Int. Rev. ges. Hydrobiol.*, 155 pp.

——. 1965. Neontological and Paleontological classification of Ostracoda. *Pubbl. Staz. zool. Napoli* [1964], **33** (suppl.), 550-87.

HOFF, C. C. 1942. The ostracods of Illinois. *Illinois biol. Monogr.*, **19** (1, 2), 1-196.

——. 1943. Seasonal changes in the ostracod fauna of temporary ponds. *Ecology*, **24** (1), 116-18.

HOWE, H. V. 1962. *Ostracod taxonomy*. Louisiana State Univ. Press, Baton Rouge. 366 pp.

KENK, R. 1949. The animal life of temporary and permanent ponds in southern Michigan. *Misc. Publs Mus. Zool. Univ. Mich.*, **71**, 5-66.

KESLING, R. V. 1951. The morphology of ostracod molt stages. *Illinois biol. Monogr.*, **21** (1-3), 1-324.

——. 1953. A slide rule for the determination of instars in ostracod species. *Contr. Mus. Paleont. Univ. Mich.*, **11** (5), 97-109.

——. 1961a. Reproduction of Ostracoda. Q17-Q19. In *Treatise on Invertebrate Paleontology* (R. C. Moore, ed.), Pt. Q., Arthropoda 3, Crustacea: Ostracoda, 442 pp, 334 Figs. University of Kansas Press, Lawrence, Kansas.

——. 1961b. Ontogeny of Ostracoda. Q19-Q20. In *Treatise on Invertebrate Paleontology* (R. C. Moore, ed.), Pt. Q., Arthropoda 3, Crustacea: Ostracoda. 442 pp., 334 Figs. University of Kansas Press, Lawrence, Kansas.

KLIE, W. 1933. Die Ostracoden der Deutschen Limnologischen Sunda-Expedition. *Arch. Hydrobiol. Suppl.*, **11**, 447-502.

——. 1935a. Ostracoda aus dem Tropischen Westafrika in Voy. Alluaud et Chappuis en Afrique Occ. Française. *Arch. Hydrobiol.*, **28**, 35-68.

——. 1935b. Süsswasser-Ostracoda aus Uruguay. *Arch. Hydrobiol.*, **29**, 282-95.

——. 1939. Ostracoden aus dem Kenia-Gebiet, vornehmlich von dessen Hochgebirgen. *Int. Rev. ges. Hydrobiol.*, **39**, 99-161.

——. 1941. Süsswasser-ostracoden aus Südosteuropa. *Zool. Anz.*, **133** (11-12), 233-44.

KORNICKER, L. S. 1959. Ecology and taxonomy of Recent marine ostracodes in the Bimini area, Great Bahama Bank. *Publs. Inst. mar. Sci. Univ. Tex.*, 1958, **5**, 194-300.

——. 1961. Ecology and taxonomy of Recent Bairdiinae (Ostracoda). *Micropaleontology*, **7** (1), 55-70.

——. 1963. Ecology and classification of Bahamian Cytherellidae (Ostracoda). *Micropaleontology*, **9** (1), 61-70.

—— AND WISE, C. D. 1960. Some environmental boundaries of a marine ostracod. *Micropaleontology*, **6** (4), 383-98.

KOZHOV, M. 1963. *Lake Baikal and its life.* Monographiae Biologicae, V. XI, W. Junk, Publishers, The Hague. 344 pp.

MCHARDY, R. A. AND BARY, B. M. 1965. Diurnal and seasonal changes in distribution of two planktonic ostracods, *Conchoecia elegans* and *Conchoecia alata minor. J. Fish. Res. Bd Can.*, **22** (3), 823-40.

MACY, R. W. AND DEMOTT, W. R. 1957. Ostracods as second intermediate hosts of *Halipegus occidualis* Stafford, 1905 (Trematoda: Hemiuridae). *J. Parasit.*, **43**, 680.

MARTIN, H. M. 1957. Outline of the geologic history of Kalamazoo County. *Rep. Mich. geol. Surv.*, 16 pp.

MÉHES, G. 1908. Beiträge zur Kenntnis der Pliozänen Ostracoden Ungarns. II. Die Darwinulidaeen und Cytheridaeen der Unterpannonischen Stufe. *Földt. Kozl.*, **38** (7-10), 601-35.

MENZEL, R. 1916. Moosbewohnende Harpaticiden und Ostracoden aus Ost-Afrika. *Arch. Hydrobiol.*, **11** (3), 486-9.

——. 1923. Beiträge zur Kenntnis der Mikrofauna von Niederländisch-Ost-Indien V. Moosbewohnende Ostracoden aus dem Urwald von Tjibodas. *Treubia*, **3**, 193-6.

MOORE, G. M. 1939. A limnological investigation of the microscopic benthic fauna of Douglas Lake, Michigan. *Ecol. Monogr.*, **9** (4), 537-82.

MORKHOVEN, F. P. C. M. VAN. 1962. *Post-Palaeozoic Ostracoda. I. Their morphology, taxonomy and economic use.* Elsevier Publishing Company, New York, Vol. 1, 204 pp.

——. 1963. *Post-Palaeozoic Ostracoda. II. Their morphology, taxonomy, and economic use. Generic descriptions.* Elsevier Publishing Company, New York. Vol. 2, 478 pp.

MÜLLER, G. W. 1912. Ostracoda. *Das Tierreich*, **31**, 1-434.

NEALE, J. W. 1965. Some factors influencing the distribution of Recent British Ostracoda. *Pubbl. Staz. zool. Napoli* [1964], **33** (suppl.), 247-307.

NICHOLS, M. M. AND ELLISON, R. L. 1967. Sedimentary patterns of microfauna in a coastal plain estuary. In *Estuaries* (G. H. Lauff, ed.), AAAS **83**, 283-8.

PETKOVSKI, T. K. 1960. Zur Kenntnis der Crustaceen des Prespasees. *Fragm. balcan.*, **3** (15), 117-31.

——. 1961. Zur Kenntnis der Crustaceen des Skadar (Scutari) Sees. *Acta Mus. maced. Sci. nat.*, **8** (2), 29-52.

PINTO, I. D. AND SANGUINETTI, Y. T. 1958. O Genótypo de *Darwinula* Brady and Robertson, 1885. *Boln Inst. Cienc. nat.*, **6**, 5-19.

—— AND KOTZIAN, S. C. B. 1961. Novos Ostracodes da Familia Darwinulidae e a variação das impressões musculares. *Boln Inst. Cienc. nat.*, **11**, 5-64.

—— AND PURPER, I. 1965. A new fresh-water ostracode *Cyprinotus trispinosus* Pinto et Purper, sp. nov., from southern Brazil, its ontogenetic carapace variation and seasonal distribution. *Esc. Geol. P. Alegre, Publ. Esp.*, **7**, 1-53.

POKORNÝ, V. 1952. The ostracods of the so-called Basal Horizon of the Subglobosa beds at Hodonin (Pliocene, Inner Alpine Basin, Czechoslovakia). *Sb. ústřed. Úst. geol.*, **19**, 229-396.

ROME, D. R. 1947. Contribution a l'étude des ostracodes de Belgique. I. Les ostracodes du Parc St. Donat à Louvain. *Bull. Mus. r. Hist. nat. Belg.*, **33** (34), 1-24.

SANDBERG, P. A. 1964. The ostracod genus *Cyprideis* in the Americas. *Stockh. Contr. Geol.*, **12**, 178 pp.

——. 1965. Notes on some Tertiary and Recent brackish-water Ostracoda. *Pubbl. Staz. zool. Napoli* [1964], **33** (suppl.), 496-514.

SARS, G. O. 1899. *An account of the Crustacea of Norway.* Vol. II. *Isopoda*, i-x, 1-270, 100 Pls. +4 supplementary Pls. Bergen, Bergen Museum.

——. 1922-28. *An account of the Crustacea of Norway.* Vol. IX. *Ostracoda*, 1-277, 119 Pls. Bergen, Bergen Museum.

SCHEERER-OSTERMEYER, E. 1940. Beitrag zur Entwicklungsgeschichte der Süsswasserostrakoden. *Zool. Jb., Abt. Anat. Ontog.*, **66** (3), 349-70.

SCOTT, T. 1891. Notes on a small collection of freshwater Crustacea from the Edinburgh District. *Proc. R. phys. Soc. Edinb. Session 1889-90*, 313-17.

SOHN, I. G. 1951. Check list of salinity tolerance of post-Paleozoic fossil Ostracoda. *J. Wash. Acad. Sci.*, **41** (2), 64-66.

——. 1965. Late Quaternary ostracodes from the southern part of the Dead Sea, Israel. *Trans. Ocean Sci., Ocean Engineering*, **1**, 82-94.

STANKOVIĆ, S. 1960. *The Balkan Lake Ohrid and its living world*. Monographiae Biologicae, 9, W. Junk, Publishers, The Hague, 357 pp.

STAPLIN, F. L. 1963. Pleistocene Ostracoda of Illinois. II. Subfamilies Cyclocyprinae, Cypridopinae, Ilyocyprinae: Families Darwinulidae and Cytheridae. Stratigraphic ranges and assemblage patterns. *J. Paleont.*, **37** (6), 1164-203.

STEPHENSEN, K. 1938. Marine Ostracoda and Cladocera. *The Zoology of Iceland* III. Part 32, 1-19, 1 map, 3 tables.

SWAIN, F. M. 1961. Superfamily Darwinulacea Brady and Norman, 1889. Q253-Q254 In *Treastise on Invertebrate Paleontology* (R. C. Moore, ed.), Pt. Q., Arthropoda 3, Crustacea: Ostracoda. 442 pp., 334 Figs. University of Kansas Press, Lawrence, Kansas.

—— AND GILBY, J. M. 1965. Ecology and taxonomy of Ostracoda and an alga from Lake Nicaragua. *Pubbl. Staz. zool. Napoli* [1964], **33** (suppl.), 361-86.

SYWULA, T. 1965. Małżoraczki (Ostracoda) Wielkopolskiego Parku Narodowego, (Ostracoda of the National Park of Great Poland). English summary. *Prace Monogr. nad Przyr. Wielkop. Parku Nar.* PTPN, **5** (2), 1-27.

TAUBE, C. M. AND BACON, E. H. 1952. Gull Lake. *Lake Inventory Summary, no. 2, Mich. Dept. Cons.*

TRESSLER, W. L. 1947. A check list of the known species of North American fresh-water Ostracoda. *Am. Midl. Nat.*, **38** (3), 698-707.

——. 1957a. The Ostracoda of Great Slave Lake. *J. Wash. Acad. Sci.*, **47** (12), 415-23.

——. 1957b. Marine Ostracoda. In *Treatise on Marine Ecology and Paleoecology. Mem. geol. Soc. Am.*, **67** (1), 1161-4.

TURNER, C. H. 1895. Fresh-water Ostracoda of the United States. *Second Report St. Zool. Minnesota*, 1895, 277-337.

VÁVRA, V. 1909. Ostracoda, Muschelkrebse. In *Die Süsswasserfauna Deutschlands* (A. Brauer, ed.), Verlag Von J. Cramer, 1961, Hafner Publishing Co., New York. **11** (2), 85-119.

VOS, A. P. C. DE. 1954. Over de oever-en bodemfauna der binnen dijkse kolken langs het Ijsselmeer. *Flora Fauna Zuiderzee*, **3**, 277-82.

—— AND STOCK, J. H. 1956. On commensal Ostracoda from the wood-infesting isopod *Limnoria*. *Beaufortia* **5** (55), 133-9.

WAGNER, C. W. 1957. *Sur les Ostracodes du Quaternaire Récent des Pays-Bas et leur utilization dans l'étude géologique des dépôts Holocènes*. Mouton and Co., The Hague. 260 pp.

——. 1965. Ostracods as environmental indicators in Recent and Holocene estuarine deposits of The Netherlands. *Pubbl. Staz. zool. Napoli* [1964], **33** (suppl.), 480-95.

WARD, H. L. 1940. Studies on the life history of *Neoechinorhynchus cylindratus* (Van Cleave, 1913) (Acanthocephala). *Trans. Am. microsc. Soc.*, **59**, 327-47.

WEYGOLDT, P. 1960. Embryologische Untersuchungen an Ostrakoden: Die Entwicklung von *Cyprideis litoralis* (G. S. Brady) (Ostracoda, Podocopa, Cytheridae). *Zool. Jb., Abt. Anat. Ont.*, **78** (3), 369-426.

WHITESIDE, E. P., SCHNEIDER, I. F. AND COOK, R. L. 1963. Soils of Michigan. *Spec. Bull. Mich. agric. Exp. Stn*, **402**, 1-52.

WINKLER, E. M. 1960. Post-Pleistocene ostracodes of Lake Nipissing Age. *J. Paleont.*, **34** (5), 923-32.

DISCUSSION

SWAIN: This paper is a significant and valuable contribution. Am I correct in thinking that the 6 m depth is either above or within the thermocline and that the 9 m depth lies below the thermocline?

MCGREGOR: During the summer and fall reproductive period the thermocline is below the 6 m depth. The 9 m depth generally is within the thermocline from about mid-June to mid-September but is below and above the thermocline, respectively, in early June and late September. The waters are well oxygenated at both depths throughout the year.

MCKENZIE: You said that the number of parasites varies with the number of young in the carapace. Can you comment further upon this interesting relationship?

MCGREGOR: This portion of the investigation is still in progress. The maximum number of parasites in any adult was nine; in ostracods with ten or more eggs and young no more than three parasites were present. One might approach this problem by statistical analysis of the field data to determine whether the incidence or level of infection influences the reproductive potential or *vice versa*. However, I believe that the better approach would be a study of the ostracod reproductive potential-parasite relationship in controlled laboratory experiments. This would include subjecting the ostracods to varying levels of parasitism to determine whether the parasite causes a reduction in reproductive potential or whether the increased number of young within the carapace forces the parasites to leave or reduces their opportunity for attachment.

HULINGS: Exactly where does this parasite occur?

MCGREGOR: The parasite is always present in the posterior portion of the ostracod carapace where the eggs and young are carried and is presumably ectoparasitic. I do not believe that the parasite derives any nutrients from the ostracod. One might suggest that the relationship is strictly commensal; however, if the animal's presence within the carapace causes a reduction of the ostracod reproductive potential, then the term parasitic may be justified.

My recent work has shown the parasite to be a rotifer, probably belonging to the Order Bdelloidea. The specific identity of this animal, its influence on reproductive potential, and the mechanism involved are currently under study.

KAESLER: It seemed from your histograms that there were just as many individuals with few numbers of young per individual as there were with large numbers of young per individual and that the same also applied for the intermediate range, the number varying from few to many. In other words the number of young per individual seemed to follow an essentially rectangular distribution rather than a normal or bell-shaped distribution. Is that so?

MCGREGOR: Yes. That is the reason I divided them into three groups and presented them in age categories. I separated them into Groups 1, 2 and 3 (refer to text) in order to obtain a biologically meaningful estimate of the mean number of young per individual (Group 1) rather than include those animals which had released all their young (Group 3) and those which had not approached maximal production (Groups 2 and 3).

SYLVESTER-BRADLEY: Parthenogenesis in *Darwinula* is coupled with a remarkably wide distribution, rather great lack of variation and an extraordinarily long geological history with very little morphological change. I would like to ask whether the males of *Darwinula stevensoni* are known anywhere in the world, and if not, whether in your future work we can gain any knowledge of the mechanism of parthenogenesis and the chromosome make-up in the species? This has considerable evolutionary interest because it would normally be regarded as impossible for a species to last through geological time if it reproduces asexually. This is against evolutionary theory where it is usually thought that the lack of flexibility given by parthenogenesis would lead to extinction. This is the exact opposite of the history of *Darwinula* so it would be very interesting indeed to know more about the reproductive capability.

MCGREGOR: In response to your second question, I plan to work with *Darwinula stevensoni* and other species of the genus for a number of years and hope to investigate some aspects of this evolutionary problem. Regarding your first question on partheno-genetic reproduction and presence of males, I know of nothing beyond a description and figure of the copulatory apparatus of a male specimen by Brady and Robertson (1870) and subsequent references to this publication. I have examined 10,000 specimens from Gull Lake and have never found a male. Juveniles reared in the laboratory through at least one reproductive cycle produced only females.

THE FRESHWATER OSTRACOD *CANDONA HARMSWORTHI* SCOTT, FROM FRANZ JOSEF LAND AND NOVAYA ZEMLYA

JOHN W. NEALE
University of Hull, England

SUMMARY

Re-examination of the type material of *Candona harmsworthi* Scott shows that the species is valid. The type material is redescribed and the appendages are figured and described for the first time. The orientation of the maxillary palp and 'masticatory' processes are noted and attention is drawn to variation in the proportions of the joints in some limbs. Comparison is made with related species.

INTRODUCTION

In his work on the crustacea of Franz Josef Land (also sometimes called Zemlya Fridtjov Nansen), published on 1 July 1899, Thomas Scott briefly described and figured the carapace of a new high latitude species of fresh-water ostracod *Candona harmsworthi*, without mentioning or figuring either limbs or soft parts. This species has never been recorded since and presents problems of interpretation. Müller in his classic survey of the ostracoda (*Das Tierreich*, 1912, **31**) lists it only cursorily in his doubtful or uncertain species, whilst Lüttig (1962) in his detailed study of the carapace shape of 117 species and subspecies of the genus omits it altogether.

Recently, while investigating Scott's material in the Royal Scottish Museum, Edinburgh, the author came across the type material from Franz Josef Land which consists of five adult carapaces and eleven immature carapaces mounted dry in a slide with the type specimens of *Herpetocypris* (?) *dubia* and *H. arctica*, all collected from ponds near Elmwood, Cape Flora (79°57'N, 50°01'E) on 30 July 1897. Another dry slide labelled 'Duplicate' from the same locality contained nine adult carapaces and one adult right valve of this species, together with 24 immature carapaces.

In addition the spirit material yielded a tube labelled 'Royal Scottish Museum, W. S. Bruce Collection. Arctic Collection 1921.145.418. *Candona harmsworthi* T. Scott. Near Kostyn Point, Novaya Zemblia 18.6.98.' This contained a few immature specimens preserved in spirit. Cape Kostin (or Kostyn) lies in south-western Novaya Zemlya at 70°55'N, 53°07'E.

By kind permission of the Museum authorities, to whom I wish to express my thanks, I was allowed to borrow Scott's material and dissect some of the spirit specimens as well as opening five of the dry adult carapaces and one immature carapace in the duplicate slide which yielded good preparations of the limbs but not of the vibratory plates of the mandibles or maxillae, or of the soft parts or genitalia. In consequence it is possible to describe the nature of this species fairly completely for the first time, to compare it with other *Candona* species and to resolve an anomaly which appears in the literature concerning the orientation of the maxillary palp with respect to the vibratory plate of that limb.

SYSTEMATIC DESCRIPTION AND COMPARISONS

Order Podocopida G. W. Müller 1894
Suborder Podocopa Sars 1866
Family Cyprididae Baird 1845
Subfamily Candoninae Kaufmann 1900
Genus *Candona* Baird 1845

Candona harmsworthi Scott 1899
Plate I, Figs 1–5

Candona Harmsworthi, sp. n., Scott 1899, p. 83, Pl. 3, Figs 16, 17.

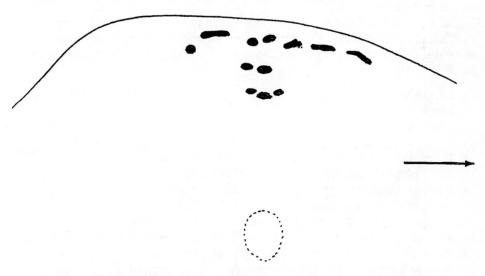

Fig. 1. *Candona harmsworthi* Scott. Paratype. Adult carapace. Bottom specimen, column 3, type slide. Dorsal muscle scars seen from right. Arrow indicates anterior. Position of main adductor scars indicated by dotted line. Freehand sketch × ca. 350.

1. *The Carapace* (Pl. I 5a, b; Figs 2, 3, 4a, b)

Scott's five adult type specimens are figured here as Plate I, and outlines of two of the immature specimens are given on Fig. 2.4, 5. His original description is concerned solely with the carapace and runs as follows:

'The shell, seen from the side, is somewhat subreniform; the dorsal margin is considerably elevated near the posterior end, where the anterior and posterior slopes meet and form an obtuse angle; the greatest height is equal to rather more than half the length; the front slope curves gently downwards to the evenly rounded anterior extremity; the posterior end is subtruncate and forms a slight curve from the obtuse dorsal angle downwards and backwards to where it meets the ventral margin; the ventral margin is distinctly incurved in front of the middle. The shell, seen from above, is ovate; the greatest width which is situated behind the middle, is equal to rather more than two-fifths of the length; extremities slightly acuminate. Length 1 mm.'

Scott's description coupled with the present figures leaves little to add except a short note on the muscle scar pattern. The main central adductor scars are typical of the genus and consist of six scars arranged as a curved anterior vertical row of four and a posterior row of two; or alternatively they may be considered as a rosette of five with one scar placed centrally above. The two sub-rounded mandibular scars are easily seen, the antennal scars less so. The dorsal series of small scars was seen fairly clearly in the bottom adult paratype in the type slide. In view of the possible taxonomic importance of these scars (Benson and MacDonald, 1963) a sketch of these is given here (Fig. 1) and may be compared with the six species of *Candona* figured by those authors. The general similarities are obvious, a centrally placed group of seven scars being disposed in three horizontal lines of 2, 2, and 3 from dorsally to ventrally, the two dorsal scars forming part of a line of generally elongate scars which runs parallel to the dorsal margin. The differences from the species figured by the aforementioned authors are too obvious to need comment. More recently Benson (1967) has returned to the whole question of discrimination (mainly at the generic level) on the basis of muscle scar pattern.

Lüttig (1962) has listed the valve characteristics of 117 species and subspecies of *Candona* under 14 different headings and for ease of comparison *C. harmsworthi* may be treated the same way. Based on the five adult type specimens the characters are as follows:

1. Length (mm) $\bar{X} - 0.991$ (Range $0.948 - 1.013$)
2. Height (mm) $\bar{X} - 0.575$ (Range $0.532 - 0.597$)
3. Length $\div$ Height $\bar{X} - 1.72$
4. Position of greatest height. $\bar{X} - +1.65$ (Range $+1.58 - +1.73$)
5. Form of dorsal margin   weakly convex
6. Form of ventral margin   moderately concave
7. Form of anterior margin   infracurvature
8. Form of posterior margin   infracurvature
9. Passage from dorsal to anterior margin   broadly rounded
10. Passage from dorsal to posterior margin   well rounded angle
11. Passage from ventral to anterior margin   well rounded
12. Passage from ventral to posterior margin   well rounded
13. Angle between ventral and dorsal margins $\bar{X} - 14°20'$ (Range $13°30' - 15°$)
14. Angle between ventral and posterior margins $\bar{X} - 43°54'$ (Range $40°30' - 52°30'$)

Among the data assembled by Lüttig (1962), three species show some correspondence but *C. acuminata* Fischer and *C. mülleri* Hartwig differ considerably in dorsal view and whilst *C. weltneri* Hartwig approaches *C. harmsworthi* in width, it is different in side view.

Scott (1899) notes that *C. harmsworthi* has a distant resemblance to *C. candida* var. *claviformis* in side view but differs in dorsal view. The latter, described by Brady and Norman (1889 p. 99, Pl. X, Figs 1, 2) was found in a pond at Sedgefield, County Durham, and is slimmer in dorsal view with the greatest width situated more posteriorly than in the present species. The latter is very wide, the greatest width being approximately equal to one half the length and situated at about the mid-point, the width being well maintained so that the extremities are bluntly pointed and the carapace has a somewhat hexagonal appearence in this view.

The specimen from the Pleistocene in the Liri Valley of Italy figured by Devoto (1965, p. 337) as *C. candida* (Müller) differs from the forms usually assigned to *C. candida* in the marked postero-dorsal 'peak' in side view and possibly belongs here.

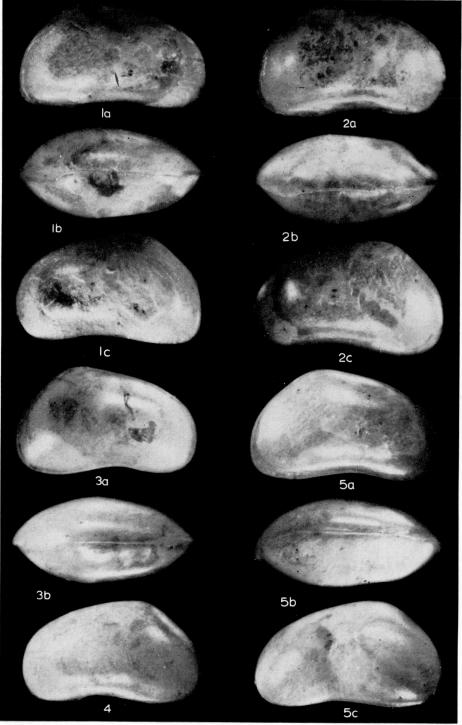

PLATE I. (Magnification in all cases × 50)

Adult type specimens of *Candona harmsworthi* Scott 1899, from freshwater ponds near Elmwood, Cape Flora, Franz Josef Land in Column 3 of the type slide in the Royal Scottish Museum, Edinburgh. Photographed dry and untreated by Mr N. Bell.

1. Lectotype. Carapace: (a) from right; (b) dorsal view; (c) from left. Type slide, top specimen.
2. Paratype. Carapace: (a) from right; (b) dorsal view; (c) from left. Second specimen.
3, Paratype. Carapace: (a) from right; (b) dorsal view. Middle specimen.
4. Paratype. Carapace: from left. Fourth specimen.
5. Paratype. Carapace: (a) from right; (b) dorsal view; (c) from left. Bottom specimen.

C. rectangulata Alm and *C. groenlandica* Brehm have been recorded from both Greenland and Novaya Zemlya and the former has also been found in Franz Josef Land, Spitzbergen and Northern Siberia. These will both be discussed in detail after the soft parts have been described and general comparisons made. Otherwise no other figured *Candona* appear to approach the present species closely.

2. *The Limbs*

Various keys have been published which make varying use of the limbs and soft parts, particularly the mandibular palp, the cleaning limb (7th pair of limbs) and the genital lobe of the female. The *C. candida* section of Müller (1900, 1912) as adopted also by Klie (1938) is characterized by the fact that 'Das vorletzte Glied des Mandibulartasters trägt an seiner medialen Seite eine gefiederte Borste'. Such a plumose seta appears to be absent in *C. harmsworthi*, the plumose seta that does occur being associated with the distal-median part of the proximal endite.

On the basis of shell shape—rounded, height typically greater than one half-length and with more than three bristles on the inner margin of the second joint of the mandible palp, *C. harmsworthi* corresponds best with Klie's *compressa*-group, but the nature of the genital lobe in the female—regarded by Klie as more important than the shell proportions, and the bundle of four setae associated with the middle joint of the mandibular palp, suggests more affinities with the *acuminata*-group. In consequence *C. harmsworthi* is referred to the *acuminata*-group as defined by Klie.

Hoff (1942) adopted Klie's broad divisions and produced a key to North American species of *Candona*, none of which are closely comparable. The closest is *C. distincta* Furtos 1933, the differences from the present species being most easily observed in the shell, which in side view has 'a distinct sinuation in the middle giving the appearence of two dorsal humps'.

In dealing with the fauna of the U.S.S.R., Bronstein (1947) uses a different basis for his key, making his primary division on the nature of the terminal joints of the first antennae. After this, varying features lead on to *C. rectangulata* Alm and *C. groenlandica* Brehm; consideration of these is deferred until after the general description and comparisons of the soft parts.

(*a*) *First pair of limbs. Antennules. (First Antennae, Upper Antennae auct.)* Fig. 2.2. The eight-jointed antennules are of typical candonid pattern and are directed forwards and upwards. The relative lengths of the joints measured along the mid-line are: 1:37, II:29, III:21, IV:11, V:16, VI:17, VII:23, VIII:23.
The three proximal joints (protopodite) are robust and almost equidimensional, being a little wider than long. Each carries a seta at, or below, the middle of the anterior side; these setae are approximately the same length as the anterior side of the joint. The middle joint carries two longer setae situated slightly to the outside at about the midpoint of the posterior margin; these setae are about as long as the respective anterior and posterior sides of these three protopodite joints.

The distal five joints (IV–VIII) are more slender, and increase in length distally, the middle joint being equidimensional, the two distal joints being much longer than wide. Joint IV is small and sometimes carries an antero-distal, slightly externally placed seta, but the development of the latter is apparently variable and depends on neither maturity nor locality. No postero-distal seta has yet been seen associated with this joint. Joint V carries a small seta, approximately equal in length to the length of the

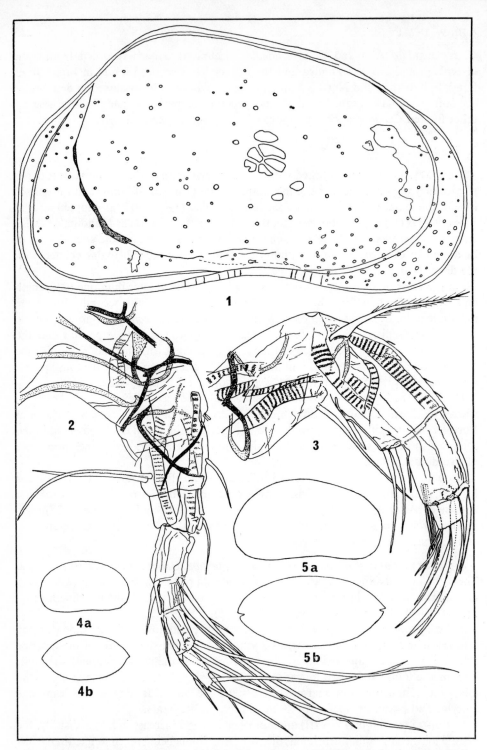

Fig. 2. *Candona harmsworthi* Scott. 1. Adult right valve. ×125. External view by transmitted light in glycerine. Dissection 3. Fresh-water pond near Elmwood, Cape Flora, Franz Josef Land. 2. Right limb No. 1. Antennule. ×380. External view. Dissection 5. From near Kostin Point, Novaya Zemlya. 3. Right limb No. 2. Antenna. ×380. External view. Dissection 5. From near Kostin Point, Novaya Zemlya. 4. Immature carapace. Paratype. ×50. (a) External lateral outline seen from right; (b) outline in dorsal view. Fresh-water pond near Elmwood, Cape Flora, Franz Josef Land. Type slide, Column 4, top-right. 5. Immature carapace. Paratype ×50. (a) Lateral outline seen from right; (b) outline in dorsal view. Fresh-water pond near Elmwood, Cape Flora, Franz Josef Land. Type slide, Column 4, top left.

joint, both anteriorly and posteriorly at its distal end. In addition, antero-distally and external of the small seta, it carries two further setae, one of medium length (about twice the length of the small setae), the other about three times as long as the small setae. Joint VI carries a postero-distal seta similar to the preceding joint, together with two very long antero-distal setae which are almost as long as the complete antennule. The penultimate joint carries one very long and one medium-length seta antero-distally, one other very long seta which leaves the inside of the joint slightly posterior of the middle of the distal edge, and a small postero-distal seta. The distal joint is armed with a terminal seta of medium length, a long antero-distal seta with a small seta outside this, and a small posterior seta.

This limb is fairly similar to the antennule of *C. candida* figured by Sars (1923, Pl. XXXII a[i]) the most obvious difference lying in the proportions of joints V and VI which are relatively shorter in the present species, and in the postero-distal setae borne on joint IV which has not been observed in *C. harmsworthi*. The long setae in *C. caudata* Kaufmann are much shorter and the distal parts of the antennules of the other species figured by Sars all appear to be more slender than *C. harmsworthi*. In the antennule of *C. candida* figured by Brady (1868, Pl. 37, Fig. 1a) joint IV (Brady's joint 3) bears a long antero-distal seta not seen in *C. harmsworthi* or Sars figure of the *C. candida* antennule.

(b) *Second pair of limbs. Antennae. (Second Antennae, Lower Antennae auct.)* Fig. 2.3. Protopodite of two robust joints (I, II), the first of which is relatively short with two well separated posterior setae, and is not figured here. The second has a long postero-distal seta which is as long as the exopodite, the latter here represented by a pilose seta situated dorso-externally at the distal end of this joint. Endopodite of three joints (III–V) of which the proximal joint is the longest and the distal the shortest. The proximal joint (III) is armed anteriorly with five minute setae or hairs, one of which overhangs joint IV. A small seta of length about half that of this joint occurs postero-proximally and two longer setae are situated postero-distally. The middle joint (IV) also has a small postero-proximal seta and a small seta situated just below the middle of the anterior margin is broken off in the figured specimen. In addition there is a bundle of three fine setae of medium length set posteriorly on the inner surface at about two-thirds the length of the joint, whilst distally two strong claw-like setae spring from the inner edge and a smaller claw occurs antero-distally. The distal (V) and smallest joint carries two claw-like setae distally, the posterior being much smaller than the anterior, with a bundle of three fine setae, which are about twice as long as this joint, postero-distally.

Relative lengths of the joints measured along the mid-line in the figured specimen are: I:—; II:15, III:12, IV:8, V:3.

In structure the antennae are closest to those of *C. candida*; the endopodite, how-ever, seems generally more robust.

(c) *Third pair of limbs. Mandibles.* Figs 3.1; 5.2, 13. The mandibles consist of a strongly chitinized *pars incisiva* with attached mandibular palp. The distal biting edge is formed by a row of seven solidly chitinized teeth, the anterior one being much the largest and the teeth generally decreasing in size posteriorly. Some of the teeth tend to be club-shaped and have a bifurcate tip. Occasionally bristles appear to be de-veloped between the more anterior teeth. The posterior margin of the distal edge carries a small pointed seta, posterior of which is a recurved pilose seta. The anterior edge carries a seta (Fig. 5.13) at about the position of maximum width. The attached

Fig. 3. Magnification in all cases ×ca. 380. *Candona harmsworthi* Scott. 1. Right limb No. 3. Mandible. External view. Dissection 5. From near Kostin Point, Novaya Zemlya. 2. Right limb No. 5. Maxilliped. External view. Dissection 5. From near Kostin Point, Novaya Zemlya. 3. Right limb No. 4. Maxilla. Seen from inside. Dissection 5. From near Kostin Point, Novaya Zemlya. 4. Left caudal ramus. Detail of distal part. External view. Dissection 3. From freshwater pond near Elmwood, Cape Flora, Franz Josef Land.

palp consists of a large protopodite with attached exopodite, and endopodite of three ill-defined endites. The exopodite is modified to form a vibratory plate with seven plumose setae and about three small setae also occur antero-dorsally. Postero-distally the protopodite carries a long, thin seta, a long pectinate seta set with rela-tively long stiff hairs, a short, smooth pointed seta (not seen in Fig. 3.1, but well seen in Fig. 5.2 [Dissection 1]), and a short, pilose, carrot-shaped organ of about the same length as the preceding seta. The proximal endite of the endopodite has two moderately long setae of differing lengths antero-distally, a pectinate seta—but with much shorter hairs than the one on the protopodite—occurs on the inside surface at the middle of the distal edge, and there is a bundle of four long, slender setae postero-distally. The middle endite has three antero-distal setae, two similar to those of the proximal endite, the third longer and either broken off or reflected back behind in Fig. 3.1; two long setae occur postero-distally, and two moderately robust spike-like setae are present distally on the inner face. The distal endite carries two terminal, claw-like setae.

This limb is poorly figured in the literature but appears fairly close to the mandible of *C. candida* figured by Sars (*op. cit.* Pl. XXXII M) apart from minor differences in shape of the *pars incisiva* in which the anterior tooth seems less well developed than in the present species.

(*d*) *Fourth pair of limbs. Maxillae.* (*Maxillules, 1st Maxillae auct.*) Figs 3.3; 5.1. This limb consists of a palp, three so-called 'masticatory' processes and a vibratory plate. The latter is well developed with a smooth, slightly concave dorsal margin and is armed posteriorly and ventrally with 18 plumose setae, whilst two setae which are apparently smooth occur anteroventrally. In *Candona* this limb is not commonly figured but Brady's *C. candida* (1868, Pl. 37, Fig. 1d) shows 16 plumose setae and no smooth setae, while Sars *C. candida* (*op. cit.* Pl. XXXII m) shows 18 plumose setae and three smooth antero-ventral setae

The relationship of the palp and 'masticatory' processes to the vibratory plate is as figured here, the setae on the endites helping to convey material to the mouth. Brady (1868), Hoff (1942) and Sars (1898) all show the parts related as figured here. On the other hand there is apparently a marked tendency for the palp and processes to rotate through 180° with respect to the vibratory plate or *vice versa* on dissection for while Sars (1923–25) shows it orientated as here in *C. candida* (Pl. XXXII m), in the half section of this species (Pl. XXXIII) and in *C. sarsi* (Pl. XXXVI m) he shows the two parts rotated 180° with respect to each other. Kesling's detailed study of *Candona suburbana* Hoff (Kesling, R. V. 1965. N.S.F. Project GB–26. *Report No. 1*, Univ. Michigan Press 56 pp., 13 Pls) which only came to hand after the present paper was completed, confirms the orientation of the various elements of the maxilla.

The palp as figured here consists of three parts. The lowest carries two relatively long pilose setae and a smooth seta of medium length. Dorsally is a bipartite endite, the ventral part carrying two blade-like setae together with a more slender seta of medium length, whilst the upper part carries four smooth radiating setae of medium length. The ventral and middle 'masticatory' processes take the form of well-defined endites each of which has half a dozen setae varying from blade-like to hair-like and short to medium in length. The dorsal 'masticatory' process is a well-defined endite with seven or eight relatively short setae and one longer one situated to the inside on the antero-dorsal edge. Dorsally, a little removed from this endite are two fairly long, smooth setae.

(e) *Fifth pair of limbs. Maxillipeds.* (*Maxillae, Second Maxillae, First Thoracic legs auct.*) Figs 3.2; 4.3. This limb, together with the cleaning limb and the caudal rami is one of the limbs most commonly figured in the literature; in the more commonly figured male (not seen) the endopodite is modified to form a clasping organ. In the female the endopodite carries three slender setae posteriorly and the dorsal surface of the endopodite has a few short hairs or spines (Fig. 4.3). The exopodite was seen in only one dissection and consists of two setae.

The protopodite has one or two relatively long, slim setae on its outer surface and the so-called masticatory process is well developed with a ventral bundle of eight setae including a pilose carrot-like seta, and a dorsal tuft of about eight setae. This 'masticatory process' is more setose and better developed than that of *C. candida* as figured by Brady (1868), Sars (1923–25) and Vávra (1891). In life the orientation of this process as seen before dissection is forwards and upwards as figured by Brady (1868, Pl. 37, Fig. 1e). This orientation also corresponds more naturally with the supposed function of passing food material forward to the maxilla and eventually to the mandibles, than the reverse orientation figured by some other authors. It would appear that the strong retaining muscles predispose this limb to rotation with regard to the endopodite on dissection out.

(f) *Sixth pair of limbs. Walking leg.* (*First thoracic leg, Second thoracic leg auct.*) Figs 4.2; 5.6–12. Protopodite with a seta in the middle of the anterodorsal side which overhangs the knee, below which are four joints and a long terminal claw. The lengths of the joints show considerable variation (v. Fig. 5) as shown by the lengths of joint II (shaded) given in Table 1, thus stressing the need for caution in selecting which limbs and which criteria to use for the discrimination of species. Using an arbitrary length of 15 for joint II, however, and expressing the length of the other joints with reference to this, the variation in the relative lengths is less than might perhaps be expected as indicated in the table below.

TABLE 1. *Candona harmsworthi* Scott. Relative lengths of joints in the sixth pair of limbs

Fig.	Length of Joint II	Relative lengths of Joints (Joint II = 15)				
		I	II	III	IV	Claw
Dissection 1. Right 5, 9	20·1	33·5	15	14·2	6·9	43·1
Dissection 1. Left 5, 10	17·6	37·9	15	12·5	7·5	—
Dissection 2. Right 5, 7	26	—	15	15	7·2	37·5
Dissection 3. Left 5, 11	22·9	30·3	15	15·3	7·3	46
Dissection 5. Left 4, 2	14·7	37	15	16	9	50
Dissection 6. Left 5, 12	21·9	33·9	15	14·5	7·2	51·1
Dissection 8. Right 5, 8	16	36·6	15	15·9	10·8	49·2
Dissection 8. Left 5, 6	16·9	35	15	13·2	9·5	50

On each joint there is an antero-distal seta which overhangs the succeeding joint. A small additional seta occurs anterodistally and to the outside on the penultimate joint and a postero-distal seta on the last joint. Posteriorly the joints have a few short hairs or spines. The claw is long and curved and smooth.

Joints II and III are more robust than those figured by Brady (1868, Pl. 37, Fig. 1f) in *C. candida*, being closer to Brady's figure of *C. detecta* (*op. cit.* Fig. 2f). They are also shorter and wider than *C. candida* as figured by Sars and most of the other species figured by him (1923–25, Pls XXXIII–XXXVII).

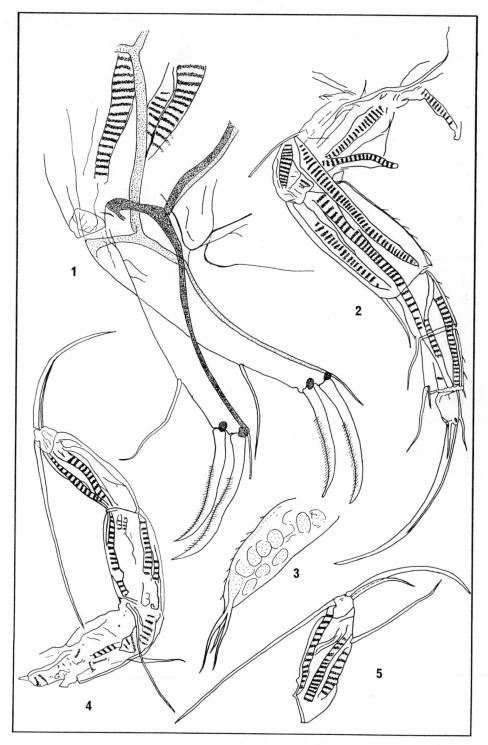

Fig. 4. Magnification in all cases × ca. 380. *Candona harmsworthi* Scott. 1. Caudal rami seen from the right. Dissection 4. Fresh-water pond near Elmwood, Cape Flora, Franz Josef Land. 2. Left limb No. 6. Walking Leg. External view. Dissection 5. From near Kostin Point, Novaya Zemlya. 3. Endopodite of right limb No. 5. Maxilliped. External view. Dissection 1. From near Kostin Point, Novaya Zemlya. 4. Left limb No. 7. Cleaning limb. External view. Dissection 5. From near Kostin Point, Novaya Zemlya. 5. Distal two segments of limb No. 7. Left limb from outside. Dissection 4. Fresh-water pond near Elmwood, Cape Flora, Franz Josef Land.

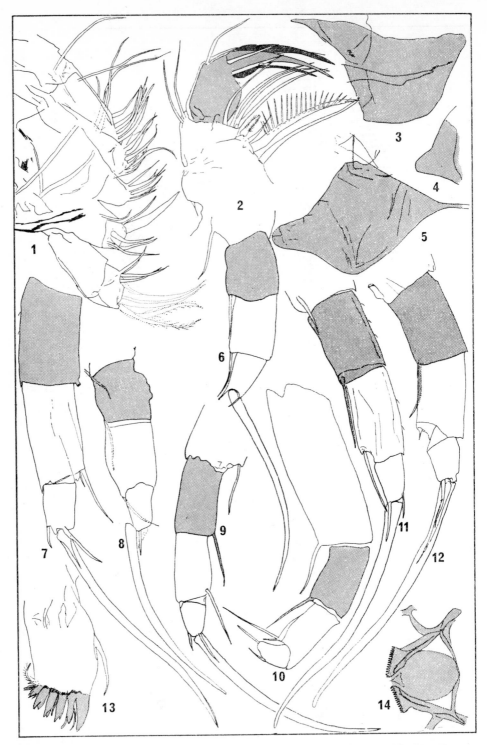

Fig. 5. Magnification of all figures X ca. 380. *Candona harmsworthi* Scott. Material from either a fresh-water pond near Elmwood, Cape Flora, Franz Josef Land (F.J.L.) or from near Kostin Point, Novaya Zemlya (N.Z.). 1. Maxillary palp and 'masticatory' processes. Dissection 2. F.J.L. 2. Mandibular palp. Dissection 1. N.Z. 3. Left genital lobe of female. Dissection 6. F.J.L. 4. Left genital lobe of female. Dissection 5. N.Z. 5. Right genital lobe of female. Dissection 2. F.J.L. 6. Left walking leg. Dissection 8. F.J.L. 7. Right walking leg. Dissection 2. F.J.L. 8. Right walking leg. Dissection 8. F.J.L. 9. Right walking leg. Dissection 1. N.Z. 10. Left walking leg. Dissection 1. N.Z. 11. Left walking leg. Dissection 3. F.J.L. 12. Left walking leg. Dissection 6. F.J.L. 13. Pars incisiva. Dissection 2. F.J.L. 14. Chitinous frame of posterior lip. Dissection 1. N.Z.

(g) *Seventh pair of limbs. Cleaning limb.* (*Second thoracic leg, Third thoracic leg auct.*) Figs 4.4, 5. A four-jointed limb which is directed dorsally, the three proximal joints being long and of similar length, the distal joint small and short. The first joint carries two setae, the penultimate joint also has a seta of medium length a little more than half way down the joint on the posterior side. The terminal joint has three setae—one anteriorly directed and very long, the other two directed posteriorly one being of medium length and the other short.

The extremely long anterior bristle and the relative lengths of the other setae are fairly distinctive.

(h) *The caudal rami.* Figs 3.4; 4.1. These are of the usual type, consisting of an elongate ramus which is gently concave anteriorly, armed with two, curved, backwardly-directed distal claws. The posterior (or subterminal) claw is a little shorter than the anterior one and is about the same length as the slender seta which occurs on the posterior side at about one-third of the length from the distal end. There is a short seta antero-distally at the 'heel' of the ramus. The outer two-thirds of the dorsal sides of the claws are pilose. In immature forms the claws tended to be slightly narrower and more curved. The only other variation noted in the rami was the presence of an additional small seta giving two setae at the 'heel' of the right ramus of Dissection 1. In this respect the left ramus of the same dissection was 'normal'.

Only differences from *Candona* species belonging to the *acuminata* group are covered here. In *C. protzi* the furcal claws apparently lack pilosity and the two claws in each ramus appear to be more widely different in size, whilst the ramus itself is rather more slender proximally and longer. In *C. hyalina* the rami are also more slender and apparently not pilose although the proportions of the rami are generally very similar. Lack of pilosity must be used with caution as a differentiating feature since its apparent absence may merely indicate insufficient examination or simply omission from the published figure.

Differences from *C. acuminata* are more marked, the furcal claws being more slender and the rami more robust in Fischer's species (v. Müller, 1900 Pl. VIII, Figs 8, 12). In *C. caudata* the dorsal seta is short and pilose as is the dorsal and distal part of the ramus, while in *C. levanderi* the ramus is more tapered and is very slender distally.

DISCUSSION

The two other *Candona* species reported from this area may now be considered more particularly.

C. rectangulata Alm 1914 was recorded by that author from Sarpiursak, Greenland; Novaya Zemlya and other Russian localities 'Jenissey, Kap Sapotschnaja, Korga', 'Nienlenka bei Ob', Olofsson added Spitzbergen, and Bronstein added Franz Josef Land, and the Basin of the R. Pechora. This species is most easily differentiated by the shell which generally has a marked antero-dorsal angle in side view (Alm, 1914, Figs 5a, b) but not in the specimens figured by Alm (*op. cit.* fig. 5c), Olofsson (1918 f. 34) or Bronstein (1947, Fig. 151.1, after Olofsson 1918). This species is also slimmer in dorsal view and in side view differs in shape posteriorly, although again Alm's figures show some variation—compare for example his Fig. 5a which approaches *C. harmsworthi* in this respect and his Figs 5b, c which are widely different. Nevertheless, as well as the differences obvious to the eye, Lüttig's measurements and data (1962) are decisively different and confirm that these are separate species.

Judging from Olofsson's Spitzbergen specimen the antennules and antennae are very similar with minor differences of detail, the walking legs of *C. rectangulata* can be matched exactly with one of the *C. harmsworthi* variants figured here, and it is in the cleaning limb that the greatest difference occurs. In the latter the penultimate joint is clearly divided in the Spitzbergen *C. rectangulata* (Olofsson *op. cit.* Fig. 34, p. 2). In passing, attention may be drawn to the variability in the shape of the genital lobe of the female (Bronstein, Fig. 6 and Olofsson, Fig. 34) and Alm (Fig. 5e) as also seen in *C. harmsworthi* (Figs 5.3–5), whilst an anomaly appears to be present in the shape of the furcal claws and length of the dorsal seta of the furca (the latter usually fairly constant) between the type figure of Alm (1914, Fig. 5e) and the figure of Bronstein (1947, Fig. 151, 6).

C. groenlandica Brehm 1911 is recorded from various localities in Greenland and the Jamal Peninsula of Novaya Zemlya, but not from Franz Josef Land. The shell clearly approaches *C. harmsworthi* in dorsal view and Alm's first figure (1914, Fig. 4a) is not far removed in side view being more evenly rounded postero-ventrally. In this respect, as indeed in general shape, it corresponds well with the immature *C. harmsworthi* figured in outline here (Figs 2, 5a, b). Alm's second lateral outline (*op. cit.* Fig. 4c) is far removed from either mature or immature *C. harmsworthi*.

Lüttig's statistics agree tolerably well with those of *C. harmsworthi* except that, according to Lüttig, the position of greatest height lies more posteriorly in *C. groenlandica* (KV = +2·12–2·16, cf. +1·58–1·73 for the present species), and the anterior and posterior margins show equicurvature not infracurvature.

The limbs are difficult to compare because of limited data in the literature (Bronstein merely repeats Alm's figures) but the cleaning limb is closely comparable with that of Scott's species and the furca and genital lobes (bearing in mind previous comments on the latter) are also similar.

The antennule, however, shows differences in the joint proportions, especially in the slender penultimate joint (Alm, 1914, Pl. 1, Fig. 10). Nevertheless, the antennules from Novaya Zemlya *C. harmsworthi* seen in Dissection 1, and those of Dissection 7 from Franz Josef Land, are virtually identical in the proportions of this joint. The other six dissections show some variation but are generally wider in proportion to the length as in the figured specimen (Fig. 2.2). Experience with variation in the walking leg proportions in *C. harmsworthi* suggests that the apparent differences between the antennules of the two species may be attributable to the same factor.

C. groenlandica is clearly very close to *C. harmsworthi* and in all probability the two species are the same, but at present the evidence is insufficient to justify the firm supression of Brehm's name as a synonym of Scott's.

CONCLUSIONS

It is concluded that *C. harmsworthi* Scott is a valid and well defined species living in high latitudes and that *C. groenlandica* Brehm is probably a synonym. Some variation is apparent in the proportion of the various limb joints especially in the case of the first and sixth pairs of limbs, and the orientation of the maxillary palp and 'masticatory' processes to the vibratory plate is as figured by Brady (1868) in *C. candida*. Perhaps more important than the simple rehabilitation of Scott's species, however, is the attention focussed on the different criteria used in the discrimination of species in the genus *Candona*. These are many and diverse, based varyingly on carapace features, limbs and soft parts. Their assessment and rationalization might provide a fertile field for further thought and discussion.

REFERENCES

ALM, G. 1914. Beiträge zur Kenntnis der nördlichen und arktischen Ostracodenfauna. *Ark. Zool.*, **9** (5), 1-20.

BAIRD, W. 1845. Arrangement of the British Entomostraca, with a list of species, particularly noticing those which have as yet been discovered within the bounds of the club. *Hist. Berwicksh. Nat. Club*, 145-8.

BENSON, R. H. 1967. Muscle-scar patterns of Pleistocene (Kansan) Ostracodes. 211-241, 15 Figs. *Essays in Paleontology and Stratigraphy Raymond C. Moore Commemorative Volume.* University of Kansas Dept. Geol. Spec. Publ. **2**, 626 pp.

BENSON, R. H. AND MACDONALD, H. C. 1963. Postglacial (Holocene) Ostracods from Lake Erie. *Paleont. Contr. Univ. Kans.*, Arthropoda Art. **4**. 1-26, Pls 1-4, 8 Figs.

BRADY, G. S. 1868. A Monograph of the Recent British Ostracoda. Part II. *Trans. Linn. Soc. Lond.*, **26**, 353-495, Pls 23-41.

BRADY, G. S. AND NORMAN, A. M. 1889. A Monograph of the marine and fresh-water Ostracoda of the North Atlantic and of Northwestern Europe. Section I. Podocopa. *Scient. Trans. R. Dubl. Soc.*, Ser. 2, **4**. 63-270, Pls 8-23.

BREHM, V. 1911. Die Entomostraken der Danmarksexpedition. Danmarksekspedition til Grönlands Nordöstkyst 1906-1908. *Meddr Grønland*, **45**, 303-17, 2 Pls.

BRONSTEIN, Z. S. 1947. Ostracoda presnii vod. *Fauna SSR Rakoobraznie* **2** (1). Akad. Nauk. SSSR, N.S. **31**, 1-339 (Russian with English summary).

DEVOTO, G. 1965. Lacustrine Pleistocene in the Lower Liri Valley (Southern Latium). *Geologica Romana*, **6**, 291-368, 61 Figs, 1 geological map.

FISCHER, S. 1851. Über das Genus *Cypris* und dessen bei Petersburg vorkommende Arten. *Zap. imp. Akad. Nauk.*, St. Petersburg, **7**, 127-67, Pls 1-11.

FURTOS, N. C. 1933. The Ostracoda of Ohio. *Bull. Ohio biol. Surv.*, **5** (6), 411-524, Pls 1-16.

HARTWIG, W. 1898. Zwei neue Candonen aus der Provinz Brandenburg. *Zool. Anz.*, **21**, 474-7, 2 Figs.

HARTWIG, W. 1899. Eine neue *Candona* aus der Provinz Brandenburg. *Zool. Anz.*, **22**, 149-51, 2 Figs.

HOFF, C. C. 1942. The Ostracods of Illinois, their biology and taxonomy. *Illinois biol. Monogr.*, **19** (1-2), 1-196, 9 Pls, 1 Fig.

KAUFMANN, A. 1900. Cypriden und Darwinuliden der Schweiz. *Revue suisse Zool.*, **8**, 209-423, Pls 15-31.

KLIE, W. 1938. In DAHL., F. *Die Tierwelt Deutschlands und der angrenzenden Meeresteile.* 34 Teil. Krebstiere oder Crustacea. III. Ostracoda, Muschelkrebse. 230 pp. Gustav Fischer, Jena.

LÜTTIG, G. 1962. Zoologische und paläontologische Ostracoden-Systematik. *Paläont. Z.*, H. Schmidt Festband. 154-84, 2 Figs, 1 table.

MÜLLER, G. W. 1894. Die Ostracoden des Golfes von Neapel und der Angrenzenden Meeres-Abschnitte. *Fauna Flora Golf. Neapel.*, **21** Monographie. i-viii, 1-404, Pls 1-40.

——. 1900. Deutschlands Süsswasser-Ostracoden. *Zoologica Stuttg.*, 112 pp., 21 Pls.

——. 1912. Ostracoda. In *Das Tierreich.* Eine Zusammenstellung und Kennzeichnung der rezenten Tierformen. Im Auftrage der *Königl. Preuß. Akad. Wiss. Berlin*, **31**. Lieferung. i-xxxiii, 1-434, 92 Figs.

OLOFSSON, O. 1918. Studien über die Süsswasserfauna Spitzbergens. *Zool. Bidr. Upps.*, V. 183-646, 69 Figs.

SARS, G. O. 1866. Oversigt af Norges marine Ostracoder. *Forh. VidenskSelsk. Krist.*, **7**, 1-130.

——. 1898. The Cladocera, Copepoda and Ostracoda of the Jana-Expedition. *Ezheg. zool. Muz.* St. Petersburg., 324-59, Pls 6-11.

——. 1923-25. *An Account of the Crustacea of Norway.* IX. *Ostracoda. Parts III-VI.* Bergen Museum. (Candonides, pp. 71-89, Pls 32-41).

SCOTT, T. 1899. Report on the marine and freshwater Crustacea from Franz Josephs Land, collected by Mr S. W. Bruce, of the Jackson Harmsworth Expedition. *J. Linn. Soc.*, **27**, 60-126, Pls 3-9.

VÁVRA, V. 1891. Monographie der Ostracoden Böhmens. *Arch. naturw. LandDurchforsch. Böhm.*, **8** (3), i-iv, 1-118, 39 Figs.

DISCUSSION

BENSON: Can you make any comment on the relationship of the frontal or antennal scar in your species. It is a problem knowing what to call this rather important scar?

NEALE: I can throw little light on the question of the frontal scar. Its principal connection appears to be with the antennae, but I did not pay particular attention to this and I am unable to say whether this is its sole connection.

BATE: In dissecting *Cyprideis torosa* from the Norman Collection, kindly supplied by Dr Harding, the V-shaped frontal scar had two muscles attached to it. The largest muscle went to the endoskeleton. The antenna was connected to the shell by several muscles, one of which was also connected to the so-called V-shaped antennal scar, so that this frontal scar serves two functions in *Cyprideis torosa*. The dissected specimen is available for examination at the British Museum (Nat. Hist.).

SWAIN: Regarding the orientation of the vibratory plate in your species, the surface of the plate is set at an angle to the plane of the maxilla as a whole, so that its orientation depends to some extent on how the cover glass is put down on the specimen. I believe that you are correct in orientating it as Kesling has done but had it been caught at the right position it would have been forced down the other way. Thus to a considerable extent it is important to try to orientate the appendages on the slide during drawing as closely as possible to their position in life.

NEALE: I think that you are absolutely right on that, and this has happened to me on a number of occasions. When cutting through the retaining tissue the vibratory plate tends to rotate, but of course one makes allowances for this. It is when you find figured half sections showing different orientations as in Sars that doubts arise. I think the only way to do this is not to dissect but to remove the outer valve and look at the limbs and their positions carefully before proceeding with the dissection. I think that once you start to separate out the limbs you run into all sorts of difficulties if you hope to work out the true orientation of some of these very delicate vibratory plates and setae.

PETKOVSKI: Habe ich Sie richtig verstanden, daß kein Männchen gefunden wurde?

NEALE: Unter diesen Exemplaren gab es kein Männchen, es waren nur Weibchen vorhanden.

PETKOVSKI: Dann weiter, ist die Gruppenzugehörigkeit dieser Art bekannt? Ist es eine Art der *candida*-Gruppe, also etwa ein Neglectoid?

NEALE: Diese *Candona* gehört zur *acuminata*-Gruppe *sensu* Klie. Sie zeigt aber gewisse Ähnlichkeiten mit der *compressa*-Gruppe in der Ausbildung der Schale und der Borsten am zweiten Glied des Mandibulartasters. Der Geschlechtshöcker und die übrigen Charaktere bestätigen die Zugehörigkeit der Art zur *acuminata*-Gruppe.

PETKOVSKI: Also gut, ich fragte, weil der Taster in der Zeichnung nicht eindeutig dargestellt war.

NEALE: Der Mandibulartaster hat vier Borsten am zweiten Glied.

PETKOVSKI: Dann kann das vorletzte Glied eine gefiederte Borste tragen.

NEALE: Es gibt überhaupt keine gefiederte Borste am medialen Rande des vorletzten Gliedes; solche Borsten sind nur an den proximalen zwei Gliedern vorhanden.

PETKOVSKI: Ja, dann ist es richtig, daß die Art zur *acuminata*-Gruppe gestellt wird, obwohl sie äußerlich an einen Neglectoid erinnert. Es ist schade, da alle diese arktischen Arten meist nur parthenogenetische Populationen haben. Die Männchen sind sonst geeigneter für die Determination.

TECHNIQUES AND APPROACHES TO TAXONOMY IN RELATION TO THE OSTRACOD SHELL

CHAIRMAN: PROFESSOR H. V. HOWE

Minutes of this session edited and prepared for publication by the Chairman

SOHN pointed out that modern developments of new techniques are about to start a new era of studying microscopic things, including ostracods. He pointed out that his slide had shown where certain elements are present in the ostracod shell and that it should be easy to determine by use of an electron microprobe all the elements in the ostracod carapace, as well as their exact location within the shell.

SYLVESTER-BRADLEY then showed a series of photographs that he had taken the week before with the new stereoscan electron microscope which had just been installed at the University of Leicester, and invited any of the delegates to visit the University and see it in operation. He pointed out that three firms had started producing these instruments; one in America, one in Japan, and one in Britain. He then went on to say: 'this machine seems to me to have brought as great an advance to micropalaeontology as did the microscope when it took over from the hand lens a century ago. With it one can see a fantastic amount of detail on an ostracod. The micrographs show much more than it is possible to see under an optical microscope. Now this has an incredible importance for taxonomy. It means that for the first time the pictures are more important than the specimens; and pictures can be duplicated, can be published. No longer will it be imperative for us to get back to the original material, because from the pictures we can actually see more than we can from the specimens. This is because the stereoscan microscope is able to give stereo pairs which show a great deal more as photographs than even the screen of the microscope itself. You can't see these things on the screen in stereo because you can only see one picture at a time. Take two pictures with a 5° tilt between them and you get an immediate 3-dimensional optical effect when you look at them with an ordinary pocket stereoscope. Let me emphasize that the machine is not fool-proof. For instance you still have to focus it. You can get your photograph out of focus. However, the depth of focus that is obtainable is much greater than that given by the optical machine. In fact, at the limits of the depth of focus, you must be getting something like 4,000 times the resolution previously obtainable on an optical microscope where, as you know, you either have to stop down when you lose resolution, or open up when you lose depth of focus.'

Sylvester-Bradley's photographs were of four specimens, each being taken at a series of magnifications and most of them in stereo pairs. The first was a whole carapace taken at ×122, followed by the muscle scar field at ×625, followed by a muscle scar at ×2,500. Other photographs were shown of hinge teeth taken at ×26,000. It was explained that one of the great advantages of the machine is that it will go down in magnification to ×20, and then without removing the specimen from the stage you can go on enlarging it up to more than ×20,000 by just moving the controls.

SOHN commented that care should be taken in coating the specimens or artifacts are likely to develop on them.

BENSON pointed out the very great expense of the stereoscan instrument will prevent many workers from obtaining one in the near future. He discussed the method originated by Triebel who used silver nitrate to coat the specimens and photograph them with certain types of more conventional optical equipment. He then showed some pictures of excellent quality, saying 'it is possible to get quite good depth of field and quite reliable photographs and quite good resolution. I published a couple of years ago methods for working with photomicrography in ostracods and have since that time modified the procedure somewhat. For instance the silver nitrate method of coating is much easier to do now than that described by Triebel, by using a rather strong parabolic reflecting lamp which some of us have used for conventional picking and examination. With a somewhat higher voltage one can generate sufficient light and heat to convert the silver nitrate to an acceptable coating. One simply applies a 5% solution of silver nitrate with a small brush and with some water to the slide, and then watching the conversion under the binocular microscope take place, one can control it by either diluting this with the water, or adding more silver nitrate to it until you get the desired effect. You get an emulsion thick layer which will allow the finest detail to be observed and in some of the pictures that I showed you could see the little exoduses of normal pore canals coming through.'

Benson went on to point out that the lenses with long working distances and insert diaphragms in the Leitz Ultropak microscope have much longer working distances than the lenses in standard biological microscopes. He discussed the importance of lighting, stating that Triebel used a diffusing screen to cut out glare from highly silvered specimens and mounted them on an opaque black disc. Benson instead preferred to use a glass mount through which light could pass permitting the background to vary from light grey to black by varying the amount of light from beneath. He concluded by saying 'Lastly the real help in this matter has been the use of an automatic camera which of course allows the light exposures to be consistent on the specimen and I have become accustomed now to be able to take 35 pictures on a 35 millimetre film and recover probably about 33 of them that are quite satisfactory. So in the case of those of us who cannot yet afford the stereoscan I think there is still hope for studying a great deal of the detail in a better fashion than has been done in the past.'

WHATLEY inquired as to the type of film Benson used and found that it was pana-tomic-X.

BATE: 'Dr Benson certainly produces some excellent photographs with the normal optical means and if everyone can produce photographs of that standard, there is probably no need to use an elaborate machine such as a stereoscan for that. Where I think the stereoscan is going to score is in determining hinge structures particularly in small species and again particularly in fossil species where specular light seems to give some indication of hinge teeth which are perhaps not in fact there. Now the stereoscan will in fact resolve this. I feel that I ought perhaps to put in a word for the British Museum stereoscan after Sylvester-Bradley has supported Leicester. The research we are proposing to do, in co-operation with Dr Harding, is sectional investigation of the ostracod shell to see what the structure is of various parts of the shell morphology. It is a new venture and how it will effect taxonomy remains to be seen. Concerning Dr Sohn's comment about artifacts, the thickness of the coating of the gold/palladium mixture is measured in angstroms, not in millimetres. Therefore, unless it is very badly coated by an amateur, artifacts should not be produced. Again I can perhaps qualify this by saying that the specimens should be clean. I think coating in the vacuum chamber would be as easy as coating with silver nitrate, it does not need to be directional. A refinement developed at the British Museum is to incorporate within the bell jar of the vacuum chamber a rotating platform in which is inserted currently 10 stubs, a stub being a metal platform on which the specimens are mounted. This platform of 10 stubs is then, through the means of an electric motor, rotated whilst the gold palladium wire is being vaporized. This results in a uniform coating over the specimen.'

SOHN mentioned having received from Dr Oertli a reprint describing a picking machine, and then went on to discuss contamination of recent material with reworked fossil shells, suggesting that the suspected fossil shell be placed in dilute acid. If no flexible residue remains after the shell is destroyed it is to be assumed that the shell was fossil.

McKENZIE: 'Last night we held an informal discussion on numerical taxonomy and agreed eventually that the purpose in classifying was to achieve a general (multipurpose) classification. As a result of this meeting I feel that we should consider the preparation of a list of shell characters which could be used by numerical taxonomists in their experimentation. Apart from purely quantitative characters, such as length and height of shell, there are many qualitative characters which can be recorded on a presence/absence basis (as pointed out to me by Dr Kornicker), e.g. the eye tubercle, also variability of the inner lamella, of the muscle scars, of various hinge elements. Perhaps we could appoint a committee to draw up an extensive list of both qualitative and quantitative shell characters.'

KAESLER pointed out that if you have 15 presence/absence characters from a hinge and 5 presence/absence characters from a muscle scar you have in effect weighted the hinge 3 times as much as the muscle scar, a situation which usually does not exist if the character is expressed by some other kind of numerical coding or by just plain numerical measurements. In the example given by Kaesler he said: 'Some of these things had to be subjective to a certain extent because I was coding for development, moderate development, or strong development.'

KILENYI: 'May I just ask one question? The whole object of numerical taxonomy is that we should get results which can be repeated. This is the whole object as far as I can see it. Now if we code things as moderate, strong or weak, these are again subjective elements. You judge how strongly spinose a form is and you describe the form accordingly. I have had this problem with *Cypridea* where the degree of overlap is supposed to be critical according to some authors. Now can we get rid of this one? Because, if we can, then I am all for it.'

KAESLER: 'Yes, we could make these operational. We can define operationally what we mean by strong caudal process, or we could in fact measure the caudal process somehow. It would take a little thought, but we certainly could make it so that it would be repeatable.'

BENSON then referred to a paper he had recently published as one approach to a shape analysis of fresh-water ostracods in which the outlines vary and are important in the identification of species. He pointed out that it is possible to break down a shape into as many as 360 characters or parts by simply getting the distance from a given point to any part of the outline.

HARDING called attention to the work of Gowers at the Rothamstead Institute, to which Kaesler replied: 'The last time I heard about anything he had done he was using factor analytic techniques and some of these have been used at the University of Kansas.'

DELORME asked if anyone had ever found shells of *Eucypris crassa* in Pleistocene or Holocene sediments. He called attention to the work of Vinogradov, who in 1953 reported that all ostracod shells were composed of calcium carbonate. Delorme then went on record to say that the shells of *Eucypris crassa* were mineralogically calcite or some similar carbonate, but with a high percentage of strontium in the carbonate lattice. X-ray fluorescence and flame photometry indicated a substantial amount of strontium in the shells. The shells of *Candona renoensis* Gutentag and Benson, 1962, obtained from the same pond as *Eucypris crassa* were also subjected to the same tests and gave negative results for strontium. X-ray powder photographs of *Eucypris crassa* shells show a calcite pattern with small deviations from the standard. It would appear then, that the large amount of strontium in the calcite lattice causes the structure to be extremely unstable above certain critical temperatures. As a consequence the shells of this peculiar species are not preserved in Pleistocene or Holocene sediments.

KORNICKER then suggested that the electron scanning microscope be used as a means of determining the presence of a ligament line in podocopid ostracods.

SYLVESTER-BRADLEY stated that he was not aware that anybody had found in the podocopids a ligament which was elastic in the sense of opening the valves externally and inquired if Kornicker had found in dead specimens of myodocopids any groove of the so-called ligament.

KORNICKER explained that when he tried to open complete specimens of the myodocopids to examine the hinge the shell broke because the ligament is stronger than the shell, saying: 'This is a problem that you don't have in the podocopid ostracods. But from drawings and photographs of podocopid ostracods, and this is supported by the evidence that Dr Benson gave us today, there is some evidence that such a ligament line is present.'

SOHN remarked that Melik (Melik, J. C. 1966. *Contr. Mus. Paleont. Univ. Mich.*, **20** (8), 195–269,) described a Palaeozoic ostracod group about a year ago that apparently had a ligament groove in both valves.

SYLVESTER-BRADLEY again expressed doubt that the podocopids have a ligament. He continued: 'I rather agree with Dr Bate that we have the valves held together by an outer layer of integument which is not in the same position as the ligament which Dr Kornicker was showing of the myodocopids, but I am not sure of that.' To which Kornicker replied: 'All that I can say is that when Mallory's Triple stain is used on sectioned podocopids and platycopids, the valves stain a light blue and in the area of the hinge, dorsal or attached margin, you get a section which stains a deep blue and very often is fibrous in nature. I am calling this the ligament. This is all the ligament is, a piece of the integument that has differentiated differently from the shell. Now it is necessary to have something there and if you wish to call it differentiated integument this is perfectly acceptable.'

PURI: 'I would like to go back to the question of loading in numerical taxonomy. All organisms have characters that are either primary, secondary or tertiary and for the classification to be meaningful it has to be subjective because it has to be used by human beings. Unless we can devise some sort of mathematical model where we can take into account the primary, secondary and subsidiary nature of some of the characters they are going to end up building a classification like a bikini. It is going to show what's apparent and will hide what's vital.'

The session then adjourned to the laboratory to examine in detail the stereo photographs which Sylvester-Bradley had brought.

TECHNIQUES AND APPROACHES TO TAXONOMY IN RELATION TO THE LIMBS AND SOFT PARTS OF OSTRACODS

CHAIRMAN: DR J. P. HARDING

This account edited and prepared for publication by the Chairman

HARDING: Dr Kornicker has given us a very clear picture of the continuous nature of the ostracod shell and other modifications of the cuticle and rightly draws attention to the great distinction between the outer hinge and the inner hinge of the calcified valves. To use the word 'ligament' for the flexible part of the cuticle which joins the valves tends to detract from its continuity with the valves; but it is not the word 'ligament' that I am not happy with but the description of the ligament as 'connected to' the shell valves, I would prefer the words 'continuous with'.

KORNICKER: I am in complete agreement with you on this. I tried to stress this continuity. When I use the words 'connected to' I do not mean that the ligament is a separate part, I am using it in the sense that a cow's leg is *connected to* the body.

FRYER: There seems to have been some confusion when the ligament of the ostracod shell was contrasted with those of bivalve molluscs. There too the ligament is continuous with the calcareous valve, both of them having been secreted by the mantle.

MCKENZIE: I would like to ask for any literature records of resistant eggs in the fresh-water cytherids and in the darwinulids, since if resistant eggs are not present in these groups, how is it that the same species occur in both Africa and South America in the Cretaceous-Jurassic?

MCGREGOR: When the eggs of *Darwinula* are developing they are constantly being revolved in the carapace by the parent. This may possibly be important during embryonic development, but I do not believe it is a valid criterion for minimizing the possibility of producing resistant eggs.

MCKENZIE: *Darwinula* from S.W. Tasmania also carries the young larvae inside the brood chamber, up to the third larval stage in the only specimen I dissected. Possibly this is a protective adaptation. The platycopids *Cytherella* and *Cytherelloidea* also have brood chambers for the larval stages, as do fresh-water cytherids.

FRYER: With regard to Dr McKenzie's question about distribution by resting eggs. I do not think we should always be willing to take these things at face value because if we think of the fresh-water Copepoda and other groups we find that organisms with resting eggs often do not have as wide a distribution as those that do not have resting eggs. Many of the Calanoida are known to have resting eggs but are very restricted in their distribution while many cyclopoids apparently have no resistant resting stages and yet are very widely dispersed.

MCGREGOR: Dr Vernon Proctor, Texas Technological College, has reported ostracods that have resistant eggs which remain viable following passage through the intestinal tracts of birds belonging to several Orders. I would be interested in trying to hatch eggs removed from the carapace of *Darwinula*, in the laboratory. I do not know how they could remain viable in the intestinal tract of birds once the ostracod appendages began to develop.

HARDING: The resting eggs in the ephippia of Cladocera are notoriously difficult to hatch, and it may be that they hatch more easily after they have passed through the gut of a duck than before.

SOHN: Bronstein has indicated that ostracods can be transported on insects. We do not know how long an ostracod, if it is in mud and clinging to a bird or some other animal, will live during transport. It might be as much as several days permitting a world-wide distribution.

FRYER: I have myself seen a species of *Cyclocypris*, attached to the setae on the limbs of water boatmen and I have no doubt that ostracods could be dispersed by this means from at least one point to another on the short flights of the water boatmen.

HOWE: Some years ago the department of wild life and fisheries in Louisiana discovered curious objects in the gullets of ducks and brought them to me. They were ostracods and it was quite easy to tell where a duck had been feeding, whether it was brackish water with *Cyprideis* in it or whether it was freshwater with such things as *Cypria* and *Candona*. I would not be surprised if those were alive at the time the duck had been examined so I would strongly suspect that ostracods may well pass through the digestive tract of migratory water fowl and I would suggest that zoologists might well examine the digestive tracts of ducks for ostracods.

PURI: The Florida Board of Conservation several years ago had a problem which centred around the migration of shrimps and their feeding grounds. I was asked to help them and we dissected eight or nine stomachs of two species of *Penaeus* and found carapaces and appendages of several cytherids in them. One of them was a brackish water form, *Cytheromorpha pascagoulensis* Mincher.

FERGUSON: Dr Vernon Proctor of the Texas Technological College and his students have been working recently on the distribution of a large number of different species by way of the alimentary canal of birds. I was requested to identify the ostracods used in these experiments; among these were several species of the genus *Cyprinotus* and a recently described new species, *Cypris puertoricoensis* Ferguson, 1967. These ostracods were among the several species of organisms found to be distributed by way of the alimentary canal of birds.

SWAIN: In contrast to the wide distribution that has been mentioned I would like to point out the very restricted distribution of certain fresh-water Ostracoda from Central Africa. Genera described by Sars have been found recently in sediments as old as Pleistocene and perhaps as old as Pliocene; some of these forms are not known anywhere else in the world.

DELORME: Perhaps it would be advisable to look as well to causes or influences which make birds migrate to certain areas. In Canada, today *Cytherissa lacustris* is not found below 54° latitude, but in the Pleistocene and Holocene sediments of the interior plains I have found this species to be extremely abundant as far south as I have worked, i.e. the 49th parallel. I think perhaps the reason is that the climatic changes since Pleistocene times have pushed the forests so far north and the type of lakes which accompany forests, that the ostracod population is forced to colonize lakes further north and become locally extinct further south.

WHATLEY: Dr McIntyre at the Torry Marine Laboratory in Aberdeen, asked me to look at the stomach content of larval plaice and quite a considerable number of the things which these larval plaice were eating off the Scottish coast were in fact ostracods, difficult to identify because they were pretty well mashed up, but I think that *Loxoconcha rhomboidea* was one of the major constituents.

KAESLER: If I may return to the economic aspects, Professor Reyment at Uppsala has found that newly hatched oyster drills (Gastropoda) in Long Island Sound feed on ostracods before they are large enough to attack oysters.

McKENZIE: We now know of a Mesozoic Lower Cretaceous bird in Australia (feathers have been found) and of course one has been reported from the Upper Cretaceous of South America. But how far would you expect these ancient birds to fly when you don't know what they looked like? Maybe they couldn't fly very far.

SOHN: Well they do not have to. They could fly a short distance on one trip and on the next one go a little farther.

BENSON: It seems to me that some of us ought to try to get to places where fish are being caught and throw some of the gut contents into water to see if they might contain living ostracods. I know of many species distributed among Pacific Islands, broadly separated one from the other. It seems to me that the only way they could arrive throughout these great distances is through transportation in the guts of fish or fowl. This really should be tested. There are too many ecologic barriers otherwise which would seem to separate members of these various faunas.

GREKOFF: We can speak only very circumspectly about Mesozoic birds. Since we do not know the configuration of the continents at that time, how can we speak about birds' m gration. In the Mesozoic we have very good faunal connections from Central or West ¡Africa with South America but rather bad connections with Europe. Moreover in the Congo basin, we have not the same species of *Cypridea* that we have in Gabon for instance, so we may speak about a barrier here as well as between Europe and Africa.

SYLVESTER-BRADLEY: I feel that there is some confusion about the similarity of species in different continents during the Mesozoic. One of the astonishing dispersals that we have in the Mesozoic is the very wide dispersal of the genus *Cypridea* and various other genera and subgenera related to it. It is not always easy to recognize species within this genus because of the extraordinary range of variation found within species. It is possible that this variation is caused by parthenogenetic reproduction, but in any case there are certain parts of the geological column where the range of variation is so great that you do not know whether you are dealing with a species complex or not and I would distrust evidence which suggested that we had the same species in widely separated continents in this particular family. What is quite clear is that we get the same genus in Japan, Africa, South America, North America and in Europe. I do not think that we can blame birds for this sort of dispersal because they would have had to evolve from one species to another on their journey across the Atlantic, but I wonder what it is that causes that particular fresh-water genus to have such a wide dispersal and yet to have species which are both important from the point of view of local occurrence and from the point of view of stratigraphical value because they altered in fairly short distances in the geological column.

SWAIN: The similarity of the facies in which the Purbeck-Wealden genera such as *Bisulcocypris* and *Theriosynoecum*, *Cypridea* and others occur are certainly very striking. They are all in pastel varicoloured mudstones or fresh-water limestones. Rather facetiously one might suggest that pterodactyls, of which there were plenty around at that time, could easily have carried a whole fauna of these Ostracoda.

KORNICKER: With regard to the weighting of characters, I think there is some general agreement that the copulatory apparatus should be relatively important for classifying taxa. If there is a good means of separating or classifying on the basis of the copulatory organ, are we wrong in not using this? I suggest that we come to some sort of agreement as to those characters that we think should be used and then try to use them.

SWAIN: A very worth while suggestion because in trying to work both with fossil shells and the appendages of living Ostracoda the characteristics we now use for taxonomic purposes appear to fall far short of our requirements. The shell features are far more diverse in many instances than would be indicated by the generally described features of the appendages, so we need something else to work with besides the appendages.

SYLVESTER-BRADLEY: I think it is one of the oddities of the taxonomy of ostracods that when we have plenty of shell features in the ornamented ostracods we find that we can divide up and classify in much finer detail than the zoologists using appendages, and yet when we have smooth ostracods with few shell features we find that the neontologists can divide up and make species far more successfully than we can. Now, clearly if we are going to get sense into the classification of ostracods we must use both the characters of the carapace and the characters of the soft parts and the mystery lying behind classification must always be which characters are evolving and which are remaining static in the phylogeny; we need to know which are the conservative characters, the ones which mark common ancestry and which characters are evolving quickly and therefore adapting themselves to environment and likely to produce convergences. Now as far as I can make out, there is not one character in the ostracods which does not sometimes behave in one way and sometimes in the other. You cannot lay down general rules of what is important in classification and what is not. One of the most interesting parts of the last session we had was the indication by two of our authors that two particular parts of the ostracod, the copulatory apparatus and the furca, taken by themselves could produce a complete classification. It seems that we need not look at anything else; we can get a classification out of just studying one part of the ostracod; this was particularly clear in Father Rome's brilliant and exciting paper describing so many different kinds of furca. Now, the danger is that we should do that; that we should be tempted to make classification from just one character like that, because there are bound to be many convergences in anything which is variable, and this applies equally to the soft parts and to every part of the shell: to the muscle scars, to the hinge, to the pore canals. None of them can be relied on absolutely, although in particular groups one can be sure that some are more reliable than others. These are the characters we are inclined to weight as taxonomists. However, numerical taxonomic colleagues say they are better not weighted because of their unreliability in certain fields.

PURI: I think it was Skogsberg who maintained that the male copulatory organ in hemicytherids and cytherids was the seat of evolution and this is perhaps the only criterion he used when he split up a lot of hemicytherids almost 20 or 30 years ago. I do not think we have been looking at soft parts in half as much detail as we have with the finer features on the carapace. If we are going to have a classification based entirely on the male copulatory organ, first we have got to find the male. About a third of the genera of fresh-water ostracods are based on the study of only females and the males are unknown. Even in a marine habitat, males are very difficult to find. G. W. Müller did not find males of half of the species he described. It has taken us five years to find males of 40 or 45 species of marine cytherids in the Gulf of Naples.

I feel that we need more seasonal work like that of Tressler and Smith where they established percentages of males which appear and disappear seasonally. I think we need studies like this to understand more fully the habitat of the males, and the ecological conditions under which they appear seasonally.

McKENZIE: There are many records of preserved soft parts from as far back as the Carboniferous, and mummified entire soft part anatomies have been found in Pleistocene or sub-Pleistocene ostracods.

HOWE: The difficulty with published classifications of recent ostracods such as Hartmann's of the Cytheracea, which I have tabulated, is that as you go through the appendages and the copulatory organ you will discover that for many genera you do not have the information. The reason for the lack of information on the part of zoologists in regard to all the genera that were involved in Hartmann's classification should be perfectly obvious. There has been an economic incentive for the palaeontologists to go to work on the hard parts of the carapace, there has been no such economic incentive for the zoologists.

McKENZIE: Taxonomic conservative characters do exist in the soft parts: we have just got to find them, and include them in Dr Kornicker's list. Dom Rome has discussed the attachments of the furca; I think the furca itself in all crustacean groups is a very conservative character and one of the things which has struck me about the furca in, say, Darwinulidae, where it is strongly reduced, is that perhaps this is an indication of the fact that Darwinulidae in an evolutionary sense are a 'stagnant' group. In Notostraca too the furca is strongly reduced and this is also a 'stagnant' group.

HULINGS: One additional comment on the use of the copulatory organs for taxonomic purposes. I have looked at a complex of genera including *Pontocythere*, *Cushmanidea* and *Hulingsina* from Cape Cod, Massachusetts, the Virginia coast, the Gulf of Mexico and Puerto Rico and I find within this complex no less than seven different forms of copulatory organ.

BENSON: In working with the copulatory organs of the hemicytherids and trachyleberids I am impressed with how complex they are morphologically and how difficult they are actually to analyse. I, a palaeontologist, would like a little morphological guidance in unravelling this anatomy, and to see more of the kind of the work that we saw from Dr Hart earlier today.

KORNICKER: I think we should think of criteria to be used in classifying in terms of their potentials for being isolating mechanisms. If we cannot use copulatory organs, what is the next best criterion? Are sex characters more important than some other characters? Should we weigh characters, or just use anything we find convenient for separating groups? I think the latter procedure is rather haphazard.

HARDING: I would like to stress the value of an understanding of the function of a feature we describe whether it be a part of the shell or a seta on an appendage. If we know what the part is for, how the animal when it is alive, uses it, we can relate this to the habits and ecology of the animal. The comparison of homologous organs in different animals can be made more intelligently, for a knowledge of the function will show which is the more primitive and indicate the direction of selection pressures. Classification becomes more certain and examples of convergent evolution can be recognized for what they are and clarify rather than cloud the issue.

III. ECOLOGY

DETERMINATION OF BIOGEOGRAPHIC BOUNDARIES OF QUATERNARY ARCTIC MARINE FAUNAS

RICHARD H. BENSON
Smithsonian Institution, Washington, D.C., U.S.A.

SUMMARY

The application of a technique basic to numerical taxonomy, also discussed elsewhere in this volume by Kaesler, was presented to demonstrate the function of a heuristic model of the structure of northern Pacific, northern Atlantic, and Arctic ostracod faunal distribution. This is the first time, known to the author, that this technique has been utilized for the study of biogeographic distribution over such a broad area. The interpretation of this model is based on the assumptions that: (*i*) the distribution of Arctic ostracod species and those of adjacent faunas is structured and this structure is polythetic and hierarchic, (*ii*) that the distribution of the Arctic fauna has progressed toward a general trophic equilibrium and has developed some recognizable inner structures, and (*iii*) that bonding between species, an expression of a tendency for mutual occurrence, can be represented in a model by clustering pairs of species, or pair-groups of species based on one of several measures of the degree of their association.

The models, in the form of a series of dendrograms, showed that the Arctic ostracod fauna (i.e., assemblage of species groups) is quantitatively more similar to the Atlantic fauna than to the Pacific fauna, even in close proximity to the Bering Strait where one might expect considerable mixing. It indicated a high degree of similarity between the fauna of northern Norway and that of the North Sea as compared with the Arctic fauna. The fauna of Iceland has two aspects, one somewhat indigenous, which may be typical of the cold temperate fauna, and the other more closely related to that of Europe. Too few samples were available from the northern Pacific for intrastructural comparisons; but considered as a whole, it is very distinct from either the Arctic or Atlantic faunas.

The results of this method were compared with more conventional intuitive evaluations of a simple data matrix, charts of concurrent range zones, and individual species maps.

The same statistical technique was applied to data compiled by Brady, Crosskey, and Robertson in the study of the 'Post-Tertiary Entomostraca of Scotland', published in 1874. The clustering of their localities, based on the relative sharing of species in common, suggested the temporary influx of a higher concentration of more northerly assemblages of species during this time than are now living in the surrounding waters. The clusters of species that might represent natural assemblages are not 'Arctic', as they suggested, but are high or cold temperate.

This study is still incomplete. The theoretical basis for the statistical procedure as applied to biogeographic and stratigraphic models is still being considered and additional distributional data are being compiled.

DISCUSSION

KAESLER: Do you think that the reason for the dispersion or strange configuration of the Jaccard coefficient in the case of the Scottish fauna is because the Scottish fauna is transitional between the Atlantic and Arctic?

BENSON: Yes, in fact the Jaccard tends to scatter units that are transitional whereas simple matching tends to amplify units that are separate.

KAESLER: Jaccard coefficients tend to be lower than simple matching coefficients, particularly if most stations contain a small proportion of the total number of species found throughout the study area.

BENSON: Most of the localities in the Firth of Clyde region tend to cluster together, as do those in the Firth of Forth region, which probably means that these environments differed from each other during the Pleistocene.

KILENYI: What was the basis of grouping on your last diagram? How did the shallow water, brackish and marine forms come up in clusters?

BENSON: I found that some of the very shallow water forms and almost littoral forms occurred together, and some of the obviously brackish forms occurred together. The others I think are eurythermal and were put in the eurythermal group. It would be interesting to look at the results of simple matching coefficients in which we deal with common absences and see how that works out.

McKENZIE: I would like to defend *Rabilimis* here. Your data indicated that *Rabilimis septentrionalis* (Brady) ranged north of about 70°N both in the Atlantic and Pacific so it is a good Arctic form. This is also the pattern for the other species we know well, *Rabilimis mirabilis* (Brady), and Dr Elofson has given considerable data on the distribution of this taxon. Furthermore there are two other species that have been assigned to the species group by Professor Swain which are also Arctic in their distribution. So, ecologically this is a good unit. Dr Elofson has also pointed out that there are anatomical distinctions which keep *Cythereis mirabilis* distinct from other trachyleberid and hemicytherid species such as those in the *Echinocythereis* group. There is also a general shape distinction; *Rabilimis* does not look much like an *Echinocythereis* or any other genus.

THE PROBLEMS OF OSTRACOD ECOLOGY IN THE THAMES ESTUARY

T. I. KILENYI
Sir John Cass College, London, England

SUMMARY

In a preliminary study of the recent ostracods of the Thames Estuary the fauna of 250 samples was quantitatively considered. The problem of biocoenosis versus thanatocoenosis is discussed and some new methods of recognizing the biocoenosis are suggested, such as state of preservation, ratio of carapaces and valves, number of juveniles, etc. Based on results achieved in this way, seven major ostracod biofacies are recognized in the Thames Estuary. These are largely controlled by salinity but in one, temperature and perhaps nature of sediment play a decisive role. The degree of mixing between bio- and thanatocoenosis is considered and a decreasing degree of mixing is found seawards in the Estuary.

The controlling factors in the distribution of Ostracoda are dealt with; the prohibiting factors being 'liquid mud' and pollution in the Inner Estuary, and 'black mud' and a certain type of grain size distribution in the Outer Estuary. The main controlling influence in the composition of the fauna is salinity, the interpretation of other factors will require further study.

INTRODUCTION

The present paper is a preliminary report on some of the results of an investigation into the ostracods of the Thames Estuary. This work is being carried out as part of a broadly based research programme, dealing with the movement, transport and deposition of sediments (Shallow Marine and Estuarine Sedimentation Research Unit). The past and present distribution of ostracods has a direct bearing on some of these questions; the presence of stenohaline forms in the only slightly brackish Upper Estuary, for instance, indicates the net landward movement of sediments within the Estuary. A positive connection also exists between ostracod faunas and sedimentary environments. Lack of ostracods in certain parts of the Estuary coincides with either toxic bottom conditions created by the peculiar conditions of the Estuary or with a certain type of grain-size distribution of marine sands.

The present study is based on approximately 250 samples which were selected from almost 1500, collected for, or by, the Shallow Marine and Estuarine Sedimentation Research Unit. The selection of these samples for microfaunal studies was guided at first by the object of achieving a more or less even coverage but as patterns of distribution were gradually recognized the choice of further samples was influenced by specific objectives. This accounts for the apparently very uneven distribution of samples (Fig. 1).

251

All the samples were taken by either a grab or cone sampler. Admittedly these methods of sampling are not ideal for ecological purposes as they disturb and sample the sediment to a depth of up to three to four inches. The sampling position was determined with a Decca Navigator or, in the Inner Estuary, by radar fixing. Most of the samples were left untreated, then washed, dried and 100 grammes sieved. The 44 and 60 mesh (B.S.) fraction (0·335 and 0·250 mm) was subsequently picked for ostracods. Naturally these were mostly in the form of empty carapaces or valves although a few specimens with appendages present were recovered. About 30 samples were taken with the view of obtaining a 'live' fauna. These were stained with Rose Bengal at the time of collecting and treated with formaldehyde. The sieving and picking followed the already described procedure. In all cases 100 grammes of sediment was sieved and the number of ostracods present in the 44 and 60 fractions was expressed in the number of valves. If less than 100 grammes of sediment was available the number of valves was corrected for 100 grammes. Similarly when the 44–60 fraction contained large quantities of sediment it was split but the number of valves was again computed for the original amount.

The picked ostracods were all mounted on micropalaeontological slides. The details of each slide were entered on large filing cards listing the species present, their numbers, the ratio of valves and carapaces, percentage of juvenile moults, nature of preservation, etc. All these results will eventually be transferred on to punched cards with a view to using the methods of numerical taxonomy.

From the above introduction it will immediately be obvious that the approach to the problem of ostracod distribution is more palaeoecological than ecological. This bias exists for various reasons, such as the methods of sampling, the type of results required, the need for a more accurate understanding of fossil estuarine ostracod faunas and finally the author's training.

ACKNOWLEDGEMENTS

The author is extremely grateful to Mr B. D'Olier, Mr M. F. Elvines and Mr R. Maddrell, Shallow Marine and Estuarine Sedimentation Research Unit, Sir John Cass College London, for generously making available their unpublished results. All the sedimentary analyses were carried out by them. Miss M. Teed and Mr C. Burchell gave valuable assistance in the preparation of diagrams.

THE HYDROGRAPHY OF THE ESTUARY

The Thames Estuary is a typical tidal estuary with a wide, triangular outer portion and a narrow, winding inner part. It is customary to talk about Inner and Outer Estuary, although the definition of these terms and the boundary between the two is not always clear. For the purposes of this study the area between London Bridge and Southend will be called the Inner Estuary, the part east of Southend the Outer Estuary. The seaward limit is taken to coincide with the sandbanks of Inner Gabbard, Galloper and South Falls (Fig. 1).

It is a positive or normal estuary (Emery et al., 1957) where the salinities are reduced upstream, average salinities, especially in the Inner Estuary, depending mainly on the volume of the fresh-water flow. The maximum ranges of variation according to seasons and state of tides are given below (Data from *Technical Paper* 11, 1964).

At London Bridge   0·01–2‰
10 miles from London Bridge   3–12‰

20 miles from London Bridge   9–20‰
30 miles from London Bridge   15–26‰
40 miles from London Bridge   20–33‰

In extreme conditions of drought or flood the range of variation can be even greater. In the Outer Estuary the salinity figures approach that of the North Sea (slightly over 34‰) although in the western parts in a wet winter this figure may drop down to well below 30‰. The Wallet and the Gore Channel are hardly affected by fresh-water flow therefore more constant salinities exist there. In the Inner Estuary some vertical grading of salinity exists, the maximum observed was 1·9‰ higher on the bottom than on the surface but on the average the figure rarely exceeds 1·0‰ (*Technical Paper* 11, 1964).

In the Inner Estuary constant dredging is carried out so only the shallow areas near the banks were sampled. In the Outer Estuary three main types of environment can be recognized; tidal flats, sandbanks and channels. The average depth of water (at low tide) is between 0–2 fathoms for the flats and sandbanks and between 8–12 fathoms for the channels. The channels carry predominantly ebb or flood currents and are termed accordingly, although there are some mixed channels as well. The tidal range is between 12–19 feet. Water temperature varies between 2–17°C but in the Inner Estuary appreciably higher temperatures prevail (see pp. 256-7).

FUNDAMENTAL FACTORS CONTROLLING THE DISTRIBUTION OF
OSTRACODA IN THE THAMES ESTUARY

The present distribution of ostracods in the Thames Estuary (biocoenosis and thanato-coenosis) is controlled by several major factors; the composite effect of which is responsible for the present-day picture. These controlling factors fall basically into two groups. The first group comprises prohibiting influences the degree and extent of which determine the distribution of the ostracod fauna as a whole, the abundance or lack of individuals and species. Some of these are due to man's artificial inter-ference in a natural environment and as such need not be taken into account in a strictly palaeoecological study. The second group of factors influence the distribution of the various species within the estuary. These are obviously the more important and also the more difficult to interpret.

Prohibitive factors

Fluid Mud. In the so-called 'Mud Reaches' of the Inner Estuary (approx. 12–30 miles from London Bridge) a peculiar environment is created by the mixing of fresh and salt water. The suspended clay minerals in the fresh-water flocculate on coming into contact with salt water and this flocculation is most intensive at or near the bottom where the salinity is higher than on the surface. The result is an approximately three feet thick layer of 'fluid mud', where the concentration of solids may be as high as 180,000 p.p.m. (Allen, 1954; Prentice and Dobson, 1967). On deposition the fluid mud produces a thixotropic soft bed of mud. The effect of the presence of fluid mud is a drastically reduced transparency and an extremely soft substratum resulting in a totally unacceptable environment for benthonic ostracods. Along the banks or where sand is carried as traction load the formation of fluid mud is reduced or may be pre-vented altogether.

Polluting discharges. Vast quantities of sewage and industrial discharges enter the

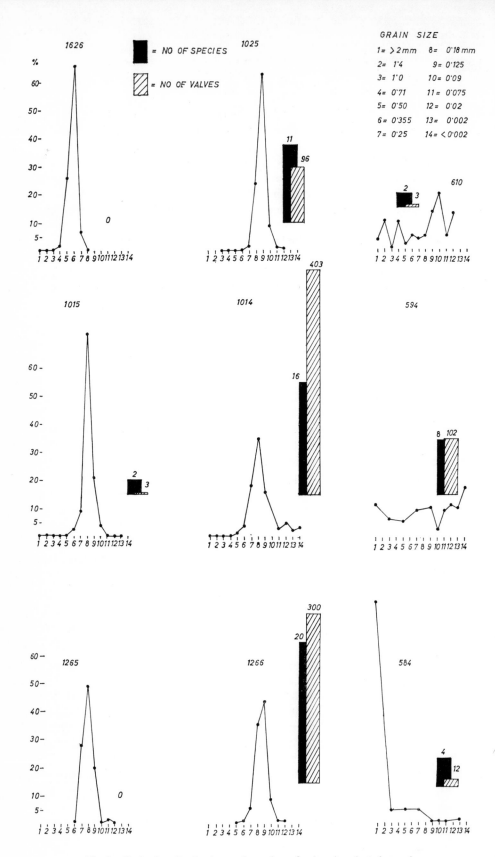

Fig. 3. Grain size distribution and number of valves in selected samples.

Inner Estuary at various points between London and Southend. Biologically, the most serious effect is caused by sewage with its high quantity of oxidizable material reducing the level of dissolved oxygen in the water, resulting at some places in completely anaerobic bottom conditions. The area affected by low oxygen levels or anaerobic conditions depends on the amount of fresh water entering the estuary (assuming, of course, on change in the level of sewage discharge). In the Autumn of 1959, for instance, a tract of the Inner Estuary from 5 to 25 miles below London Bridge was anaerobic, and even 40 miles down the water was only 60 per cent saturated with oxygen (The Effect of Polluting Discharges, *Technical Paper* 11, 1964). The degree of oxygenation is, however, very variable along the length and width of the Inner Estuary and in recent years great advances have been achieved in reducing anaerobic conditions.

Black Mud. This term is used to describe a black, thixotropic muddy deposit, found generally in channels, East of the Black Deep (Warp-Oaze area, Outer Estuary). It is in fact an accumulation of dumped material originating from the Inner Estuary and thus is closely akin to the consolidated fluid mud and its influence on benthonic ostracods must be the same. Populations from samples with a high black mud content tend to show the presence of thanatocoenose only.

Nature of Sediment. The above mentioned limiting factors are more or less restricted to the Inner Estuary or a relatively small area of the Outer Estuary. In the rest of the area studied the nature of the substratum seems to have the most decisive influence. The effect of bottom sediments on the ostracod fauna is either prohibitive or selective, the former is discussed here, the latter in the next part.

The deposits in the Outer Estuary are remarkably uniform, well-sorted marine sands. There is very little difference between the sandbanks and channels in this respect. The mean grain size is between 0·12–0·2 mm but the histograms of sandbank samples show a higher degree of sorting; samples from the channels generally show slightly right-skewed histograms with a small silt-clay fraction. In spite of this rather uniform nature of the sediments, very slight changes in grain-size distribution have a profound effect on the ostracod fauna. The 'critical grain size' in this respect is between 0·25–0·50 mm. If the quantity of this fraction exceeds about 10 per cent of the total it is rare to find any ostracods at all (Fig. 2). The presence of grains 0·5 mm and larger in any quantity has little influence on the abundance of ostracods. Right-skewness of the histogram has invariably a positive influence on the numbers of ostracods present.

This close connection between sediments and populations cannot be recognized in the Blackwater Estuary and Wallet (samples 610, 594 and 584, Fig. 3), which from a sedimentological point of view represent a different environment. Here the sediments are very poorly sorted and the histograms of various samples do not show any particular trend.

The interpretation of this close connection between bottom sediment and abundance of ostracods is a complex question and basically two lines of argument are open.

1. The lack of ostracods in sediments with a high percentage of 0·25–0·5 mm grain size is due to a critical current velocity at or near the bottom. This current might be too fast for the existence of ostracods. A strong argument against this interpretation seems to be the presence of often large numbers of foraminifera in these ostracod-barren samples. Even if one takes into account the different hydrodynamic properties of the ostracod carapace and the test of foraminifera it is difficult to see that such a difference could result in this kind of clear cut separation. In any case there are some ostracods and foraminifera of comparable size and not all that different shape

(e.g. moults of *Cushmanidea elongata* and *Quinqueloculina* sp.) which do not occur together if the 'critical fraction' is present.

2. It seems therefore more logical to correlate the nature of the substratum with the ability of the ostracod to exist over (or in) such a milieu. Wieser (1959) studying the effect of grain-size on the distribution of small invertebrates, including ostracods, came to the conclusion that the grain-size distribution exerts a profound influence on the epi- and infauna. He found that the 0·2 mm median grain size acts as a sort of barrier, separating the interstitial sliders which inhabit sediments with higher grain-size from the burrowing or surface crawling organisms. The majority of the benthonic podocopid ostracods can live either interstitially or on the surface (or burrowing).[1] The latter mode of life is the most commonly mentioned in the literature. Williams (1969 and personal communication) paid special attention to the interstitial fauna in some beach sands on the coast of Anglesey and found a rich ostracod fauna in the interstices although mainly in the form of young instars. He came to the conclusion that the shape and especially the size of the voids in the sediment have an important limiting influence on the ostracod fauna.

In the Thames Estuary most of the sampled sediments are too fine grained for the interstitial existence of most of the ostracod species, although it is quite possible that younger instars of all species do live this way. The effect of grain size distribution on bottom dwelling or burrowing ostracods has received scant attention so far.

Only laboratory experiments would provide the answer as to exactly how the presence of large quantities of the 'critical fraction' influences ostracods, but it is most likely that their locomotion is adversely affected. Kornicker (1958, 1965b) reports the absence of ostracods from the highly sorted oolite sands of the Bahamas although the average grain size of the oolites is considerably bigger (1–3 mm) than our 'critical fraction'. Very possibly the shape of the grains also has a bearing on this; well rounded grains of more or less equal size would provide a very unstable floor (Kornicker, 1961). This supposition seems to be further supported by cases where although the 'critical fraction' is present in large quantities the number of ostracods is still high. On further examination it is found that instead of the usual well-rounded quartz grains the material consists of shell fragments.

In the Blackwater Estuary and the Wallet the effect of the 'critical fraction' is minimized or non-existent because of the poor sorting of the sediments.

Factors influencing the composition of the ostracod fauna

Salinity. Neale (1965) considered salinity one of the most important factors influencing the distribution of various species and genera of ostracods and this is well borne out by the ostracod fauna of the Thames Estuary. Of the seven major ostracod biotopes recognized, six are definitely controlled by variation in salinity. The changes in fauna due to salinity are described later.

Temperature. The effect of temperature on the ostracod faunas is difficult to see in the Thames Estuary, largely because it is superimposed on the salinity gradient. Temperatures in the Inner Estuary are abnormally high owing to the heat rejection of

---

[1] Wieser (1959) introduced the term 'critical grain size', taking the 0·2 mm median diameter as such. Some objections can be raised against the use of median grain size and the figure mentioned. Median grain size does not reflect the composition of the sediment representatively, it can also be very misleading as in the case of strongly skewed distribution. The number of fractions considered has also an important bearing on the question. The suitability of a sediment for a burrowing or interstitial mode of life cannot be expressed by median grain size alone, the proportion of the various fractions is certainly more important.

several large power stations (in the region of 350×10^9 BTU/day). This 'overheating' of the Inner Estuary is however most pronounced in the liquid mud affected and polluted 'mud reaches' (down to 25–30 miles from London Bridge) where there are either no ostracods present or only fresh-water forms (sample 2034).

It appears, however, that the distribution of one species, '*Cythereis*' *tuberculata* is controlled or at least influenced by temperature. This form is the dominant species in the Blackwater Estuary (Fig. 5), an area where the water temperature is raised by the cooling water of the Bradwell power station. A further, but much less pronounced concentration of '*C.*' *tuberculata* is found in Sea Reach (Fig. 5), an area of polyhaline water with still slightly higher average temperatures than the rest of the Outer Estuary.

Depth. A more detailed analysis of the material available is needed to establish close connection between depth and the distribution of the various species. Certain patterns are however discernible at this stage. Some phytal species, e.g. *Sclerochilus contortus, Paradoxostoma normani*, etc., are most common in depths from two feet to two fathoms, beyond this depth they decline gradually to about 8 fathoms and in deeper waters they are extremely rare.

Fig. 6 shows the quantity of valves found at various depths in the Outer Estuary. About 200 samples are considered in this diagram, chosen in such a way that each depth is represented by about an equal number of samples. There seem to be appreciable peaks at $2\frac{1}{2}$, $4\frac{1}{2}$ and 8 fathoms.

Factors such as food supply, shelter, other chemical parameters, etc., have not yet been evaluated.

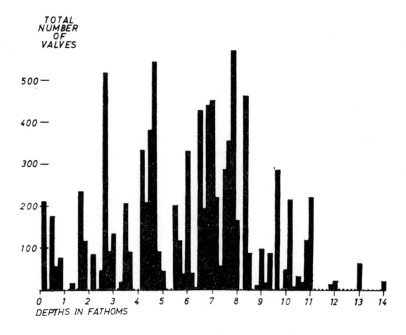

Fig. 6. Total number of valves found at various depths.

BIOCOENOSIS OR THANATOCOENOSIS

As with other small aquatic organisms with hard exoskeletons, ostracods present a difficult problem in ecological-palaeoecological studies. Post-mortem transport may be very intensive, masking or completely destroying the original ecological picture. Ostracods seem particularly vulnerable to post-mortem displacement due to the hydrodynamic properties of the shell and the possibility of gas being trapped in the carapace.

It appears from papers dealing with ostracod ecology that according to the circumstances the thanatocoenosis may or may not be representative of the biocoenosis. Two factors may have a decisive influence here; the energy level of the environment, and population density.

1. Shallow marine or tidal estuarine and deltaic areas with high current velocities would tend to produce unrepresentative thanatocoenoses (Wagner, 1957), whereas lakes, tidal pools, lagoons or deeper seas would fall into the other category (Kornicker, 1965a; Puri et al., 1964).

2. High population densities—all other things being equal—would tend to produce a more representative thanatocoenosis and therefore it may be suggested tentatively that tropical death assemblages may be truer representatives of the living fauna than temperate ones.

There is a considerable amount of confusion in the literature on the meaning of the term biocoenosis or 'living forms'. Studies with a zoological background consider only those specimens which are actually alive at the time of sampling, whereas the micropalaeontological-palaeoecological approach seems to accept all those carapaces to be 'living' which show traces of organic matter still adhering to the exoskeleton. The differences between the two methods are those of technique, purpose and philosophy, and it is obvious therefore that the parallel usage of the two methods is often incompatible. Maddocks (1966) has demonstrated very effectively the difficulties involved in such a parallel use of the two methods.

In the Thames Estuary the problem of thanatocoenosis versus biocoenosis is extremely acute. Faunas of several distinct ecological environments are displaced and mixed together by a system of strong tidal currents and already deposited sediments and their enclosed fauna are redeposited. The number of living specimens (in the strict sense of the word) is low and the difficulties involved in sampling make the zoological-ecological approach impractical. In the present study various ways have been tried to establish the composition of the biocoenosis. It must be stressed, however, that the term biocoenosis is interpreted in the widest sense, including all that live or lived at the place of sampling. In fact our biocoenosis is the equivalent of Wagner's biocoenosis plus thanatocoenosis I (Wagner, 1957, 1964).

Standard staining technique

This seems to be the only method in use to determine whether empty carapaces or valves are 'living' or dead ones. It is based on the assumption that any trace of protoplasm present in or on the surface of the valve or carapace is part of the original soft body and therefore the animal must have died recently, although there is no record in the literature of how long after the animal's death the stain may still be taken up. In the present writer's opinion staining techniques can be very unreliable, especially in estuarine environments where the possibility of organic pollution exists. It was often found for instance that sand grains, or even pebbles, took up the stain, mainly in samples from the Inner Estuary.

About 12 per cent of the samples collected were stained with Rose Bengal, the majority of them from the Inner Estuary. A great deal of difficulty was experienced in trying to decide which carapace or valve should be considered 'stained' as complete gradation existed between specimens with no sign of stain to bright red ones. Those specimens where weak staining affected the outside of the carapace only were not considered to be 'stained'.

Valves-carapaces ratio

It is reasonable to assume that in the course of transportation the carapaces tend to open and therefore an increasing number of valves as against carapaces will be found the further the specimens have been displaced. This is true however for certain species only. For example *Leptocythere pellucida*, *L. castanea*, various species of *Hemicytherura* and others were found mostly as carapaces, whereas very large, smooth forms such as *Candona neglecta* occurred always as separated valves. Two species seemed especially well suited in this respect; *Cyprideis torosa* and *Cushmanidea elongata*. In the case of the latter, for instance, two main conclusions can be drawn (Fig. 8): *C. elongata* is found more in the form of complete carapaces than isolated valves on sandbanks, the reverse being true for channels; and the ratio of closed carapaces is highest between the Oaze and Kentish Knock, the ratio rapidly decreasing to nil upriver, and N and S of the area mentioned. In our view these two statements roughly delineate the area populated by *C. elongata*.

Mode of preservation

Generally speaking, in papers dealing with the ecology of recent ostracods three 'levels' of preservation are considered; living, dead but with all soft parts present, and empty carapaces (= subfossil).

The vast bulk of the material dredged from the Thames Estuary falls into the third category. No actually living specimen was found but this was due to the methods of sampling. A small percentage of carapaces and valves did contain all or some of the soft parts but since most of the specimens were 'subfossil' special attention was paid to the state of preservation. The following series shows the various stages and the corresponding preservation, from living to subfossil.

TIME (↓)	Biocoenosis s.l.	1. Carapace with soft parts. 2. Carapace or valve with some appendages or traces of cuticle. 3. Transparent carapaces and valves.
	Thanatocoenosis	4. Opaque (white, yellow or brown) carapaces or valves. 5. Opaque valves, filled with various matrices (quartz grains, shell fragments, etc.).

A further distinctive type of preservation is a black coloured carapace or valve; this is no doubt due to it coming into contact with black mud and stages 2–5 may be involved.

It is not suggested here that the above sequence is necessarily the same elsewhere, but in the Thames Estuary, its use produced unequivocal results that corresponded closely to those of other observations.

Juvenile-Adult ratio

It is generally accepted that a high ratio of young moults in a population indicates a natural, living population. Two species, *Cushmanidea elongata* and *Cyprideis torosa*, were examined from this point of view with rather disappointing results. Although a tendency for young moults in the fauna to occur on certain sandbanks was discernible, too many samples showed contradictory results to accept this as a satisfactory method.

Derived material

It was accepted at the beginning of this study that great difficulties would be encountered in connection with derived ostracod valves. Upper Cretaceous and Eocene rocks form the banks and the floor of the Estuary and deposits ranging from Lias to Pliocene are found within the catchment area. Particularly serious contamination was expected from the Eocene London Clay and Woolwich Beds (both rich in ostracods) which floor a large part of the Estuary. It was therefore very surprising to find that contamination from pre-existing sediments was negligible and was almost entirely confined to Upper Cretaceous forms.

Altogether 14 samples contained derived ostracods; seven of these were closely grouped together (1119, 1036, 1073, 1103, 1105 and 1140) in an area north-west of Margate (Fig. 9). All these samples contained various species of *Cytherella* and *Cytherelloidea*, mainly in the form of chalk infilled valves. These valves were transported into this area by the strong flood currents which sweep into this area from the Straits of Dover. Three samples (4, 2029 and 2039) contained the same derived fauna but were transported into their present position (Sea Reach and entrance to Medway) from the Inner Estuary, where the Chalk forms the floor of the channels.

A very surprising find was a single well-preserved valve of *Theriosynoecum* sp., very likely to be of Wealden age, from the Long Sand (1252).

THE DISTRIBUTION OF OSTRACODA

At the present stage only the major features of ostracod distribution are examined; only those species are dealt with which occur in fairly large numbers and thus are suitable for statistical treatment.

Seven major ostracod biofacies were recognized in the estuary each with a characteristic assemblage (Fig. 14). The boundaries between these biofacies are in some cases rather vague and in others arbitrary but surprisingly often they are very clearly defined. Each of these characteristic assemblages is based on the broadly interpreted biocoenosis of the most common species. Whether a carapace or valve was the representative of a 'living' or 'dead' specimen was decided by using the various criteria described in the section on biocoenosis and thanatocoenosis. The degree of transportation was qualitatively assessed by the ratio of 'living/dead' specimens and carapace/valve ratio, simply assuming that a lower figure in both cases will indicate a higher degree of transportation. The biocoenosis of the various samples was then compared and the degree of similarity judged, again qualitatively only. Obviously, numerical methods would give a more objective solution and also a much more detailed picture but we believe that the broad results would be substantially the same.

For easy comparison the dominance of a species in the various samples was used as the first coefficient of similarity. A further comparison of the samples on the basis of the second dominant species gave substantially the same results.

Biofacies I

This biofacies is based on the fauna of one sample (2034) only and consequently its boundaries are ill-defined. This sample was taken near the south bank of Halfway Reach, approximately 14 miles from London Bridge. Salinities here vary enormously according to the volume of fresh-water flow; minimum-maximum values recorded between 1951–54 range from 1·82‰ to 17·8‰ at half-tide (*Technical Paper 11*, 1964).

Ostracod assemblage (sample 2034):

Ilyocypris gibba	47%	
Candona neglecta[1]	18%	
Limnocythere inopinata	12%	
Candona sp	4%	
Cypria opthalmica	3%	Biocoenosis s.l.
Candona compressa	3%	
Cypridopsis aculeata	3%	
Cypridopsis vidua.	3%	
Cyprinotus salinus	3%	
Cyprideis torosa[2]	32%	
Leptocythere castanea	13%	
Loxoconcha elliptica	20%	
Hemicythere villosa	8%	Thanatocoenosis
Hirschmannia viridis	6%	
Cushmanidea elongata.	10%	
Heterocythereis albomaculata . . .	5%	
Aurila convexa	3%	

Ratio of transported specimens very high (about 40%).

Characteristic features of assemblage: Oligohaline forms, *Ilyocypris gibba* dominating.

Biofacies II

The area occupied by this biofacies extends from Cliff to Canvey Island (approximately 33–39 miles from London Bridge). It is in fact the entrance to the Inner Estuary and it is separated from Biofacies I by 10 miles of ostracod-barren area, a heavily polluted stretch of water. Salinity figures range approximately between 20–30‰. Bottom conditions are very variable as in the rest of the Inner Estuary but the effect of the 'critical fraction' is already noticeable in the well sorted sands.

The biocoenosis is richest on the shallow (1 fathom) flats of the south side. The central navigable channel is constantly dredged and it rapidly fills up again with material entering from the Outer Estuary; samples obtained from here contain transported specimens from biofacies III-IV.

[1] Some may have been transported.
[2] A small percentage may belong to the biocoenosis.

Ostracod assemblage (composite result of 10 samples):

Cyprideis torosa . . . . . . very common ⎫
Leptocythere pellucida. . . . . very common ⎪
Sclerochilus contortus . . . . . common ⎪
Loxoconcha rhomboidea . . . . common ⎪
Ilyocypris gibba . . . . . . common ⎪
Cushmanidea elongata. . . . . rare ⎬ Biocoenosis s.l.
Cyprinotus salinus . . . . . rare ⎪
Candona neglecta. . . . . . rare ⎪
Limnocythere inopinata . . . . rare ⎪
Hirschmannia viridis . . . . . rare ⎭

Hemicythere villosa . . . . ⎫
Heterocythereis albomaculata . . . ⎪
'Cythereis' whiteii. . . . . . ⎬ Thanatocoenosis
Cytheropteron sp.. . . . . . ⎪
Aurila convexa . . . . . . ⎪
Hemicytherura sp. . . . . . ⎭

Ratio of transported specimens: very high to medium.

Characteristic features of assemblage. This biofacies is probably the least valid and the most difficult to assess as it represents a strongly transitional area with marked horizontal salinity gradient and fast tidal currents. The dominance of *C. torosa* with the continued presence of many of the oligohaline species of Biofacies I seems to be characteristic.

Biofacies III

The area included stretches from the approximate seaward limit of the Inner Estuary to include the Sea Reach, Warp area and part of the Maplin Sands and the Cant. Salinity is in the region of 30‰ dropping much below this figure during periods of increased fresh-water flow, to as low as 25‰ or under. The fauna is rich, up to 23 species and the approximate composition, taking only the most common species into account, is given below.

Cyprideis torosa . . . . . . 40% ⎫
Cushmanidea elongata. . . . . 20–25% ⎪
Hemicythere villosa . . . . . 10–15% ⎪
Leptocythere pellucida. . . . . 5–10% ⎪
Loxoconcha elliptica . . . . . 5% ⎬ Biocoenosis s.l.
Loxoconcha rhomboidea . . . . 5% ⎪
Heterocythereis albomaculata . . . 5% ⎪
Aurila convexa . . . . . . 3% ⎭

Candona neglecta. . . . . . ⎫
Ilyocypris gibba . . . . . . ⎬ Thanatocoenosis

Ratio of transported specimens: Low.

Characteristic features of assemblage: The dominance of *Cyprideis torosa* with *Cushmanidea elongata* as second dominant species in association with forms given above.

Biofacies IV

Covers the largest part of the Outer Estuary, with a very uniform ostracod fauna. Salinity is near 34‰ but slight seasonal fluctuations can still be expected.

A typical assemblage (sample 1326):

Cushmanidea elongata.	29%	
Hemicythere villosa	25%	
Loxoconcha rhomboidea	8%	
Loxoconcha granulata	7·5%	Biocoenosis s.l.
Cyprideis torosa	7%	
Leptocythere pellucida.	3%	
Aurila convexa	5%	

plus 23 species, each less than 3%.

Candona neglecta   Thanatocoenosis

Ratio of transported specimens: Very low.

Characteristic features of assemblage: Dominance of *Cushmanidea elongata*. The second dominant species is *Cyprideis torosa* in the west slowly giving way to *Hemicythere villosa* in the east.

Biofacies V

A very narrow area of channels and flats south of the Margate Sands; salinity is around 34‰ with very little variation. Water circulation is much less intense than in areas discussed so far and the average depth of channels is markedly less, being in the region of 6 fathoms.

A typical assemblage (sample 1148):

Hemicythere villosa	23%	
Loxoconcha rhomboidea	22%	
Cyprideis torosa	16%	
Heterocythereis albomaculata . . .	9·5%	
Leptocythere pellucida.	8%	Biocoenosis s.l.
Hirschmannia tamarindus	7%	
Loxoconcha elliptica	3·5%	

plus 15 more species, each less than 3%.

Candona neglecta   Thanatocoenosis

Ratio of transported specimens: Almost nil.

Characteristic features of assemblage: Dominance of *Hemicythere villosa* with high percentages of *Loxoconcha rhomboidea* and *Heterocythereis albomaculata*.

S

Biofacies VI

The extreme eastern area of the estuary, it is in point of fact more closely related to the North Sea. Many of the samples in this area are barren owing to the unfavourable nature of the substratum but those which contain large numbers of ostracods gave a very uniform fauna.

A typical assemblage (sample 1179):

Hemicythere villosa	32%	⎫
Cushmanidea elongata.	28%	⎪
Leptocythere pellucida.	10%	⎬ Biocoenosis s.l.
Hirschmannia tamarindus	5%	⎪
Loxoconcha rhomboidea	5%	⎭

plus 12 more species, all less than 5%.

Ratio of transported specimens: Nil.

Characteristic features of assemblage: Dominance of *Hemicythere villosa* followed by *Cushmanidea elongata*. *Loxoconcha rhomboidea* in insignificant quantities.

Biofacies VII

The Blackwater Estuary and the Wallet are characterized by a completely differen ostracod assemblage. As yet it is not clear what factor is responsible for such a sharp distinction in the fauna. No salinity figures are available but because of the very restricted fresh-water flow from the River Blackwater it cannot be significantly below 34‰. The poorly sorted sediments contrast sharply with the rest of the estuary and slightly higher water temperatures exist due to the proximity of the power station at Bradwell and restricted water circulation.

A typical assemblage (sample 577):

Hemicythere villosa	26%	⎫
'*Cythereis*' *tuberculata.*	25%	⎪
'*Cythereis*' *whiteii.*	15%	⎬ Biocoenosis s.l.
Loxoconcha granulata	12%	⎪
Cythere lutea.	10%	⎭

plus 5 more species.

Ratio of transported specimens: Nil.

Characteristic features of assemblage: The high percentage of '*Cythereis*' *tuberculata* coupled with that of '*Cythereis*' *whiteii* and *Cythere lutea* produce a very distinctive assemblage. A further difference from the other biofacies is the low number of species; approximately half of that of the rest of the estuary.

Within each of these major biofacies further differentiation occurs, the main division being between the channels on the one hand and the shallower sandbanks and flats on the other. One immediately obvious fact is the always higher portion of thanato-coenosis in channel samples. In the Warp-Oaze area, for example, up to 80 per cent of the individuals in the channels may belong to the thanatocoenosis, and in a few cases the figure may be almost 100 per cent. On the banks and flats the biocoenosis is more numerous. Further seawards this difference between channels and sandbanks in the ratio of thanatocoenosis/biocoenosis diminishes although it continues to exist.

Differences in the 'living' fauna between the various sub-environments within one biofacies exist, but are difficult to recognize, as a statistical study of the whole fauna is necessary since the major features of faunal composition are constant within the given biofacies.

REFERENCES

ALLEN, F. H. 1954. The Thames Model Investigation. A study of siltation problems. *J. Instn Wat. Engrs*, **8** (3), 232-42.

BRADY, G. S. AND ROBERTSON, D. 1870. The Ostracoda and Foraminifera of Tidal Rivers. *Ann. Mag. nat. Hist.*, (4) **6**, 1-33.

——. AND NORMAN, A. M. 1889. A monograph of the marine and freshwater Ostracoda of the North Atlantic and North-Western Europe. Section 1. Podocopa. *Scient. Trans. R. Dubl. Soc.*, Ser. 2, **4**, 63-270.

CLOET, R. L. 1967. Determining the dimensions of marine sediment circulations and the effect of spoil dumping and dredging upon them and on navigation. *British National Conference on the Technology of the Sea and Sea-Bed, III.* Ministry of Technology.

DOBSON, M. R. 1965. The Sediments of the Thames Estuary. Unpublished Ph.D. thesis. Univ. of London.

ELOFSON, O. 1941. Zur Kenntnis der marinen Ostracoden Schwedens mit besonderer Berücksichtigung des Skageraks. *Zool. Bidr. Upps.*, **19**, 215-534.

EMERY, K. O. AND STEVENSON, R. E. 1957. Estuaries and Lagoons. I. Physical and Chemical Characteristics. *Mem. geol. Soc. Am.*, No. 67, **1**, 673-93.

HEDGPETH, J. W. 1957. Estuaries and Lagoons. II. Biological Aspects. *Mem. geol. Soc. Am.*, No. 67, **1**, 693-750.

IMBRIE, J. AND NEWELL, N. D. (ed.) 1964. *Approaches to Paleoecology.* Wiley, New York.

JONES, T. R. 1850. Description of the Entomostraca of the late Pleistocene beds of Newbury, Copford, Clacton and Grays. *Ann. Mag. nat. Hist.* (2), **6**, 25-28.

KORNICKER, L. S. 1961. Ecology and taxonomy of Recent Bairdiinae (Ostracoda). *Micropaleontology*, **7**, 55-70.

——. 1965a. A seasonal study of living Ostracoda in a Texas bay (Redfish Bay) adjoining the Gulf of Mexico. *Pubbl. Staz. zool. Napoli* [1964], **33** (suppl.), 45-60.

——. 1965b. Ecology of Ostracoda in the northwestern part of the Great Bahama Bank. *Pubbl. Staz. zool. Napoli* [1964], **33** (suppl.), 345-60.

MADDOCKS, R. F. 1966. Distribution patterns of living and subfossil podocopid ostracodes in the Nosy Bé area, Northern Madagascar. *Paleont. Contr. Univ. Kans.*, Arthropoda, Paper 12, 72 pp., 63 Figs.

McKENZIE, K. G., 1965. The ecologic associations of an ostracode fauna from Oyster Harbour, a marginal environment near Albany, Western Australia. *Pubbl. Staz. zool. Napoli* [1964], **33** (suppl.), 421-61.

NEALE, J. W. 1965. Some factors influencing the distribution of Recent British Ostracoda. *Pubbl. Staz. zool. Napoli* [1964], **33** (suppl.), 247-307.

PRENTICE, J. E. AND DOBSON, M. R. 1967. Sedimentation in Halfway Reach, Thames, England. *Preprints, Seventh Int. Sed. Congr.*

PURI, H. S., BONADUCE, G. AND MALLOY, J. 1965. Ecology of the Gulf of Naples. *Pubbl. Staz. zool. Napoli* [1964], **33** (suppl.), 87-199.

SANDBERG, P. 1965. Notes on some Tertiary and Recent brackish-water Ostracoda. *Pubbl. Staz. zool. Napoli* [1964], **33** (suppl.), 496-514.

WAGNER, C. W. 1957. *Sur les Ostracodes du Quaternaire récent des Pays-Bas et leur utilisation dans l'étude géologique des dépôts holocènes.* 259 pp. Mouton et Co., The Hague.

——. 1965. Ostracods as environmental indicators in Recent and Holocene estuarine deposits of the Netherlands. *Pubbl. Staz. zool. Napoli* [1964], **33** (suppl.), 480-95.

WATER POLLUTION RESEARCH. 1964. The Effect of Polluting Discharges on the Thames Estuary. *Technical Papers* 11. H.M.S.O., London.

WIESER, W. 1959. The effect of grain size on the distribution of small invertebrates inhabiting the beaches of Puget Sand. *Limnol. Oceanogr.*, **4** (2), 181-94.

WILLIAMS, R. B. 1966. Recent marine podocopid Ostracoda of Narragansett Bay, Rhode Island. *Paleont. Contr. Univ. Kans.*, Arthropoda, Paper 11, 36 pp., 27 Figs.

WILLIAMS, R. 1969. Ecology of the Ostracoda from selected marine intertidal localities on the coast of Anglesey. 299-329. *The Taxonomy, Morphology and Ecology of Recent Ostracoda.* Oliver and Boyd, Edinburgh. 532 pp.

DISCUSSION

HART: Can you say which species might be sensitive to thermal organic pollution and which might be particularly indicative of such conditions?

KILENYI: No, because the polluted part of the estuary is the fresh-water part of the estuary and thus falls into my first biofacies which is fresh water. One cannot test this on the marine ostracods.

GREKOFF: I was very interested by the whole of this very instructive paper. Concerning contamination from older sediments I have my own example from a fresh-water pond in the Rambouillet Forest, Paris Basin. The surrounding country is composed of Oligocene rocks and in the mud of this fresh-water pond I found Oligocene species of *Cytheridea* mixed in with Recent species of *Ilyocypris*. This bears out the importance of distinguishing between additional fauna from older sediments and the indigenous fauna.

KILENYI: The Thames is a very high energy environment where all the derived valves are obviously worn by the constant water movement. These sediments are all worked and re-worked glacial sands; the chalk fossils are mainly *Cytherella* and *Cytherelloidea* which are much harder, thicker and more resistant to erosion which is why they are preserved.

BATE: The fresh-water ostracods found out in the estuary must obviously have come through the polluted zone of the river. Did they appear normal or had the shell surface any peculiar growth or encrustation? The reason I ask is that in a paper which appeared in *Nature* about 1959 by the late Dr L. R. Cox, I think in conjunction with a Japanese mollusc worker, it was stated that gastropods living in areas where there was sewage pollution developed quite strong tubercular growths on the shell.

KILENYI: I don't know this paper but it is true that the *Ilyocypris gibba* and *Limnocythere inopinata* are tuberculate. There are both tuberculate and non-tuberculate forms but these do not come from the polluted zone but from tidal marshes, etc., along the edge of the river where the pollution is very restricted. The actual polluted part of the river is mainly the centre where you get the de-oxygenation.

HASKINS: It is interesting that Mr Kilenyi finds chalk forms but very few Tertiary forms since, as the Thames flows over a lot of Tertiary deposits, one would expect to find re-worked Tertiary fossils in the Thames Estuary. This is also the case in the North Sea itself. We do quite a lot of work on oil exploration at our laboratories and we found that in the Upper Tertiary deposits there are very few re-worked Lower Tertiary forms, but in certain cases, there are considerable numbers of chalk fossils. This may be something to do with the chalk standing out as scarps and being more liable to attack by current and wave action than the softer, low-lying Tertiary deposits.

OERTLI: I was particularly interested by the black valves. First, have you any idea what is the chemical component that makes them turn black? Secondly, do all the species that you have found in the black mud show this colour? Thirdly, have you found the black colour also in living species?

KILENYI: I do not know the exact cause of the black coloration, but I assume that it is caused by some valves coming into contact with the 'Black Mud'. This is a highly organic mud which is partly sewage sludge, partly just extremely fine silt and clay. The answer to your second question is that as far as I remember all species had some black coloured valves. In some areas the number of black valves and carapaces was extremely high, as for instance in the Warp-Oaze area; in others, generally in the outer parts of the Estuary, there were hardly any. In answer to your third question I have never found a living ostracod with a black stain on it.

WHATLEY: Why do you, and Dr Bate earlier in the discussion, assume that these fresh-water forms have come down through the polluted zone of the Thames? Draining the chalk of the north Kent coast are a number of small rivers, such as the Darent, all of which discharge into the estuary. These rivers and the Medway contain large populations of *Ilyocypris* and *Limnocythere* living much closer to where you have found them than to where such forms can live in the Thames. Alternatively, could one not invoke a submarine chalk spring somewhere in the estuary?

KILENYI: Of course many of the fresh-water forms may have come from the rivers you mention and then been transported by tidal streams to the Outer Estuary. The Medway is, however, an exception as practically no sediment from this river reaches the Outer Estuary. Submarine chalk springs may exist but we found no evidence of these. Furthermore the presence of the thick layer of sand and gravel that forms the bed of the estuary would almost certainly eliminate the effect of such springs.

HOWE: From an oil company's standpoint, and they are the ones that are really interested in ecology, I think this is perhaps the most significant ecologic paper that I have encountered. Here, you may not consider that this is a delta but I think that it has the form of a more or less submerged delta and this study goes all the way from freshwater to marine conditions. I believe that, at least as regards petroleum exploration, this is the most significant paper from an economic geology standpoint that has been presented by anyone.

A PRELIMINARY ACCOUNT OF THE ECOLOGY AND DISTRIBUTION OF RECENT OSTRACODA IN THE SOUTHERN IRISH SEA

R. C. WHATLEY and D. R. WALL
University College of Wales, Aberystwyth, Wales

SUMMARY

Preliminary results concerning the distribution of 65 species of Ostracoda are presented. Four assemblages of ostracods are recognized within the area, inhabiting four distinct environments. The provisional results of seasonal studies, and an examination of the relationship of Ostracoda to algae are discussed. Tentative conclusions, and an outline of future projects to substantiate these and to investigate other aspects of the faunas are given.

INTRODUCTION

In this paper, preliminary results from nine months of a considerably longer study are presented. The examination of the ostracod faunas in this area forms part of a much larger project being carried out by the Department of Geology at Aberystwyth to investigate the geological history, superficial sediments and the Quaternary and Recent microfaunas of the southern part of the Irish Sea.

The objects of the present study are multiple and are enumerated below:

(*i*) To produce a map of the distribution of living and dead forms in the area under consideration and to attempt to account for the observed discrepancy between the biocoenosis and the thanatocoenosis.

(*ii*) To examine the relationship between the distribution of living ostracods and such factors as substrate and bathymetry.

(*iii*) To investigate the possible response of living forms to seasonal fluctuations in dissolved oxygen, pH, temperature, salinity, etc.

(*iv*) To establish the nature of, and to attempt to explain, the relationship of Ostracoda to seaweeds on this extremely exposed coast.

(*v*) To investigate, under laboratory conditions, the life histories of selected ostracod species.

(*vi*) To review the taxonomic status, particularly at the generic level, of many Recent British marine ostracods.

(*vii*) To attempt to arrive at conclusions which can be applied to problems of a palaeoecological nature.

Much of the work is necessarily incomplete and only those parts of the study which are furthest advanced are discussed here.

A network of samples has been collected, out to approximately 20 fathoms, on both the Irish and Welsh sides of the Irish Sea. In addition, a series of traverses have been taken across St. George's Channel. It is intended eventually to produce results based upon all these samples. The present account is confined to the area marked with vertical lines, to the west and south-west of Aberystwyth, shown in Fig. 1. The location of the sample stations within this area is given in Fig. 2.

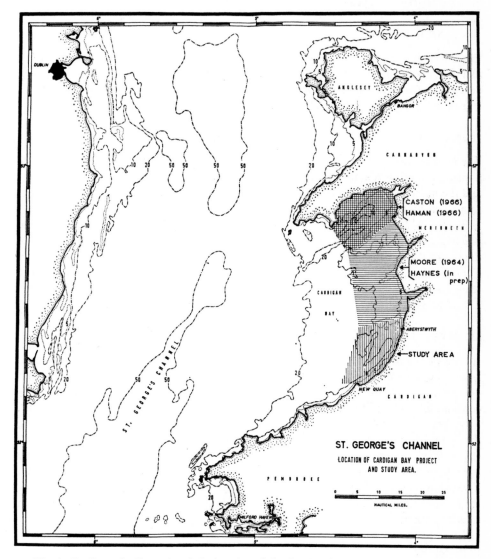

Fig. 1. St. George's Channel and the location of the Cardigan Bay project and study area.

CLIMATE

The climatic conditions of the area are discussed by Evans (1947). A summary of Evans's account is given below. The climate is typically cold-temperate, with a mean annual temperature range (1943–45) of 14°C, and with a minimum of 0°C and a maximum of 20°C. Mean annual rainfall, (1936–45) is approximately 35 inches, with rainfall being fairly evenly distributed throughout the year, with a minimum in March-April. Twelve miles inland the rainfall rises to over 100 inches.

Approximately 70 per cent of the winds (1943–45) blow from between NNE and SSW. Wind strengths (Beaufort Scale) during this period were mostly of strength 1 to 5, although on 20 per cent of the days winds of strength 6–12 occurred. Gale force winds (8–12) most frequently occur during the autumn and winter but are not uncommon throughout the year.

TIDES AND CURRENTS

The tidal range on the coast is considerable with a maximum spring tide range of 17 ft, and a minimum neap tide range of 5 ft. The tide floods from south to north, yet, despite the considerable tide range, tidal currents are not strong, being of the order of 1 knot at springs and $\frac{1}{2}$ knot at neaps.

TEMPERATURE AND SALINITY

The annual variation in sea-water temperature is approximately 11°C with (1943–45) a maximum of 18°C and a minimum of 2·5°C. The maximum occurs in late August and the minimum between late December and early March. The water temperature is somewhat higher in the autumn than in the spring.

Evans (1947) records a mean salinity, in the inshore waters of the Aberystwyth area, of 34·5‰ and also mentions low salinities of 32‰ at Aberystwyth, probably in times of flood. In the present study, a large number of salinity readings has been taken in Cardigan Bay, the mean of which is approximately 33‰ (range 31·8–35·1). Inshore readings, especially off the mouths of rivers, have given readings which range between 27‰ and 34‰. Selected temperature and salinity profiles are given in Fig. 5.

LIGHT PENETRATION

In times of calm weather and strong sunlight, at a distance of some 5 miles offshore, a Secchi disc can be seen to a depth of over 5 fathoms (30 feet). At most other times however, the turbidity of the water, due to the plankton and suspended sediment, is considerable. Often it is not possible to see the disc at more than $1\frac{1}{2}$ fathoms (9 feet).

In times of west or south-west gales, coupled with flooding rivers and spring tides, the degree of turbidity is very high, especially nearshore. This must have an inimical effect on plants and animals in the area. The position is aggravated by the large amount of boulder clay available for erosion on the cliffs to the south of Aberystwyth. The measurements of transparency, which range between $1\frac{1}{2}$ and $5\frac{1}{2}$ fathoms, do not record the minimum value as sampling is not possible in a small vessel on this coast in rough weather.

BATHYMETRY

The bathymetry of the area is shown in Fig. 3. The area sampled in detail ranging from the high water mark, down to 14·5 fm. Between Aberystwyth and New Quay there is a fairly steep slope culminating in the linear trough parallel to the coast known as the Trawling Ground. The undulating area to the west and north-west of this feature is to a large extent covered by mobile sand.

North of Aberystwyth, extending for some 7 miles at right angles to the coast, is a shallow linear area, Sarn Wallog, which is thought to be a morainic feature (Wood, 1966). The Sarn deflects the northward flowing flood tide westwards and causes considerable turbulence in this area, particularly at its shallow western extremity.

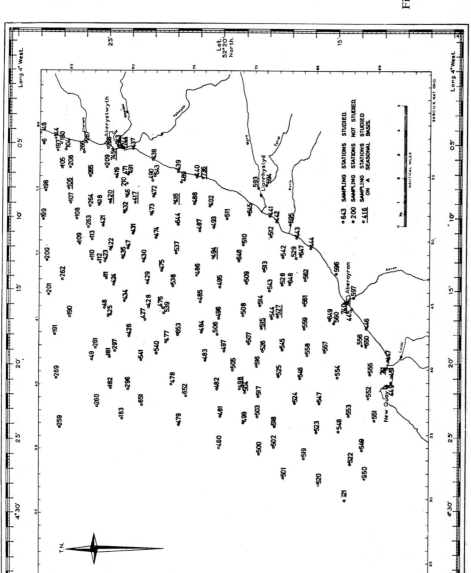

Fig. 2. Sampling stations.

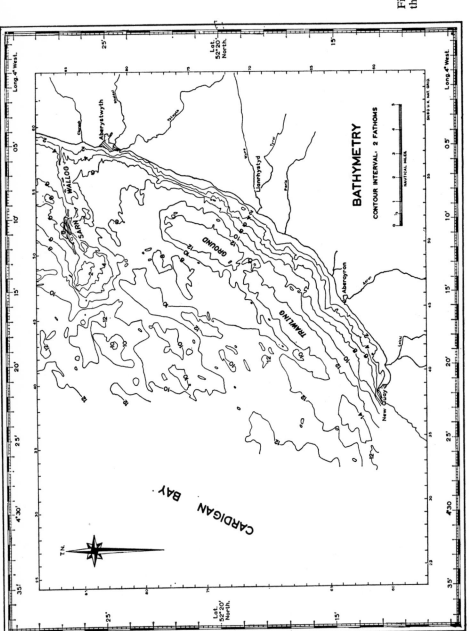

Fig. 3. The bathymetry of the study area.

SUBSTRATE

The area is thought to be floored by the Irish Sea Boulder Clay of Saale age, which has been collected in cores taken principally in the seaward part of the area. In most places, this boulder clay platform is covered by a veneer of superficial sediment. This ranges in size from the large boulders and cobbles of the Sarn and the sublittoral zone, through gravel and sand, to the coarse silt of the Trawling Ground. The clay recorded to the north of New Quay is now thought to be of Pleistocene origin. The work of Jones (MS) indicates that little primary sedimentation is taking place in Cardigan Bay although the mobility of the existing sediments is considerable. The existing superficial cover is thought to have been brought into the area by the transgression which has occurred in this area since the end of the Pleistocene.

There is considerable mobility in the western extremities of the area where large sand waves occur. Work on sediment mobility is being undertaken in the department, and it is hoped to produce a mobility map in the near future.

METHODS

The samples have been collected from the Department of Geology's motor launch 'Antur', using a Holme Vacuum Grab, a Van Veen Grab and a 'home-made' dredge, yet to be described as the 'Arklow Dredge'. Positions are fixed using the Decca Navigation system. Samples for microfaunal examination are treated with buffered formalin and 10 ml is picked. Conductivity, salinity and temperature are recorded on the Model R55–3 Salinometer, Industrial Instruments Inc., New Jersey. Transparency is recorded using a Secchi Disc. Bottom water is collected in a Knudsen bottle and analysed in the laboratory. Whilst the authors recognize the inherent inaccuracies of the standard Winkler method of measuring dissolved oxygen concentration, this is still considered to give the best approximation, and has the advantage of world-wide usage. The method used in the present study is that outlined by Barnes (1959). Dissolved oxygen concentrations are quoted in ml/l and in saturation values using the conversion tables of Green and Carrit (1967). The determination of pH was made using the Pye Dynacap pH meter. With the recent aquisition of the Unicam atomic absorption spectrophotometer and the Technicon auto-analyser, studies of sea-water chemistry will be expanded, both in large scale and in micro-environmental work.

THE OSTRACODA

Fig. 6 shows the distribution of living ostracods within the area, based on samples which were collected during the period November 1963 to September 1964.

Individual specimens are considered to be living if they contain appendages or protoplasm. In certain species, such as *Cythere lutea*, *Loxoconcha rhomboidea* and *Heterocythereis albomaculata*, living forms are readily distinguishable with their conspicuous eyes and externally visible appendages and setae. More difficult to distinguish however, are many members of the Cytheruridae. *Semicytherura* species, for example, commonly exhibit post-mortem closure of the valves without any append- ages being visible externally. With such forms, it is often necessary to open the cara- pace to determine whether or not it contains soft parts.

The samples used in this study had been previously stained with Rose Bengal for foraminiferal work. In many cases, the protoplasm of foraminifera, particularly that in the last chamber, can be seen to have taken up the stain. In the living ostracod,

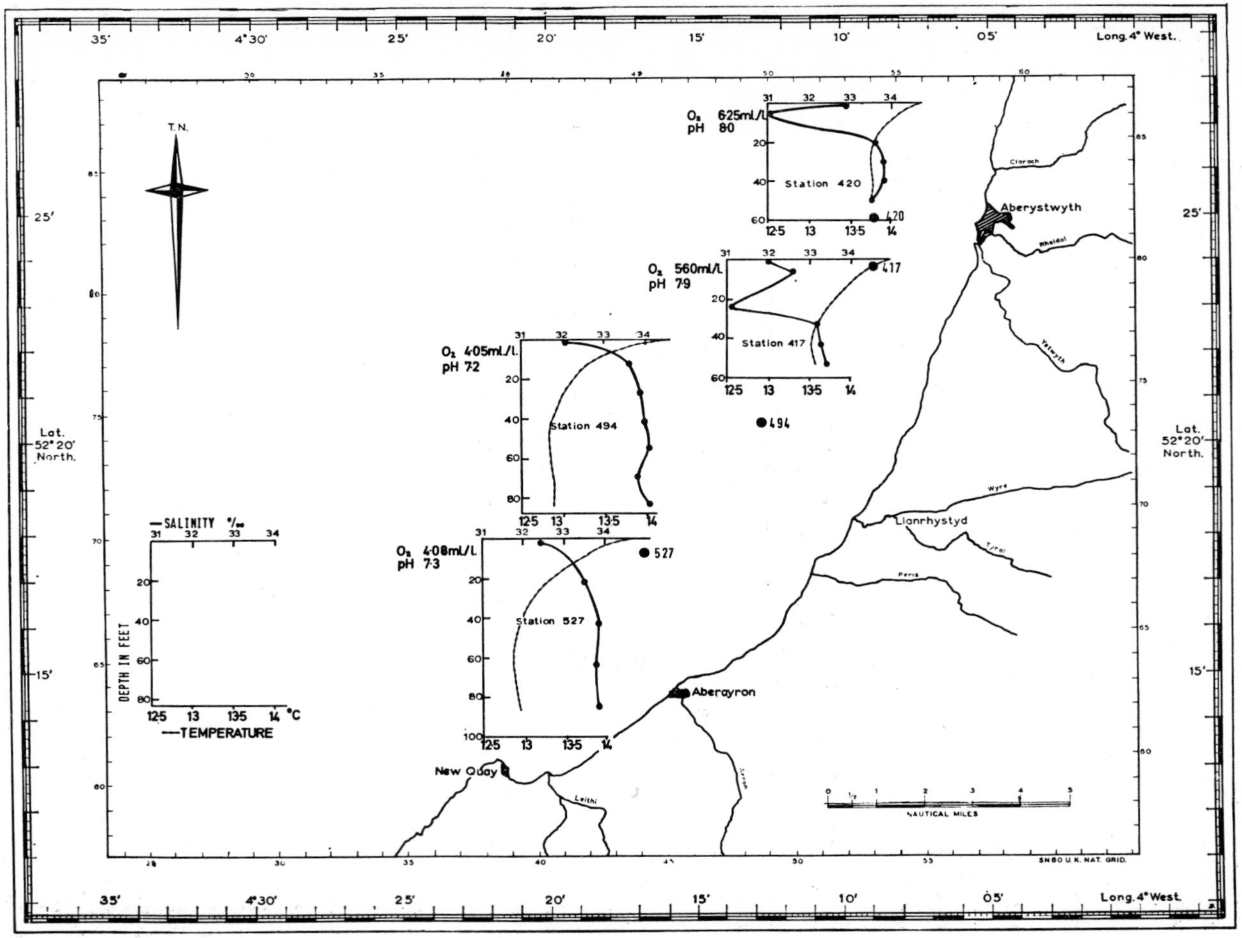

Fig. 5. Selected temperature—salinity profiles of readings taken on 11 June 1967.

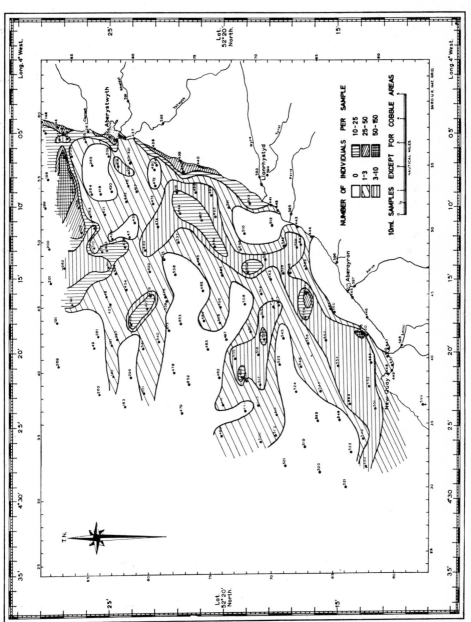

Fig. 6. Distribution of total live Ostracoda.

however, this is precluded by the total enclosure of the animal in chitin which is not stained by Rose Bengal. Indeed, in the examination of many thousands of ostracods from the Irish Sea and the Dovey Estuary, neither author has observed a single ostracod, demonstrably alive because of the presence of soft parts, to be stained by Rose Bengal. In fact, in the present study Rose Bengal has proved to be of value as an indicator of dead rather than of living forms. Many empty valves and carapaces have taken up the stain, particularly in the normal pores, and it is thought that this is due to the stain having been taken up by proteinaceous material or micro-organisms which have become aggregated in the empty shell.

Experiments are continuing in an attempt to find a stain which will be taken up by the protoplasm through the chitin of preserved ostracods. A number of stains, such as Lissamine Green, have been tried but they have only proved successful when access to the protoplasm has been artificially contrived.

The distribution of the living Ostracoda (Fig. 6) can be correlated to a large extent with three principal factors; substrate, depth and the distribution of algae. These three factors are interrelated in that the nature of the substrate is often a function of depth, and the presence of dense weed growth is largely dependent upon shallow water and a substrate suitable for attachment.

The area of highest concentration occurs on Sarn Wallog, where shallow water and a substrate of large cobbles and boulders, combine to produce conditions conducive to dense weed growth. It is difficult to compare quantitatively the results from areas of cobble and boulder with those of sand and finer sediments. In the latter material, 10 ml samples have been picked. In the areas of coarser substrate, the sample examined may be from large weeds or from small algae attached to the boulders. The anomalous high concentration obtained from algae encrusting a single boulder dredged up at station 556, two miles north-west of New Quay, can be cited as an example.

The second highest concentration of living forms, occurring from the littoral zone out to a depth of 5 fm, is recorded off Aberystwyth and both north and south of the town. Here rock platforms and coarse sediments support a dense weed population dominated by *Laminaria*. The nature of the relationship of the Ostracoda to the algae is discussed later in the paper.

In the areas where depth, or the nature and mobility of the substrate preclude dense algal growth, the number of living individuals declines. In these offshore areas there seems to be little selectivity of sediment type and the highest concentrations may occur in silt or sand. The coarser sand grades however, usually contain smaller numbers. The large number of barren samples in the western part of the region is thought to be due to the high mobility of the substrate as evidenced by the occurrence of sand waves in this area. Smaller barren areas off Aberystwyth, which appear to be adequately sampled, are also thought to be due to high mobility. The fact that no ostracods have been recorded from a large area off Aberaeron, however, is almost certainly due to inadequate sampling. Further collecting is being carried out in these areas although samples are difficult to obtain as the bottom is largely of large boulders.

The distribution of dead Ostracoda is shown in Fig. 7. The areas of greatest concentration occur in the silt of the Trawling Ground and in the fine sediments flanking it. This area is obviously a trap for the finer sediment grades, and it is noticeable that a substantial proportion of the dead valves and carapaces occurring here are immature. A considerable number of species occurring dead have yet to be found living in the area. Many of these are species with delicate shells which seem not to have been transported far. They have almost certainly been brought into the area

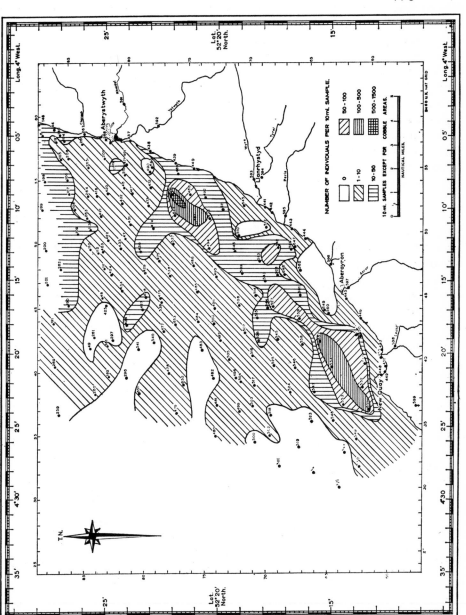

Fig. 7. Distribution of total dead Ostracoda.

from the south by the currents of the flood tide. It is intended to sample the rocky and weed-rich coast of Pembrokeshire and south Cardiganshire, to the south-west, in order to ascertain the provenance of these forms.

In attempting to relate the biocoenosis to the thanatocoenosis one is confronted, in this area, with a further consideration. The superficial sediments in Cardigan Bay are largely of Pleistocene and early post-glacial origin, rather than Recent. They have been brought into this area, during the last 10,000 years, by an irregularly encroaching marine transgression. These deposits contain ostracods which must contribute considerably to the total of dead forms. Additionally, the Irish Sea Boulder Clay which is known to floor a large part of the area, contains Ostracoda many of which live in the area today. Ostracods obtained from this boulder clay, and from Holocene estuarine deposits exposed at Borth, 7 miles north of Aberystwyth, are indistinguishable from forms only recently dead.

By investigating the ecological preference of the living forms in greater detail and by studying the sediment transport properties of the dead shells, it is hoped to be able to arrive at conclusions applicable to palaeoecology.

Fig. 8 shows the distribution pattern of the total ostracods and is analagous to the data with which the palaeoecologist has to work. It is doubtful whether even the most careful worker in this field would be able to distinguish between concentrations due to animals living in this area or having accumulated there by sedimentary processes.

SPECIES LIST

In the list given below, those species found living are prefixed with an (L) and those in which males have been found are prefixed with an (S). A total of 65 species are listed, 22 are syngamic. All are members of the Podocopina, and all but 3 belong to the Cytheracea.

(L) *Propontocypris trigonella* (Sars)
(L) *Pontocypris frequens* (Müller)
(L) *Paracypris polita* Sars
(S) (L) *Cythere lutea* Müller
(S) (L) *Pterygocythereis antiquata* (Baird)
    Pterygocythereis emaciata (Brady)
    Bythocythere constricta Sars
    Bythocythere cf. B. bradyi Sars
(L) *Heterocyprideis sorbyana* (Jones)
(L) *Cuneocythere semipunctata* (Brady)
  ? *Cuneocythere* sp.
(S) (L) *Eucythere argus* (Sars)
(S) (L) *Eucythere declivis* (Norman)
(L) *Neocytherideis subulata* (Brady)
(L) *Neocytherideis* sp.
(L) *Cushmanidea elongata* (Brady)
(L) ? *Cytherura cuneata* Brady
(S) (L) *Hemicytherura cellulosa* (Norman)
(S) (L) *Hemicytherura clathrata* (Sars)
(L) *Microcytherura fulva* (Brady and Robertson)
(S) (L) *Semicytherura sella* (Sars)
(S) (L) *Semicytherura striata* (Sars)
(L) *Semicytherura acuticostata* (Sars)
    Semicytherura angulata (Brady)
    Semicytherura simplex (Brady and Norman)

(S) *Semicytherura nigrescens* (Baird)
    Semicytherura cf. S. producta (Brady)
(L) *Cytheropteron nodosum* Brady
(L) *Cytheropteron latissimum* (Norman)
    Cytheropteron subcircinatum Sars
    Cytheropteron alatum Sars
    Cytheropteron punctatum Brady
    Cytheropteron crassipinnatum Brady and Norman
(S) (L) *Hemicythere villosa* (Sars)
(L) *Aurila convexa* (Baird)
(S) (L) *Heterocythereis albomaculata* (Baird)
    Urocythereis cf. U. oblonga (Brady)
(S) (L) *Leptocythere pellucida* (Baird)
(S) (L) *Leptocythere castanea* (Sars)
(S) (L) *Leptocythere tenera* (Brady)
(S) (L) *Leptocythere macallana* (Brady and Robertson)
(S) (L) *Callistocythere crispata* (Brady)
(S) (L) *Loxoconcha rhomboidea* (Fischer)
(S) (L) *Loxoconcha guttata* (Norman)
(S) (L) *Cytheromorpha fuscata* (Brady)
(L) *Elofsonia* sp.
    Elofsonia pusilla (Brady and Robertson)
(L) *Hirschmannia viridis* (Müller)
(S) (L) *Hirschmannia tamarindus* (Jones)

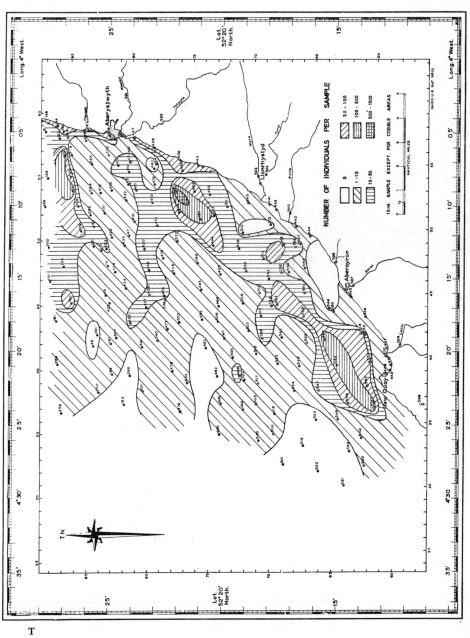

Fig. 8. Distribution of total Ostracoda.

(L) *Phlyctocythere fragilis* (Sars)
(S) (L) *Paradoxostoma variabile* (Baird)
(S) (L) *Paradoxostoma normani* Brady
(L) *Paradoxostoma ensiforme* Brady
(L) *Paradoxostoma bradyi* Sars
(L) *Paradoxostoma hibernicum* Brady
    Paradoxostoma abbreviatum Sars
    Paradoxostoma rostratum Sars

(L) *Paradoxostoma* sp. 1
(L) *Paradoxostoma* sp. 2
(L) *Paracytherois flexuosa* (Brady)
(L) *Paracytherura arcuata* (Brady)
(L) *Cytherois fischeri* (Sars)
(L) *Sclerochilus contortus* (Norman)
    Trachyleberis dunelmensis (Norman)
(L) *Xenocythere cuneiformis* (Brady)

OSTRACOD ASSEMBLAGES

Four distinct ostracod assemblages are recognized within the area. These are out-lined in Fig. 9. Whereas the Littoral and Laminarian assemblages are recognized largely on the basis of living forms, the majority of individuals within the 'Sand' and 'Silt' assemblages are dead. The distribution of the four assemblages is shown in Fig. 10.

The Littoral assemblage is dominated by *Cythere lutea* and *Heterocythereis albomaculata*. This assemblage is recognized, principally in rock pools, on the weed-rich wave cut platform composed of Silurian Aberystwyth Grits, both north and south of Aberystwyth. The rock pools containing this fauna vary in size from small clefts, with a capacity of 2–3 gallons up to very large pools with a capacity of many hundreds of gallons. The restriction of this assemblage to the areas shown in Fig. 10, is due to the absence of rock platforms and pools in the littoral zone to the south.

Using the nomenclature of Lewis (1964), these pools are situated principally in the littoral fringe and in the eulittoral zone and as such are above mean low water mark, and thus above the normal range of *Laminaria*.

The ostracods in these pools occur principally on the Chlorophyceae, particularly *Cladophora* and *Enteromorpha*; also on the Ulvaceae, and to some extent on the Rhodophyceae such as *Corallina*. In the deeper, more sheltered pools, some individuals have been found living in the sandy sediments on the bottom.

In those pools nearest the seaward boundary of the eulittoral zone, *Laminaria* occasionally occurs. In this case there is often an overlap, within the one pool, of the Littoral and Laminarian assemblages. There appears to be no mixing of the two Assemblages however, as *Cythere lutea* has never been found in a *Laminaria* holdfast in these pools.

Those species found living in this assemblage are listed below:

Cythere lutea	49%	*Aurila convexa*	3%
Heterocythereis albomaculata	24%	*Loxoconcha rhomboidea*	1%
Hirschmannia viridis	8%	*Semicytherura striata*	1%
Paradoxostoma bradyi	4%	*Paradoxostoma* sp. 1.	1%
Hemicythere villosa	4%		

Others, approximately 5 per cent:

Microcytherura fulva	*Leptocythere castanea*
Hirschmannia tamarindus	*Paradoxostoma normani*
Paradoxostoma sp. 2	*Hemicytherura cellulosa*

The Laminarian assemblage, as its name suggests, is closely associated with *Laminaria* and has in fact been isolated exclusively from the holdfasts of *Laminaria digitata* and *L. hyperborea*. This association occurs in three principal areas. The first is in the sublittoral zone, below low water mark off the rocky shores immediately to the SSW of Aberystwyth; the second, again just below low water mark, between

Aberaeron and the mouth of the Afon Peris, and the third on Sarn Wallog. Other
small patches of *Laminaria* occur off, and north of Aberystwyth; they are too small,
however, to be represented in Fig. 10.

This assemblage has been found only in the holdfasts of *Laminaria*, there being little
other shelter for ostracods, and most other small invertebrates in this high energy
zone. Other algae, the fronds of *Laminaria* and the fine sediments between the rocks
and boulders have been examined but with rare exceptions have failed to produce
ostracods.

The assemblage is dominated by *Loxoconcha rhomboidea* although *Hemicythere*

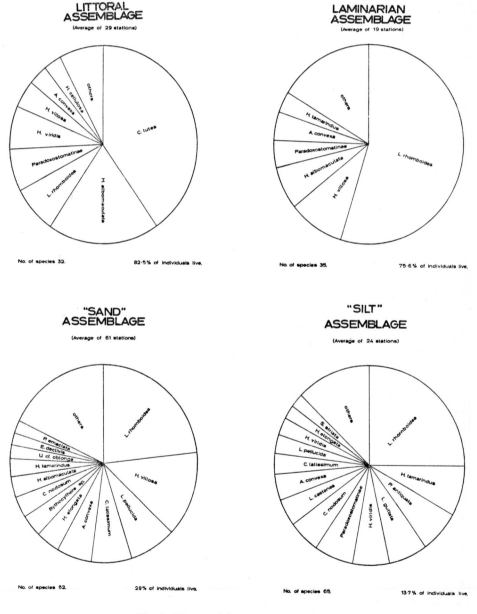

Fig. 9. Nature of the ostracod assemblages.

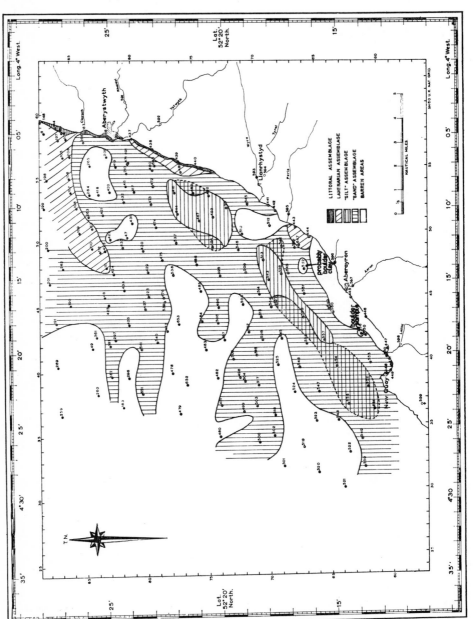

Fig. 10. Distribution of the ostracod assemblages.

villosa and *Heterocythereis albomaculata* are also important. It differs from the Littoral assemblage in the absence of *Cythere lutea*, the greater abundance of *L. rhomboidea* and in the reduced importance of *H. albomaculata*.

In the Littoral assemblages, 32 species occur of which 14, and 82·5 per cent of all individuals were living. In the Laminarian assemblage 35 species have been recorded of which 15 are living. 75·6 per cent of individuals in this assemblage are live.

The 15 living species are listed below:

Loxoconcha rhomboidea	65%	*Hirschmannia viridis*	2%
Heterocythereis albomaculata	10%	*Paradoxostoma ensiforme*	2%
Hemicythere villosa	8%	*Paradoxostoma* sp. 1	1%
Aurila convexa	4%	*Hirschmannia tamarindus*	1%
Hemicytherura cellulosa	3%	*Elofsonia* sp.	1%

Others, approximately 3 per cent:

Cushmanidea elongata	*Cuneocythere semipunctata*
Paradoxostoma variabile	*Paradoxostoma bradyi*
Semicytherura acuticostata	

Experiments, including artificial holdfasts, are in preparation to investigate the true nature of the relationship between the environment and the animal. For example, does the holdfast merely provide shelter or does it also provide a primary or secondary food supply?

The 'Sand' Assemblage, which covers the larger part of the area studied, contains 52 species of which 25 species and 29 per cent of individuals are living. Taking the total figure, *Loxoconcha rhomboidea* is the dominant species, followed by *Hemicythere villosa*. This position is reversed when the living forms only are considered. The exact nature of the relationship of the ostracods to the environment is difficult to assess in this assemblage. The samples have been collected in many cases with a dredge which takes sediment from several centimetres below the surface. It is thought that many of the ostracods, particularly those from coarse shelly sands, are living interstitially. Plans are in hand to investigate this possibility in the most likely areas, by collecting small core samples with SCUBA gear.

A large number of sand samples from the western fringes of the area have failed to yield ostracods, living or dead. These samples are, for the most part, in areas of sand waves where the mobility of the sand is probably too great for the surface, or the interstitial habitat to be tenable. It is probable that the destructive power of these high mobility sands also precludes the preservation of any dead shells.

The 'Sand' assemblage occurs at all depths from the intertidal zone down to 12 fathoms, and thus it is thought that some selective relationship exists between its members and the substrate. The nature of this relationship, however, requires further investigation.

The species which are found living in this assemblage are given as percentages below:

Hemicythere villosa	24%	*Leptocythere tenera*	2%
Loxoconcha rhomboidea	16%	*Hemicytherura cellulosa*	2%
Elofsonia sp.	9%	*Hirschmannia tamarindus*	2%
Aurila convexa	9%	*Eucythere declivis*	2%
Leptocythere pellucida	8%	*Semicytherura sella*	1%
Cytheropteron latissimum	3%	*Cushmanidea elongata*	1%
Cytheropteron nodosum	3%	*Xenocythere cuneiformis*	1%
Leptocythere castanea	3%		

Others, approximately 14 per cent:

Heterocythereis albomaculata	*Paradoxostoma ensiforme*
Hirschmannia viridis	*Loxoconcha guttata*
Semicytherura acuticostata	*?Cytherura cuneata*
Callistocythere crispata	*Pterygocythereis emaciata*
Sclerochilus contortus	*Paradoxostoma normani*

The 'Silt' assemblage is confined to two deep linear areas in the Trawling Ground. This assemblage contains 65 species, 14 of which have been found living. Only 13·7 per cent of the individuals in this assemblage are live. A large number of species has not been found living within the area as a whole, and is thought to have been swept into the Trawling Ground from the south. *Loxoconcha rhomboidea* is the most common species but it occurs only as instars and has not been recorded live. *Hirschmannia viridis, H. tamarindus, Pterygocythereis antiquata* and *Loxoconcha guttata* are also important, all but the first occurring live. It is recognized that this assemblage is more the product of sedimentary processes than of selection of the silt environment by the species concerned. The forms found living in this assemblage are listed below:

Leptocythere pellucida	20%	*Hirschmannia tamarindus*	7%
Loxoconcha guttata	14%	*Elofsonia* sp.	6%
Pterygocythereis antiquata	13%	*Hemicytherura cellulosa*	5%
Leptocythere castanea	12%	*Aurila convexa*	4%
Leptocythere tenera	8%		

Others, approximately 11 per cent:

Cytheropteron nodosum	*Neocytherideis subulata*
Hemicythere villosa	*Propontocypris trigonella*
Semicytherura sella	

This assemblage differs from the 'Sand' assemblage not only in comprising fewer living species and individuals but in its distinctly different character. *Loxoconcha rhomboidea* does not occur living and *Hemicythere villosa* is much less important. *Pterygocythereis antiquata* does not occur living in any other part of the area studied and is assumed to prefer the silt substrate to any other.

Fringing the two areas from which the 'Silt' assemblage has been identified, occur mixed populations of the 'Sand' and 'Silt' assemblages. Further sampling, to be carried out over the next two years, will be required to investigate the factors governing the existence of these assemblages.

THE DISTRIBUTION OF INDIVIDUAL SPECIES

It is intended, when the study is complete, to have investigated in detail, the individual distributions of all those species which occur living in the area. One such distribution is presented here, and Figs 11–13 show the distribution and population structure of *Loxoconcha rhomboidea*.

Fig. 11, illustrates the live distribution of the species and shows living forms occurring in all but the 'Silt' assemblage, although it does occur live in the zone of mixing between the 'Silt' and 'Sand' assemblages.

The areas of greatest abundance are in the Laminarian assemblage, on Sarn Wallog, and in two sublittoral areas south of Aberystwyth. Living forms occur consistently throughout the Littoral assemblage but its distribution in the 'Sand' assemblage is extremely irregular.

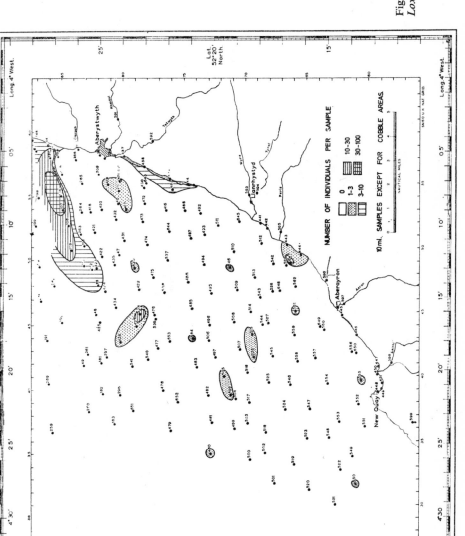

Fig. 11. Distribution of live
Loxoconcha rhomboidea.

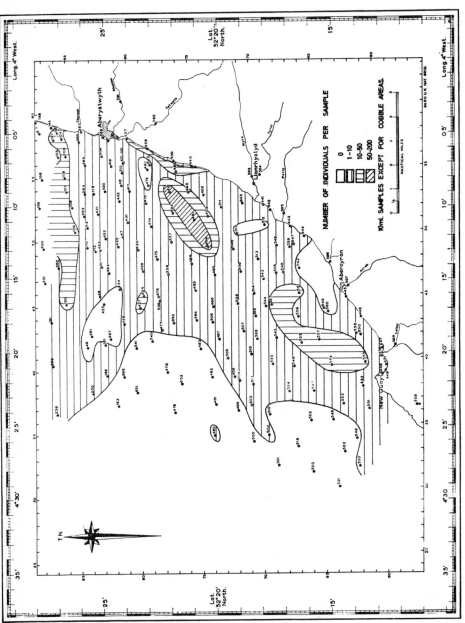

Fig. 12. Distribution of dead
Loxoconcha rhomboidea.

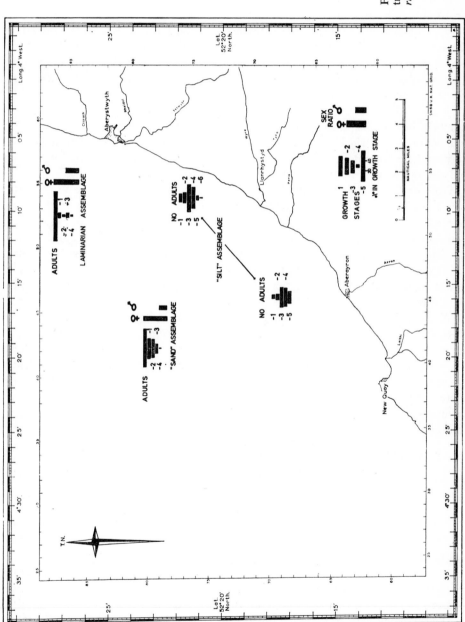

Fig. 13. Population-age distribution of *Loxoconcha rhomboidea*.

The distribution of dead *Loxoconcha rhomboidea* (Fig. 12) differs markedly from that of the living. There are two principal concentrations, one in the silt of the Trawling Grounds, where the species does not occur living, and the other on Sarn Wallog and in the Laminarian assemblage along the coast. It occurs, almost ubiquitously, throughout the area of the 'Sand' assemblage.

The discrepancy between the living and dead distribution of this species is an example, in miniscule, of the difficulty encountered in trying to correlate the biocoenosis and the thanatocoenosis. By comparing the live and dead distribution of a large number of individual species it is hoped that some means of resolving this problem may be arrived at.

Fig. 13 shows the population age structure of total *Loxoconcha rhomboidea* within the area. The Laminarian assemblage (which is almost identical with the Littoral) has a large number of adults proportional to the number of instars. In the 'Sand' assemblage the total number of instars is greater than that of adults, and in the 'Silt' assemblage, where the species does not occur living, the population is composed entirely of instars. Of these instars, stages 3 and 4 are the most abundant. Most of the dead ostracod numbers in the 'Silt' assemblage are, in fact, made up by instars and adults of a number of species which have yet to be recorded there. It is suggested that the bottom currents here are not strong enough to transport the adults of these species and can only bring the immature valves and carapaces into the area.

The population age structure of *L. rhomboidea* in the Laminarian assemblage, where adults greatly outnumber instars, poses a number of problems. In this area, where the greatest density of living forms occurs, one might expect to find the reverse with instars outnumbering adults. Exhaustive seasonal sampling coupled with a detailed study of the life history of the species (similar to those conducted by Theisen, 1966) are in hand to investigate the possibility of a seasonal change in the ratio of adults to instars. At different times of the year, various growth stages may preponderate depending upon whether the species winters as eggs, various instars, or as adults. Alternatively, as suggested by Hagerman (personal communication, 1967), migration in marine benthonic ostracods may be more important than realized hitherto. It is possible that development may take place in one environment with a late stage migration to an environment more favourable to the adults. Presumably the eggs would be deposited in the environment favoured by the adult and this concept therefore begs the important question of how development can take place in one environment if the eggs are deposited in another. This problem is considered again later in the paper in connection with *Cythere lutea*.

THE SEASONAL DISTRIBUTION OF THE OSTRACODA

A number of stations at sea and in the littoral zone have been collected seasonally. The coastal stations are sampled monthly but those in the bay, due to inclement weather during the winter months, have been collected less regularly.

The eight stations collected at sea (Table 1) are underlined in Fig. 2. They were originally selected to encompass as many different substrates as possible and yet to be collected during a single day.

At each of the stations, which have been collected on four occasions to date, various physical and chemical factors of the bottom water are measured and the ostracods are picked from a 10 ml sample. The results are tabulated in Table 2.

The results, in that so few living forms were collected, are disappointing. Sampling will be continued, however, for a further year, and expanded to include more stations

TABLE 1. Data of offshore collecting stations

Station	Lat.	Long.	Depth (fm)	Bottom
106	52°26′59″N	4°7′14″W	6·75	Boulders, cobbles and algae
420	52°24′53″N	4°09′05″W	8·5	Coarse sand
417	52°23′50″N	4°09′06″W	8·17	Cobbles in silty matrix
416	52°22′00″N	4°09′05″W	11	Coarse silt
494	52°20′25″N	4°12′45″W	14	Fine sand
527	52°17′22″N	4°16′52″W	14·5	Coarse silt
515	52°18′05″N	4°17′41″W	11	Small cobbles
498	52°19′00″N	4°21′55″W	10·5	Coarse sand

TABLE 2. Physical, chemical and ostracod data of the eight stations collected at sea

Stn.	Date	Sal. ‰	pH	T °C	Trans-parency fms	O_2 ml/l	O_2 % Sat.	Ostracods Live No.	Live No. of sp.	Dead No.	Dead No. of sp.
106	21. 9.66	34·6	6	16·5	—	—	—	1	1	5	2
	22.11.66	32·5	7·0	8·5	2·3	5·5	84	2	2	17	9
	14. 4.67	31·8	7·9	7·2	1·5	7·1	102	0	0	2	2
	11. 6.67	33·3	7·9	13·9	4·5	6·2	106·5		Barren		
420	21. 9.66	34·8	—	16·9	—	—	—	0	0	1	1
	22.11.66	33·2	7·0	8·9	2	5·6	86		Barren		
	14. 4.67	32·1	—	7·2	—	—	—		No Sample		
	11. 6.67	33·5	7·9	13·8	4·5	—	—		Barren		
417	21. 9.66	34·6	—	16·5	—	—	—		Barren		
	22.11.66	32·9	7·0	8·4	1·5	5·6	86·5	1	1	10	5
	13. 4.67	32·0	7·75	7·3	1·65	7·0	101	1	1	2	2
416	21. 9.66	34·7	—	16·5	—	—	3	3	3	878	53
	22.11.66	33·4	7·0	9·0	1·8	5·5	85	9	8	1374	45
	13. 4.67	32·0	7·4	7·4	1·7	6·2	89	28	14	1052	47
	11. 6.67	33·7	7·7	12·9	3·7	3·9	66	19	12	1290	47
494	21. 9.66	34·8	—	16·4	—	—	—	9	5	90	25
	22.11.66	33·4	7·0	9·1	2·5	5·4	84	4	3	147	34
	13. 4.67	32·5	7·3	7·2	2·3	7·2	107	8	3	71	23
	11. 6.67	34·1	7·7	12·8	5·0	4·5	76	1	1	91	26
527	21. 9.66	34·9	—	16·4	—	—	—	4	4	247	34
	22.11.66	33·4	7·0	9·0	—	5·7	87·7	5	4	253	36
	13. 4.67	33·1	7·55	7·2	1·5	6·5	—	7	4	149	32
	11. 6.67	33·8	7·9	12·8	5·0	4·1	68	3	2	343	38
515	21. 9.66	34·8	—	16·5	—	—	—	1	1	23	10
	22.11.66	33·5	7·0	9·0	2	5·4	83·5	1	1	2	1
	13. 4.67	32·5	7·7	7·2	1·5	7·1	105	1	1	5	3
	11. 6.67	33·6	7·9	13·15	5	6·2	104		Barren		
498	21. 9.66	35·1	—	16·5	—	—	—	19	4	13	6
	22.11.66	33·4	7·0	9·5	1·7	5·7	89·5	0	0	1	1
	13. 4.67	32·4	7·8	7·2	1·5	6·9	—	0	0	2	2
	11. 6.67	33·9	7·8	12·9	5·5	7·0	110		Barren		

within the Laminarian zone. In addition, sampling with SCUBA gear is being initiated in those areas where, because of a rocky bottom, samples are difficult to obtain by conventional means.

Although the seasonal changes in temperature, oxygen concentration and transparency are considerable they seem to exert little effect upon the numbers of living ostracods. Stations 416, in the 'Silt' assemblages, and 494 and 527 in the mixed 'Sand' and 'Silt' area, do show an increase in living forms in April but there is little other correlation between stations.

Station 106 is in the Laminarian assemblage and is particularly disappointing in that large numbers of living forms were collected here prior to the commencement of the seasonal sampling. However, even with a large dredge, sampling in an area of large boulders is very selective and a different type of sample may be collected on each occasion. To resolve this, SCUBA gear will be used in future. Station 420 is in an area of coarse mobile sand which is largely barren of ostracods; 417, 515 and 498 are in the 'Sand' assemblage.

Along the coast, between Aberystwyth and New Quay, five stations have been sampled at monthly intervals (Table 3). These stations are underlined in Fig. 2 and are:

741. Sandy beach just south of the mouth of the Afon Leithi at G.R. 392598.
740. Silt from Aberaeron Harbour, G.R. 455629.
743. Silt from beneath Trefechan Bridge at the throat of Aberystwyth harbour, where the Afon Rheidol enters the harbour. G.R. 577801.
744. Weed covered rocks at the entrance to Aberystwyth harbour G.R. 576805.
745. Weed rich rock pool at the seaward extremity of College Rocks, Aberystwyth, G.R. 576818.

The results from 10 months sampling at these stations are given in Table 3.

All the samples have been collected at low tide so that the salinity values are the minimum. Ostracoda have been picked from 10 ml of sediment and from 100 g (wet) of weed.

At station 741, a very exposed sandy beach, only one live ostracod was recorded during the ten months. The beach is of fine sand which probably precludes the interstitial habitat, and in this exposed intertidal zone the surface habitat would be equally untenable.

The silt samples from the more sheltered harbour at station 740 contain rather few living forms and no seasonal effect is apparent. Those species most commonly recorded live from this station are *Aurila convexa*, *Hemicythere villosa*, *Heterocythereis albomaculata* and *Loxoconcha rhomboidea*.

Station 743, from the silty most fluvial part of Aberystwyth harbour has only one ostracod species living, *Leptocythere pellucida* (Baird). The seasonal distribution of this form is shown in Fig. 14. After a winter when low numbers have been consistently recorded at this station, a considerable increase in numbers, largely instars, occurs in March 1967, to be immediately followed by a decline to the previous level. This rise does not seem to be correlated with any significant change in the physical parameters also shown in Fig. 14. One of the authors (R.C.W.) has observed a similar increase in numbers, in March and April, in the same species from a nearby estuary. Here, however, it is not followed by a decline, the same numbers being maintained until September. It is assumed that this sudden increase in numbers at this station in March represents the normal spring increase in numbers once the temperature becomes sufficiently warm for development to take place (Theisen, 1966). No explanation

TABLE 3. Sample data for stations between Aberystwyth and New Quay.

Stn.	Date	Time	Sal. ‰	T °C	O_2 ml/l	O_2 ‰	pH	Ostracods Live No.	Live No. of sp.	Dead No.	Dead No. of sp.
741	30. 9.66	1050	34·4	15·2	—	—	—			Barren	
	31.10.66	1112	33·7	11·5	5·7	93	7·9			Barren	
	5.12.66	1352	31·4	7·3	6·4	96	7·2			Barren	
	12. 1.67	1020	32·6	6·8	6·7	98·5	6·8			Barren	
	3. 2.67	1100	30·8	8·5	5·5	85·5	7·4	0	0	1	1
	3. 3.67	0915	30·0	7·0	6·5	94	7·8	0	0	2	2
	6. 4.67	1300	32·6	9·0	7·7	117	7·7	0	0	30	12
	10. 5.67	1500	31·5	14·5	7·3	118·5	7·5	1	1	1	1
	7. 6.67	1425	32·4	15·8	6·3	114	7·9			Barren	
	27. 7.67	0715	31·0	17·0	5·5	101	7·8			Barren	
740	30. 9.66	1005	22·0	13·4	—	—	—	1	1	5	4
	31.10.66	1033	3·1	8·1	7·4	95	7·4	9	4	36	10
	5.12.66	1322	3·2	6·2	7·7	94·5	6·9	1	1	3	3
	12. 1.67	0950	3·1	5·5	8·4	102	6·9			Barren	
	3. 2.67	1133	0·3	8·7	6·6	85	7·6	1	1	3	3
	3. 3.67	0945	0·8	7·0	7·9	97	8·4	3	3	18	11
	6. 4.67	1230	6·0	8·0	6·4	82	7·2	2	2	108	27
	10. 5.67	1430	0·5	14·0	8·5	115	7·0	0	0	31	10
	7. 6.67	1345	0·2	15·5	8·1	121	8·1	0	0	32	12
	27. 7.67	0630	0·0	16·0	6·5	99	8·0			Barren	
743	30. 9.66	1355	0·6	14·8	—	—	—	4	1	0	0
	31.10.66	1435	0·9	9·0	7·0	89	7·6	2	1	0	0
	5.12.66	1451	0·7	5·5	8·1	96	7·3	6	1	0	0
	12. 1.67	1135	2·6	5·0	8·2	96·5	7·5	3	1	0	0
	3. 2.67	0950	0·2	8·0	6·7	83·5	4·3	16	1	5	2
	3. 3.67	0805	0·3	6·5	7·9	95	7·2	6	1	0	0
	6. 4.67	1405	0·4	8·5	7·7	98·5	8·7	53	1	3	1
	10. 5.67	1605	16·0	16·5	6·7	102	8·6	3	1	0	0
	7. 6.67	0807	0·3	16·0	6·8	102	7·2	5	1	0	0
	27. 7.67	0800	0·0	17·0	6·6	100	7·0	7	1	0	0
744	30. 9.66	1330	5·0	14·8	—	—	—	4	4	83	24
	31.10.66	1414	2·2	9·2	7·0	91	7·7	1	1	42	17
	6.12.66	1357	1·4	6·0	8·0	95	7·6	2	2	18	11
	12. 1.67	1120	4·3	5·1	8·7	103	7·5	1	1	12	8
	3. 2.67	1000	2·1	8·5	6·8	87	7·1	2	2	12	8
	3. 3.67	0830	1·2	7·0	7·9	97	7·2	0	0	16	9
	6. 4.67	1415	2·6	9·0	9·0	117	8·9	0	0	82	21
	10. 5.67	1555	27·8	16·0	8·6	117	8·5	0	0	60	19
	7. 6.67	1535	17·2	20·0	9·5	169	8·1	1	1	16	10
	27. 7.67	0820	16·0	17·5	6·2	102	7·4	2	1	27	9
745	30. 9.66	1425	34·4	17·2	—	—	—	1	1	7	6
	31.10.66	1455	33·5	12·0	5·8	97	8·0	0	0	4	4
	6.12.66	0838	27·5	6·4	7·0	99	6·5	0	0	1	1
	21. 1.67	1150	30·1	6·7	7·1	102	7·0	1	1	0	0
	3. 2.67	0930	31·8	7·5	5·1	76	6·9	0	0	3	2
	3. 3.67	0735	29·3	6·5	6·6	96	7·6	4	3	7	6
	6. 4.67	1430	30·4	9·0	6·3	95·5	7·7	9	5	9	6
	10. 5.67	1620	33·2	13·5	8·5	135	8·4	135	6	12	7
	7. 6.67	1610	31·6	17·0	7·3	134	8·0	140	10	27	16
	27. 7.67	0855	30·5	17·0	6·2	112	7·6	20	11	7	5

other than the suggestion that the population was killed off by pollution or migrated elsewhere, can be put forward to account for the decline after March.

Station 744, collected from algae (*Fucus*, *Cladophora* and *Corallina*) attached to rocks at the entrance to Aberystwyth harbour, produced few living ostracods. No seasonal effect is apparent in their distribution and it is thought that the high degree of turbulence and turbidity at this station is not conducive to the existence of large populations.

Station 745 is a rock pool on the lower part of the eulittoral zone. From this pool, bottom sediment and algae (*Ulva*, *Cladophora*, *Corallina* and *Enteromorpha*) have been examined. Throughout the autumn and winter months, very few living ostracods were recorded at this station. There is a marked increase as shown in Fig. 15, in the numbers collected in April and this is sustained into May with a decline into June.

The forms which appear at this station are, with the exception of a very few penultimate instars, all adults. *Cythere lutea*, not recorded at the station during the autumn

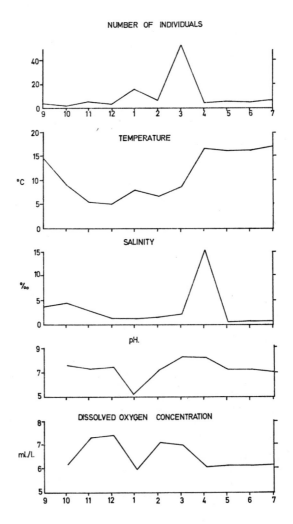

Fig. 14.   Seasonal distribution of living Ostracoda, Aberystwyth Harbour (743). 1966-67.

or winter, appears in April, shows a slight decline into May, and then a great reduction in numbers in June and July. Samples taken in August 1967, and not represented in Fig. 15, failed to reveal the species.

There are two main possibilities concerning the sudden appearance of living forms at this station. The first is that they may have wintered as eggs, on the weed or in the sediment of the station. It is well known that certain species winter entirely or principally as eggs. It seems unlikely, however, that these eggs could have hatched and reached maturity between the 3rd of March and the 6th of April, 1967. In the light of the possibility that very early instars might have been missed when picking the March samples, they have been carefully re-examined without a single live instar being recovered. The second possibility is that the adults could have suddenly migrated into the area, having wintered in deeper water to avoid the very low winter temperatures of the littoral zone. If this is the case, why have they left so soon, long before the temperature falls below the figure for April when they first appeared? It is obvious

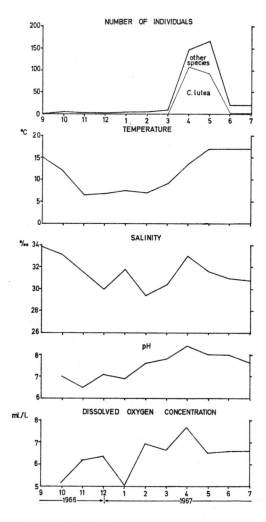

Fig. 15. Seasonal distribution of living Ostracoda College Rocks (745).

that closer sampling over the critical period, which is projected for 1968, coupled with a detailed study of the life history of *Cythere lutea*, is required to resolve this problem.

THE RELATIONSHIP OF OSTRACODA TO ALGAE

Colman (1940) and Wieser (1952) working on weed rich reefs or shores in the Plymouth area, and Chapman (1955) in the Azores, have demonstrated the importance of the relationship between the distribution and abundance of ostracods and that of certain seaweeds.

A study of this relationship has been made at Monks Cave, in the area of sample 736 (Fig. 2), 4½ miles SSW of Aberystwyth. Here (at G.R. 555745) is a substantial wave cut platform with numerous rock pools. Both pools and platform support dense weed growths. Three pools have been selected for investigation. The first (using the nomenclature of Lewis, 1964) is on the littoral fringe and does not receive marine water during neap tides. The second is a mid-eulittoral pool, and the third a sub-littoral pool which, even at lowest spring tides, is always within reach of the waves. The results from February 1966 and April 1967, both collected at low spring tides, are presented in Fig. 16.

The pool on the littoral fringe has a capacity of some 250 gallons and a maximum depth of 3 ft 6 in. It is floored by cobbles and sand. Fourteen species of algae have been recorded in this pool, of which *Ulva* sp., *Cladophora rupestris* and *Corallina officinalis* are the most abundant. From each weed, 100 g (wet) is collected and placed separately in a strong formalin solution and vigorously agitated. The solution is then passed through a 200 inch mesh sieve. To check the validity of this method, representative weed samples have been picked wet under a low powered microscope. No significant difference was observed between the two methods and, as the method involving formalin is much quicker when a large number of samples require to be processed, it has been adhered to in this study.

The largest number of ostracods in the first pool were found on *Ulva* sp., *Cladophora rupestris* and *Enteromorpha clathrata*. The latter two species have dense intergrowths and provide considerable shelter. Other species, such as *Chrondus crispus* and *Halidrys siliquosa*, being long and filamentous provide little shelter, and ostracods were not found on them. Of the 16 species of algae in this pool, 9 have been found to have ostracods living on them.

There seems to be some selectivity on the part of the ostracods in that most species are restricted to a very few weed types. Of the six most common species, *Cythere lutea* and *Paradoxostoma bradyi* occur on only one weed; *Hemicythere cellulosa* occurs on two; *Aurila convexa* on three; *Heterocythereis albomaculata* on four; and *Hirschmannia viridis* on five.

From the winter samples, only 5 ostracod species are recorded, whereas 13 occur in the spring samples. Numbers of individuals are also somewhat higher in the spring.

At the edge of this pool, and on much of the surrounding rock platform, is a dense growth of *Fucus spiralis* and *Fucus serratus*. Ostracods are very rare on these weeds and although large amounts of *Fucus* have been examined, nothing like the maximum number recorded by Colman (1,480 individuals per 100 g) have been found. This large figure was recorded from *Fucus spiralis* although the mean recorded for this weed by Colman was only three individuals. Hagerman (personal communication, 1967) has found several thousand living ostracods on *Fucus* in the Baltic, which are usually concentrated on the epiphytes associated with this genus. *Fucus* is abundant on the west coast of Wales but, probably because of the high energy nature of the littoral zone, epiphytes and ostracods are very rare.

The mean salinity in this pool is 31·5‰. The temperature in November was 8°C and in April 9°C, the pH on each occasion was 7·1.

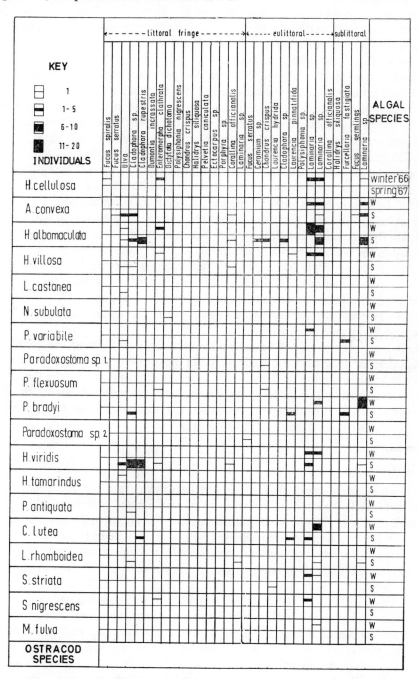

Fig. 16. Relationship of living ostracods to algae at Monk's Cave (736).

U

The mean salinity of the second pool is 32·5‰ and the third pool 33‰. Temperature and pH show very little variation between the three pools. This pool is a linear cleft extending parallel to the coast and is approximately 3 ft deep. Unlike the first pool, it has water, draining off from higher up on the littoral fringe and eulittoral zone, flowing gently through it throughout the time it is uncovered. There are fewer species of algae in this pool, and of the 10 recorded, *Corallina officinalis*, *Cladophora* sp., and *Ceramium* sp. are dominant. *Fucus serratus* is also common but occurs principally on the rock surface surrounding the pool.

The winter sampling in this pool produced 11 species and the spring sampling only 9. The largest number of living ostracods was from the holdfasts of the few *Laminaria* which occur here. These holdfasts produced 11 species in the winter and 5 in the spring. Of the twelve ostracod species found in the holdfasts, six are restricted to this environment, within this pool. Of the other eight algae present, ostracods occur on five of them, but always in lower numbers than in the *Laminaria* holdfasts.

The third pool is large and shallow with a mean depth of approximately $\frac{1}{2}$ m and an area of approximately 30 square metres. Even at lowest spring tides, it is connected to the sea. *Laminaria* forms some 95 per cent of the flora occurring here and the three other algae are all rare. Only seven ostracod species have been recorded from this pool, six from *Laminaria* holdfasts and three on other weeds. Three species were recorded in the winter, and six in the spring.

The absence of *Ascophyllum nodosum* in this area may be responsible for the absence of *Xestoleberis aurantia* which has not yet been recorded in Cardigan Bay. Colman's work seems to indicate an association between the two, although one of the authors (R. C. W.) working on the Ostracoda of the Fleet in Dorset, has found this ostracod to be principally associated with *Fucus*.

These three rock pools will be sampled continually over the next two years as it is evident that the present data is inadequate. In the course of this work, artificial holdfasts will be placed in the sublittoral zone in an attempt to discover why this environment is so frequently populated by large numbers of ostracods of varying species.

In addition, tank experiments are continuing to investigate the nature of the relationship of certain ostracod species to other algae.

DISCUSSION

The most apparent conclusion which can be drawn from the preliminary results presented here is the large amount of further sampling and experimentation which is required in this small area to substantiate any of the suggested divisions or relationships. Only when this has been achieved will it be possible to properly evaluate the four ostracod assemblages which are provisionally recognized or equally to assess the true nature of the apparent seasonal changes in ostracod distribution in the various environments.

One firm conclusion is that ostracod numbers on this very exposed coast, with its low fertility and often extremely turbid waters are, compared to such areas as Plymouth Sound and the Menai Straits (Williams, personal communication 1967), extremely low.

The particular value of this study is felt to be in its comparisons with more sheltered areas such as those quoted above and those in the Baltic where a large number of ecological studies have been made. It is hoped that when this work is completed it will provide a valuable standard of comparison with the results achieved in these more favourable areas.

ACKNOWLEDGEMENTS

The authors wish to express their gratitude to the following for assistance in the course of this work. Dr M. R. Dobson and the late Mr W. Lucas for assistance in collecting at sea; Mr A. S. G. Jones whose MS work on the sediments of the area has been invaluable; Dr A. V. Bromley and Mr R. Fuge for instruction and assistance in chemical analysis; Dr P. Cattermole for critically reading the MS. Mr H. Williams and many others of the Department of Geology, U.C.W., whose help is gratefully appreciated.

The original network of samples was collected by Drs J. R. Haynes and H. Jones and the ostracods picked from these samples by Dr K. Atkinson. Dr Haynes was also responsible for the fixing of all stations before 248 by sextant prior to the aquisition of Decca. The authors wish to gratefully acknowledge this work undertaken before their arrival at the University College of Wales.

We wish particularly to thank Dr A. D. Boney of the Department of Botany, U.C.W., and his students, for valuable advice and discussions concerning the algae, and also Mr R. Williams of the Menai Bridge Marine Station, University College of North Wales, and Dr Lars Hagerman of the University of Lund, for stimulating discussions.

SELECTED BIBLIOGRAPHY

BAIRD, W. 1850. *The Natural History of the British Entomostraca.* The Ray Soc., London, i-vii, 1-364, Pls 1-36. (Ostracoda, 138-82, Pls 18-23).

BARNES, H. 1959. *Apparatus and Methods of Oceanography.* Part one. *Chemical.* George Allen and Unwin Ltd., London.

BRADY, G. S. 1865. Deep-sea dredging on the coasts of Northumberland and Durham. *Rep. Br. Assn Advmt Sci.,* 189-93.

——. 1868. A Monograph of the Recent British Ostracoda. *Trans. Linn. Soc. London,* **26**, 353-495, Pls 23-41.

——. 1869. Ostracoda from the River Scheldt and the Grecian Archipelago. *Ann. Mag. nat. Hist.,* (4), **2** 45-50.

—— AND NORMAN, A. M. 1889. A Monograph of the Marine and Freshwater Ostracoda of the North Atlantic and of North-Western Europe. Section 1. Podocopa. *Sci. Trans. R. Dubl. Soc.,* Ser. 2, **4**, 63-270. Pls 8-23.

——. CROSSKEY, H. W. AND ROBERTSON, D. 1874. A Monograph of the Post-Tertiary Entomostraca of Scotland including species from England and Ireland. *Palaeontologr. Soc.,* 1-232, 16 Pls.

—— AND ROBERTSON, D. 1869. Notes on a weeks dredging in the west of Ireland. *Ann. Mag. nat. Hist.,* (4), **3** 353-74.

——. AND ROBERTSON, D. 1870. The Ostracoda and Foraminifera of Tidal Rivers. *Ann. Mag. nat. Hist.,* (4), **6** 1-33, Pls 4-10.

CARPENTER, J. H. 1965. The accuracy of the Winkler method for dissolved oxygen analysis. *Limnol. Oceanogr.,* **10**, 135-40.

CHAPMAN, G. 1955. Aspects of the Fauna and Flora of the Azores. VI. The density of animal life in the coralline alga zone. *Ann. Mag. nat. Hist.,* (12) **8**, 801-5.

COLMAN, J. 1940. On the faunas inhabitating intertidal seaweeds. *J. mar. biol. Assn U.K.,* **24**, n.s. 129-83, 23 tables.

ELOFSON, O. 1941. Zur Kenntnis der marinen Ostracoden Schwedens mit besonderer Berücksichtigung des Skageraks. *Zool. Bidr. Upps.,* **19**, 215-534, 52 Figs, 42 maps.

EVANS, R. G. 1947. The intertidal ecology of Cardigan Bay. *J. Ecol.,* **34**, 273-309, 10 Figs, 5 tables.

GREEN, E. J. AND CARRITT, D. E. 1967. New tables for oxygen saturation of seawater. *J. mar. Res.,* **25** (2), 140-7, tables 1-2, Fig. 1.

HANAI, T. 1957. Studies on the Ostracoda from Japan. 1. Subfamily Leptocytherinae new subfamily. *J. Fac. Sci. Tokyo Univ.,* Section 2, **10**, 431-68.

HELLAND-HANSEN, B., JACOBSEN, J. P. AND THOMPSON, T. G. 1948. Chemical methods and units. *Publ. Sci. Assoc. Oceanogr. Phys.,* No. 9, 1-28.

HIRSCHMANN, N. 1909. Beiträg zur Kenntnis der Ostracoden-fauna des Finnischen Meeresbusens. *Meddn Soc. Fauna Flora fenn.,* **35**, 282-96.

JONES, T. R. 1857. A Monograph of the Tertiary Entomostraca of England. *Palaeontogr. Soc.,* **9** (1856), 1-68, Pls 1-6.

Lewis, J. R. 1964. *The Ecology of Rocky Shores*. Biological Science texts. The English University Press, London. 1-323.

Müller, G. W. 1894. Die Ostracoden des Golfes von Neapel und der angrenzenden Meeres-Abschnitte. *Fauna Flora Golf. Neapel.*, **21**, Monographie i-viii, 1-404, Pls 1-40.

Müller, O. F. 1785. *Entomostraca seu insecta testacea, quae in aquis daniae et norvegiae reperit, descripsit et iconibus illustravit.* Lipsiae et Havniae, Frankfurt. 1-135.

Neale, J. W. 1965. Some factors influencing the distribution of Recent British Ostracoda. *Pubbl. Staz. zool. Napoli* [1964], **33** (suppl.), 247-307. Pl. 1, Figs 1-11, tables 1-5.

Norman, A. M. 1862. On species of Ostracoda new to Great Britain. *Ann. Mag. nat. Hist.*, (3) **9**, 43-52.

Puri, H. S. 1953. The ostracode genus *Hemicythere* and its allies. *J. Wash. Acad. Sci.*, **43** (6), 169-79, Pls 1-2.

——. 1957. Notes on the ostracode subfamily Cytherideidinae Puri, 1952. *J. Wash. Acad. Sci.*, **47** (9), 305-6.

Remane, A. 1933. Verteilung und Organisation der benthonischen Mikrofauna der Kieler Bucht. *Wiss. Meeresunters.*, N.F. **21**, Heft 2, 16-221, 10 tables, 6 Figs.

Sars, G. O. 1866. Oversigt af Norges marine ostracoder. *Forh. VidenskSelsk. Krist.*, **7**, (1865), 1-130.

——. 1922-28. *An account of the Crustacea of Norway. Vol. IX, Ostracoda.* 1-277, 119 Pls. Bergen Museum, Bergen.

Theisen, B. F. 1966. The life history of seven species of ostracods from a Danish brackish-water locality. *Meddr Kommn Danm. Fisk.-og Havunders.*, N.S. **4** (8), 215-70.

Wagner, C. W. 1957. *Sur les Ostracodes du Quaternaire récent des Pays-Bas et leur utilisation dans l'étude Géologique des dépôts holocènes.* Mouton et Co. The Hague. 1-259, 50 Pls, 21 Figs.

Walton, C. L. 1915. The shore fauna of Cardigan Bay. *J. mar. biol. Assn U.K.*, **10**, 102-13.

Wieser, W. 1952. Investigations on the microfauna inhabiting seaweeds on rocky coasts. IV. Studies on the vertical distribution of the fauna inhabiting seaweeds below the Plymouth Laboratory. *J. mar. biol. Assn U.K.*, **31**, n.s. 145-74, 12 tables, 4 Figs.

——. 1959. The Effect of Grain Size on the Distribution of Small Invertebrates Inhabiting the Beaches of Puget Sound. *Limnol. Oceanogr.*, **4** (2), 181-94, 10 Figs, 2 tables.

Wood, A. 1966. Archwilio Bae Aberteifi. *Y Gwyddonydd*, **IV** (4), 1-7. Figs 1-4.

DISCUSSION

Kornicker: I think that the distribution of the dead abundances relative to the type of sediment is more than coincidental. Possibly it is a function of the kind of sample or measurement that was taken. The only true measurement, which is one that we can seldom get, is between two time planes. If you are comparing the dead population in say a clay as against a sand, you will naturally get major differences. I would welcome your comments on this.

Wall: I agree that this is a factor to take into account as is also the fact that we may be dealing here with sub-Recent material. In this area we have carapaces of species of the subfamily Paradoxostomatinae preserved, species which do not occur living, or at least that I have not yet found living, in Cardigan Bay. I do not think that this could explain the difference between finding three carapaces in a mobile sand area and over a thousand carapaces in a muddy area.

Kornicker: Was this based on equal weight or equal volume samples?

Wall: Equal volume samples.

McKenzie: During your account of the numbers of living ostracods obtained you illustrated counts per 10 ml sample. How did you collect a 10 ml sample in the areas of cobble grade sediments?

Whatley: The numbers of ostracods in these areas do not bear a strict relationship to the numbers obtained from other areas. In the areas of cobble and boulder, sampling with dredge or grab is very much a 'hit or miss' process. In consequence, this type of substrate cannot, by these methods, be sampled quantitatively. Usually the ostracods in these areas are obtained from encrusting green algae scraped from the stones or from *Laminaria* holdfasts. In either case all the ostracods are taken into account.

ECOLOGY OF THE OSTRACODA FROM SELECTED MARINE INTERTIDAL LOCALITIES ON THE COAST OF ANGLESEY

R. WILLIAMS[1,2]

Marine Science Laboratories, Menai Bridge, Anglesey

SUMMARY

The distribution of the ostracod fauna from five localities on the coast of Anglesey was investigated. No representatives of the suborder Mydocopa were found. Species belonging to the family Polycopidae were *Polycope orbicularis* (G. O. Sars) and *Polycopsis compressa* (Brady and Robertson) and to the family Cyprididae, *Pontocypris hispida* (G. O. Sars) and *Bythocypris obtusata* (G. O. Sars). All other species belonged to the family Cytheridae. The total population was found to consist of 58 species. The fauna from each of these localities was divided into two assemblages; the population occurring on the beach sediment and the population on the algae adjoining the beach. A number of species were found to be specific to these separate habitats. The populations inhabiting the intertidal sediments were contaminated with a number of species from deeper water localities, these forms were generally represented by single valves and carapaces.

A list of all the Foraminifera found in the habitats investigated is included in this work.

INTRODUCTION

During an intensive study of the habitat relationships of the interstitial and associated meiofauna of selected intertidal beaches, the Ostracoda was one of the groups studied in detail. Hitherto, this group has received little attention in the area, the records of species being extremely sparse. A quantitative and qualitative study of the ostracod fauna was carried out with selective observations to detect seasonal fluctuations in the population. The occurrence and influence of the ostracod species inhabiting the algal zones adjoining these habitats were also investigated.

The beach localities chosen for this investigation exhibited the following range of deposits: gravel, coarse sand, medium sand, fine sand, and fine sand with varying proportions of silt. The various environmental characteristics of each of these habitats were studied, e.g. particle size analysis, the shape and the size of the interstices in the sediment, porosity, water content, salinity, temperature. The linking of these measurements with the general distribution of the Ostracoda in the habitats was attempted to ascertain whether the faunal and environmental elements corresponded to the general pattern of distribution. The results obtained from this section are beyond the scope of the present paper and will be reported in a subsequent publication.

[1] This work was carried out during the tenure of a S.R.C. Research Student grant.
[2] Now at: Oceanographic Laboratory, Edinburgh, Scotland.

299

DESCRIPTION OF THE LOCALITIES

The five intertidal beach localities (Fig. 1) displayed varying gradations in their physical and environmental characters. The most important of these related characters are: sediment composition, exposure to wave action, phytal influence and the effects of salinity change on the beach due to land drainage.

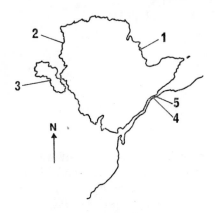

Fig. 1. Map of Anglesey showing the positions of the localities: 1. Traeth Bychan; 2. Porth Swtan; 3. Porth Castell; 4. Church Island; 5. Menai Bridge.

Locality 1. Traeth Bychan

Situated on the east coast of the island this sheltered beach faces ENE. The tidal range from high water to low water is 22 feet. From a small pebble beach at high water neap tide level (H.W.N.) down to the level of low water neap tides (L.W.N.) there are 220 yards of gently sloping sand. The sediment consists of medium size quartz particles with the larger fraction composed of broken lamellibranch shell. Due to the slope of the beach a greater part of it is saturated at ebb tide. The water table at the minimum is only five centimetres from the surface of the sand. At the north end of the beach there is a fresh-water inflow which affects the salinity of a large section of the shore. Even in June the flow of water is sufficient to give a salinity reading of 4‰ on the surface of the sand at low water. The rocky boulder beach close to the sand supports a relatively sparse localized population of algae. The dominant weeds are *Ascophyllum nodosum* (L.) Le Jol. and *Fucus serratus* (L.).

Locality 2. Porth Swtan

This locality faces due west. On the exposure scale, defined by Ballantine (1961), it would be in category 2, i.e. a very exposed shore. The main algae on the shore are *Enteromorpha compressa* (L) Grev., *Porphyra umbilicalis* (L.) J.Ag. and *F. serratus*, with the notable absence of *A. nodosum*. The littoral zone is 150 yards of gently sloping fine sand, the tidal rise and fall being 16·5 feet. The sediment is poorly graded (Hough, 1957) composed in the main of particles of the same size. There is a fresh-water seepage on to the shore at the high water mark which causes a reduction in salinity to 27‰ from 33·5‰ on certain areas of the beach.

Locality 3. Porth Castell

Although on the west coast of the island this locality faces north-west. It is protected from the direction of the prevailing seas by a headland. As a result of this the beach is fairly sheltered, possessing a tidal range of 17 feet. The length of the beach is 70 yards, hence the slope is greater than the two previously described localities. A fresh-water stream flowing on to the beach at high water causes salinity reductions down to 22‰ at mean tide level (M.T.L.). The rocks adjoining the beach support the typical phytal zonation of a sheltered shore (Stephenson and Stephenson, 1949; Lewis, 1953, 1955; Ballantine, 1961), the dominant weed being *A. nodosum*. The sediment is a medium to coarse sand.

Locality 4. Church Island

At extreme low water spring tides—2 feet below C.D., a large gravel bank composed of barnacle shell fragments is exposed in this area. This sediment was included in this study because of the large nature of the particles and hence the large interstices in the deposit. The median diameter of the particles is in the region of 1400 microns. It is a very sheltered locality surrounded by a profusion of algae, dominated by *Laminaria* spp.

Locality 5. Menai Bridge

The beach at Menai Bridge is a very sheltered area in the Menai Straits with a tidal rise and fall of 21·6 feet for spring tides. The beach sediment is composed mainly of very fine sand particles with a varying fraction of silt. The particle size analyses of all the localities studied were based on sievings with British Standard sieves related to the Wentworth grain size scale for sediments. The phytal cover on the shore is limited to outcrops of rock occurring from high water to mean tide level. The main weed attached to the rocks is *A. nodosum*. The individual specimens of this weed have dense tufted growths of a number of epiphytic weeds. There is a high proportion of silt and organic debris found trapped in the interstices of this dense growth, increasing the 'habitat potential' for a number of species of the cryptofauna. This was noticed by Dahl (1948), and Wieser (1952, 1959), who correlated the high densities of Nematoda with the amount of silt trapped in these growths.

Materials and Methods

Samples were collected at five stations (a, b, c, d, e) along a line traverse at H.W.N., M.T.L. and L.W.N. on the shores of localities 1, 2 and 3. Each of the samples was taken by means of a 0·1 m² frame which was pressed into the sand to a depth of 2–3 cm. The samples were then scooped into small screw-topped jars, which were marked with the date, location, sample number and tidal level. The jars were then stored in a cold room at 1°C and worked on as soon as possible. The depth of two to three centimetres was chosen because of the concentration of the interstital and associated fauna in the top surface centimetres of the sand in the localities studied. Each sample jar contained approximately 200 ml of damp sediment. Two samples were taken at each station, one for the estimation of the fauna and the other for the measurement of the physical parameters of the sediment. The fauna samples were subsampled, the respective aliquots being 50 ml of damp sand. Every animal in one of these subsamples was counted and the population structure determined. The other aliquots of the original sample were examined to see if the sample was a representative portion and that all the species had been identified. The fauna was also identified from samples taken down to 20 cm below the surface of the sand at selected stations along the traverses. Samples of the weeds were collected from the phytal zones at each locality. These samples were obtained from the corresponding tidal levels of the traverses. The algal samples were sealed in polythene bags and placed in a cold room.

The cryptofauna was maintained in a living condition at the temperature of 1°C, but the fauna was identified as soon as possible from the date of sampling. The sediment and the algal regions at the smaller localities 4 and 5 were selectively sampled by a similar procedure to the one outlined above.

The fauna was removed from the weed and sediment by washing twice with an isotonic solution of $MgCl_2$ and finally with a solution of 1% formalin. The solutions containing the dead and anaesthetized animals were filtered through a nylon sieve with a mesh size of 150 microns. The sieved animals were washed into petri dishes and then transferred in solution to plankton counting trays for quantitative and qualitative analysis. The animal counts from the sediments were expressed in numbers per 50 ml of wet deposit. This number multiplied by 400 gave the approximate numbers of animals per square metre of sediment down to a depth of two centimetres. For similar comparative reasons the cryptofauna counts were expressed in numbers per 100 g of damp weed.

LISTS OF SPECIES

Ostracoda; 58 *species*

Species	Localities
1. *Cythere lutea* O. F. Müller	5sa*, 4s*a*, 1s*a*, 2s*a*, 3s*a*
2. *Cyamocytheridea papillosa* (Bosquet)	4s
3. *Cyamocytheridea punctillata* (Brady)	4sa, 5s
4. *Heterocyprideis sorbyana* (Jones)	3s
5. *Eucythere declivis* (Norman)	3s
6. *Eucythere argus* (G. O. Sars)	1s
7. *Cushmanidea elongata* (Brady)	1s*
8. *Neocytherideis subulata* (Brady)	4s, 5s
9. *Cytherura gibba* (O. F. Müller)	3s*, 4s, 5a*
10. *Semicytherura sella* (G. O. Sars)	1s, 2s, 3s*, 4s, 5s
11. *Semicytherura nigrescens* (Baird)	5sa*, 1a*, 3s*a*, 4s*a*
12. *Semicytherura acuticostata* (G. O. Sars)	3s
13. *Semicytherura striata* (G. O. Sars)	1s*, 4s*a, 5s
14. *Semicytherura angulata* (Brady)	4a, 5s
15. *Microcytherura fulva* (Brady and Robertson)	4s
16. *Hemicytherura clathrata* (G. O. Sars)	1s
17. *Hemicytherura cellulosa* (Norman)	5sa, 1a*, 2sa*, 3s, 4sa*
18. *Cytheropteron latissimum* (Norman)	1sa, 3s, 5s
19. *Cytheropteron nodosum* Brady	1s, 2s, 3s, 4sa, 5s
20. *Cytheropteron inflatum* Brady	3s
21. *Hemicythere villosa* (G. O. Sars)	5sa, 1s*a, 2s, 3sa*, 4sa*
22. *Hemicythere emarginata* (G. O. Sars)	3s
23. *Hemicythere angulata* (G. O. Sars)	1s
24. *Elofsonella concinna* (Jones)	1s, 2sa, 3s, 5s
25. *Aurila convexa* (Baird)	1s, 3s, 4s, 5s
26. *Heterocythereis albomaculata* (Baird)	4s*, 1s*a*, 2s*a*, 3sa*
27. *Leptocythere pellucida* (Baird)	5s*a, 1s*a, 2s, 3sa, 4s*
28. *Leptocythere macallana* (Brady and Robertson)	1s
29. *Leptocythere castanea* (G. O. Sars)	1s, 4s*, 5s*
30. *Leptocythere tenera* (Brady)	5a
31. *Cytheromorpha fuscata* (Brady)	5s, 1s, 2sa, 3s, 4sa*
32. *Hirschmannia viridis* (O. F. Müller)	5s*a*, 4s*a*, 1as, 2sa, 3sa*
33. *Hirschmannia tamarindus* (Jones)	4sa, 1s, 2s, 3s
34. *Loxoconcha impressa* (Baird)	5s*a*, 4s*a*, 1a*s, 2s*a*, 3s*a*
35. *Loxoconcha granulata* G. O. Sars	1s
36. *Paradoxostoma variabile* (Baird)	5s*, 1s, 2s, 3a*, 4s*a*

37. *Paradoxostoma ensiforme* Brady	4s
38. *Paradoxostoma bradyi* G. O. Sars	4a
39. *Paradoxostoma hibernicum* Brady	2sa
40. *Paradoxostoma abbreviatum* G. O. Sars	4a, 5s
41. *Paradoxostoma normani* Brady	5sa
42. *Paradoxostoma sarniense* Brady	3a*
43. *Cytherois fischeri* (G. O. Sars)	4s*a
44. *Cytherois vitrea* (G. O. Sars)	1s*a*, 3s*a, 4a
45. *Cytherois pusilla* G. O. Sars	3s*a*, 4a, 5s
46. *Paracytherois arcuata* (Brady)	1s, 4s, 5s
47. *Paracytherois flexuosa* (Brady)	5s
48. *Sclerochilus contortus* (Norman)	1s, 4a, 5s
49. *Trachyleberis dunelmensis* (Norman)	⋅ 1s, 3s, 4s
50. *?Cythereis tuberculata* (G. O. Sars)	3s
51. *Xestoleberis aurantia* (Baird)	1a*, 2s*a*, 3s*a*
52. *Xestoleberis depressa* (G. O. Sars)	1s
53. *?Cythere globulifera* Brady	4s
54. *Xenocythere cuneiformis* (Brady)	1s, 2s, 3s
55. *Bythocypris obtusata* (G. O. Sars)	5s
56. *Pontocypris hispida* G. O. Sars	5s
57. *Polycope orbicularis* G. O. Sars	4s*
58. *Polycopsis compressa* (Brady and Robertson)	4s*

Localities 1—5 as in text. s—sediment, a—algae, *—live individuals present.

Foraminifera; 28 species

Species	Localities
1. *Miliolinella subrotunda* (Montagu)	4s*a*, 2s
2. *Spirillina vivipara* Ehrenberg	4s*
3. *Rosalina globularis* d'Orbigny	4s*a*, 2s, 3
4. *Ammonia beccarii* (Linné)	4s, 5a, 2s, 3s, 4s
5. *Lagena clavata* (d'Orbigny)	4s
6. *Lagena sulcata* (Walker and Jacob)	4s, 2s
7. *Oolina hexagona* (Williamson)	4s
8. *Oolina williamsoni* (Alcock)	4s, 5s
9. *Oolina melo* d'Orbigny	5s
10. *Bulimina elongata* d'Orbigny	4s
11. *Elphidium incertum* (Williamson)	4s, 5a
12. *Elphidium macellum* (Fichtel and Moll)	4a, 5a*s*, 2s
13. *Elphidium selseyense* Heron-Allen and Earland	4a, 5a, 3s
14. *Elphidium magellanicum* Heron-Allen and Earland	4a
15. *Elphidium crispum* (Linné)	3s
16. *Elphidium discoidale* (d'Orbigny)	2s, 4a*, 5a*
17. *Nonion depressulum* (Walker and Jacob)	4a
18. *Nonion pompilioides* (Fichtel and Moll)	4s, 5a
19. *Nonion* cf. *N. granosum* (d'Orbigny)	4s
20. *Trochammina squamata* Jones and Parker	4s
21. *Trochammina nana* (Brady)	4a
22. *Massilina secans* (d'Orbigny)	4s, 2s, 3s
23. *Discorbis bradyi* Cushman	4a
24. *Cibicides lobatulus* Walker and Jacob	4a*, 2s
25. *Pateoris hauerinoides* (Rhumbler)	2s
26. *Quinqueloculina seminulum* (Linné)	2s, 3s, 4s
27. *Quinqueloculina auberiana* d'Orbigny	2s
28. *Quinqueloculina bicornis* (Walker and Jacob)	2s

Localities 1—5 as in text, s—sediment, a—algae, *—live species (recognized by feeding cyst).

RESULTS OF THE INVESTIGATION

Traeth Bychan. Locality 1

The algae. The algae from this locality supported a sparse cryptofauna. From Tables 1a and 1b it is noticed that the ostracods formed a small percentage of the population, the highest density recorded per 100 g of weed was only twelve individuals. The numbers occurring on the algae were fairly constant as shown by repeating the sampling throughout the respective algal zones. Where a fluctuation of numbers occurred a direct correlation could be made with the amount of epiphytic growth on the *A. nodosum*. The main epiphytes on this weed were: *Polysiphonia lanosa* (L) Tandy; *Pilayella littoralis* (L) Kjellm; *Ceramium rubrum* (Huds) J. Agardh; and *Ulva lactuca* (L). It is the *P. littoralis* which is responsible for the dense mats of epiphytic growth. The living ostracods found on *A. nodosum* were:

Cythere lutea	*Heterocythereis albomaculata*
*Semicytherura nigrescens**	*Loxoconcha impressa*
*Hemicytherura cellulosa**	*Xestoleberis aurantia**

with the occasional specimen of *Cytherois vitrea*. Eleven species are found on the weed, the three above marked * were found to be specific to the algae.

The sediment. The results of the quantitative and qualitative work on the three traverse lines are shown in Tables 1c, 1d and 1e. Twenty-seven species were found on the sediment of the shore, representatives of eight species were found alive.

H.W.N. traverse. Nineteen species occurred at this level on the beach, six of which were repeatedly found alive in the samples. These were:

Cythere lutea	*Heterocythereis albomaculata*
Cushmanidea elongata	*Leptocythere pellucida*
Hemicythere villosa	*Xestoleberis depressa*

The larval stages of *L. pellucida* contributed to the majority of the species found.

M.T.L. traverse. Twenty-one species occurred at this level, individuals of only four species were alive. These were:

Cushmanidea elongata	*Leptocythere pellucida*
Semicytherura striata	*Hemicythere villosa*

The majority of these twenty-one species were represented in the fauna by dead forms or empty carapaces and valves. Some of these species were typical algal forms, while others occur on sublittoral bottom sediments. *L. pellucida* is the most abundant living species forming 25–95 per cent of the living population.

L.W.N. traverse. The assemblage at low water consisted of sixteen species, representatives of three of these were found alive:

Cushmanidea elongata	*Leptocythere pellucida*
Cytherois vitrea	

Again *L. pellucida* was the most abundant species forming 42–100 per cent of the live population. *Cytherois vitrea* was found at station Le, the nearest area sampled to the algal zones. This species was the only living form to be found on *F. serratus* and it probably accounts for the occasional individual being found on the sediment adjoining the algal zone.

TABLE 1a. The various groups of animals inhabiting three species of weed from Traeth Bychan

Results expressed in numbers of animals per 100g
of damp weed and in percentage of total population.
Ascophyllum nodosum—epiphytes—Polysiphonia lanosa
Pilayella littoralis

Species of algae	*Fucus spiralis*		*Ascophyllum nodosum*		*Fucus serratus*	
Weight in g	59		173		170	
Sample No.	HF_1		MA_1		LF_1	
Tidal level	HWN		MTL		LWN	
Date	17.5.67		17.5.67		17.5.67	
Acarina	26	25·0	139	28·2	41	25·9
Polychaeta	—	—	2	0·5	—	—
Oligochaeta	—	—	8	1·5	—	—
Copepoda						
Harpacticoida	9	8·3	50	10·1	12	7·8
Insecta						
Diptera Larvae	—	—	2	0·3	—	—
Isopoda	3	3·3	4	0·7	—	—
Amphipoda	24	23·3	3	0·6	—	—
Gastropoda	17	16·7	25	5·1	1	0·7
Lamellibranchia	—	—	2	0·3	—	—
Nemertini	—	—	5	1·1	—	—
Nematoda	17	16·7	237	48·2	101	63·3
Turbellaria	—	—	5	1·1	3	1·9
Ostracoda	7	6·7	12	2·3	1	0·4
Total	103	100	494	100	159	100

TABLE 1b. Ostracods inhabiting three species of algae from Traeth Bychan

Counts of living (a), *dead* (d) *and right* (v) *valves of adult and juvenile species*

Species of algae	*Fucus spiralis*			*Ascophyllum nodosum*			*Fucus serratus*		
Weight in g	59			173			170		
% of ostracods of total pop.	6·7			2·3			0·4		
% of living ostracods of tot. pop.	0·0			1·2			0·4		
Species	a.	d.	v.	a.	d.	v.	a.	d.	v.
Cythere lutea ad. juv.				3					
Semicytherura nigrescens				2					
Hemicytherura cellulosa				2					
Cytheropteron latissimum					1				
Hemicythere villosa						1			
Heterocythereis albomaculata				1					
Leptocythere pellucida		3		1					
Hirschmannia viridis				1					
Loxoconcha impressa				1	3				
Cytherois vitrea		1			2	1	1		
Xestoleberis aurantia				1					
Total	0	4	0	10	8	2	1	0	0

TABLE 1c. Ostracods from seven samples of beach sand taken along a line transect at H.W.N. Traeth Bychan, 17.5.67

Counts of living (a), *dead* (d) *and right valves* (v) *of adult and juvenile species per 50 ml of wet sand (approx. 65 g)*

Sample No.	Ha	Hb	Hc₁	Hc₂*	Hd₁	Hd₂*	He
Tot. pop. of animals in sample	1006	789	176	68	36	105	155
% of ostracods of the total pop.	3·6	8·2	24·4	28·0	52·7	11·4	7·8
% of living ostracods of the tot. pop.	3·3	3·0	5·1	10·3	2·8	0·0	0·6
Salinity ‰ surface	29·5	23·5	23·0	—	0·5	—	16·0
15 cm below surface	29·5	21·7	—	22·0	—	5·8	17·5

Species	Ha			Hb			Hc₁			Hc₂*			Hd₁			Hd₂*			He		
	a	d	v	a	d	v	a	d	v	a	d	v	a	d	v	a	d	v	a	d	v
1. *Cythere lutea†* ad.					2	1														1	
juv.				2																	
2. *Eucythere argus*				1																	
3. *Cushmanidea elongata*		1	1		2	10		3			2	2		3	3		1	1			3
juv.	8			4																	
4. *Semicytherura striata*				2										1						1	
5. *Cytheropteron nodosum*											1										
6. *Hemicythere villosa†*					1	1		1	1					2	1		2	3			
juv.	6			1																	
7. *Hemicythere angulata*																					2
8. *Elofsonella concinna*																					1
9. *Aurila convexa*				1						1	1										
10. *Heterocythereis albomaculata†*	2			1						1	1										
juv.	8																				
11. *Leptocythere pellucida†*				5	7	5	10	2		2				1	1		1	1		1	2
juv.	8			12			9			7			1								
12. *Leptocythere macallana*				2	2		1	3					3	2			1				
13. *Hirschmannia tamarindus*				2			3														1
14. *Loxoconcha impressa†*																					1
15. *Loxoconcha granulata*							2			2	2		2	2			1		1	1	
16. *Sclerochilus contortus*											1										
17. *Trachyleberis dunelmensis*											1										
18. *Xestoleberis depressa*	1																				
19. *Xenocythere cuneiformis*											1										
Total	33	1	2	24	20	21	9	18	16	7	7	5	1	10	8	0	5	7	1	4	

	Ha	Hb	Hc₁	Hc₂*	Hd₁	Hd₂*	He
Sum total of species	36	65	43	19	19	12	13
Number of ostracods per m²	14,400	28,000	17,200	7,600	7,600	4,800	5,200

* Samples taken 15cm below the surface of the sand.
† Species occurring on the algae adjoining the beach.

TABLE 1d. Ostracods from seven samples of beach sand taken along a line transect at M.T.L. Traeth Bychan, 7.5.67

Counts of living (a), dead (d) and right valves (v) of adult and juvenile species per 50 ml of wet sand (approx. 65 g)

Sample No.	Ma	Mb	Mc1	Mc2*	Md1	Md2*	Me
Tot. pop. of animals in sample	622	338	606	148	21	88	154
% of ostracods of the total population	7·2	14·5	9·8	11·5	61·9	14·0	46·8
% of living ostracods of the tot. pop.	6·4	13·6	8·6	3·4	14·3	2·2	13·6
Salinity ‰ surface	30·8	32·4	30·6	—	1·5	—	32·4
15 cm below surface	30·1	31·8	—	32·0	—	26·6	31·1

Species	Ma			Mb			Mc1			Mc2*			Md1			Md2*			Me		
	a	d	v	a	d	v	a	d	v	a	d	v	a	d	v	a	d	v	a	d	v
1. Cushmanidea elongata ad.		1		2			2	2		2			1	1					3	1	7
juv.	5			3			7									2				2	
2. Semicytherura sella																				2	
3. Semicytherura striata				1															1	1	
4. Hemicytherura clathrata																				2	1
5. Cytheropteron nodosum																					2
6. Cytheropteron latissimum†																					2
7. Hemicythere villosa†										1	2			1		1					
8. Hemicythere angulata														1							
9. Elofsonella concinna									2			3		1			2				
10. Aurila convexa										1				1							
11. Heterocythereis albomaculata†														1					1	1	
12. Leptocythere pellucida†	35	1		43			45			5	2	2	3	1	1	2	1	1	11	7	6
13. Leptocythere macallana																			2		
14. Leptocythere castanea																			3		
15. Cytheromorpha fuscata										1											
16. Hirschmannia viridis																					1
17. Hirschmannia tamarindus																				2	
18. Loxoconcha granulata				1	1					1				1		1				2	8
19. Paradoxostoma variabile							1														
20. Paracytherois arcuata																					1
21. Trachyleberis dunelmensis																					2
Total	40	3	2	46	1	2	52	5	2	5	7	5	3	8	2	2	7	4	21	21	30

Sum total of species	45	49	59	17	13	13	72
Number of ostracods per m²	18,000	19,600	23,600	6,800	5,200	5,200	28,800

* Samples taken 15 cm below the surface of the sand.
† Species occurring on the algae adjoining the beach.

TABLE 1e. Ostracods from seven samples of beach sand taken along a line transect at L.W.N. Traeth Bychan, 17.5.67

Counts of living (a), *dead* (d) *and right valves* (v) *of adult and juvenile species per 50 ml of wet sand (approx. 65 g)*

Sample No.	La	Lb	Lc$_1$	Lc$_2$*	Ld$_1$	Ld$_2$*	Le
Tot. pop. of animals in sample	601	476	425	157	251	315	377
% of ostracods of the total pop.	8·6	9·4	6·6	9·6	5·2	4·5	5·0
% of living ostracods of the tot. pop.	6·7	7·1	3·0	2·0	1·6	0·6	1·6
Salinity ‰ surface	32·1	31·9	31·5	—	3·2	—	32·5
15 cm below surface	32·5	31·1	—	32·0	—	31·2	32·5

Species	La a	La d	La v	Lb a	Lb d	Lb v	Lc$_1$ a	Lc$_1$ d	Lc$_1$ v	Lc$_2$* a	Lc$_2$* d	Lc$_2$* v	Ld$_1$ a	Ld$_1$ d	Ld$_1$ v	Ld$_2$* a	Ld$_2$* d	Ld$_2$* v	Le a	Le d	Le v
1. *Cythere .lutea*† ad.				1			1														
juv.																					
2. *Cushmanidea elongata*	1	1								4	1		1			2					
(juv.)	15			14			8														
3. *Semicytherura striata*																				1	
4. *Cytheropteron latissimum*†																				1	
5. *Elofsonella concinna*	1			2	1								1			1					
6. *Aurila convexa*							1							1		1					
7. *Leptocythere pellucida*†		6		4	2		3	2		7	3		1			1	3		1	2	1
(juv.)	25			20			5			3			4			2			4		
8. *Leptocythere macallana*	2									1									1		
9. *Leptocythere castanea*							2						1	2		1					
10. *Cytheromorpha fuscata*				1									1								
11. *Hirschmannia tamarindus*																			1		
12. *Loxoconcha granulata*	1						1						1				2			2	2
13. *Paradoxostoma variabile*																				1	
14. *Cytherois vitrea*†																			1		
15. *Paracytherois arcuata*																			1		
16. *Sclerochilus contortus*							1														
Total	40	11	1	34	8	3	13	7	8	3	9	3	4	5	4	2	4	7	6	10	3
Sum total of species	52			45			28			15			13			13			19		
Number of ostracods per m²	20,800			18,000			11,200			6,000			5,200			5,200			7,600		

* Samples taken 15 cm below the surface of the sand.
† Species occurring on the algae adjoining the beach.

From the three traverses it can be seen that the dominant ostracod species is *L. pellucida*. Larval stages of this ostracod were even found 15 cm below the surface at stations Md$_2$* and Mc$_2$* at mean tide level. One live juvenile was found at station Hd$_1$ where the salinity was only 0·5‰. In all cases the line traverses cross the fresh-water stream at the station d$_1$. Stations c$_1$ and e$_1$ were sampled either side of the stream. A decrease in numbers was observed as one progressed towards the areas of lower salinity. The larval forms of *C. elongata* were never found where the salinity was below 23‰. The intertidal animals living in the areas affected by the fresh-water stream will be subjected, at the maximum, to a salinity gradient of 0–33·5‰ every tidal cycle of twelve hours, but there are an infinite number of constantly changing gradations present in the area. The species living at the higher levels of the shore are obviously subjected to the salinity drop caused by the fresh water for greater periods of time.

Porth Swtan. Locality 2

The algae. The dominant algae on the shore are *P. umbilicalis* and *F. serratus.* The fauna occurring on these species (Tables 2a and 2b) is found in higher densities than in the previous locality even though the exposure is far greater. A factor which contributes to this difference is desiccation. The broad flat fronds of *E. compressa* and *P. umbilicalis* form dense mats on the surface of the rocks maintaining a moist environment beneath them at ebb tide; even when the sun is strong in summer there is little fluctuation of the temperature. The *F. spiralis* and other brown weeds at Traeth Bychan occur in localized growths on the boulders of the shore. Drainage and air circulation causes greater desiccation of these weeds than at Porth Swtan, affording a less favourable environment for the cryptofauna.

Nine species were found on the weeds, individuals of six of these species occurred alive, these were:

Cythere lutea	*Hirschmannia viridis**
Heterocythereis albomaculata	*Loxoconcha impressa*
*Hemicytherura cellulosa**	*Xestoleberis aurantia**

The species marked * were found to be specific to the algae of this locality. Occasional live specimens of the other species were found on the sediment, probably these individuals originally came from the algal zones being detached by wave action.

The sediment. The constant wave action on the shore would make this an unfavourable environment for colonization by surface living ostracods. The occasional species found on the beach are not burrowing forms and are too large to utilize the interstitial pore system formed by the spatial relationships of the individual quartz particles. Seventeen species were found on the beach, only individuals of three species were alive, and these, as stated above, probably originated from the algae.

H.W.N. traverse. Three stations were allocated to this traverse due to the proximity of a pebble beach. The sand in the samples was desiccated, the water table being 35 cm below the surface of the sand. The ostracods were represented by the occasional valves of *L. impressa.*

M.T.L. traverse. Seventeen species (Table 2c) occurred at this level. Only two individuals out of the 191 ostracods determined for the traverse were alive. These were *C. lutea* and *L. impressa.* The majority of the species were represented as valves.

L.W.N. traverse. Fifteen species were recorded from this assemblage. One individual *H. albomaculata*, was found alive. All the other species were represented by valves and carapaces. Sampling in January 1967 gave even fewer numbers of ostracods. The seas were rough during this month, on one occasion removing approximately one foot of sediment from the beach during a tidal cycle. Very few species occurred in the algal zones due to wave action and to the loss of the *P. umbilicalis* belt which occurs on the rocks only during the summer months.

Table 2a. The various groups of animals inhabiting three species of weed from Porth Swtan

Results expressed in numbers of animals per 100 g of damp weed and in percentage of total population

Species of algae	*Enteromorpha compressa*		*Porphyra umbilicalis*		*Fucus serratus*	
Sample No.	HE$_1$		LP$_1$		LF$_1$	
Weight in g	49		57		142	
Tidal level	HWN		LWN		LWN	
Date	25.6.67		25.6.67		25.6.67	
Acarina	184	19·8	14	3·2	88	16·5
Coelenterata	—	—	28	6·35	—	—
Polychaeta	—	—	—	—	3	0·5
Copepoda						
Harpacticoida	246	26·4	321	73·0	76	14·4
Isopoda	—	—	—	—	9	1·6
Amphipoda	—	—	—	—	3	0·5
Gastropoda	8	0·9	14	3·2	9	1·6
Lamellibranchia	37	4·0	—	—	9	1·6
Insecta						
Diptera larvae	—	—	—	—	9	1·6
Decapoda						
Megalopa	8	0·9	—	—	—	—
Nematoda	246	26·4	35	7·9	288	54·3
Turbellaria	25	2·7	—	—	—	—
Ostracoda	176	18·9	28	6·35	40	7·4
Total	930	100	440	100	534	100

Table 2b. Ostracods inhabiting three species of algae from Porth Swtan

Counts of living (a), dead (d) and right valves (v) of adult and juvenile species

Species of algae		*Enteromorpha compressa*			*Porphyra umbilicalis*			*Fucus serratus*		
Weight in g		49			57			142		
% of ostracods of total pop.		18·9			6·35			7·4		
% of living ostracods of tot. pop.		12·8			0·8			5·9		
Species		a.	d.	v.	a.	d.	v.	a.	d.	v.
Cythere lutea	ad.	36		6	4		4			
	juv.	12						16		4
Hemicytherura cellulosa								4		
Elofsonella concinna			4			4				
Heterocythereis albomaculata		4								
Cytheromorpha fuscata									4	
Hirschmannia viridis									16	
Loxoconcha impressa			4	8			4	4		
Paradoxostoma hibernicum				6					4	4
Xestoleberis aurantia		6								
Total		58	8	20	4	4	8	24	24	8

TABLE 2c. Ostracods from seven samples of beach sand taken along a line transect at M.T.L. Porth Swtan, .6.67

Counts of living (a), dead (d) and right valves (v) of adult and juvenile species per 50 ml of wet sand (approx. 65 g)

Sample No.	Ma	Mb	Mc$_1$	Mc$_2$*	Md$_1$	Md$_2$*	Me
Tot. pop. of animals in sample	57	59	132	41	171	37	596
% of ostracods of the total population	21·1	30·5	70·2	24·0	8·2	13·5	3·9
% of living ostracods of the tot. pop.	0	0	0	0	0	0	0·3
Salinity ‰ surface	33·3	33·2	33·5	—	28·5	—	33·5
15 cm below surface	33·3	33·2	—	33·5	—	32·3	33·5

Species		Ma			Mb			Mc$_1$			Mc$_2$*			Md$_1$			Md$_2$*			Me		
		a.	d.	v.	a.	d.	v.	a.	d.	v.	a.	d.	v.	a.	d.	v.	a.	d.	v.	a.	d.	v.
Cythere lutea†	ad.			2			3		10	2					1	3		2				4
	juv.								11			4						1				
Semicytherura sella									1													
Hemicytherura cellulosa†									1													
Cytheropteron nodosum																		2				
Hemicythere villosa						2			2													2
Elofsonella concinna†				2		2			2			1			1						1	2
Heterocythereis albomaculata†									2													
Leptocythere pellucida																						1
Cytheromorpha fuscata†				2		3	1		3	2		1			1			1				
Hirschmannia viridis†									5			1										3
Hirschmannia tamarindus							1		1						1							3
Loxoconcha impressa†			1	3		1	5		11	10		3			1	4	2	1				3
									9													1
Paradoxostoma variabile																		1				
Paradoxostoma hibernicum†				2					8						1	1						1
									2													
Sclerochilus contortus									1													
Xestoleberis aurantia†									7													
									3													
Xenocythere cuneiformis									1													
Total		0	1	11	0	6	12	25		69		10		5		9		5		2	4	17
Sum total of species			12			18			94			10			14			5			23	
Number of ostracods per m²			4,800			7,200			37,600			4,000			5,600			2,000			9,200	

* Samples taken 15 cm below the surface of the sand.
† Species occurring on the algae adjoining the beach.

X

Porth Castell. Locality 3

The algae. A dense cover of brown weeds occurred on the rocks at this locality. The majority of *A. nodosum* plants had large growths of epiphytes, the dominant species being *P. lanosa.* At low water a number of filamentous red algae were found, such as *Cystoclonium purpureum* (Huds) Batt. The cryptofauna occurred in very high densities on these small algae. Tables 3a and 3b illustrate the populations on one sample of *C. purpureum,* over 6000 ind./100 g of damp weed. The majority of the weight of this weed is taken up by water. It is one of the few examples where the Ostracoda is equally dominant with the Copepoda, together they form 70 per cent of the population. It is amazing that such a population density occurs on this type of alga. There are no epiphytic diatoms or adhering detritus on the surface of the weed.

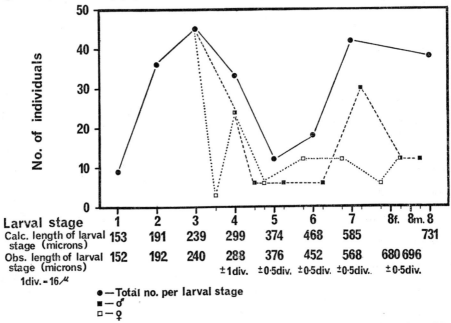

Fig. 2. The larval population of *Cythere lutea* O. F. Müller collected from the red alga *Cystoclonium purpureum* (Huds) Batt. 2 July 1967. Stage 8 was regarded as the adult, the first stage was not sampled. (Calc. growth factor—1·25.)

It seems likely that the animals are grazing on some cellular exudate from the plant itself which is found on the surface epithelium. A large number of animals was also noted on the *A. nodosum* plants, the majority of the species being localized to the areas of epiphytic growth. Twelve species were recorded from the algae (Table 3b). Representatives of eleven of these were living. The majority of these living forms were larval stages of *H. viridis, S. nigrescens* and *C. lutea* in fact, seven of the eight larval stages of *C. lutea* were found on specimens of *Cystoclonium purpureum* (Fig. 2). Five of the eleven living species were found to be specific to the algae, the other six occurred in the beach areas adjoining the algal zones.

The ostracods specific to the algae were:

Hemicythere villosa	*Paradoxostoma variabile*
Heterocythereis albomaculata	*Paradoxostoma sarniense*
Hirschmannia viridis	

TABLE 3a. The various groups of animals inhabiting four species of weed from Porth Castell

Results expressed in numbers of animals per 100 g of damp weed and in percentage of total population

Species of algae	Fucus spiralis		Ascophyllum nodosum		Fucus serratus		Cystoclonium purpureum	
Sample No.	HF₁		MA₁		LF₁		EL₁	
Weight in g	96		186		90		26	
Tidal level	HWN		MTL		LWN		ELWS	
Date	2.7.67		2.7.67		2.7.67		2.7.67	
Acarina	1	0·9	242	4·8	450	19·3	181	2·9
Polychaeta	—	—	—	—	1	—	—	—
Oligochaeta	—	—	—	—	3	0·2	—	—
Copepoda	—	—	3,763	74·3	1,317	56·3	2,462	39·3
Isopoda	—	—	8	0·2	10	0·4	423	6·8
Amphipoda	29	25	16	0·3	30	1·3	96	1·5
Gastropoda	73	62·5	151	3·0	163	7·0	327	5·2
Lamellibranchia	—	—	—	—	3	0·1	—	—
Insecta								
Diptera larvae	—	—	53	1·0	3	0·1	—	—
Nematoda	12	9·8	161	3·2	97	4·2	600	9·6
Turbellaria	—	—	331	6·5	229	9·4	131	2·1
Ostracoda	2	1·8	339	6·7	40	1·7	2,039	32·6
Total	117	100	5,064	100	2,346	100	6,259	100

TABLE 3b. Ostracoda inhabiting four species of algae from Porth Castell

Counts of living (a), dead (d) and right valves (v) of adult and juvenile species

Species of algae	Fucus spiralis			Ascophyllum nodosum			Fucus serratus			Cystoclonium purpureum		
Weight in g	96			186			90			26		
% of ostracods of total pop.	1·8			6·7			1·7			32·6		
% of living ostracods of tot. pop.	0·0			6·5			1·7			32·6		
Date	2.6.67			2.6.67			2.6.67			2.6.67		
Species	a.	d.	v.	a.	d.	v.	a.	d.	v.	a.	d.	v.
there lutea ad.				50						47		
juv.				260			16			343		
micytherura nigrescens				85			17			14		
				135			3			16		
emicythere villosa				14		14						
eterocythereis albomaculata										5		
ptocythere pellucida			1									
rschmannia viridis												
										70		
xoconcha impressa										6		
radoxostoma variabile										18		
radoxostoma sarniense										2		
therois vitrea			1									
therois pusilla				24								
stoleberis aurantia				40	8					9		
Total	0	0	2	608	8	14	36	0	0	530	0	0

The sediment

H.W.N. traverse. The ostracod assemblage at this tidal level consisted of carapaces and valves of six ostracods. In order of dominance they were, *H. villosa, L. impressa, S. nigrescens, C. lutea, L. pellucida* and *X. aurantia.* Station Ha of this traverse was desiccated, the other stations were saturated by the fresh-water stream entering the beach at this level. The salinities at these stations were in the range of 1–5‰. The anaerobic layer, the 'black layer', at its maximum was only 7 cm from the surface at these stations. There was a general increase in the intensity of the blackness from the surface down to this layer.

M.T.L. traverse. Seventeen species were found in this traverse (Table 3c) seven of which were repeatedly found alive in the samples. These species were:

*Cythere lutea**	*Paracytherois arcuata*
Semicytherura gibba	*Cytherois vitrea*
*Semicytherura nigrescens**	
*Cytherois pusilla**	

The species marked * have large populations on the adjoining algae. All the species above are either algal living forms or occur on plant debris. There is usually an accumulation of plant debris at the sides of the beaches near the algal zones, and it is here that these species which live on plant debris are found in greatest numbers. This is represented in the traverses by station *e* and in this locality by station *a* as well. The width of the beach is such that the traverse passes right across the beach.

L.W.N. traverse. There were 24 species found along this traverse (Table 3d); individuals of 8 of these species occurred alive. These were:

*Cythere lutea**	*Loxoconcha impressa*
Cytherura gibba	*Cytherois pusilla**
*Semicytherura nigrescens**	*Cytherois vitrea*
Semicytherura sella	*Xestoleberis aurantia*

The species marked * have populations on the algae adjoining the beach and are represented on the sediment by the occasional specimen. The dominant species is *S. sella* which lives on plant debris, *S. gibba* is also classified in this category. *C. vitrea* and *L. impressa* are found living in the algae as well as on plant debris. The width of the beach at this level is only fifteen yards, the sediment being coarse sand with large amounts of plant debris accumulated on the surface. This accounts for the large numbers of algal and plant debris living ostracods inhabiting this level of the shore.

Altogether 34 species were recorded from this locality, 29 on the sediment and 12 on the algae, 5 of which were specific to this substrate.

Church Island. Locality 4

This locality was chosen because of the nature of the sediment and for the large varied fauna which inhabits the interstices of the deposit. It is situated at E.L.W.S., hence any fluctuation of temperature and salinity can be discounted in comparison with the previously described localities. Thirty-three species were found in this locality, 27 in the sediment and 21 on the algae, 6 of the latter species were only found on the algae but they were represented by carapaces and valves.

TABLE 3c. Ostracods taken from seven samples of beach sand taken along a line transect at M.T.L. Porth Castell, 2.6.67

Counts of living (a), dead (d) and right valves (v) of adult and juvenile species per 50 ml of wet sand

Sample No.	Ma_1			Ma_2			Mb_1			Mb_2			Mc			Md			Me		
Tot. pop. of animals in sample	825			366			592			240			379			386			711		
% of ostracoda of the total pop.	5·9			16·4			12·5			14·2			13·5			10·4			7·2		
% of living ostracods of the tot. pop.	0·4			0·0			0·5			0·0			2·4			2·3			0·4		
Salinity ‰ surface	29·2			—			21·2			—			27·9			30·7			32·6		
15 cm below surface	—			24·1			—			21·4			23·4			28·5			31·9		
Species	a.	d.	v.	a.	d.	v.	a.	d.	v.	a.	d.	v.	a.	d.	v.	a.	d.	v.	a.	d.	v.
1. Cythere lutea† ad.		2				13			6			2			2			2		1	
juv.						3															
2. Heterocyprideis sorbyana														1							1
3. Eucythere declivis														1							
4. Cytherura gibba							2		4			2									
5. Semicytherura nigrescens†													2	6	2	3					
6. Semicytherura sella																6			2		
7. Hemicytherura cellulosa						2			2												
8. Hemicythere villosa†		2	2						4	2		4		4	4		2	3			2
9. Hemicythere emarginata			2			6												2			
10. Aurila convexa																					1
11. Leptocythere pellucida†	2											2			2			2		1	2
12. Cytheromorpha fuscata						6			4			12			4			2		10	5
13. Loxoconcha impressa†		10	16		8	22		8	30		2	18	2	2	22		12	8		8	16
14. Cytherois pusilla†							1						2								
15. Paracytherois arcuata	1												2								
16. Cytherois vitrea†													1								
17. Cythereis tuberculata												1									2
Total	3	14	32	0	8	52	3	16	55	0	6	28	9	12	30	9	17	14	2	20	29
Sum total of species		49			60			74			34			51			40			51	
Number of ostracods per m²		19,600			24,000			29,600			13,600			20,400			16,000			20,400	

† Species occurring on the algae adjoining the beach.

TABLE 3d. Ostracods taken from five samples of beach sand taken along a line transect at L.W.N. Porth Castell, 2.6.67

Counts of living (a), *dead* (d) *and right valves* (v) *of adult and juvenile species per 50 ml of wet sand*

	La	Lb	Lc	Ld	Le
Sample No.	La	Lb	Lc	Ld	Le
Tot. pop. of animals in sample	1820	824	397	704	1004
% of ostracods of the total pop.	8·6	19·9	29·0	15·3	15·3
% of living ostracods of the tot. pop.	4·8	12·1	6·6	5·4	2·4
Salinity ‰ surface	33·4	33·5	33·0	33·2	32·5
5 cm below surface	31·5	33·1	32·2	32·4	32·6

Species	La a	La d	La v	Lb a	Lb d	Lb v	Lc a	Lc d	Lc v	Ld a	Ld d	Ld v	Le a	Le d	Le v
1. *Cythere lutea*† ad.		12					4	2	6				4	2	18
juv.				4			2	2		4			6	4	
2. *Cytherura gibba*	20						8								
3. *Semicytherura sella*	9			52			12			12			10		
(juv.)										2					
4. *Semicytherura nigrescens*†	17			16											2
(juv.)				4											2
5. *Semicytherura acuticostata*										2					
6. *Hemicytherura cellulosa*										2	2				
7. *Cytheropteron inflatum*												2			
8. *Cytheropteron nodosum*															1
9. *Cytheropteron latissimum*										2					
10. *Hemicythere villosa*†	4	8		8						16		8	4	10	
11. *Elofsonella concinna*										4					
12. *Aurila convexa*								2			2	1			4
13. *Heterocythereis albomaculata*†							2			2					
14. *Heterocyprideis sorbyana*										1					
15. *Leptocythere pellucida*†	4			2						2			3		
16. *Cytheromorpha fuscata*							2	4		10	2		10	4	
(juv.)													8	3	
17. *Hirschmannia viridis*†															12
(juv.)															4
18. *Hirschmannia tamarindus*														2	2
19. *Loxoconcha impressa*†	20	8	4	14	18	4	16	26		14	10	21		10	31
20. *Trachyleberis dunelmensis*										2					
21. *Cytherois pusilla*†	2						2								
22. *Cytherois vitrea*†	16	5		24	4					6					
(juv.)	22	3													
23. *Xestoleberis aurantia*†							2						6		
24. *Xenocythere cuneiformis*															2
Total	86	28	40	100	28	38	26	28	61	38	25	45	24	29	91

	La	Lb	Lc	Ld	Le
Sum total of species	154	166	115	108	144
Numbers of ostracods per m²	61,600	66,400	46,000	43,200	57,600

† Species occurring on the algae adjoining the beach.

TABLE 4a. The various groups of animals inhabiting five species of weed from Church Island

Results expressed in numbers of animals per 100 g of damp weed and in percentage of total population

Species of algae	Cladophora sericea		Fucus serratus		Halidrys siliquosa		Griffithsia flosculosa		Laminaria saccharina	
Weight in g	4.9		30.7		22.4		10.0		28.0	
Sample No.	EC_1		EF_1		EH_1		EG_1		EL_1	
Tidal level	ELWS		ELWS		ELWS		ELWS		ELWS	
Date	23.6.67		23.6.67		23.6.67		23.6.67		23.6.67	
Acarina	41⎱41	0.3	29	1.0	89	0.4	180⎱360	1.4	—	—
Pycnogonida	—		—		—	0.1			—	
Oligochaeta	143	0.5	—		36	0.1	—		—	3.3
Polychaeta	143	0.5	72	2.4	232	1.0	660	1.6	50	3.3
Copepoda	15,939	58.7	1,492	50.1	15,571	64.3	22,800	57.3	379	25.0
Isopoda	82	0.3	—		—		30	0.1	—	
Amphipoda	286	1.0	59	2.0	911	3.8	480	1.2	21	1.4
Gastropoda	1,408	5.2	46	1.5	411	1.7	1,020	2.6	7	0.5
Lamellibranchia	245	0.9	68	2.3	214	0.9	450	1.1	29	1.9
Insecta Diptera larvae	82	0.3	10	0.3	18	0.1	—		—	
Decapoda Megalopa	—		—		18	0.1	30	0.1	7	0.5
Nematoda	918	3.4	365	12.3	1,321	5.4	8,640	21.7	593	39.1
Nemertini	20	0.1	10	0.3	161	0.6	30	0.1	—	
Turbellaria	—		26	0.9	89	0.4	150	0.4	—	
Ophiuroidea	—		3	0.1	18	0.1	180	0.4	—	
Ostracoda	7,959	29.3	798	26.8	5,121	21.1	4,800	12.0	429	28.3
Total	27,164	100	2,978	100	24,210	100	39,810	100	1,515	100

TABLE 4b. Ostracods inhabiting five species of algae from Church Island

Counts of living (a), dead (d) and right valves (v) of adult and juvenile species

Species of algae	Cladophora sericea			Fucus serratus			Halidrys siliquosa			Griffithsia flosculosa			Laminaria saccharina		
	a.	d.	v.	a.	d.	v.	a.	d.	v.	a.	d.	v.	a.	d.	v.
Weight in g	4·9			30·7			22·4			10·0			28·0		
% of ostracods of total pop.	29·3			26·8			21·1			12·0			28·3		
% of living ostracods of tot. pop.	28·7			25·8			19·2			10·1			25·4		
1. Cythere lutea ad.	4						12			9					
juv.							24								
2. Cyamocytheridea punctillata			2			1									
3. Semicytherura nigrescens	7			4			36			4	8		2	2	
4. Semicytherura striata	2			2			12							2	
5. Semicytherura angulata								2							
6. Hemicytherura cellulosa	4				1		4	2			3				
juv.							4								
7. Cytheropteron latissinum			1												
8. Cytheropteron nodosum			2												

9. *Hemicythere villosa*							1				3				
10. *Leptocythere pellucida*					1		4				12				
11. *Cytheromorpha fuscata*	1		2		1										
12. *Hirschmannia viridis*	96			167			388						4	4	
13. *Hirschmannia tamarindus*	152						332			207			80	1	
14. *Loxoconcha impressa*	27						51			7	20	6	2	1	
15. *Paradoxostoma variabile*	85			60			225			175	14		20	1	
16. *Paradoxostoma abbreviatum*	3						12	12		2	7				1
17. *Paradoxostoma bradyi*			1		1		20								
18. *Cytherois vitrea*									1						
19. *Cytherois fischeri*			1												
20. *Cytherois pusilla*					1										
21. *Sclerochilus contortus*	2					2		2			3		2		
Total	382	0	8	235	8	2	1,128	18	1	404	70	6	108	11	1

TABLE 4c. Ostracods from four samples of barnacle shell gravel taken at E.L.W.S. Church Island

Counts of living (a), dead (d) and right valves (v) of adult and juvenile species per 50 ml of saturated gravel (approx. 100 g)

	CIc$_9$	CIc$_6$	CIf$_8$	CIf$_4$
Sample No.	CIc$_9$	CIc$_6$	CIf$_8$	CIf$_4$
Tot. pop. of animals in sample	827	663	919	718
% of ostracods of the total pop.	12·9	8·6	10·0	5·2
% of living ostracods of the tot. pop.	1·6	6·3	1·0	1·4
Date	28.6.67	1.4.67	28.6.67	1.4.67

Species	CIc$_9$ a.	d.	v.	CIc$_6$ a.	d.	v.	CIf$_8$ a.	d.	v.	CIf$_4$ a.	d.	v.
1. Cythere lutea† (ad./juv.)	ad.	1	2	1			1	2	3			2
2. Cyamocytheridea punctillata†		1										
3. Cyamocytheridea papillosa		1	1						1			
4. Neocytherideis subulata			1									
5. Cytherura gibba		1	1			2						1
6. Semicytherura nigrescens†		1	1	1								
7. Semicytherura sella	2						1	2	2			
8. Semicytherura striata†	1		1	1				2	1		1	
9. Hemicytherura cellulosa†		1	1	1								2
10. Cytheropteron nodosum†			5						3			2
11. Hemicythere villosa†		5	2					3	5		1	2
12. Aurila convexa		2	1					2	1		1	2
13. Heterocythereis albomaculata				3						2		
14. Leptocythere pellucida†	2	4	6	6	1			7	4	3		
15. Leptocythere castanea	2 / 3	9	2	13	1		4	2	4	4	1	3
16. Cytheromorpha fuscata†		3	2		1			4	4			2
17. Hirschmannia viridis†		2	10	3	1		3	3	4		1	
18. Hirschmannia tamarindus†		2	2					1				
19. Loxoconcha impressa†	4		12	6 / 3	2	2		9	3 / 4	1		
20. Paradoxostoma variabile†		2	1	1	1			1				3
21. Paradoxostoma ensiforme			1			2		2	1			1
22. Cytherois fischeri†	1			1							1	
23. Paracytherois arcuata	1											
24. Sclerochilus contortus†		1	1			1			1			1
25. Trachyleberis dunelmensis			1						1			
26. Cythere ? globulifera								1				
27. Polycopsis compressa				3								
Total	13	39	54	42	8	7	9	41	42	10	6	21
Sum total of species	106			57			92			37		
Number of ostracods per m²	42,400			22,800			36,800			14,800		

† Species occurring on the algae adjoining the beach.

The algae. The main species in the phytal belt surrounding the sediment were sampled to ascertain the ostracod populations inhabiting them. The algae were *Cladophora sericea* (Huds) Kutz., *F. serratus*, *Halidrys siliquosa* (L.) Lyngb., *Griffithsia flosculosa* (Ellis) Batt. and *Laminaria saccharina* (L.) Lamour. (holdfasts). The main epiphytes occurring on these weeds were *Ceramium rubrum* (Huds) J. Agardh, on *F. serratus*, and *Sphacelaria pennata* (Huds) Lyngb. on *H. siliquosa*. The last epiphyte occurs in dense clumps at intervals along the thallus of *H. siliquosa*. The results of the examinations of these species for their cryptofauna are seen in Tables 4a and 4b. The epiphyte of *H. siliquosa* is definitely responsible for the high count of live species, the vast majority of the fauna being localized in the clumps. Nine species had living representatives on the algae, four of these did not occur in the sediment. They were:

Hemicytherura cellulosa	*Cytheromorpha fuscata*
Hemicythere villosa	*Hirschmannia tamarindus*

Not one of the above four species could be placed in the category of living specifically on algae. Species which could be placed in this category are found on the sediment because of the proximity of the dense algae to the beach. An example of this is seen in the distribution of *H. viridis*. Its maximum density is found on *Halidrys siliquosa* at 720 live ind./100 g of damp weed; in comparison, only three were found alive in 50 ml of damp sediment. The dominant species inhabiting the algae were *H. viridis*, *L. impressa*, *S. nigrescens*, *C. lutea*, and *P. variabile*.

The sediment. The assemblage consisted of 27 species (Table 4c), 12 of which had living representatives in the sediment. Five of these species were individuals of the above algal living species. The other seven were

*Semicytherura striata**	*Leptocythere castanea**
*Polycope orbicularis**	*Leptocythere pellucida**
*Polycopsis compressa**	*Cytherois fischeri*
Heterocythereis albomaculata.	

The species marked * are specific to bottom sediments, they are found very rarely, if ever, living in the algae. *C. fischeri* and *H. albomaculata* are species which live among the algae and plant debris.

Menai Bridge. Locality 5

In this locality a vertical traverse was taken down the shore to ascertain the numbers and distribution of the ostracod population. Selective samples were collected at H.W.N., M.T.L., M.T.L./L.W.N. and L.W.N., corresponding algal samples were taken at these levels. The results of this traverse are shown in Tables 5a, 5b and 5c. Thirty species of ostracods were found, eight of which had living representatives in the locality. The other species were present in the counts as carapaces and valves.

The algae. The main weed was *A. nodosum*. The four samples of this alga, 1–4, illustrate a progression of the percentage of epiphytic growth and sedimentation. The difference in numbers in Copepoda and Nematoda are in the region of a ×15 increase while the increase in Ostracoda from samples 1–4 was ×28. Eleven species were found on the algae, five of which had living populations, these were:

Cythere lutea	*Hirschmannia viridis*
Semicytherura nigrescens	*Loxoconcha impressa*
Cytherura gibba	

TABLE 5a. The various groups of animals inhabiting five species of weed from Menai Bridge

Results expressed in numbers of animals per 100 g of damp weed and in percentage of total population

Species of algae	*Fucus spiralis*		*Ascophyllum nodosum*		*Ascophyllum nodosum*		*Ascophyllum nodosum*		*Ascophyllum nodosum*	
Sample No.	HF₁		LA₂		HA₁		MA₁		LA₁	
Weight in g	11·6		41·0		19·7		18·7		28·6	
Tidal level	HWN		MTL/LWN		HWN/MTL		MTL		MTL/LWN	
Date	14.6.67		14.6.67		14.6.67		14.6.67		14.6.67	
Acarina	405	6·4	24	3·2	823	31·9	80	6·7	151	0·9
Pycnogonida	—		5		15		—		17	
Oligochaeta	—		20	1·8	—		11	0·2	339	1·8
Polychaeta	—		—		—		—		28	0·1
Copepoda	362	5·8	273	25·1	909	34·6	2,310	36·5	5,699	30·9
Isopoda	129	2·0	45	4·0	142	5·4	11	0·2	206	1·1
Amphipoda	112	1·8	10	0·9	15	0·6	—		273	1·5
Gastropoda	—		20	1·8	56	2·1	43	0·7	178	1·0
Lamellibranchia	—		15	1·4	46	1·8	75	1·2	346	1·9
Insecta										
Diptera larvae	224	3·6	29	2·7	61	2·3	625	10·3	850	4·6
Nematoda	5,026	80·0	527	48·4	416	15·9	2,524	40·0	8,252	44·7
Nemertini	—		—		—		75	1·2	49	0·3
Turbellaria	—		45	4·0	56	2·1	128	2·0	31	0·2
Sipunculoidea	—		—		10	0·4	—		—	
Ostracoda	26	0·4	73	6·7	76	2·9	64	1·0	2,038	11·0
Total	6,284	100	1,086	100	2,625	100	5,946	100	18,457	100

TABLE 5b. Ostracods inhabiting five samples (2 species) of algae from Menai Bridge

Counts of living (a), dead (d) and right valves (v) of adult and juvenile species

Species of algae Weight in g % of ostracods of total pop. % of living ostracods of tot. pop.	Fucus spiralis 11·6 0·4 0·4			Ascophyllum nodosum 41·0 6·7 4·9			Ascophyllum nodosum 19·7 2·9 2·1			Ascophyllum nodosum 18·7 1·0 0·5			Ascophyllum nodosum 28·6 11·0 6·7		
Species	a.	d.	v.	a.	d.	v.	a.	d.	v.	a.	d.	v.	a.	d.	v.
1. Cythere lutea (ad. / juv.)							1 4			2	3			14	
2. Semicytherura nigrescens				2									115 60		
3. Cytherura gibba				2		2	1			3					
4. Hemicytherura cellulosa						2									
5. Hemicythere villosa						2						4			
6. Leptocythere pellucida								1	1					10	
7. Leptocythere tenera														2	
8. Hirschmannia viridis	2			2 16		2	5		2				27 249		17 71
9. Loxoconcha impressa	1												4		
10. Cytherois pusilla														3	
11. Paradoxostoma normani															11
Total	3	0	0	22	0	8	11	1	3	5	3	4	455	29	99

TABLE 5c. Ostracods from four samples of 'mud' (very fine sand $+16-33\%$ silt) taken from the intertidal region at Menai Bridge. 14.6.67

Counts of living (a), dead (d) and right valves (v) of adult and juvenile species per 10 ml of damp sediment (approx. 26 g)

Sample No.	MBh_2	MBm_1	MBl_1	MBl_4
Tot. pop. of animals in sample	932	2892	1771	1464
% of ostracods of the total pop.	1·2	3·0	2·5	6·1
% of living ostracods of the tot. pop.	0·3	0·4	0·2	0·8
Tidal level	HWN	MTL	MTL/LWN	LWN

Species		MBh_2			MBm_1			MBl_1			MBl_4	
	a.	d.	v.	a.	d.	v.	a.	d.	v.	a.	d.	v.
1. *Cythere lutea*† (ad. / juv.)												1
2. *Cyamocytheridea punctillata*									2		2	
3. *Neocytherideis subulata*											2	5
4. *Semicytherura nigrescens*†		1							4			
5. *Semicytherura sella*					4						2	2
6. *Semicytherura striata*					2						2	2
7. *Semicytherura angulata*						5						
8. *Hemicytherura cellulosa*†												2
9. *Cytheropteron nodosum*												2 2
10. *Cytheropteron latissimum*									2			
11. *Hemicythere villosa*†						2			1			4
12. *Elofsonella concinna*									2			
13. *Aurila convexa*									2		2	
14. *Leptocythere pellucida*†		1		1 1	4	4			2		3	1
15. *Leptocythere castanea*	1	1		4	4	3	4	1	4	5 2	10	4
16. *Cytheromorpha fuscata*					2	4		2	6			2
17. *Hirschmannia viridis*†	1		2		2	14			4		1	5
18. *Loxoconcha impressa*†					2	4			6	1 3	1	5
19. *Paradoxostoma variabile*				4 1	6 1	6			2		4	2
20. *Paradoxostoma abbreviatum*											3	1
21. *Paradoxostoma normani*†												2
22. *Cytherois pusilla*			1									
23. *Paracytherois arcuata*			2								2	
24. *Paracytherois flexuosa*						4						
25. *Sclerochilus contortus*						4						
26. *Bythocypris obtusata*												1
27. *Pontocypris hispida*			1									
Total	2	3	6	11	27	50	4	3	37	11	34	43
Sum total of species		11			88			44			88	
Number of ostracods per m²		22,000			176,000			88,000			176,000	

†Species occurring on the algae adjoining the beach.

The dominant species were *H. viridis* and *S. nigrescens*. All the above species are algal living ostracods, *C. gibba* is often found associated with plant debris.

The sediment. Twenty-seven species were recorded from the sediment, the majority of these ostracods are dead individuals, empty carapaces and valves. Five of the species had living forms on the sediment, these were:

Leptocythere pellucida	*Hirschmannia viridis**
Leptocythere castanea	*Loxoconcha impressa**
*Paradoxostoma variabile**	

The species marked * are algal and plant detritus living ostracods. The two species of *Leptocythere* are specific to this type of sediment. From the distribution of these two species in the five localities it can be seen that *L. pellucida* shows a preference for the clean sand type of sediment. while *L. castanea* occurs on fine sediments with silt fractions and a high percentage of organic detritus. The localities in which *L. castanea* occur, show little salinity variation from 33·5‰, while *L. pellucida* is present at Traeth Bychan at salinities of 0·5‰. This in fact is the reverse of what might be expected. It may be due to a sediment preference over-riding the effects of other physical factors.

DISCUSSION

The investigation of the five intertidal localities revealed two basic populations of ostracods. One occurring on the sediment and the other on the phytal zones adjoining the sediment. The populations taken as whole units were by no means specifically confined within these habitats. There was a great deal of mixing of the two populations especially at the boundaries of the two habitats where a great deal of algal detritus accumulated. The 58 species could be divided into four groups: the algal living species, like *Xestoleberis aurantia* and *Paradoxostoma sarniense*, which are specific to this type of habitat; the algal, plant detritus living species, representatives of these species are *Cytherois vitrea*, *Cytherois fischeri* and *Cytherura gibba*. These forms occur in the algal zones and usually in the stations next to the phytal belts. The sediment living species such as *Leptocythere pellucida*, *Leptocythere castanea* and the rarer *Polycope orbicularis* and *Polycopsis compressa* which are specific to the sediments, *L. pellucida* showing a preference for the cleaner types of sands. The last group are the species from deeper water localities, such as *Xenocythere cuneiformis*. These species are usually represented in the assemblages as single valves and carapaces.

The adults of *L. pellucida* were found living down to a depth of a few centimetres from the surface at Traeth Bychan. The larval stages were collected down to a depth of 15 cm. The larval instars are small enough to utilize the interstitial pore system of the sediment, but the adults are far too large. This is illustrated in Fig. 3. The sizes of the adults, with small projections to show their general shape, are shown relative to the sizes of the interstices of the sediment of Traeth Bychan. This photograph was taken by projecting thin slides of the actual sediment, which were obtained by impregnating undisturbed samples with unsaturated polyester resins. In habitat 4, Church Island, the particles are very large and angular forming a large interstitial system. These interstices are large enough in this particular sediment to be utilized by the ostracods, as seen in Fig. 4. This example is an exceptional case and it was chosen for this investigation purely for that purpose. It can be seen that the sediment type of ostracods found in all the other localities cannot utilize the interstitial pore system and therefore must be classified as surface dwellers, living in the top few centimetres of the deposit.

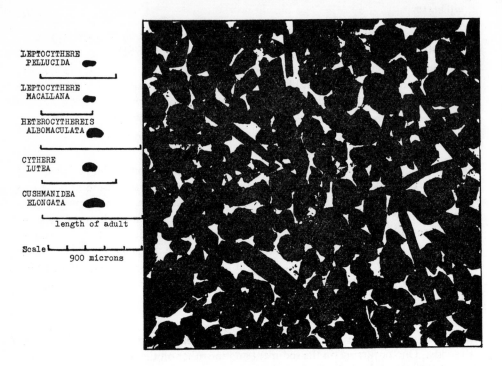

Fig. 3. Traeth Bychan shell sand showing the relative sizes of the Ostracoda to the interstices of the deposit (the interstices are white).

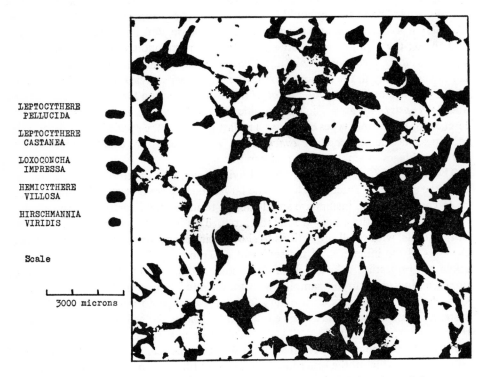

Fig. 4. Church Island barnacle shell gravel showing the relative sizes of the Ostracoda to the interstices of the deposit (the interstices are black).

ACKNOWLEDGEMENTS

I wish to thank Mr R. Smith for identifying certain of the algae and especially Dr K. Atkinson for his identification of the Foraminifera species.

SELECTED BIBLIOGRAPHY

BALLANTINE, W. J. 1961. A biologically-defined exposure scale for the comparative description of rocky shores. *Field Studies*, **1** (3), 1-19.

BRADY, G. S. 1868. A Monograph of the Recent British Ostracoda. *Trans. Linn. Soc. London*, **26**, 353-495, Pls 23-41.

—— AND NORMAN, A. M. 1889. A Monograph of the Marine and Freshwater Ostracoda of the North Atlantic and of North-Western Europe. Section I. Podocopa. *Sci. Trans. R. Dubl. Soc.*, **4**, 63-270, Pls 8-23.

—— AND NORMAN, A. M. 1896. A Monograph of the Marine and Freshwater Ostracoda of the North Atlantic and of North-Western Europe. Part II. Sections II-IV. Mydocopa, Cladocopa and Platycopa. *Sci. Trans. R. Dubl. Soc.*, **5** (12), 621-746, Pls 50-68.

COLMAN, J. 1940. On the faunas inhabiting intertidal seaweeds. *J. mar. biol. Assn U.K.*, **24**, n.s. 129-83, 23 tables.

DAHL, E. 1948. On the smaller Arthropoda of marine algae, especially in the polyhaline waters off the Swedish west coast. Dissertation, Lund. *Undersökn. över Öresund*, **35**, 193.

ELOFSON, O. 1941. Zur Kenntnis der marinen Ostracoden Schwedens, mit besonderer Berücksichtigung des Skageraks. *Zool. Bidr. Upps.*, **19**, 215-534.

——. 1940. Notes on the ostracod fauna of Plymouth. *J. mar. biol. Assn U.K.*, **24**, 495-504, 13 Figs.

HAGERMAN, L. 1965. The ostracods of the Øresund, with special reference to the bottom-living species. *Ophelia*, **2** (1), 49-70.

——. 1966. The macro- and microfauna associated with *Fucus serratus* L., with some ecological remarks. *Ophelia*, **3**, 1-43.

HARTMANN, G. 1965. Neontological and Paleontological classification of Ostracoda. *Pubbl. Staz. zool. Napoli* [1964], **33** (suppl.), 550-87.

HOUGH, B. K. 1957. *Basic Soils Engineering*. Ronald Press Comp., New York.

KLIE, W. 1929. In Grimpe, G. Ostracoda. *Tierwelt der Nord- und Ostsee*. Lief. **16**, xb1-xb50, 51 Figs.

——. 1938. In Dahl, F. *Die Tierwelt Deutschlands und der angrenzenden Meeresteile*. 34 *Teil. Krebstiere oder Crustacea. III. Ostracoda, Muschelkrebse*, i-iv, 230 pp. 768 Figs. Gustav Fischer, Jena.

LEWIS, J. R. 1953. The ecology of rocky shores around Anglesey. *Proc. zool. Soc. Lond.*, **123** (3), 481-549.

——. 1955. The mode of occurrence of the universal intertidal zone in Great Britain. *J. Ecol.*, **43**, 270-90.

REYS, S. 1961. Recherches sur la systématique et la distribution des ostracodes de la région de Marseille. *Recl. Trav. Stn mar. Endoume*, **22** (36), 53-109.

SARS, G. O. 1928. *An account of the Crustacea of Norway*. Vol. IX, *Ostracoda*, 1-277, 119 Pls. Bergen Museum, Bergen.

SKOGSBERG, T. 1920. Studies on marine ostracods. Part 1, (Cypridinids, Halocyprids, and Polycopids). *Zool. Bidr. Upps.*, *Suppl. Bd.* **1**, 1-784, 153 Figs.

——. 1928. Studies on marine ostracods. Part II. External morphology of the genus *Cythereis* with descriptions of twenty-one new species. *Occ. Pap. Calif. Acad. Sci.*, No. **15**, 1-155, Pls I-VI.

STEPHENSON, T. A. AND STEPHENSON, A. 1949. The universal feature of zonation between tide-marks on rocky coasts. *J. Ecol.*, **38**, 289-305.

THEISEN, B. F. 1966. The life history of seven species of ostracods from a Danish brackish water locality. *Meddr Kommn Danm. Fisk. -og Havunders.*, N.S. **4**, 215-70.

WIESER, W. 1952. Investigations on the microfauna inhabiting seaweeds on rocky coasts. IV. Studies on the vertical distribution of the fauna inhabiting seaweeds below the Plymouth Laboratory. *J. mar. biol. Assn U.K.*, **31**, 145-74.

——. 1959. Zur Ökologie der Fauna mariner Algen mit besondere Berücksichtigung des Mittelmeeres. *Int. Rev. Hydrobiol.*, **44**, 137-80.

DISCUSSION

SOHN: Did you also find juveniles during the winter months?

WILLIAMS: I found juveniles from April to early November, the highest numbers recorded were in June and July. I have not sampled sufficiently to give a definite answer regarding their occurrence in the winter months.

BENSON: How large were your samples?

WILLIAMS: The size of the sediment samples was approximately 50 ml of damp sand. This sample size gave population densities of 1000 animals in some cases and as low as 20 animals in others. The ostracod population ranged from 1–60 per cent in these populations. The size of the algal samples varied. In my tables I have for comparative purposes expressed the results as the number of animals found in 100 g of damp weed. A weed like *Cystoclonium purpureum*, 26 g of which gave a population of 1,627 animals, 33 per cent of which were ostracods, contrasts with *Ascophyllum nodosum* where 173 g gave 851 animals, 2·3 per cent of which were ostracods. I have dealt with the total animal populations in these samples, therefore, I have adjusted the size of the sample accordingly. In most cases I have duplicated and triplicated these samples and analysed them statistically.

BENSON: Concerning dead populations, the more I have seen around the coast here and in studying Brady, Crosskey and Robertson, the more impressed I am with the possibilities of pollution from older material in the beach cliffs. Have you any comments or knowledge of this?

WILLIAMS: There could be no Pleistocene contamination from the cliffs on the beach populations that I have studied. A number of valves that I have recorded are of deeper water forms and these have been carried on to the beaches by tidal movements, etc.

KAESLER: How do you sample the number per square metre and to what depth have you sampled?

WILLIAMS: The interstitial fauna and the ostracod fauna has its maximum density in the first few centimetres from the surface. Therefore I sampled with a 0·1 square metre former down to a depth of 2 cm, with selective sampling down to 15 cm. The 'black layer' usually occurred a little beyond this depth. I then took 50 ml sub-samples from this sample and extracted and counted the total populations. The numbers are then expressed in terms of population/square metre to a depth of 2 cm. It is very arbitrary to extrapolate data from a number of 50 ml samples but it gives a comparative measure of the population density.

McKENZIE: Hartmann has described interstitial species in several papers. Furthermore, Marinov has noted that species live interstitially in the Black Sea area. Would you please comment on this?

WILLIAMS: Although Hartmann has found species like *Microcythere subterranea* in the interstices of sands, I have found no species that I can say are truly interstitial forms. The ostracods that I have found which are living in the interstices are the juveniles of the species which occur on the surface layers of the sediment.

McKENZIE: You stated that your counts were mostly of juveniles. Does this not imply that you also found some adults living interstitially?

WILLIAMS: I recorded live adults in habitats 1, 2, 3, down to a depth of 2 cm. Since all these beaches are saturated to some degree at low water, then the properties of these sediments are completely changed. There is a lowering of the shear strength, enabling these animals to burrow down to this level, but I doubt if the adults of these species could burrow down to 15 cm. I have recorded juveniles down to 15 cm but they are small enough to utilize the interstitial pore system, the adults are far too big.

McKENZIE: Some small ostracods seem to be adapted in shell shape for interstitial life in that they possess a pointed anterior. This would be advantageous for moving through sediment interstices as long as the sediment was sufficiently 'open'—and one of your slides showed that this condition was met in part of your study area. Does this not suggest that the interstitial habitat *is* characteristic for some ostracods?

WILLIAMS: *Leptocythere pellucida* is one of the species that I have found in greatest abundance on the sandy habitats (1, 2, 3) and this has a blunt anterior end. Examining the void size distributions of these sediments it is clear that there are insufficient voids of the diameter required by this animal for it to move through the interstices of the deposit.

As I stated previously the only ostracods able to utilize the pore systems in these deposits are the juveniles. If you have voids of a certain diameter exceeding a certain percentage in a deposit then they will be of such a frequency as to form a continuous system in that sediment thus allowing the animals, which require that void diameter as a minimum, to move through the system. The sediment would have to have particles of the granule size to allow *L. pellucida* to move through the interstices. The only sediment examined that approaches this size is the Church Island Barnacle shell gravel, habitat 4. This has voids up to 1,500 microns, the mean for the void distribution being 360 microns. This sediment possesses a high enough frequency of large voids to allow the movement of species like *Leptocythere castanea*, *Loxoconcha impressa* and *Leptocythere pellucida* through the pore spaces.

McKENZIE: Your ostracod populations are rather high compared with interstitial ostracod populations that have been obtained by Madame Reys for example.

WILLIAMS: It is what you would expect when a comparison is made of the population densities of surface detritus feeders with the densities of more specialized interstitial living forms. In fact my ostracod populations are low compared with the populations found in the Øresund (Hagerman, personal communication).

KILENYI: Grain size and grain shape has a very close bearing on the interstitial ostracods because if you have hexagonal close fitting between rounded grains you have no space between them. This might easily account for the differences between Anglesey and the French coast. Another feature is the grain size distribution itself. If you get a particular grain size distribution you may have no gaps left between the grains, so there might be a critical grain size; I found this in dealing with the Thames Estuary ostracods. There you get no interstitial forms. Your slides showed rather angular grains which would allow these species to develop.

WILLIAMS: The packing of the particles of a sediment greatly affects the porosity of the deposit which in turn would affect the faunal distribution. The voids of habitats 1, 2 and 3 are too small to allow the surface living ostracods to live interstitially. In habitat 4, the particles are very angular. A lot of bridging occurs between the particles forming large voids, thus allowing the surface forms to penetrate into the sediment, but as I pointed out previously these species are not typical interstitial forms.

KILENYI: There was a great decrease in the number of ostracods towards the mouths of inflowing streams. Could this not be associated with the amount of silt or clay brought in by the streams, thereby filling these voids?

WILLIAMS: The small fraction portion of any sediment is important to the interstitial habitat, because as you said it fills up the interstices. In fact, when it reaches a certain percentage in a sediment it can exclude certain interstitial species yet give rise to a new habitat for smaller invertebrate species. I have measured the particle size distribution at every station I have sampled. Where the streams enter the beaches at high water, the means and medians obtained from the cumulative curves are usually higher than the surrounding sediment. Lower down the shore the converse is true. In the habitats I have studied it is not the sediment composition which is responsible for the low numbers of ostracods in these streams but the salinities.

HASKINS: Did you find living representatives of all the ostracod species that you listed and identified?

WILLIAMS: No, I did not, but in my paper I have listed the species I have found alive, dead with the intact animal inside, empty carapaces and valves. These are further subdivided into adults and juveniles, because I think that it is very important to count all specimens when dealing with total animal composition of any habitat.

HULINGS: Did you distinguish between the sexes and did you find any seasonality as far as males and females were concerned?

WILLIAMS: I did. I found that more females than males occurred in all my adult species. In the juvenile population of *Cythere lutea* I found approximately equal numbers of both male and female.

OSTRACODES DE GROTTES SOUS-MARINES SEMI-OBSCURES DES CÔTES DE PROVENCE

SIMONE REYS
Station Marine d'Endoume, Marseille, France

SUMMARY

Forty-one samples, each collected by scraping $\frac{1}{4}$ m² of rocky substrate, were obtained from submarine caves near Marseille. Seventeen species of ostracod were found, essentially species known from photophilic algae or even the rhizomes of *Posidonia*, although some appear to have a preference for the cave biotopes. About a dozen species occur in each of the five recognized cave biotopes of which one species dominates the particular biotope. *Paracytherois striata* dominates the population associated with *Parazoanthus axinellae* and *Paramuricea clavata*, *Sclerochilus* sp. with *Corallium rubrum*, *Sclerochilus aequus* with sponges, and *Xestoleberis parva* the population associated with *Eunicella cavolini*.

A nine-metre length of nylon string, suspended in one cave for 235 days, yielded an extraordinarily rich ostracod fauna of 1575 individuals, 1286 of which were *Loxoconcha decipiens*.

Alors que les peuplements algaux et les fonds sableux ou détritiques ont souvent fait l'objet de recherches concernant la connaissance des Ostracodes, il semble que les grottes sous-marines n'ont jamais été étudiées en ce qui concerne ces crustacés. Aussi ai-je trouvé intéressant de profiter de prélèvements faits en plongée, dans des grottes de la région de Marseille essentiellement, par M. True et J. G. Harmelin, que je remercie ici très vivement. Les prélèvements ont été effectués par grattages du substrat rocheux sur une surface de 25 cm sur 25 cm; le tri de la macrofaune ayant été préalablement effectué, j'ai pu disposer de la totalité de leurs fonds de tamis dans lesquels j'ai pu récupérer tous les Ostracodes. De la sorte, j'ai pu disposer de 41 prélèvements, intéressant 5 peuplements différents : à *Eunicella cavolini*

à *Parazoanthus axinellae*
à *Paramuricea clavata*
à *Corallium rubrum*
à Spongiaires
(pour plus de détail, cf. la liste de Stations)

D'autre part, dans une des grottes sous-marines, une expérience a été effectuée par M. True. Une ficelle de nylon de 9 m de long a été suspendue dans l'axe principal de la grotte (−10 à −19 m).

Elle a été immergée en mars 1966 et ressortie au bout de 235 jours. La microfaune de cette installation m'a été confiée et s'est trouvée être extraordinairement riche en Ostracodes (1575 individus, dont 1286 de *Loxoconcha decipiens*).

LISTE DES STATIONS ÉTUDIÉES

Peuplement à *Eunicella cavolini*:
EC0: 15.3.66 −15 m Plane (tombant extérieur)
EC1: 1.4.66 −11 m Grand Salaman (Frioul)

Peuplement à *Parazoanthus axinellae*:
PA1:
PA2:
PA3: } 4.4.63 au 9.5.63 −8 m Niolon (paroi latérale)
PA4:
PA5:
PA6:
PA7: } 4.4.63 au 9.5.63 −5 à −8 m Niolon (plafond)
PA8:
PA9:
PA10:
PA11: } 4.4.63 au 9.5.63 −6 m Niolon (arche)
PA12:
PA1 Vill.: 27.3.65 Villefranche
PA2 Vill.: 26.3.65 Villefranche

Peuplement à *Paramuricea clavata*:
MC3:
MC4: } 23.5.63 au 29.5.63 −18 m Plane (gauche entrée)
MC5:
MC6:
MC7: } 23.5.63 au 29.5.63 −16 m Plane (en haut de l'arche)
MC8:
MC10:
MC11: } 23.5.63 au 29.5.63 −12 à −16 m Plane (plafond)
MC12:
MC3 réinstallation: 29.10.66 −18 m Plane (bord gauche arche)
MC4 réinstallation: 29.10.66 −18 m Plane (bord gauche arche)
PC2: 14.5.66 −25 m Farillon du milieu
PC3: 31.5.66 −16 m Plane (entrée gauche grotte)

Peuplement à *Corallium rubrum*:
CR1
CR2
CR3 } réinstallation: 6.12.66 −13 m Plane
CR4
CR2: 20.5.66 −11 m Cap Sormiou
CR3: 8.6.66 −9 m Plane
CR4: 15.6.66 −25 m Sud de Riou
CR6: 10.1.67 −23 m Moyade (boyau)
CR7: 17.1.67 −23 m Moyade (chambre)

Peuplement d'éponges:
EG1: 5.5.66 −23 m Moyade (chambre)
EG2: 17.5.66 −23 m Moyade (intérieur du tunnel)
EG3: 31.5.66 −20 m Figuier (30 m à l'intérieur de la grotte)

Peuplement expérimental sur ficelle de nylon, immergée le 8.3.66 et prélevée le 29.10.66 profondeur: −10 à −19 m. Plane.

LISTE DES ESPÈCES RENCONTRÉES DANS CES PEUPLEMENTS

Propontocypris levis (Müller)
Bairdia corpulenta Müller
Bairdia frequens Müller
Bairdia longevaginata Müller
Cytherois frequens Müller
Paracytherois acuminata Müller
Paracytherois striata Müller
Paradoxostoma angustum Müller
Paradoxostoma caecum Müller
Paradoxostoma simile Müller
Paradoxostoma versicolor Müller
Sclerochilus abbreviatus Brady et Robertson
Sclerochilus aequus Müller
Sclerochilus contortus (Norman)
Sclerochilus sp.
Xestoleberis parva Müller
Loxoconcha decipiens Müller

REMARQUES SYSTÉMATIQUES

Les espèces d'Ostracodes rencontrées dans ces peuplements de grottes ne posent pas de grands problèmes systématiques, sauf pour une espèce de *Sclerochilus*, que j'ai rencontrée en assez grand nombre dans ce biotope.

15 exemplaires dans les *Parazoanthus axinellae*
41 exemplaires dans les *Paramuricea clavata*
13 exemplaires dans les *Corallium*
17 exemplaires dans les Spongiaires
7 exemplaires sur la ficelle expérimentale.

Cette espèce de *Sclerochilus* ressemble beaucoup au *Sclerochilus gewemülleri dubowski* Marinov 1962; mais en attendant de pouvoir le comparer avec un paratype de ce dernier, je l'ai laissé sous la dénomination de *Sclerochilus* sp.

REMARQUES ÉCOLOGIQUES

Nous ne pouvons pas faire de différenciation entre les populations d'Ostracodes des différents peuplements étudiés. Il semble y avoir pour chaque peuplement, une douzaine d'espèces, dont une espèce qui prolifère et domine en nombre le reste du stock: nous avons ainsi *Paracytherois striata* dans les peuplements à *Parazoanthus axinellae* et à *Paramuricea clavata*, *Sclerochilus* sp. dans les peuplements à *Corallium rubrum*, *Sclerochilus aequus* dans les peuplements à Spongiaires et *Xestoleberis parva* dans le peuplement à *Eunicella cavolini*.

Précisons que, dans le peuplement à *Eunicella cavolini*, qui n'est pas précisément un peuplement de grotte, mais plutôt de paroi extérieure, le stock spécifique des Ostracodes est réduit de moitié et qu'on y trouve plus que 6 espèces (cf. tableau).

Les espèces rencontrées dans ces peuplements de grottes sont essentiellement des espèces connues, jusqu'à présent, de peuplements d'algues photophiles ou même de rhizomes de Posidonies; notamment *Bairdia frequens*, *Sclerochilus abbreviatus*, *Sclerochilus aequus*, *Cytherois frequens*, *Paradoxostoma angustum*, *Paradoxostoma caecum*, *Paradoxostoma versicolor*, *Paracytherois striata*, *Xestoleberis parva*.

Mais quelques espèces semblent avoir une prédilection plus marquée pour les biotopes de grotte. Il s'agit de *Bairdia longevaginata*, *Bairdia corpulenta*, *Sclerochilus contortus*, *Sclerochilus* sp., *Paracytherois acuminata* et *Loxoconcha decipiens* que je n'ai trouvé que dans ce biotope; mais Müller et Puri signalent leur présence en Méditerranée dans les algues calcaires et les Posidonies. Signalons, en ce qui concerne les deux espèces de *Bairdia*, que je les avais déjà déterminées des grottes de la région provençale et de la Grèce.

En ce qui concerne l'aspect quantitatif numéral, la moyenne des individus par peuplement laisse apparaître une importante diminution dans le peuplement à *Corallium rubrum*, alors qu'elle est la plus haute dans celui des Spongiaires et des *Parazoanthus axinellae*.

Ces moyennes sont, pour 625 cm², de 17 dans les *Parazoanthus axinellae*.

de 27,8 dans les *Eunicella cavolini*.

de 14,7 dans les *Paramuricea clavata*.

de 5 dans le Corail.

de 27 dans les Spongiaires.

de 768 sur la ficelle.

On ne peut, dans l'état actuel des recherches, trouver une explication à ces différences. Pour le cas d'un peuplement expérimental sur ficelle on peut envisager que le nombre d'individus particulièrement élevé, tient au fait que, lors du prélèvement, la ficelle a été mise dans un sac de plastique sans aucune perte de matériel, alors que pour des grattages on ne peut envisager de récolter la totalité du matériel.

Enfin, certaines stations (MC3, MC4 et CR2) ont fait l'objet d'un second grattage après avoir laissé le peuplement se réinstaller durant trois ans. En ce qui concerne les Ostracodes, on constate alors que toutes les espèces des peuplements normaux environnants sont présentes mais avec quelques variations dans le nombre des individus.

BIBLIOGRAPHIE SOMMAIRE

CARAION, F. E. 1962. Cytheridae noi pentru fauna Pontica Romineasca. *Studii Cerc. Biol. Seria 'biologie animala'*, **14** (1), 111-21.

DUBOWSKI, N. 1939. Zur Kenntnis der Ostracoden Fauna des Schwarzen Meeres. *Trudy karadah. nauch. Sta. T.I. Vyazems'koho*, **5**, 3-68.

MARINOV, T. 1962. Über die Muschelkrebs-Fauna des Westlichen Schwarzmeerstrandes. *Bull. Inst. cent. Rech. Sci. pisc. pech. Varna*, **2**, 81-108 [en russe].

MÜLLER, G. W. 1894. Die Ostracoden des Golfes von Neapel und der angrenzenden Meeres-Abschitte. *Fauna Flora Golf. Neapel*, Monogr. **21**, i-viii, 1-404, 40 Pls.

PURI, H. S. 1963. Preliminary Notes on the Ostracoda of the Gulf of Naples. *Experientia*, **19**, 368-73.

REYS, S. 1961. Recherches sur la systématique et la distribution des Ostracodes de la région de Marseille. *Recl Trav. Stn mar. Endoume*, **22** fasc. (36), 53-109.

——. 1963. Ostracodes des peuplements algaux de l'étage infralittoral de substrat rocheux. *Recl Trav. Stn mar. Endoume*, **28**, fasc. 43, 33-47.

——. 1964. Note sur les Ostracodes des Phanérogames marines des côtes de Provence. *Recl Trav. Stn mar. Endoume*, **32**, fasc. 48, 183-202.

TRUE, M. A. Étude quantitative de quatre peuplements sciaphiles sur substrat rocheux dans la région marseillaise. (Sous presse.) *Bull. Inst. océanogr. Monaco*.

DISTRIBUTION OF SPECIES OF THE GENUS *SEMICYTHERURA* IN THE NORTHERN ADRIATIC SEA BETWEEN VENICE AND TRIESTE, ITALY[1]

MARIO MASOLI
University of Trieste, Italy

SUMMARY

A study has been made of the distribution of the genus *Semicytherura* in the Northern Adriatic Sea. The distribution of each species belonging to this genus has been ascertained and distribution maps have been drawn. Environmental factors which can have influenced the species distribution have been examined. Since the depth, transparency, temperature and salinity of the water in this basin is practically constant or changes very little, the factor which seems to influence the species distribution is the type of bottom. Examination of the species distributions shows that they develop following the palaeocoastline which has been reconstructed by means of sedimentological studies.

RIASSUNTO

E' stata esaminata la distribuzione del genere *Semicytherura* reppresentato nello Adriatico Settentrionale da 11 specie. Sono state definite le aree di distribuzione del genere e delle singole specie e sono state disegnate le carte ad esse relative. Sono stati presi in esame i fattori ambientali che possono avere influenzato la diffusione delle specie. Dato che batimetria, trasparenza, temperatura e salinita hanno in questo bacino caratteristiche costanti o presentano variazioni di lieve entita, il fattore che maggiormente influenza la distribuzione delle specie sembra essere quello sedimentologico. Esaminando l'andamento delle aree di distribuzione delle specie si nota infatti che queste si sviluppano con modalita che rispecchiano l'andamento delle paleocoste che sono state ricostruite per mezzo di studi sedimentologici.

During 1965 and 1966 the Geological and Palaeontological Institute of Trieste University under the direction of Professor G. A. Venzo made two marine geology surveys in the northern Adriatic Sea. The area studied lay between latitudes 45°47′N and 45°22′N and longitudes 13°45′E and 12°40′E. The lithological nature of the bottom varied but there was little variation in depth, the maximum depth being 31·5 m. In this area 90 offshore stations and 36 near-shore grab samples have been collected. For each station a bottom sample was collected by a modified Van Veen grab sampler and a bottom core was also taken. At the same time data on chemical and physical factors were collected at the surface and near the bottom. Further details on the

[1] The present research was supported by the National Research Council, Italy, Committee for Oceanography and Limnology.

distribution of the sediments and their grain size characteristics are given in Brambati and Venzo (1967), and Venzo and Stefanini (1967). In the present work only the off-shore bottom samples have been considered since in the near shore samples, pollution and other environment modifications are too numerous to be significant for valid ecological purposes in relation to the number of samples available. The samples were washed with highly diluted H_2O_2 using 200 mesh sieves and the residues were split by means of a microsplitter.

Examination of the bottom samples collected in the area showed rich faunas of Ostracoda and Foraminifera. The genus *Semicytherura* was well represented both as regards number of species and abundance of individuals and detailed study was restricted to this genus. Eleven species of *Semicytherura* were found, ten of which have been classified, whilst the eleventh is probably new. The purpose of the present study is to define the distribution areas and analyse the ecological factors which have influenced the distribution of the genus and the various species in the area. Figures and descriptions of the species are given in Masoli 1968.

The following species were found in the samples examined:

1. *Semicytherura acuticostata* (Sars, 1886)
2. *Semicytherura alifera* (Ruggieri, 1952)
3. *Semicytherura brachyptera* (Ruggieri, 1952)
4. *Semicytherura incongruens* (G. W. Müller, 1894)
5. *Semicytherura inversa* (Seguenza, 1879)
6. *Semicytherura nigrescens* (G. W. Müller, 1894)
7. *Semicytherura paradoxa* (G. W. Müller, 1894)
8. *Semicytherura rara* (G. W. Müller, 1894)
9. *Semicytherura sulcata* (G. W. Müller, 1894)
10. *Semicytherura tergestina* Masoli 1968
11. *Semicytherura sp. 1.*

These species are generally associated with different representatives of other genera of Ostracoda. Sometimes (in the places indicated * on the maps) associated brackish- and fresh-water ostracods have also been found. These ostracods have evidently been carried by currents to the place where the sample was taken. It is clear that these have no significance as regards the interpretation of the ecology of the genus *Semicytherura*.

AREAS OF DISTRIBUTION

The genus

The genus *Semicytherura* is found all over the area considered (Fig. 1) except for a small area delimited on the east by the coast of the Gulf of Trieste and on the west by a line joining stations 37–60–50. Specimens of *Semicytherura* have not been found at stations 39, 60, 62, 46 and 49 which are located east of the latter line. It is thought that this absence is related to the large number of fresh-water submarine springs which can cause a sensible, if somewhat circumscribed, variation in the salinity and in the temperature of the water. Over the remaining area the genus is variously represented both as regards numbers and preservation. The area bounded by stations 69, 70, 83, 82 and 71 is the richest in number of ostracod species, and numbers of individuals are also very abundant in this area, where the depth varies between 14 and 23 metres and the bottom sediment is an argillaceous, shelly, medium to coarse sand.

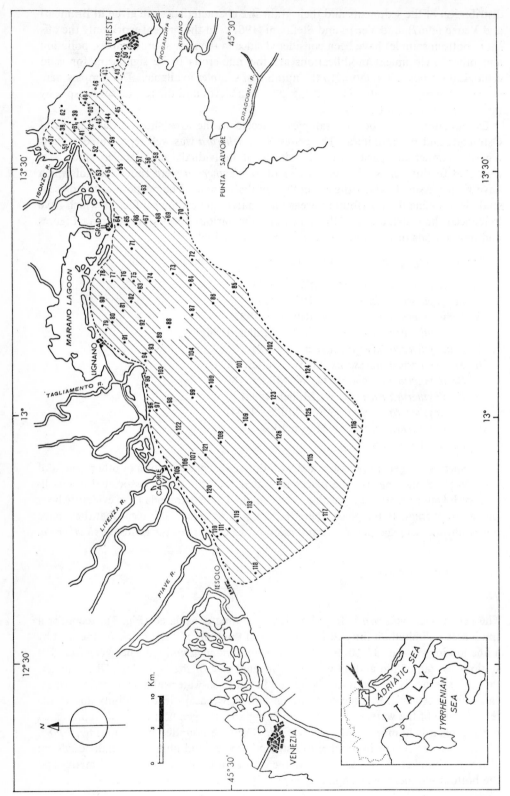

Fig. 1. Distribution of the genus *Semicytherura*.

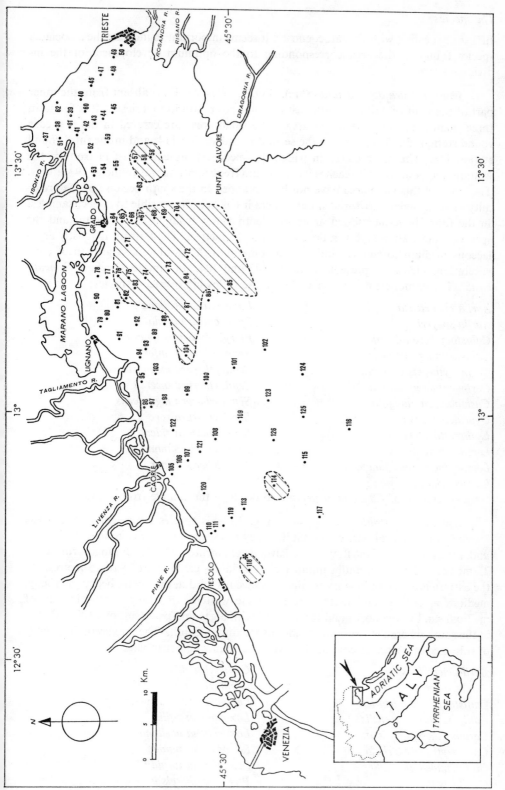

Fig. 2. Distribution of *S. acuticostata*.

The species

In a study dealing with thanatocoenoses, it seems important to indicate the associated species found in the area corresponding to the optimum development of the individual species.

1. *Semicytherura acuticostata* (Sars). Fig. 2. This species is absent from the inner part of the Gulf of Trieste, north-east of a line from Grado to Trieste. It is found in three small areas and one larger area. The small areas are centred in the first case round stations 56, 57 and 58; in the second round station 114; and in the third round station 118. The biggest area in which the species is present is just in front of the Marano Lagoon. Here it reaches its maximum frequency and was found at 19 stations. The limits of this large area have not been mapped in the south because of the proximity of Jugoslavian territorial waters where it has been impossible to collect samples. In the four above-mentioned areas the depth varies from 13 to 23 metres and the species is most abundant at a depth of 14 to 15 metres. The bottom is mostly argillaceous medium to coarse sand, which sometimes shows abundant growth of algal vegetation, such as is present at stations 73 and 104. In sample 82 where *S. acuticostata* is commonest it is associated with the following ostracod species:

Aurila cicatricosa
Aurila speyeri
Callistocythere adriatica
Callistocythere littoralis
Carinocythereis carinata
Carinocythereis turbida
Cushmanidea elongata
Cyprideis torosa
Cytheretta subradiosa
Leptocythere bacescoi
Leptocythere multipunctata
Loxoconcha agilis
Loxoconcha avellana
Loxoconcha tumida
Pterygocythereis jonesii
Semicytherura alifera
Semicytherura incongruens
Semicytherura inversa
Semicytherura nigrescens
Semicytherura rara
Semicytherura sulcata
Xestoleberis communis
Xestoleberis dispar

Rare specimens of *Candona neglecta* also occur in the sample from station 118.

2. *Semicytherura alifera* (Ruggieri). Fig. 3. This species is present in two areas located in the central part of the Gulf. They are separated by the zone which comes under the influence of the flow of the River Tagliamento. The depth ranges from 14 to 25 metres and the maximum number of individuals came from 14 to 15 metres. In the eastern area the sediments are fine argillaceous sand in the north, becoming shelly, medium to coarse sand in the south; in the western area they consist instead of medium sand in the north and sand mixed with algae and peat (stations 100 and 101) in the south where the sea floor reaches its maximum depth. The maximum frequency of this species occurs in conjunction with the following assemblage:

Aurila cicatricosa
Aurila speyeri
Callistocythere adriatica
Callistocythere littoralis
Carinocythereis carinata
Carinocythereis turbida
Cushmanidea elongata
Cyprideis torosa
Cytheretta subradiosa
Leptocythere bacescoi
Leptocythere multipunctata
Loxoconcha agilis
Loxoconcha avellana
Loxoconcha tumida
Loxoconcha turbida
Pterygocythereis jonesii

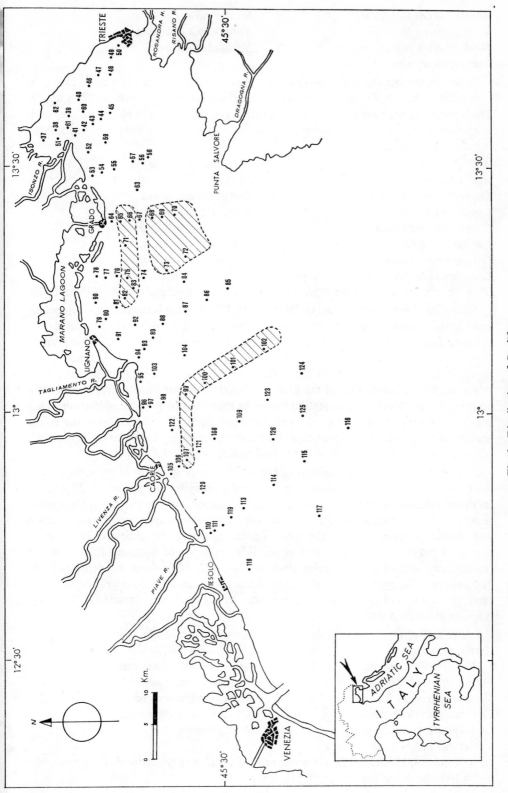

Fig. 3. Distribution of *S. alifera*.

Semicytherura incongruens   *Semicytherura sulcata*
Semicytherura inversa   *Xestoleberis communis*
Semicytherura nigrescens   *Xestoleberis dispar*
Semicytherura rara

3. *Semicytherura brachyptera* (Ruggieri), Fig. 4. This species has been found at isolated stations (76, 87, 98, 119) and in the small area containing stations 125 and 126. The bottom sediments are very different in grain size from place to place and the depth is between 14 and 28 metres. *Semicytherura brachyptera* is associated with the following ostracod species:

Aurila convexa   *Loxoconcha tumida*
Aurila speyeri   *Paracytheridea bovettensis*
Carinocythereis turbida   *Quadracythere prava*
Cyprideis torosa   *Semicytherura acuticostata*
Cytheretta subradiosa   *Semicytherura incongruens*
Loxoconcha agilis   *Semicytherura inversa*
Loxoconcha avellana   *Xestoleberis communis*
Loxoconcha bairdi   *Xestoleberis dispar*

Some specimens of *Candona neglecta* have been found at stations 98, 125 and 126.

4. *Semicytherura incongruens* (G. W. Müller), Fig. 5. This is the commonest species and has the widest distribution of all the species. In the area sampled, this species has not been found in a narrow zone located in the innermost part of the Gulf of Trieste (stations 62, 39, 60, 46, 49) and at station 88. The latter is located 8 kilometres offshore, in front of the delta of the River Tagliamento. It is believed that the absence of this species in the inner part of the Gulf is related to the presence of fresh-water submarine springs. Its absence at station 88 can be explained by the type of bottom which consists of rounded medium sand. This species is generally frequent over the whole area but it reaches its maximum frequency between the Rivers Isonzo and Tagliamento where the sea floor varies between fine and coarse sand.

5. *Semicytherura inversa* (Seguenza), Fig. 6. *S. inversa* is absent in the inner part of the Gulf of Trieste and has been found west of a line joining stations 55, 57 and 58. Its distribution area extends westward as far as station 118. In the north the species is present further inshore between the Rivers Isonzo and Tagliamento, and occurs a little further offshore between the River Tagliamento and Iesolo beach. It occurs at depths of between 7 and 31·5 metres on different types of bottom. Its abundance reaches a maximum at the deepest stations (116, 123, 125) where the depth is 27 to 31·5 metres. The type of bottom at these stations is a medium to coarse argillaceous sand; at station 123 green algae are present. At its maximum abundance *S. inversa* is found associated with:

Aurila cicatricosa   *Loxoconcha avellana*
Aurila convexa   *Loxoconcha tumida*
Callistocythere adriatica   *Paracytheridea bovettensis*
Carinocythereis carinata   *Pterygocythereis jonesii*
Carinocythereis turbida   *Semicytherura inversa*
Cushmanidea elongata   *Semicytherura nigrescens*
Cytheretta subradiosa   *Semicytherura sp. 1.*
Cytheridea neapolitana

At stations 118, 123, 124, 125, 126 specimens of *Candona neglecta* and *Ilyocypris gibba* which have been carried in by currents are present.

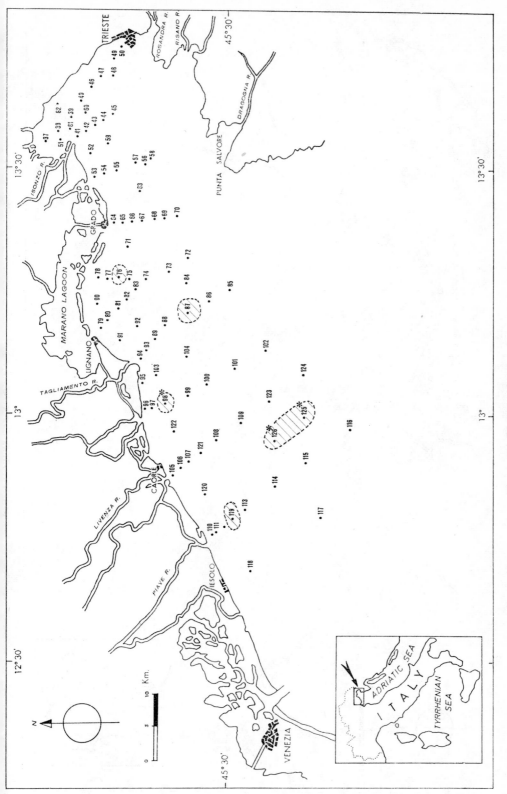

Fig. 4. Distribution of *S. brachyptera*.

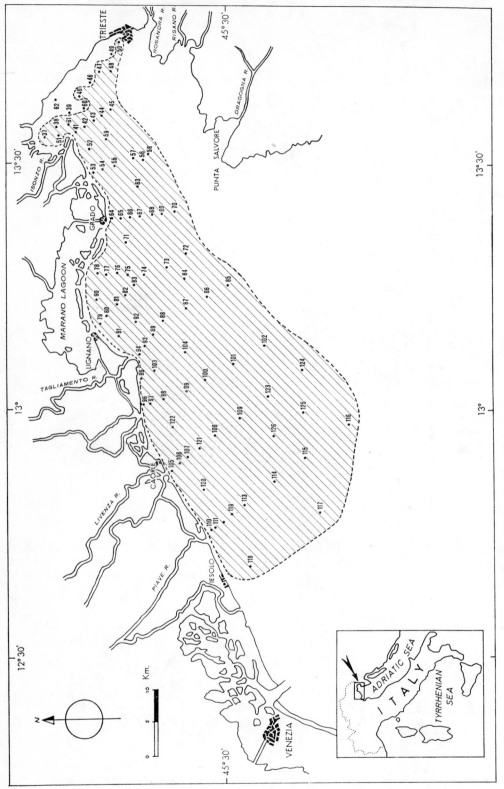

Fig. 5. Distribution of *S. incongruens*

343

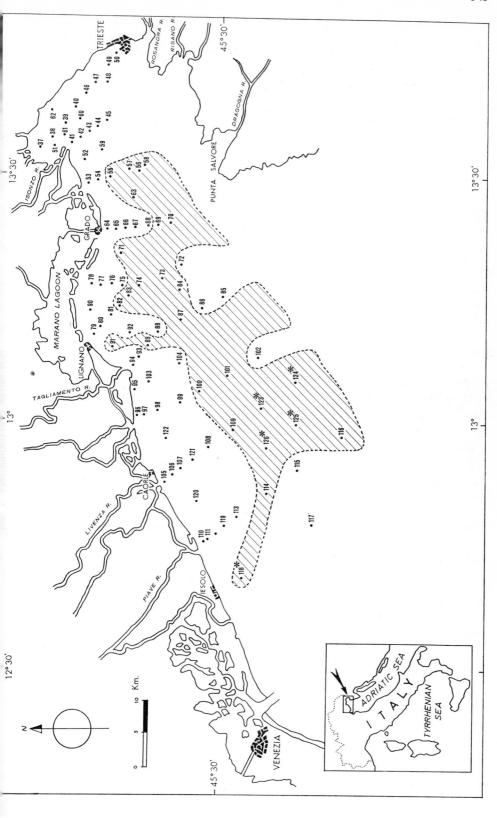

Fig. 6. Distribution of *S. inversa*.

6. *Semicytherura nigrescens* (G. W. Müller), Fig. 7. This species is rare and occurs in only four small areas whose distribution is shown in Fig. 7. It ranges in depth from 14 to 20 metres and the greatest number of specimens was found at station 58 where the sea floor consists of medium argillaceous sand and mud and the depth is 20 metres. At this station the associated ostracod fauna is:

Aurila speyeri	*Loxoconcha tumida*
Callistocythere adriatica	*Loxoconcha turbida*
Carinocythereis carinata	*Paracytheridea bovettensis*
Carinocythereis turbida	*Paradoxostoma simile*
Cushmanidea elongata	*Pterygocythereis jonesii*
Cyprideis torosa	*Semicytherura acuticostata*
Cytheretta subradiosa	*Semicytherura incongruens*
Cytheridea neapolitana	*Xestoleberis communis*
Leptocythere multipunctata	*Xestoleberis dispar*

A few specimens of *Candona neglecta* and *Ilyocypris gibba* were present at stations 89, 112 and 114.

7. *Semicytherura paradoxa* (G. W. Müller), Fig. 8. This species is restricted to five small areas and the number of specimens found is small. The depth distribution range is 11 to 23 metres and the substrate consists of medium sand and very sandy mud, sometimes associated with green algae. The maximum number of specimens occurs in association with the following ostracod assemblage:

Carinocythereis carinata	*Loxoconcha tumida*
Carinocythereis turbida	*Loxoconcha turbida*
Cushmanidea elongata	*Neocytherideis fasciata*
Cyprideis torosa	*Pseudocytherura calcarata*
Cytheretta subradiosa	*Quadracythere prava*
Cytheridea neapolitana	*Semicytherura acuticostata*
Leptocythere multipunctata	*Semicytherura alifera*
Loxoconcha agilis	*Semicytherura incongruens*
Loxoconcha avellana	*Semicytherura sulcata*
Loxoconcha bairdi	*Semicytherura tergestina*
Loxoconcha napoliana	*Xestoleberis dispar*
Loxoconcha stellifera	

Station 89 also yielded a few *Candona neglecta*.

8. *Semicytherura rara* (G. W. Müller), Fig. 9. The distribution of this species is very restricted and it is present in a small area in front of the Marano Lagoon and at station 124. The corresponding sediments are fine sand and very sandy mud. *S. rara* occurs from depths of 10 to 28·5 metres and the maximum number of specimens was found at a depth of 14 to 15 metres where the bottom was sandy with some mud content. In this last environment the species is associated with the following ostracod species:

Aurila cicatricosa	*Cushmanidea elongata*
Aurila speyeri	*Cyprideis torosa*
Callistocythere adriatica	*Cytheretta subradiosa*
Callistocythere flavidofusca	*Cytheridea neapolitana*
Carinocythereis carinata	*Leptocythere bacescoi*
Carinocythereis turbida	*Leptocythere multipunctata*

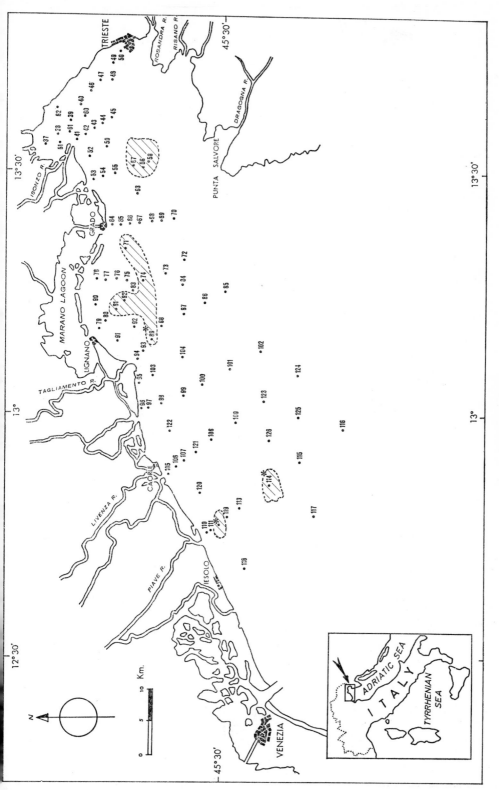

Fig. 7. Distribution of *S. nigrescens*.

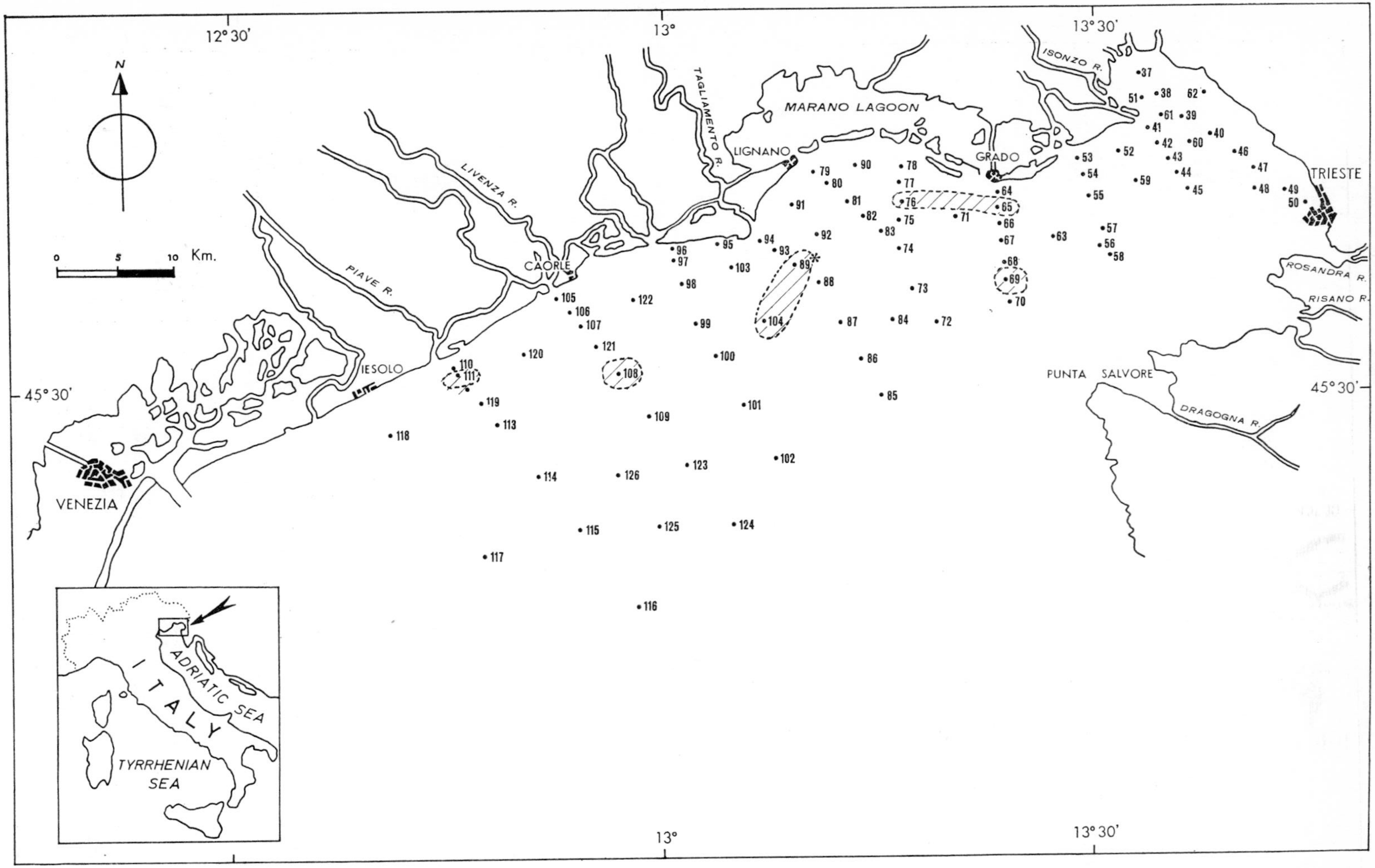

Fig. 3. Distribution of *S. aenedora*

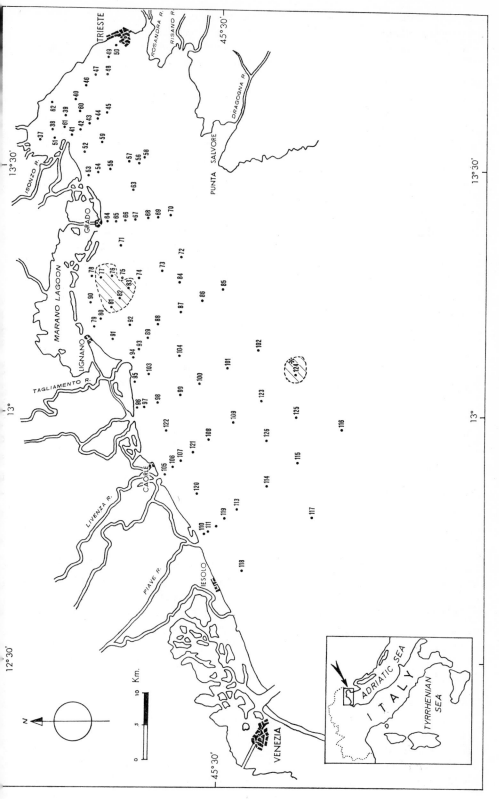

Fig. 9. Distribution of *S. rara*.

Loxoconcha agilis
Loxoconcha avellana
Loxoconcha tumida
Loxoconcha turbida
Paracytheridea bovettensis
Pseudocytherura calcarata
Pterygocythereis jonesii
Semicytherura acuticostata

Semicytherura alifera
Semicytherura incongruens
Semicytherura inversa
Semicytherura nigrescens
Semicytherura tergestina
Xestoleberis communis
Xestoleberis dispar

At station 124 *Candona neglecta* and *Ilyocypris gibba* are also present.

9. *Semicytherura sulcata* (G. W. Müller), Fig. 10. The distribution area of this species consists of a linear belt with irregular form, especially in its northern part. It lies relatively close inshore and is limited on the east by Grado and on the west by the River Piave. This species is also present at station 118. In the northern part of this area the sediments are very sandy mud, fine sand, and argillaceous and sandy silts; in the central part they consist of medium to coarse sand, argillaceous and sometimes shelly. In the southern part of this area medium sand is present. *S. sulcata* occurs at depths between 3 and 20 metres, the maximum number of specimens having been found at 15 metres. The associated ostracod assemblage at the station at which this species was most abundant is:

Carinocythereis carinata
Carinocythereis turbida
Cushmanidea elongata
Cyprideis torosa
Cytheretta subradiosa
Cytheridea neapolitana
Leptocythere multipunctata
Loxoconcha agilis
Loxoconcha avellana
Loxoconcha bairdi
Loxoconcha napoliana
Loxoconcha stellifera

Loxoconcha tumida
Loxoconcha turbida
Neocytherideis fasciata
Pseudocytherura calcarata
Quadracythere prava
Semicytherura acuticostata
Semicytherura alifera
Semicytherura incongruens
Semicytherura paradoxa
Semicytherura tergestina
Xestoleberis dispar

Some specimens of *Candona neglecta* and *Ilyocypris gibba* were present at stations 80 and 118.

10. *Semicytherura tergestina* Masoli, Fig. 11. This species has a wide distribution and is present in the area well defined by stations 64, 70, 100, 113, 118, 122 and 82. In the centre of this area there is a small region where the species has not been found (stations 99 and 104). The presence of *Semicytherura tergestina* seems to be linked with medium argillaceous sand and very sandy mud. At stations 99 and 104, where it is absent, the bottom is medium sand. It is found in depths ranging from 7 to 23 metres, and has a maximum frequency at a depth of about 20 metres where it is associated with the following:

Callistocythere adriatica
Callistocythere flavidofusca
Carinocythereis carinata
Carinocythereis turbida
Cushmanidea elongata
Cyprideis torosa

Cytheretta subradiosa
Cytheridea neapolitana
Leptocythere multipunctata
Loxoconcha avellana
Loxoconcha napoliana
Loxoconcha stellifera

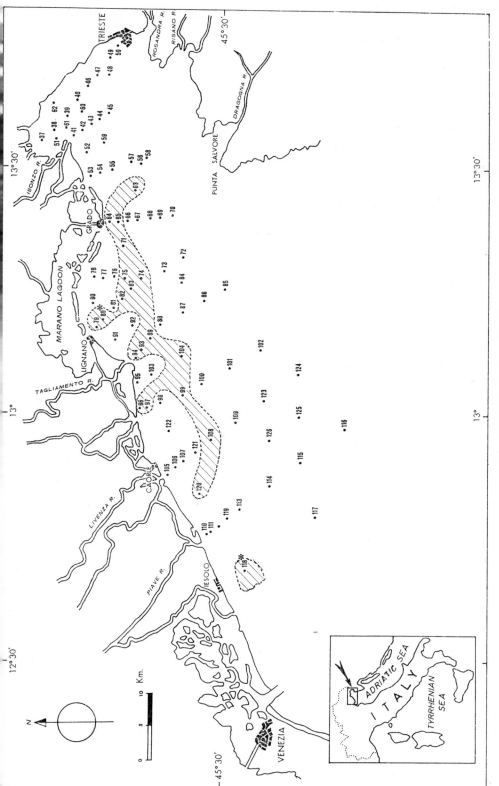

Fig. 10. Distribution of *S. sulcata*.

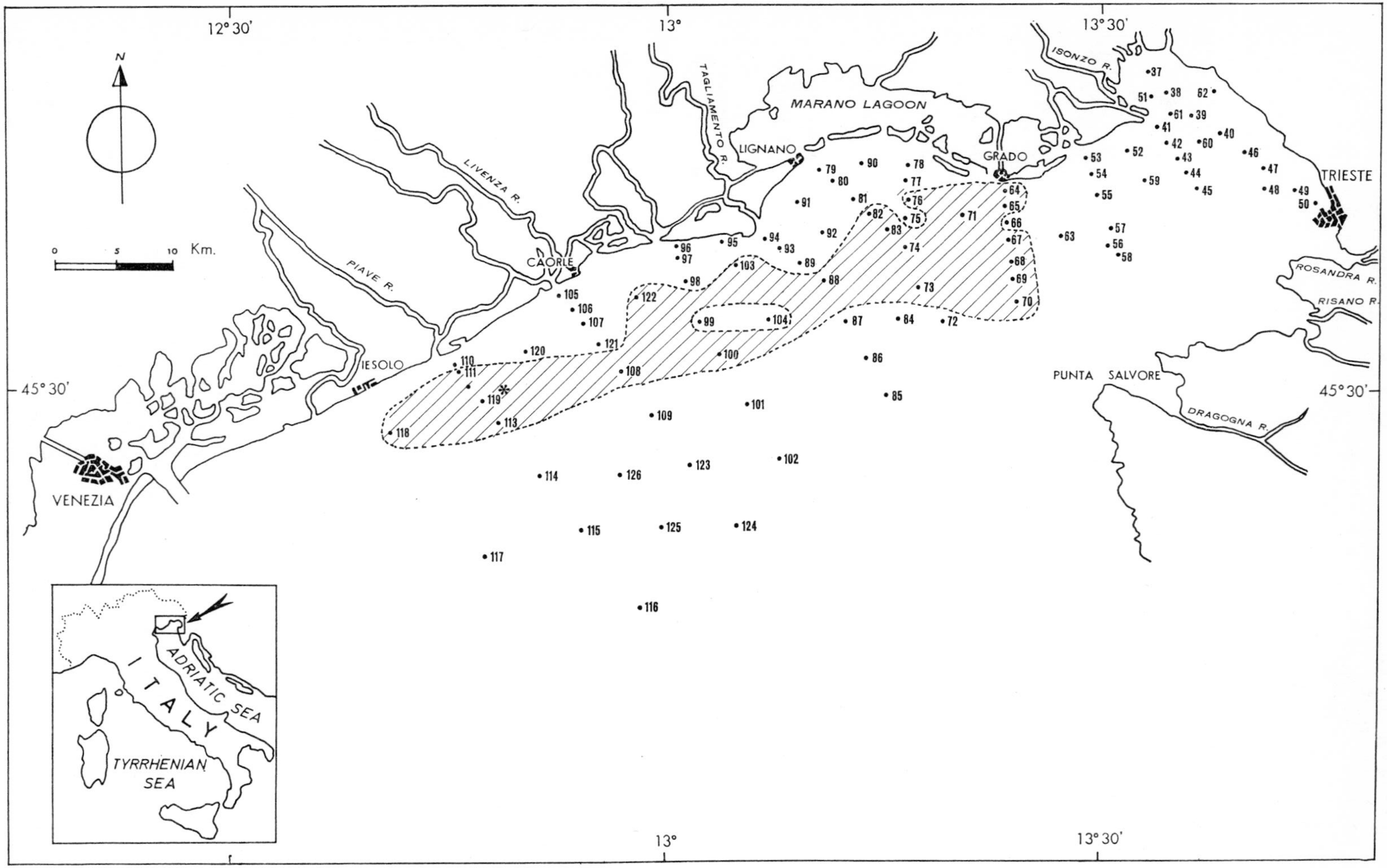

Fig. 11. Distribution of Stations.

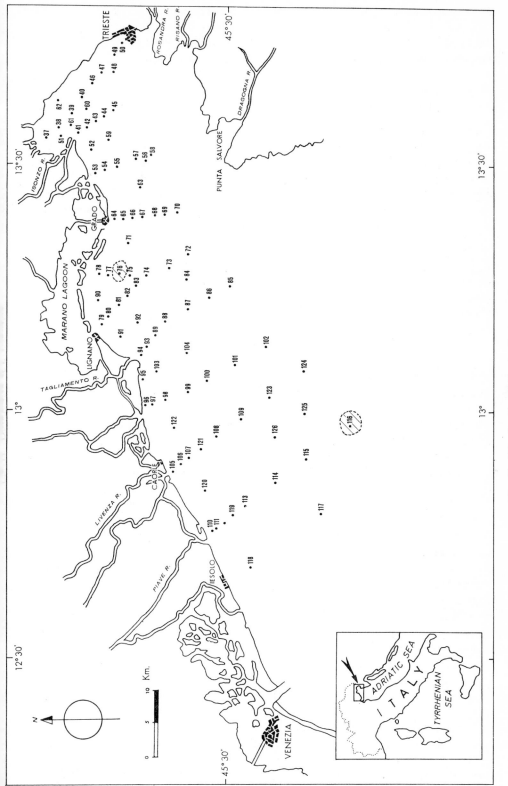

Fig. 12. Distribution of *Semicytherura* sp. 1.

Loxoconcha tumida
Loxoconcha turbida
Neocytherideis fasciata
Paracytheridea bovettensis
Paradoxostoma simile
Propontocypris cf. *P. setosa*

Pseudocytherura calcarata
Semicytherura brachyptera
Semicytherura incongruens
Xestoleberis communis
Xestoleberis dispar

At station 119, Ostracod 'A' which is probably not a marine species, is also present.

11. *Semicytherura* sp. 1, Fig. 12. This species has been found at two stations only (76 and 116) which are widely separated and in which the depth and type of bottom are very different. At station 76 the bottom is fine argillaceous sand, and the depth is 14 metres; at station 116 the bottom consists of medium argillaceous sand and the depth is 31·5 metres. The only environmental factor in common at the two stations is the presence of mud. *Semicytherura* sp. 1. is associated with the following species:

Aurila cicatricosa
Aurila convexa
Callistocythere adriatica
Carinocythereis carinata
Carinocythereis turbida
Cushmanidea elongata
Cytheretta subradiosa

Cytheridea neapolitana
Loxoconcha bairdi
Loxoconcha tumida
Paracytheridea bovettensis
Pterygocythereis jonesii
Semicytherura incongruens
Semicytherura inversa

ECOLOGICAL REMARKS

In the area of research, some environmental factors, such as depth, transparency and temperature, vary very little. Temperature varies appreciably only in a restricted area confined to the innermost part of the Gulf where submarine springs are recorded. In this respect it is interesting to note that while the mean annual temperature of the water near the bottom is between 14° and 16°C, in the area affected by these springs the average is about 10°C. Salinity decreases only near the mouths of the rivers entering the Gulf and there the genus *Semicytherura* was found to be scarce or absent. The only species found in this environment is *Semicytherura incongruens* which shows good adaptability to changes of salinity and temperature. The above-mentioned factors do not explain the differences in the distribution and frequency of the 11 species encountered. The major environmental factor which seems to influence their distribution is the nature of the bottom. It was found that the sediments of the area show appreciable variations and that the distribution areas of the species were related to this. It was also noted that the genus *Semicytherura* was most abundant, both as regards number of species and specimens, in an area where the prevailing sediments were argillaceous, medium-coarse shelly sands and argillaceous very fine grained sands. Both these types of sediments are well represented in front of the Marano Lagoon.

For some species a variation in grain size of the sediments corresponds with a reduction in the number of specimens; other species disappear when the grain size changes. It is reasonable to suppose that the decrease of ostracod specimens in areas of relatively coarse deposition (e.g. medium sands) is caused by mechanical factors such as sorting of the sediments, mechanical destruction of the carapaces by the shifting coarse materials, and the impossibility of the formation of organic calcium carbonate (one of the basic requirements of ostracods). At the same time it is difficult for lime-secreting plants to settle and gain a foothold. In relation to this last point it has not been possible to establish a direct relationship and this is only a hypothesis. Areas of fine grained sedimentation such as mud and sandy mud form unfavourable areas either because of the decreasing transparency of the sea water, or because of the poor support afforded by the bottom, especially in the case of the poorly ornamented forms. The area of maximum concentration of species and specimens lies close and subparallel to the coastline, but is well removed from areas influenced by land drainage. The whole area is characterized by moderate depths of 13 to 18 metres. Contour lines of the distribution areas of each species follow the near-by palaeocoastline which has been reconstructed by sedimentological methods.

In short it may be said that in the basin examined, the ecological factors which appear to influence the distribution of the species of *Semicytherura* are sedimentological. In the basin as a whole, other environmental factors change very little and only affect small areas.

ACKNOWLEDGEMENTS

I wish to thank Professor G. A. Venzo for his valuable advice, Professor G. Ruggieri, Director of the Geological and Palaeontological Institute of Palermo University who revised the classification and the manuscript, and Dr G. Bonaduce of the Stazione Zoologica, Naples, for his help.

REFERENCES

ASCOLI, P. 1965. Ecological study on Ostracoda from bottom cores of the Adriatic Sea. *Pubbl. Staz. zool. Napoli* [1964], **33** (suppl.), 213-46, 3 Figs, 4 Pls.
BENSON, R. H., *et al.* in MOORE, R. C. 1961. *Treatise on Invertebrate Paleontology, Pt. Q. Arthropoda. 3. Crustacea: Ostracoda*. 442 pp., 334 Figs. Lawrence, Kansas, University of Kansas and Geological Society of America.
BRAMBATI, A. AND VENZO, G. A. 1967. Recent sedimentation in the Northern Adriatic Sea between Venice and Trieste. *Studi trent. Sci. nat., A.* **44** (2), 202-74.
ELOFSON, O. 1941. Zur Kenntniss der marinen Ostracoden Schwedens mit besonders Berücksichtigung des Skageraks. *Zool. Bidr. Upps.*, **19**, 215-534, 42 kart.
GRAF, H. 1940. Marine Ostracoden von Arbe (Adria). *Zool. Anz.*, **130** (1/2), 25-30.
KEIJ, A. J. 1957. Eocene and Oligocene Ostracoda of Belgium. *Mém. Inst. r. Sci. nat. Belg.*, **136**, 3-210, 23 Pls.
KLIE, W. 1942. Adriatische Ostracoden. I. *Zool. Anz.*, **138** (3/4), 86-89.
——. 1942. Adriatische Ostracoden. II. *Zool. Anz.*, **138** (9/10), 197-210.
——. 1942. Adriatische Ostracoden. III. *Zool. Anz.*, **139** (3/4), 67-73.
MASOLI, M. 1968. Ostracodi recenti dell'Adriatico Settentrionale, tra Venezia e Trieste. *Memorie Mus. Stor. nat. Venezia trident.* **17** (1), 1-100 13 Pls.
MOSETTI, F. 1966/67. Caratteristiche idrologiche dell'Adriatico Settentrionale. Situazione estiva. *Memorie R. Ist veneto Sci.*, **125**, 147-75, 6 Figs, 8 tables.
NEALE, J. W. 1965. Some factors influencing the distribution of Recent British Ostracoda. *Pubbl. Staz. zool. Napoli* [1964], **33** (suppl.), 247-307, 11 Figs, 1 Pl., 5 tables.
PURI, H., BONADUCE, G. AND MALLOY, J. 1965. Ecology of the Gulf of Naples. *Pubbl. Staz. zool. Napoli* [1964], **33** (suppl.), 87-199, 67 Figs, 1 table.
ROME, D. R. 1965. Ostracodes des environs de Monaco, leur distribution en profondeur, nature des fonds marins explorés. *Pubbl. Staz. zool. Napoli* [1964], **33** (suppl.), 200-12, 1 carte, 1 table.
RUGGIERI, G. 1952. Nota preliminare sugli Ostracodi di alcune spiaggie adriatiche. *Note Lab. Biol. mar. Fano*, **1** (8), 57-64, 4 Figs.
——. 1953. Iconografia degli Ostracodi marini del Pliocene e del Pleistocene italiani. *Atti Soc. ital. Sci. nat.*, **92**, 40-56, 16 Figs.
——. 1959. Enumerazione degli Ostracodi marini del Neogene, Quaternario e Recente italiani descritti o elencati nell'ultimo decennio. *Atti Soc. ital. Sci. nat.*, **98**, 183-208.
——. 1962. Gli Ostracodi marini del Tortoniano (Miocene medio superiore) di Enna, nella Sicilia centrale. *Palaeontogr. ital., Mem.* **2**, 1-68, tav. XI-XVII, 15 Figs.
——. 1965. Ecological remarks on the present and past distribution of four species of *Loxoconcha* in the Mediterranean. *Pubbl. Staz. zool. Napoli* [1964], **33** (suppl.), 515-23, 11 Figs.

VENZO, G. A. AND STEFANINI, S. 1967. Distribuzione dei carbonati nei sedimenti di spiaggia e marini dell'Adriatico Settentrionale tra Venezia e Trieste. *Studi trent. Sci. nat., A.* **44** (2), 178-201.

WAGNER, C. W. 1957. *Sur les Ostracodes du Quaternaire récent des Pays-Bas et leur utilisation dans l'étude géologique des dépôts Holocènes.* 259 pp. Mon., Mouton et C. ed., 'S-Gravenhage.

——. 1965. Ostracods as environmental indicators in Recent and Holocene estuarine deposits of The Netherlands. *Pubbl. Staz. zool. Napoli* [1964], **33** (suppl.), 480-95, 6 Figs.

DISCUSSION

OERTLI: It would be very interesting if, in his further investigations, Dr Masoli would study the near-shore faunas of Trieste. The highlands of Trieste are the classical karst region, in fact the name karst comes from nearby Yugoslavia, and we have many superficial rivers which disappear in the calcareous massifs. Some very important resurgences can be observed in the Gulf of Trieste, and the effects are sometimes so great as to alter the sea level. It would be interesting to see the faunas in this region, and especially if these resurgences take up subterranean species.

ASCOLI (for MASOLI): The area was probably not sampled because it is very crowded with shipping, Trieste being an important harbour, and perhaps also because of very strong currents. It would, however, be extremely interesting and I will make this point to Dr Masoli. It would be interesting to see if there are any karstic subterranean currents coming from the land by dying the water. This has been done inland but I do not know whether it has been done with waters that are supposed to flow into the sea. It might be proved that some ostracods living along the coast line, live also in the continental waters nearby. We do know that some waters from inland disappear and enter the open sea.

DELORME: Has Dr Masoli made a study of the currents? I would suggest that the presence of *Ilyocypris gibba* indicates derivation from the rivers which empty into the Bay.

ASCOLI (for MASOLI): He points out which species are supposedly being carried in by rivers and which species are carried in from continental waters by means of subterranean currents.

KILENYI: There seems to be some sort of correlation in the position of the entrance of the rivers and the presence of *Ilyocypris gibba* and *Candona neglecta* and I think that hey both come in from the rivers. Dr Masoli very rightly emphasizes the correlation between sediments and fauna and in this respect it is extremely important that we should be very exact about the nature of the sediment. To say sand or clay means absolutely nothing because these are subjective terms. It is not even enough to give the median grain size because this can be interpreted in various ways. The most useful is a very accurate grain size analysis, preferably sticking to the Phi's or Phi analysis, which will then give some sort of method by which we can correlate faunas and sediments.

PURI: How much of this study was based on living forms?

ASCOLI (for MASOLI): This study is based on thanatocoenosis only because it aims to be a palaeoecological study and to point out the geographical distribution of *Semicytherura* thanatocoenosis and not of biocoenosis in the Gulf of Trieste.

KILENYI: Regarding Dr Masoli's comments on the mechanical destruction of the valves; I have worked in an equally high energy environment, an estuary with tidal currents in the region of three to four knots, and I found very little evidence of valves, even relatively old and long dead valves having been destroyed by current action. In spite of the large traction load present in this particular area there was no evidence for this.

WHATLEY: On Dr Masoli's maps, *Semicytherura* did not seem to occur close to the coast. In the Irish Sea area, this genus occurs frequently, mostly as dead carapaces, well up the Welsh estuaries. Do you think that its absence from the near-shore waters and lagoons in the Mediterranean is due to the very low tidal range in the Mediterranean relative to that of the Welsh coast?

ASCOLI (for MASOLI): In the Mediterranean the tidal range is extremely low. As far as I remember it is less than half a metre in the Gulf of Trieste. On the whole, in the genus *Semicytherura* the majority of species are not present along the coast except for *S. incongruens*, which is a very typical shallow-water indicator over the whole Italian coastline.

PETKOVSKI: Ich habe eine Frage zur Sammeltechnik: Sind außer quantitativen Proben auch qualitative genommen worden, und wie groß ist die Probenfläche?

ASCOLI (for MASOLI): The amount of sediment collected is not stated. For each station a bottom sample has been collected by a big, modified Van Veen grab sampler together with a bottom core, so there is a bottom core and a grab sample for each station.

PETKOVSKI: Bei den qualitativen Proben wird mit einem Bodenzugnetz oder einer Dredge stets mehr Material gesammelt als mit einem Bodengreifer. So weißi ch einen Fall vom Bodensee, wo die quantitativen Proben allein keinen vollständigen Überblick über die Bodenfauna boten. Es ist am besten, wenn man gleichzeitig quantitative und qualitative Proben nimmt.

ASCOLI (for MASOLI): Dr Masoli does not state whether the results are based on grab samples, on core samples, or both.

BENSON: Would Dr Ascoli care to comment on the distinctiveness of the species dealt with? In my experience there is considerable variability within species of *Semicytherura*. Could you give some indication of the problems of recognizing these species in order to plot their distribution.

ASCOLI: As far as I recall from my study of samples from the Adriatic Sea, the genus *Semicytherura* is generally extremely rare except *S. incongruens* and *S. sulcata* which are quite good indicators of shallow-water conditions. All other species such as *S. paradoxa* and *S. acuticostata* are very rare and generally found at depths greater than 15 metres. Only *S. incongruens* is very abundant and characteristic in shallow-water samples from an ecological point of view.

From a palaeontological or zoological point of view you can describe many new species from the Trieste area and the whole Adriatic area in general. I am sure that besides *S. tergestina* you could find many other new species. As far as I remember, however, such new species are very rare and confined to very restricted areas.

DISTRIBUTION OF OSTRACODA
IN THE MEDITERRANEAN

HARBANS S. PURI, G. BONADUCE and A. M. GERVASIO
Florida Geological Survey, Tallahassee, U.S.A., and
Stazione Zoologica, Napoli, Italy

SUMMARY

The surface temperature of the water during summer in the Western Mediterranean is 20°–25°C (subtropical), with a greater change in seasonal temperature than the eastern part which is tropical (summer temperature 25°–27°C). Salinity in the western part is +38‰ while in the eastern tropical part it is 40‰. The Mediterranean Sea has a much lower fertility than the neighbouring Atlantic. The bottom sediments consist of sands, clays, muds, *Globigerina* ooze and calcarenites. Vegetation of the coastal areas in depths up to 100 m consists of *Posidonia*, calcareous algae and fibrous algae.

This uniformity in bottom conditions has resulted in a remarkably similar shallow shelf (depth up to 100 m) fauna from the Balearic Sea in the west to the Levantine Coast in the east. This assemblage consists of species like *Buntonia giesbrechti* (Müller), *Neocytherideis foveolata* (Brady), *Cytheridea neapolitana* Kollmann, *Carinocythereis runcinata* (Baird), *C. quadridentata* (Baird), '*Cythereis*' *polygonata* (Rome), *Cushmanidea elongata* (Brady), *Hemicytherura videns* (Müller), *Basslerites teres* (Brady), and *Aurila convexa* (Baird). The phytal fauna associated with bottom vegetation consists predominantly of species of *Paradoxostoma, Xestoleberis, Cytherois, Sclerochilus* and *Paracytherois*. Calcareous algae support an assemblage that consists of *Hemicytherura videns* (Müller), *Loxoconcha rhomboidea* (Fischer) and some other forms.

An ecologically stable fresh-water (limnetic) assemblage exists in the upper reaches of the rivers, which is characterized by species of *Candona, Ilyocypris, Herpetocypris, Ilyodromus, Cyprinotus* and other fresh-water genera. This fresh-water assemblage gives way to a more restricted shallow littoral lagoon assemblage (mixomesohaline), with typical brackish-water species like *Loxoconcha elliptica, L. dorsosalina, Cyprideis littoralis, Aurila woodwardi* and *Xestoleberis aurantia*. In the vicinity of Korfu, an oligohaline fauna consisting of *Cytherois stephanidesi* Klie and *Leptocythere* ? *ilyophila* Hirschmann is known to exist.

Over five off-shore inner neritic assemblages exist in the Tyrrhenian and Ionian seas. The abyssal assemblages are characterized by *Globigerina* ooze bottom with sparse ostracod fauna comprising species of *Cytheropteron* and *Pseudocythere caudata* Sars.

The present study is based on 70 bottom cores in an east-west traverse from Port Said on the east through the continental shelf, Nile Cone, Herodotus Abyssal Plain, Mediterranean Ridge and Sicilia Basin in the Eastern Mediterranean to Gibraltar in the west.

INTRODUCTION

During our ecologic studies of the ostracoda of the Gulf of Naples (Puri *et al.*, 1965), it was suspected that the eight major assemblages defined in the Gulf of Naples area extended around the western Mediterranean to perhaps as far west as Gibraltar. There is a remarkable similarity between faunas around Baleares, Banyuls, Monaco, Marseille and the Gulf of Naples. The association of ostracods with certain types of vegetation and substrate is well established from studies by G. W. Müller (1894), De Buen (1916), Rome (1939, 1942, 1944), Reys (1961, 1963), Hartmann (1953), Puri *et al.* (1965). Except for Rome's samples, which were collected at various depths to below 400 m, the majority of the rest of the studies have been based on samples collected from shallow waters of less than 30 m depth. The purpose of this study is to supplement the present knowledge by examination of deep-water assemblages in the Mediterranean from 70 sediment cores collected by *Vema* (Cruise 10) from Gibraltar in the west to Port Said in the east.

The Mediterranean Basin in this report includes the so-called 'Mediterranean-Atlantic faunas' and consists of the Western Mediterranean (Balearic Sea, the Tyrrhenian Sea and the Adriatic Sea), the Eastern Mediterranean (the Ionian Sea, the Aegean Sea, the Sea of Candia and the Nile Cone), the Black Sea, the Sea of Azov, the Caspian Sea and the Red Sea. Our current knowledge of the ostracod assemblages from these areas is summarized together with new data.

BATHYMETRY

Although soundings have been made in the Mediterranean by the ancients since the days of Aristotle, circa 330 B.C., and Posidonius, circa 85 B.C. (who remarked on the deepest point, about 1000 fathoms off Sardinia), it was not until 1947 that the first modern data on the Mediterranean were collected by the Swedish research ship *Albatross* (Petterson, 1957; Koczy, 1956). In the last ten years, several surveys have been made chiefly by American and Russian ships and a serious attempt has been made to study bathymetry, physiography and sedimentation of the Mediterranean. Maurice Ewing, in 1956, and C. L. Drake in 1959, made several traverses by research vessel *Vema* of the Lamont Geological Observatory (Cruise 10 and Cruise 14) and conducted bathymetric, gravity and sediment surveys. Later on, additional data were gathered by the Lamont Research ship *Conrad*. J. L. Worzel, of the Lamont Geological Observatory conducted gravity and echo-sounding surveys on board U.S.S. *Compass Island* (Worzel, 1959). In 1959, J. B. Hersey conducted surveys by Woods Hole research ship *Atlantis* (Cruise 42) and a later Cruise 151; Dr Hersey also obtained further data from Cruise 7, Cruise 21 and Cruise 43 of the Woods Hole ship *Chain* (Hersey, 1965). Hersey also directed surveys of the Woods Hole vessel *Yamacraw* in 1958. Emery and Benton in 1959 ran several traverses off the coast of Israel (Emery and Benton, 1960) and Emery, in 1962, conducted bathymetric survey in the Eastern Mediterranean on board *Aragonese* (Allan *et al.*, 1964). Emery, Heezen and Allan (1966) recently summarized this new data, and also data provided by the Soviet research vessels *Ob*, *Vavilov* and *Vityaz*, and published the first modern physiographic map of the Eastern Mediterranean. Data obtained from the *Chain*, *Atlantis*, *Yamacraw*, *Vema* and *Compass Island* formed the basis of studies of the sediments of the Tyrrhenian Abyssal Plain (Ryan *et al.*, 1965) and the Ionian Sea submarine canyons and the turbidity currents (Ryan and Heezen, 1965). Data obtained from these reports was used to prepare a physiographic map of the Mediterranean (Fig. 2). Data on the

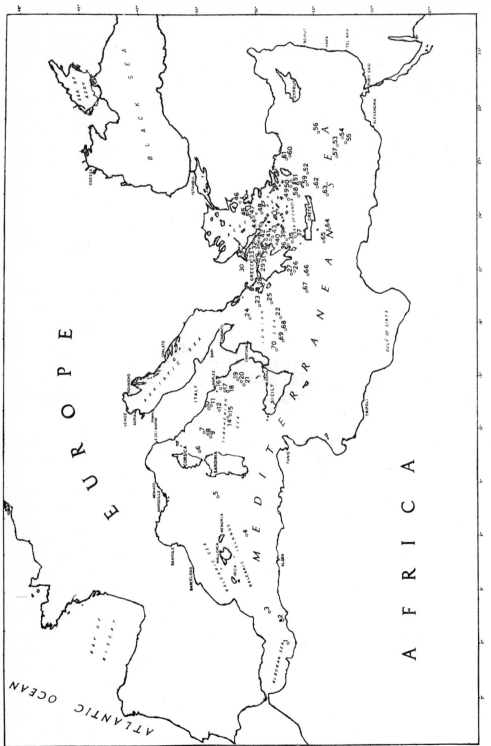

Fig. 1. Map of the Mediterranean Sea showing location of *Vema* (Cruise 10) stations.

position and depth of the cores used in this study was provided by the Lamont Geological Observatory and formed the basis of the location map (Fig. 1).

THE MEDITERRANEAN-ATLANTIC FAUNAS

Mediterranean Sea

The Mediterranean Sea (Fig. 2) is treated here as a unit although it is realized that the boreal North Atlantic fauna extends into the Mediterranean through the Straits of Gibraltar.

The surface temperature of the water during summer in the western Mediterranean is 20°–25°C (subtropical), with a greater change in a seasonal temperature than the eastern part which is tropical (summer temperature 25°–27°C). Salinity in the western part is $+38\%_0$ while in the eastern tropical part it is $40\%_0$. Subsurface salinity is contoured on Fig. 3. The Mediterranean Sea has a much lower fertility than the neighbouring Atlantic. The bottom sediments consist of sands, clays, muds and shell sands. Vegetation of the coastal areas in depths up to 100 m consists of *Posidonia*, calcareous algae and fibrous algae. This uniformity in bottom conditions has resulted in a remarkably similar fauna from the Baleares on the west to the Levantine Coast on the east. Ecologically the ostracod fauna has been studied in much greater detail than anywhere else in the world.

The region has provided classic works, such as G. W. Müller (1894), Rome (1939, 1942, 1964), Reys (1961, 1963) and Hartmann (1954).

The Western Mediterranean

Phytal

The fauna associated with bottom vegetation, east of the Balearic Islands, consists predominantly of species of *Paradoxostoma*, *Xestoleberis*, *Cytherois*, *Sclerochilus* and *Paracytherois*. The following species commonly occur from Marseille on the west (Reys, 1961, 1963) through Monaco (Rome, 1939, 1942, 1965) to the Naples region Puri *et al.*, 1965) on the east:

Paradoxostoma intermedium Müller (Fig. 4), *P. taeniatum* Müller (Fig. 5), *P. caecum* Müller (Fig. 6), *P. rarum* Müller (Fig. 7), *P. breve* Müller (Fig. 8), *P. incongruens* Müller (Fig. 9), *Xestoleberis parva* Müller, *X. fuscomaculata* Müller, *X. communis* Müller (Fig. 10), *X. labiata* Brady and Robertson, *X. margaritea* (Brady), *X. decipiens* Müller, *Cytherois frequens* Müller, *Sclerochilus abbreviatus* Brady and Robertson, *S. aequus* Müller (Fig. 11) and *Paracytherois striata* (Müller).

Calcareous Algae: The following fauna associated with the coralline algae in the Marseille area also occurs as far east as the Gulf of Naples:

Aurila convexa (Baird) (Fig. 12), *Hemicytherura videns* (Müller) (Fig. 13), *Loxoconcha rhomboidea* (Fischer), *Xestoleberis parva* Müller, *X. pellucida* Müller (Fig. 14), *Paradoxostoma intermedium* Müller, *P. rarum* Müller, and *Cytherois frequens* Müller.

The faunal uniformity of the Western Mediterranean is shown by the following species which commonly occur in Marseille, Monaco and the Gulf of Naples:

Aurila convexa (Baird), *Hemicytherura videns* (Müller), *Xestoleberis parva* Müller, *X. margaritea* Brady, *X. decipiens* Müller, *X. fuscomaculata* Müller, *X. communis* Müller, *Paradoxostoma intermedium* Müller, *P. taeniatum* Müller, *P. caecum* Müller, *P. rarum* Müller, *P. breve* Müller, *P. incongruens* Müller, *Cytherois frequens* Müller, *Sclerochilus abbreviatus* Brady and Robertson, *S. aequus* Müller and *Paracytherois striata* Müller. [Text continued on p. 373]

2 A

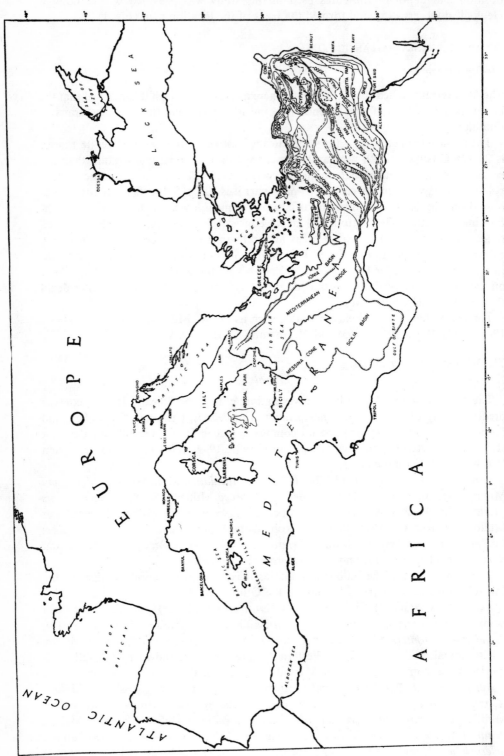

Fig. 2. Physiographic map of the Mediterranean Sea showing boundaries of physiographic provinces and bottom contours at 500 m intervals (data from Emery, Heezen and Allan, 1966; Ryan,

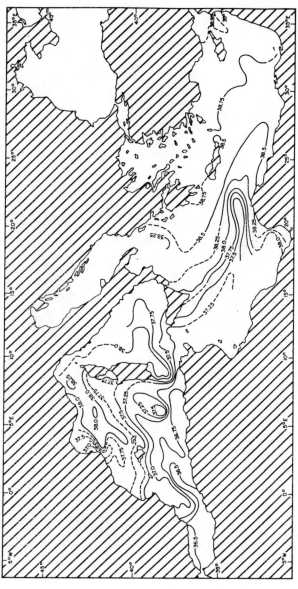

Fig. 3. Map of the Mediterranean Sea showing subsurface salinity (after Lacombe and Tchernia, 1960).

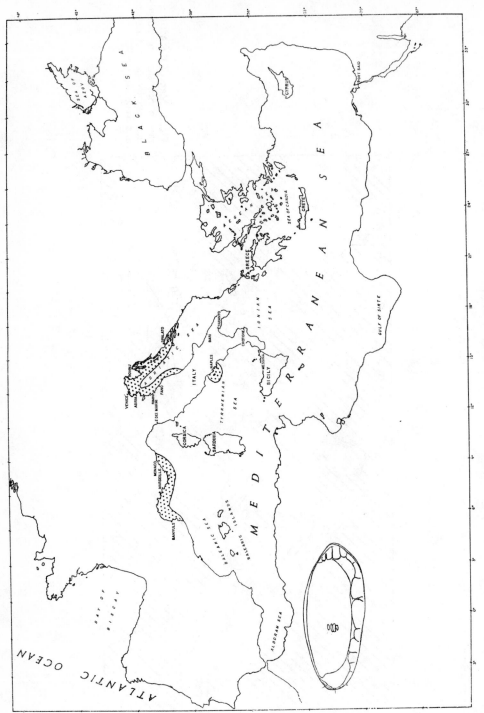

Fig. 4. Distribution of *Paradoxostoma intermedium* Müller in the Mediterranean.

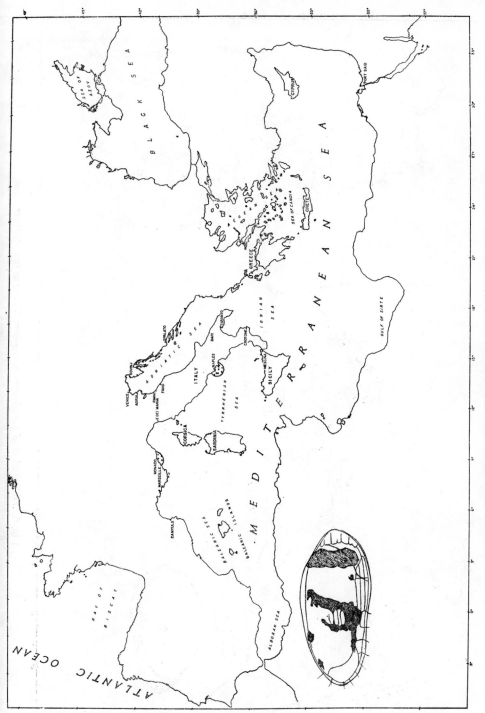

Fig. 5. Distribution of *Paradoxostoma taeniatum* Müller in the Mediterranean.

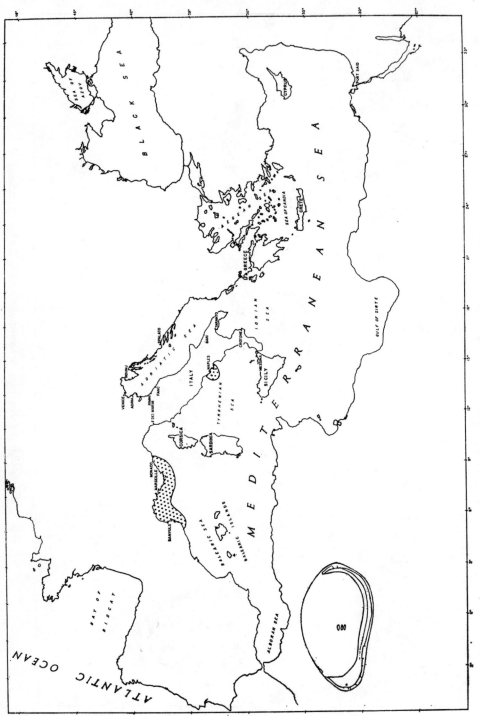

Fig. 6. Distribution of *Paradoxostoma caecum* Müller in the Mediterranean.

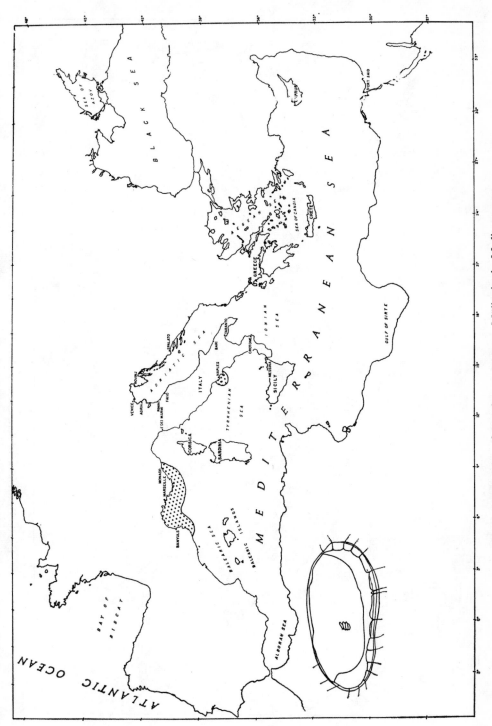

Fig. 7. Distribution of *Paradoxostoma rarum* Müller in the Mediterranean.

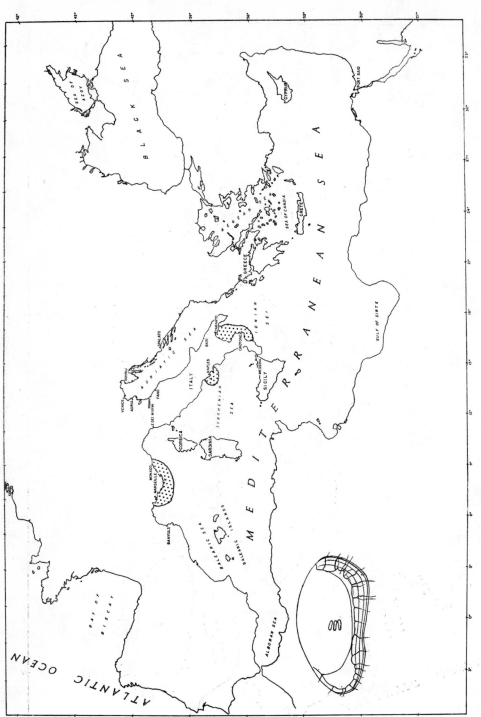

Fig. 8. Distribution of *Paradoxostoma breve* Müller in the Mediterranean.

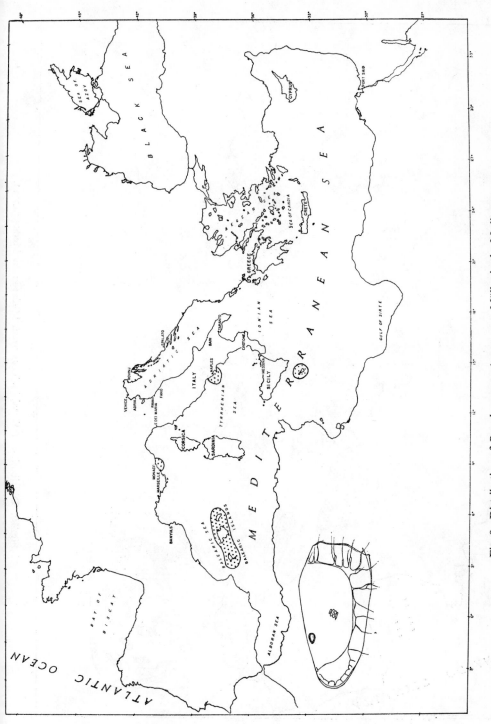

Fig. 9. Distribution of *Paradoxostoma incongruens* Müller in the Mediterranean.

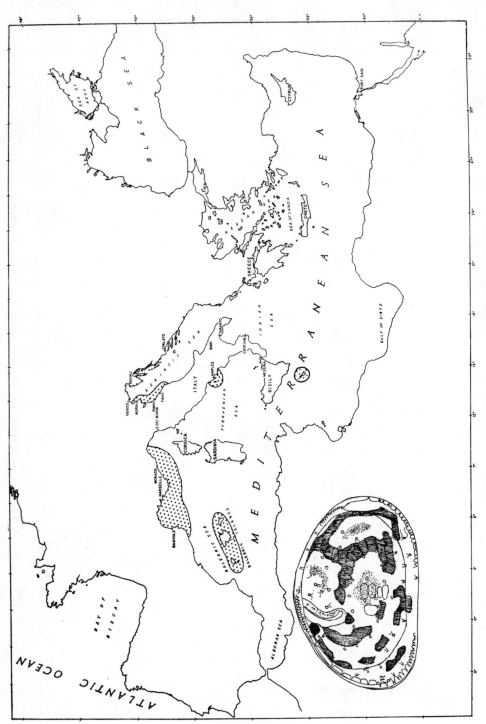

Fig. 10. Distribution of *Xestoleberis communis* Müller in the Mediterranean.

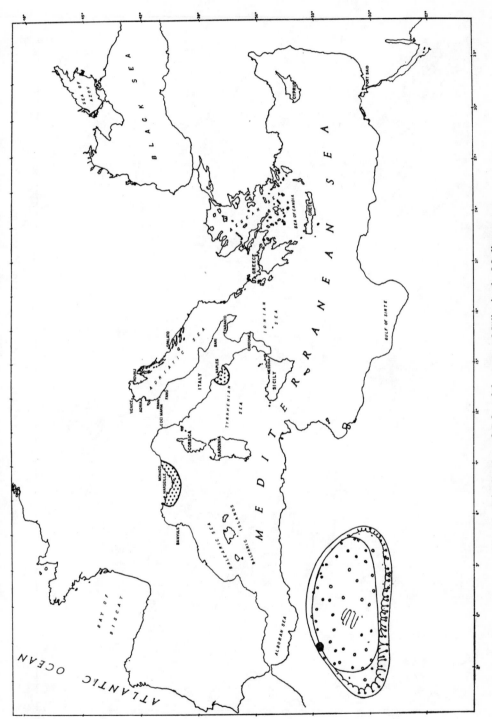

Fig. 11. Distribution of *Sclerochilus aequus* Müller in the Mediterranean.

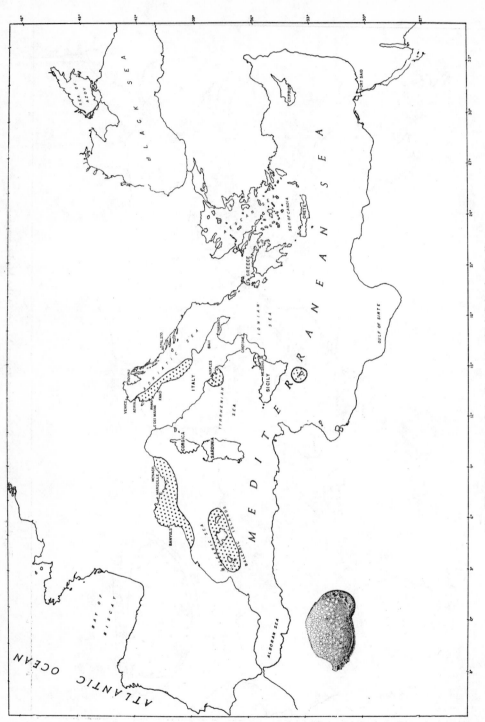

Fig. 12. Distribution of *Aurila convexa* (Baird) in the Mediterranean.

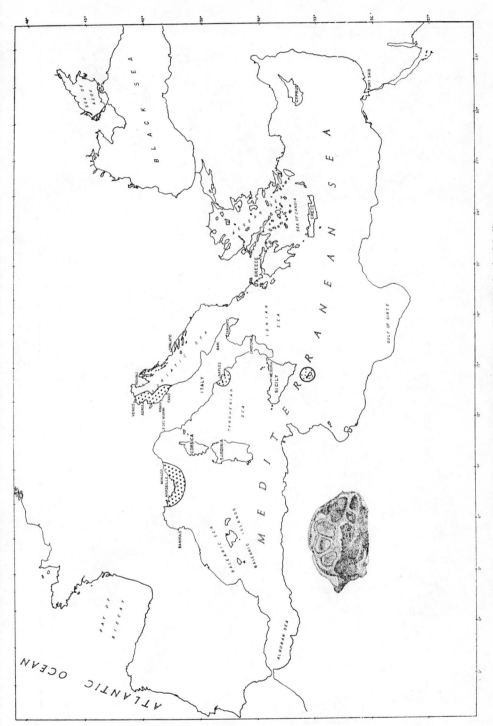

Fig. 13. Distribution of *Hemicytherura videns* (Müller) in the Mediterranean.

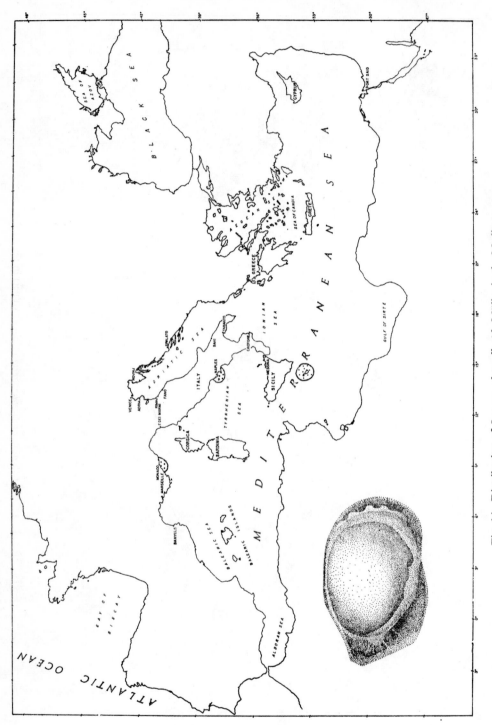

Fig. 14. Distribution of *Loxoconcha pellucida* Müller in the Mediterranean.

For discussion in this paper, the western Mediterranean is divided into Baleares, Tyrrhenian Sea, Adriatic Sea and Ionian Sea.

Baleares. In the French Riviera, Rome (1939, 1942, 1965) studied the ostracod fauna near Monaco and discovered a marked similarity between the fauna of that area and that of the Gulf of Naples. Eighty-six species, representing 26 genera, were found to be common to the two areas. Rome's samples were collected at various depths to below 400 m. The majority of these samples, however, were collected from shallow waters of less than 30 m depth. Bottom conditions in the Monaco area are similar to those of the Gulf of Naples, with sands and plants giving way to clays with increasing depth.

In a recent study (Reys, 1961, 1963) in the vicinity of Marseille, in the Golfe du Lion, based on 37 samples from 5 biocoenoses, 85 species, distributed over 22 genera were reported.

There is remarkable similarity between faunas around the Baleares, Banyuls, Monaco, Marseille and the Gulf of Naples. The association of ostracods with certain types of vegetation and substrate is well established from studies by G. W. Müller (1894), De Buen (1916), Rome (1939, 1942, 1965), Reys (1961), Hartmann (1953) and Puri *et al.*, (1965).

The following species occur in all the five areas (Baleares, Banyuls, Monaco, Gulf of Naples and Marseille):

Loxoconcha rhomboidea (Fischer), *Xestoleberis labiata* Brady and Robertson, *X. margaritea* Brady, *X. communis* Müller.

The following species occur in the above four areas except Banyuls:

Xestoleberis decipiens Müller, *Paradoxostoma incongruens* Müller, *Cytherois frequens* Müller and *Paracytherois striata* Müller.

The following species occur in all the above areas except Baleares:

Xestoleberis parva Müller, *X. fuscomaculata* Müller, *X. communis* Müller, *Paradoxostoma intermedium* Müller, *P. caecum* Müller and *P. rarum* Müller.

The following 20 species occur in Monaco, Gulf of Naples and Marseille:

Aurila convexa (Baird), *Hemicytherura videns* (Müller), *Xestoleberis parva* Müller, *X. labiata* Brady and Robertson, *X. margaritea* Brady, *X. decipiens* Müller, *X. fuscomaculata* Müller, *X. communis* Müller, *Paradoxostoma intermedium* Müller, *P. taeniatum* Müller, *P. caecum* Müller, *P. rarum* Müller, *P. breve* Müller, *P. incongruens* Müller, *Cytherois frequens* Müller, *Sclerochilus abbreviatus* Brady and Robertson (= *S.* (?) *levis* Müller), *S. aequus* Müller and *Paracytherois striata* Müller.

The following species are associated with coralline algae in the Marseille area:

Bairdia raripila Müller, *Aurila convexa* (Baird), *Hemicytherura videns* (Müller), *Loxoconcha rhomboidea* (Fischer), *Xestoleberis parva* Müller, *X. labiata* Brady and Robertson, *X. pellucida* Müller, *X. margaritea* Brady, *X. fuscomaculata* Müller, *Paradoxostoma intermedium* Müller, *P. parallelum* Müller, *P. fuscum* Müller, *P. taeniatum* Müller, *P. caecum* Müller, *P. rarum* Müller, *P. atrum* Müller, *P. breve* Müller, *Cytherois frequens* Müller.

In the Baleares, 5 cores (V10–1, depth 1365 m; V10–2, V10–3, depth 2556 m; V10–4, depth 2260 m; and V10–5, depth 2723 m) were examined of which one core (V10–3, depth 2556 m) did not contain ostracods. The abyssal assemblage represented by these samples is dominated by species of *Krithe* and *Bythocypris obtusata*.

A mixture of both abyssal and shallow marine forms was encountered in V10–2, where such shallow-water forms as species of *Callistocythere* and *Aurila* occur at a much greater depth. The following species were encountered at the various stations:

V10–1   (1365 m)   *V10–4*   (2260 m)
Krithe sp. *1*   *Bythocypris obtusata*
                                      Polycopsis n. sp. 1
V10–2   *Propontocypris* sp.
Aurila sp.   *?Trachyleberidea* n. sp.
Callistocythere sp.
Cytheropteron aff. *C. alatum*   *V10–5*   (2723 m)
Hemicytherura videns   *Bythocypris obtusata*
Krithe sp. 1   *Polycope* sp. 1
Loxoconcha sp.   *Pontocypris* sp.
Paracypris sp.

V10–3   (2556 m)
No ostracods

Tyrrhenian Sea. The Gulf of Naples (Fig. 15) has been studied in detail by Müller (1894); and Puri *et al.* (1965). Because of very little rainfall in this region, surface run off has very little influence on the water of the Gulf. Consequently the open-sea character due to the connection with the Mediterranean is even more pronounced. The bottom sediments are derived mostly from volcanic tuff except in the Sorrentine Peninsula and Capri, where erosion of sedimentary rocks provides some clastic material. The bottom temperature at depths greater than 50 m is uniform (14°–15°C).

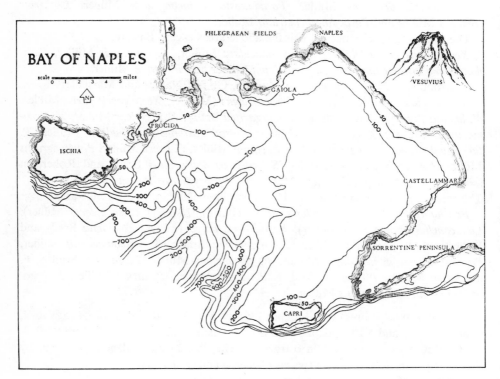

Fig. 15. Gulf of Naples showing depth (contour interval 100 m, 50 m contour also shown).

Hydrogen ion concentration (pH) is between 7·7 and 8·20. Bottom salinity varies between 37‰ around the coast to 38·50‰ at the connection with the Mediterranean. Transparency of the water is as high as 37 m below surface in the middle of the Gulf.

Lagoons

Lago di Patria is a shallow water (average depth 2 m) littoral lagoon: salinity seldom exceeds 13 to 14‰, minimum temperature is 7° to 8°C and maximum temperature is 30°C (Sacchi, 1961). Distribution of ostracods from this area consists of 16 species of a typical brackish-assemblage which is shown on Fig. 16 (data from McKenzie, 1963).

Shelf

Three shallow shelf (depth up to 100 m) inner neritic assemblages are recognized (Puri *et al.*, 1965); these consist of:

Aurila convexa (Baird), *Buntonia sublatissima* (Neviani), *Carinocythereis carinata* (Roemer), *C. antiquata* (Baird), *C. quadridentata* (Baird), *Costa runcinata* (Brady), *Cuneocythere* n. sp. *A*, *Cytheridea neapolitana* Kollmann, '*Cythereis*' *polygonata* Rome, *Cytheridea* n. sp. *A*, *Eucythere declivis* (Norman), *Hemicytherura videns* (Müller), *Krithe similis* Müller, *Leptocythere bacescoi* (Rome), *L.* n. sp. *C.*, *Mutilus speyeri* (Brady), *Neocytherideis foveolata* (Brady), *Quadracythere* (*?*) *prava* (Baird) and *Urocythereis margaritifera* (Müller).

The following five species are restricted to depths up to 50 m or attain their maximum development in these depths and may represent a distinct assemblage:

Costa batei (Brady), *Cushmanidea elongata* (Brady), *C. turbida* (Müller), *Krithe reniformis* (Brady), and *Semicytherura incongruens* Müller.

Ostracods associated with bottom vegetation like *Posidonia*, calcareous algae, algae, seaweeds, *Phyllochaetopterus socialis*, and bottom animals, like sponges, are here considered as separate assemblages although it is realized that these assemblages are confined to depths less than 100 m. In most cases *Posidonia* occurs on shallow banks around Ischia, Procida, Naples and the Sorrento Peninsula; while calcareous algae are confined to Ischia bank, Gulf of Pozzuoli, south of Naples, north of Castellammare and Bocca Piccola. In some places, however, *Posidonia* and calcareous algae occur together and it is impossible to delineate their boundaries either by bottom vegetation or by their ostracod assemblage. The boundaries of *Posidonia*, calcareous algae and fibrous algae ostracod assemblages coincide with the limits of their vegetation pattern.

The following species are associated with *Posidonia*:

Basslerites teres (Brady)
Cylindroleberis mariae (Baird)
Cylindroleberis teres Norman
Cytherelloidea sordida (Müller)
Cytheretta rubra Müller
Macrocypris succinea Müller
Neocytherideis foveolata (Brady)
Paracypris complanata (Brady and Robertson)
Philomedes aspera Müller
Polycope dentata Brady
Polycope dispar Müller
Polycope fragilis Müller

Polycope reticulata Müller
Polycope serrata Müller
Polycope striata Müller
Polycope tuberosa Müller
Pontocypria spinosa Müller
Pontocypris declivis Müller
Pontocypris levis Müller
Pontocypris mediterranea Müller
Pontocypris setosa Müller
Pontocypris succinea Müller
Pontocypris subfusca Müller
?Sarsiella capsula Norman

2 B

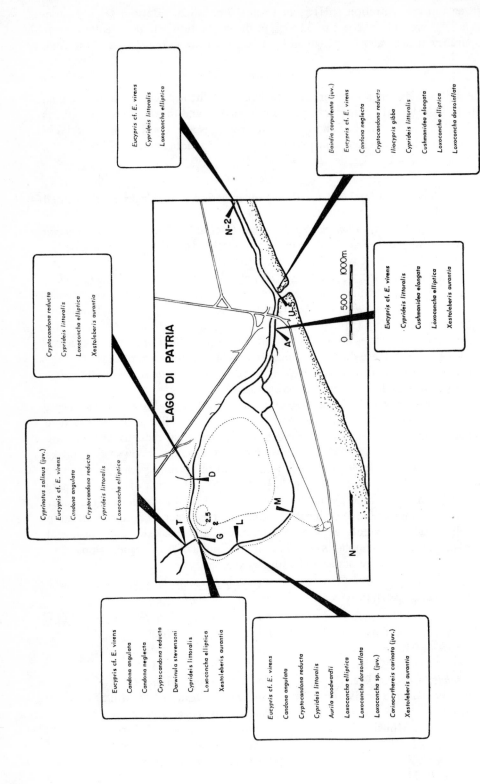

Fig. 16. Distribution of brackish-water assemblage in the Lago di Patria (data from MacKenzie, 1963).

The following species are associated with calcareous algae:

Cylindroleberis mariae (Baird)

Cylindroleberis teres Norman

Cypridina mediterranea Costa

Macrocypris succinea Müller

Philomedes interpuncta (Baird)

Philomedes levis Müller

Polycope maculata Müller

Polycope rostrata Müller

Polycope striata Müller

Polycope tuberosa Müller

Polycopsis compressa Müller

Pseudophilomedes angulata Müller

Pseudophilomedes foveolata Müller

?Sarsiella capsula Norman

Off-Shore Assemblage and Depth Zonation

Off-shore inner neritic assemblages (depth over 100 m) can be separated into three depth zones: 100 m to 200 m; 200 m to 300 m; and over 300 m.

The following are indicative of 100 m to 200 m depth zone:

Buntonia giesbrechti (Müller)

Buntonia sublatissima dertonensis Ruggieri

Cuneocythere n. sp. *C*

Cytherella n. sp.

Cytheropteron ? crassipinnatum Brady and Norman

Cytheropteron latum Müller

Cytheropteron n. sp. *H.*

Cytherura dispar Müller

Loxoconcha decipiens Müller

Loxoconcha napoliana Puri

Loxoconcha n. sp. *A*

Loxoconcha n. sp. *D*

Loxoconcha versicolor Müller

Pterygocythereis jonesii (Baird) and subspecies *ceratoptera* and *fimbriata.*

Semicytherura acuminata (Müller)

Semicytherura acuticostata (Sars)

Tetracytherura angulosa (Seguenza)

The following species are indicative of 200 m to 300 m depth zone:

Bosquetina carinella (Reuss)

Cytheropteron alatum Sars

Henryhowella sarsi (Müller)

Krithe n. sp. *B*

Pseudocythere caudata Sars

Semicytherura paradoxa (Müller)

Trachyleberid n. gen. *D.* n. sp. *A*

The following species occur in depths over 300 m:

Cytheropteron rotundatum Müller

Cytherura rara Müller

Kangarina abyssicola (Müller)

From beach sand at Forte dei Marmi, Tuscany, Ruggieri (1955) described *Tyrrhenocythere pignattii* Ruggieri.

In the Tyrrhenian Sea, 14 cores, V10–6 (depth 818 m), V10–7 (depth 1387 m), V10–8 (depth 1339 m), V10–9 (depth 1043 m), V10–10 (depth 543 m), V10–11 (depth 192 m), V10–12 (depth 1884 m), V10–15 (depth 3439 m), V10–17 (depth 1033 m), V10–18 (depth 265 m), V10–19 (depth 150 m), V10–20 (depth 256 m) and V10–21 (depth 1266 m), were examined and one core (V10–15) did not yield any ostracods although a foraminiferal fauna was present. Core V10–19 (depth 150 m) and V10–11 (depth 192 m) have yielded an offshore shelf assemblage. The following species are present:

V10–19   (150 m)

Argilloecia acuminata
Argilloecia bulbifera
Bairdia? 'mediterranea'
Bosquetina dentata
Bythocypris bosquetiana
Bythocypris n. sp. (= *B. ?bosquetiana*)
Bythocypris obtusata
Bythocythere ? insignis
Carinocythereis laticarinata
Costa edwardsi
Cytherois sp.
Cytheropteron alatum
Cytheropteron aff. *C. alatum* (with
   punctations)
Eucytherura complexa
Eucytherura n. sp. *1*
Henryhowella sarsi
Loxoconcha decipiens
Loxoconcha ovulata

Loxoconcha n. sp. *F*
Loxoconcha n. sp. *X*
Microcythere hians
Microcythere inflexa
Microcythere n. sp. *X*
?Microcythere sp.
Paracytherois oblonga
Paracytherois cf. *P. oblonga*
Paracytherois striata
Paradoxostoma n. sp. *X*
Polycope cf. *P. dentata*
Polycope dispar
Polycope frequens
Polycope reticulata
Polycope rostrata
Polycope ? striata
Pontocypris n. sp. *B*
Pseudocythere caudata

V10–11   (192 m)

Argilloecia sp. *1*
Aurila convexa
Aurila sp. *H*
Bairdia ? 'mediterranea'
Bosquetina ? dentata
Callistocythere cf. *C. diffusa*
Cythere sp.
Cytheropteron sp.
Eucythere declivis
Hemicytherura defiorei

Henryhowella sarsi
Leptocythere sp.
Loxoconcha decipiens
Loxoconcha aff. *L. impressa* (= *?L.
   obliquata* Seguenza)
Loxoconcha sp. *1*
Paracytherois oblonga
Pseudocythere caudata
Semicytherura acuticostata
Xestoleberis sp.

The above assemblages are mixed with both shallow-water forms (e.g. *Aurila convexa*) and deeper water species (e.g. *Bythocypris obtusata*, species of *Argilloecia*).

   The following bathyal assemblage is represented by three cores (V10–10, depth 543 m, V10–18, depth 265 m, and V10–20, depth 256 m):

V10–20   (256 m)

Bairdia ? 'mediterranea'
Bythocypris bosquetiana (fossil)

V10–10   (543 m)

Argilloecia acuminata
Henryhowella sarsi
Polycope n. sp. *1*

V10–18   (265 m)

Argilloecia acuminata
Argilloecia bulbifera
Argilloecia minor
Bosquetina dentata
Bythocypris bosquetiana

Bythocypris obtusata
Bythocypris n. sp. *B*
Carinocythereis laticarinata
Cytheropteron? punctatum
Erythrocypris cf. *E. serrata*

Loxoconcha cf. *L. impressa*
Loxoconcha n. sp. *F*
Loxoconcha n. sp. *X*
Microcythere gibba
Microcythere *? hians*
Microcycyere *? inflexa*
Microcythere cf. *M. nana*
Monoceratina n. sp.
Paracytherois *? oblonga*
Paracytherois sp.
Paradoxostoma cylindricum

Polycope cf. *P. dentata*
Polycope *? dispar*
Polycope fragilis
Polycope frequens
Polycope n. sp. *A*
Polycope reticulata
Polycope rostrata
Pontocypris n. sp. *B*
Pseudocythere caudata
Trachyleberid n. g. D, n. sp. A
Xestoleberis n. sp. A

Abyssal Fauna

The abyssal fauna in the Tyrrhenian Sea is represented by two distinct assemblages, one west of the Tyrrhenian Abyssal Plain (represented by C10–6, depth 818 m, C10–7, depth 1387 m, and C10–8, depth 1337 m) and the other east of this abyssal plain (represented by C10–9, depth 1043 m, V10–12, depth 1884 m, V10–16, depth 986 m V10–17, depth 1033 m and V10–21, depth 1266 m). A new species of a deep water *Cytherella* (n. sp. *A*) was found to be restricted to the assemblages west of the Tyrrhenian Abyssal Plain. The following species were encountered at the various stations:

V10–7 (1387 m)
Argilloecia acuminata
Bythocypris obtusata
Caudites n. sp. *A*
Cytherella n. sp. *A*
Polycope n. sp. *1*

V10–8 (1337 m)
Argilloecia acuminata
Bythocypris obtusata
Cytherella n. sp. *A*
Polycope cf. *P. dentata*
Pterygocythereis jonesii
Xestoleberis *? rara*

V10–6 (818 m)
Argilloecia acuminata
Argilloecia *? clavata*
Argilloecia sp. *M*
Bythocypris obtusata
Cytherella n. sp. *A*
Erythrocypris cf. *E. serrata*
Polycope n. sp. *3*
Polycope reticulata
Polycope n. sp. *1*

The following abyssal fauna was encountered in the east of the Tyrrhenian Abyssal Plain:

V10–9 (1043 m)
Bosquetina sp.
Bythocypris obtusata
Henryhowella sarsi
Polycope n. sp. *1*
Pterygocythereis sp.

V10–15 (3439 m)
No ostracods

V10–12 (1884 m)
Argilloecia acuminata
Argilloecia sp. *1*
Aurila convexa
Bosquetina sp.
Bythocypris obtusata
Cytheropteron alatum
Cytherura aff. *C. costata*
Cytherura dispar
Krithe sp. 1
Loxoconcha cf. *L. impressa*
Polycope *? frequens*
Semicytherura sp.

V10–16   (986 m)
Argilloecia acuminata
Bythocypris obtusata
Xestoleberis sp.

V10–17   (1033 m)
Argilloecia acuminata
Bythocypris bosquetiana
Bythocypris obtusata
Cytherella ? vulgata
Cytherura rara
Cytherura ? rara
Polycope n. sp. *1*

V10–21   (1266 m)
Argilloecia acuminata
Argilloecia ? caudata
Argilloecia ? minor
Bythocypris obtusata
Krithe sp. *1*
Polycope cf. *P. dentata*
Polycope ? fragilis
Polycope frequens (broken)
Polycopsis n. sp. *1*

The Adriatic Sea. The Adriatic is bordered on the east by Yugoslavia and Albania and on the west by the Italian coast; in the south it connects with the Ionian Sea. In the northern part, shallow depths (up to 80 m) are encountered, and the middle is moderately deep (80 to 200 m); the greatest depths (up to 1200 m) are encountered in the southern part. There is very little drainage into the sea except for the Po River, which flows into the sea south of Venice. Surface and salinity are normal open sea (35 to 38‰) and are fairly uniform. Bottom sediments in the northern part are variable and consist of mostly sand and argillaceous sands and clay. The central part shows a predominantly muddy bottom with occasional sand and clay; the southern part is mostly clay except in the southernmost part which exhibits a mixture of sand, mud and clay (Ascoli, 1965).

Ostracods of the Adriatic have been studied by Klie (1942), Ruggieri (1952) and Ascoli (1965).

Data from bottom cores throughout the Adriatic have enabled Ascoli (1965) to recognize the following 4 depth zones:

'*Inner Sublittoral*' (31 to 42 m) is a shallow inner neritic assemblage which is dominated by *Cytheridea neapolitana* Kollmann, *Cushmanidea elongata* (Brady) and *Semicytherura incongruens* (Müller). This assemblage inhabits a predominantly sandy bottom, with clay and mud.

'*Transitional*' (42 to 74 m) zone consists of species which are common in the 31 to 42 m assemblage together with an increase in the population of *Pterygocythereis*, *Callistocythere diffusa* (Müller) and *Henryhowella*. This assemblage does not show a preference for a definite sediment type.

'*Outer Sublittoral*' (74 to 243 m) is characterized by *Bairdia*, *Loxocythere* and *Loxoconcha tamarindus* (Baird). *Bosquetina dentata* (Müller), *Henryhowella* and *Cytherella* are frequent.

'*Bathyal*' (243 to 1192 m) is characterized by species of *Macrocypris*, *Argilloecia* and *Krithe*.

The following species are common between the Adriatic, Gulf of Naples, Monaco, and Marseille:

Hemicytherura videns (Müller), *Loxoconcha rhomboidea* (Fischer), *Xestoleberis labiata* Brady and Robertson, *Xestoleberis pellucida* Müller, *Xestoleberis decipiens* Müller, *Xestoleberis communis* Müller, *Paradoxostoma intermedium* Müller, *Paradoxostoma fuscum* Müller, *Paradoxostoma rotundatum* Müller, *Paradoxostoma atrum*

Müller, *Sclerochilus abbreviatus* Brady and Robertson, *Pontocypris pirifera* Müller, *Callistocythere lobiancoi* (Müller) and *Neocytherideis foveolata* (Brady).

Ionio Basin. In the Ionio Basin east of the Mediterranean Ridge, three cores, V10–23 (depth 3466 m), V10–24 (depth 2250 m), and V10–25 (depth 3101 m) were studied. These sediments have yielded an abyssal fauna which as a composite assemblage consists of species of *Polycope, Loxoconcha ? impressa, Bythocypris obtusata, Krithe* n. sp., and *Argilloecia acuminata.* The following species were encountered at the various stations:

V10–23   (3466 m)
Polycope sp. (broken)

V10–25   (3101 m)
Loxoconcha cf. *L. impressa*
Polycope sp.

V10–24   (2250 m)

Argilloecia acuminata
Bairdia serrata
Bythocypris obtusata
Cytheropteron alatum
Erythrocypris cf. *E. serrata*
Krithe sp. *1*

Loxoconcha cf. *L. impressa*
Monoceratina n. sp. *2*
Polycope cf. *P. dentata*
Polycope frequens
Polycope reticulata

Strait of Messina. This strait with a narrow passage only $1\frac{1}{2}$ miles wide in the north, widens to almost 8 miles in the south. Waters from the Tyrrhenian and Ionian Seas mix through these straits and cause strong currents. Three traverses were made through the strait in the spring 1961 from the Port of Messina to a depth of 50 m. Ryan and Heezen (1965) give a sill depth of less than 200 m.

Sediments eroded from the coasts of Calabria and Sicily are swept through the strait by northward flowing under current and deposited at various depths. *Posidonia* commonly occurs at depths of 35 m and lower depths.

Seguenza (1883–86) described the following species from the Port of Messina area:

Asterope gracilis Seguenza
Argilloecia messanensis Seguenza
Bairdia complanata Brady
Bairdia expansa Brady
Bairdia formosa Brady
Bairdia messanensis Seguenza
Bairdia subdeltoidea Münster
Cypridina messanensis Claus
Cythere albomaculata Baird
Cythere convexa Baird
Cythere crispata Brady
Cythere edwardsii Roemer
Cythere emaciata Brady
Cythere jonesii Baird
Cythere prava Brady
Cythere speyeri Brady
Cythere stimpsoni Brady
Cythere tenera Brady
Cythere tuberculata Sars
Cythere venus Seguenza

Cythere venus messanensis Seguenza
Cythere woodwardi Brady
Cytherella areolata Seguenza
Cytherella calabra Seguenza
Cytherella punctata Brady
Cytheridea punctillata Brady
Cytheridea torosa (Jones)
Cytherideis gracilis (Reuss)
Cytherura acuticostata Sars
Cytherura biproducta Seguenza
Cytherura calcarata Seguenza
Cytherura exagonalis Seguenza
Cytherura ornata Seguenza
Cytherura quadrata Norman
Cytherura speciosa Seguenza
Cytherura striata Sars
Loxoconcha avellana Brady
Loxoconcha lata Brady
Loxoconcha seminulum Seguenza
Loxoconcha tenuis Seguenza

Loxoconcha tumida Brady
Macrocypris elongata Seguenza
Macrocypris gracilis Seguenza
Macrocypris setigera Brady
Macrocypris trigona Seguenza
Paracypris polita Sars
Pontocypris faba? (Reuss) Brady
Pontocypris interposita Seguenza
Pontocypris mytiloides (Norman)
Pontocypris polita Seguenza

Pontocypris punctata Seguenza
Pontocypris trigonella Sars
Xestoleberis compressa Seguenza
Xestoleberis curta? Brady
Xestoleberis depressa Sars
Xestoleberis intermedia Brady
Xestoleberis labiata Brady and Robertson
Xestoleberis margaritea Brady
Xestoleberis producta Seguenza
Xestoleberis saccata Seguenza

The lectotypes of Seguenza's species were destroyed in the 1908 earthquake. P. Ascoli is currently engaged in designation of neotypes of the new species described by Seguenza. A complete analysis of the ostracod fauna in this area must await until the taxonomic and nomenclatural problems in the Seguenza study are resolved.

Sicilia Basin. The Mediterranean Ridge divides the Ionian Sea into two basins; the Ionio Basin and Sicilia Basin (Ryan and Heezen, 1965). Five cores, V10–22 (depth 3427 m), V10–67 (depth 2906 m), V10–68 (depth 3538 m), V10–69 (3089 m) and V10–70 (2499 m) have yielded an abyssal assemblage which consists of species of *Polycope*, *Bythocypris obtusata* and *Bythocypris bosquetiana*. The following species are present at the various stations:

V10–22 (3427 m)
Bythocypris bosquetiana
Cytheropteron ? rotundatum
Polycope cf. *P. dentata*

V10–68 (3538 m)
Polycope cf. *P. dentata*
Polycope frequens
Polycope reticulata

V10–67 (2906 m)
Bythocypris obtusata
Polycope n. sp. *4*

V10–69 (3089 m)
Polycope frequens

V10–70 (2499 m)
? ?Conchoecia sp.
Polycope frequens

Malta Area. A sample of clay dredged from a depth of 100 fathoms south of the Island of Malta, has yielded an assemblage in which the following species are common:

Bosquetina carinella (Reuss)
Cytheropteron alatum Sars
Cytheropteron rotundatum Müller
Cytherura dispar Müller
Eucytherura gibbera Müller
Hemicytherura videns (Müller)
Henryhowella sarsi (Müller)
Loxoconcha decipiens Müller
Loxoconcha pellucida Müller
Mutilus (*?*) *speyeri* (Brady)
Polycope cf. *P. dentata* Brady

Polycope dispar Müller
Polycope reticulata Müller
Polycope rostrata Müller
Polycope striata Müller
Pterygocythereis fimbriata (Münster)
Pterygocythereis jonesii (Baird)
Semicytherura acuticostata (Sars)
Semicytherura mediterranea (Müller)
Semicytherura paradoxa (Müller)
Semicytherura punctata (Müller)

During the summer of 1966, 8 grab samples were collected north-east of the Isle of Malta, in two traverses, one from Valetta Harbour and the other from St Paul's Bay (see Fig. 17).

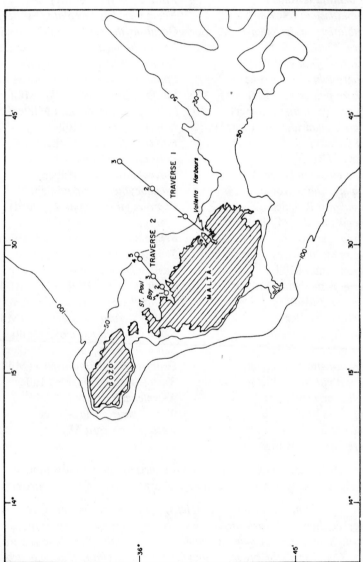

Fig. 17. Hydrographic map of Malta and Gozo area showing location of samples. Soundings in fathoms (from British Hydrographic Office, 1966, Chart 194).

The ostracod fauna from this area is very close to that of the Bay of Naples.
The following species commonly occur in the Malta Sill:

Argilloecia acuminata Müller
Argilloecia bulbifera Müller
Argilloecia caudata Müller
Argilloecia levis Müller
Aurila convexa (Baird)
Bairdia frequens Müller
Bairdia longevaginata Müller
Bairdia serrata Müller
Bosquetina dentata (Müller)
Buntonia sublatissima (Neviani)
Buntonia sublatissima dertonensis Ruggieri
Bythocythere ? insignis Sars
Callistocythere diffusa (Müller)
Callistocythere lobiancoi (Rome)
Callistocythere pallida (Müller)
Carinocythereis carinata (Roemer)
Carinocythereis quadridentata (Baird)
Costa batei (Brady)
Costa edwardsi (Roemer)
Cushmanidea elongata (Brady)
'*Cythereis*' *polygonata* Rome
Cytherella vulgata Ruggieri
Cytherelloidea sordida (Müller)
Cytherois frequens Müller
Cytheropteron alatum Sars
Cytheropteron latum Müller
Cytheropteron rotundatum Müller
Cytherura costata Müller
Cytherura dispar Müller
Eucythere declivis Norman
Eucytherura gibbera Müller
Hemicytherura defiorei Ruggieri
Hemicytherura videns (Müller)
Henryhowella sarsi (Müller)
Krithe reniformis Müller
Krithe similis Müller
Leptocythere bacescoi Rome

Loxoconcha decipiens Müller
Loxoconcha impressa Baird
Loxoconcha levis Müller
Loxoconcha ovulata Costa
Loxoconcha pellucida Müller
Loxoconcha versicolor Müller
Loxocythere angulosa (Seguenza)
Macrocypris succinea Müller
Mutilus speyeri (Brady)
Neocytherideis foveolata (Müller)
Occultocythereis lineata (Müller)
Paracytheridea bovettensis (Seguenza)
Paracytherois striata Müller
Paradoxostoma incongruens Müller
Paradoxostoma versicolor Müller
Polycope cf. *P. dentata* Müller
Polycope fragilis Müller
Polycope frequens Müller
Polycope reticulata Müller
Polycope rostrata Müller
Pseudocythere caudata Sars
Pterygocythereis jonesii (Baird)
Pterygocythereis n. sp. *1*
Semicytherura acuminata (Müller)
Semicytherura acuticostata (Sars)
Semicytherura alifera Ruggieri
Semicytherura cribriformis (Müller)
Semicytherura inversa (Seguenza)
Semicytherura mediterranea (Müller)
Semicytherura paradoxa (Müller)
Semicytherura punctata (Müller)
Urocythereis margaritifera (Müller)
Xestoleberis communis Müller
Xestoleberis dispar Müller
Xestoleberis plana Müller
Xestoleberis rara Müller

A shallow shelf (less than 108 m) is characterized by a sandy mud bottom with
calcareous algae. The ostracod assemblage characteristic of this environment is:

*Loxoconcha ovulata, Loxoconcha decipiens, Loxoconcha impressa, Aurila convexa,
Mutilus speyeri, Semicytherura incongruens, Semicytherura cribriformis, Urocythereis
margaritifera, Xestoleberis dispar, Xestoleberis communis, Xestoleberis rara, Xestoleberis plana, Quadracythere prava, Neocytherideis foveolata, Cushmanidea elongata,
'Cythereis' polygonata* and *Costa batei.*

An off-shore area (from 108 m to below 132 m) is characterized by a muddy bottom.

The ostracod assemblage characteristic of this environment consists of the following species:

Cytheropteron alatum, Cytheropteron rotundatum, Cytheropteron latum, Bythocypris ? insignis, Eucythere declivis, Pterygocythereis jonesii, Pterygocythereis sp. *1, Henryhowella sarsi, Buntonia sublatissima, Buntonia sublatissima dertonensis, Bosquetina carinella, Polycope reticulata, Polycope* cf. *P. dentata, Polycope rostrata, Polycope frequens, Argilloecia bulbifera, Argilloecia acuminata, Henryhowella sarsi, Cytherella vulgata.*

The distribution of species in this area is related to the type of bottom and depth. The above assemblages are related to the Gulf of Naples assemblages from similar bottom types and corresponding depths.

The Eastern Mediterranean

Aegean Province. This physiographic feature (see Fig. 2) as defined by Emery, Heezen and Allan (1966, p. 181) extends seaward from Greece and western Turkey towards the south and south-east, where it is bounded by the Pliny and Strabo trenches. So defined, it includes the Islands of Crete, Rhodes and countless small islands with insular shelves. Besides the small islands, this province is marked with several abyssal plains over 1500 m in depth and five of these appear on the bathymetric map.

Brady (1866) described 19 species and varieties from this area, 14 of them as new. Most of his assemblage came from Sponge sands at Levant, and consists of the following forms:

Bairdia crosskeiana Brady	*Cytherella punctata* Brady
Costa batei (Brady)	*Cytheridea margaritea* Brady
Cythere hodgei Brady	*Echinocythereis cribriformis* (Brady)
Cythere jurinei Von Münster	*Loculicytheretta pavonia* (Brady)
Cythere jurinei costata Brady	*Loxoconcha affinis* (Brady)
Cythere oblonga Brady (not *Cythere*	*Loxoconcha glabra* (Brady)
oblonga M'Coy 1844)	*Neocytherideis gracilis* (Reuss)
Cythere plicatula (Reuss 1850)	

Two species, *Neocytherideis nobilis* (Brady) and '*Cythere*' *scabra* Von Münster, were reported by Brady off Crete in 40 and 360 fathoms respectively. Brady also listed the following three species from shallow water from Smyrna (Izmir) on the western coast of Turkey: *Loxoconcha grisea* (Brady), *L. modesta* (Brady) and '*Cythereis*' *subcoronata* (Speyer).

In the Aegean Sea 15 cores were studied, with a depth range of 73 m to 860 m. Cores V10–29 to V10–34 are located in the vicinity of Athens. The shallowest core V10–29 (73 m) has yielded the following assemblage which consists of both mixohaline and euhaline species:

V10–30 (73 m)

Bairdia ? '*subdeltoidea*'	*Pterygocythereis jonesii*
Bosquetina dentata	*Pterygocythereis* n. sp. *1*
Candona candida	*Tyrrhenocythere* n. sp. *1*
Candona sp. *1*	*Xestoleberis dispar*
Cyprideis torosa	*Xestoleberis fuscomaculata*
Cytherella ? n. sp. A.	*Xestoleberis ? margaritea*

Core V10–34 (depth 422 m) has yielded an open shelf assemblage, which consists of the following species:

V10–34 (422 m)

Bosquetina dentata	*Microcythere* sp. *2*
Buntonia sublatissima dertonensis	*Paracytherois striata*
Cytherois ? n. sp. *A*	*Paradoxostoma versicolor*
Cytherura dispar	*Semicytherura acuticostata*
Henryhowella sarsi	*Xestoleberis pellucida*
Leptocythere n. sp. *L*	*Xestoleberis ? rara*
Microcythere sp. *1*	*Xestoleberis* sp.

The following bathyal assemblage is represented by cores V10–29 (depth 649 m), V10–31 (depth 790 m), V10–32 (depth 768 m) and V10–33 (depth 823 m) and this assemblage consists of the following species:

V10–29 (649 m)
Cytherella vulgata
Leptocythere n. sp. *L*
Loxoconcha sp.
Xestoleberis dispar

V10–32 (768 m)
Bosquetina aff. *B. dentata*
?Bythocypris sp.
Cytheropteron latum
Henryhowella sarsi
Leptocythere n. sp. *L*

V10–31 (790 m)
Argilloecia sp. 1
Bosquetina aff. *B. dentata*
Buntonia sublatissima dertonensis
Loxoconcha cf. *L. impressa*

V10–33 (823 m)
Eucytherura n. sp. *1*
Henryhowella sarsi

Cores V10–40, V10–41, V10–42, V10–43 are located south-east of Athens and have yielded two distinct assemblages. A near-shore assemblage is present in V10–41 (depth 135 m) which consists of following shallow shelf forms:

V10–41 (135 m)

Aurila convexa	*Loxoconcha ovulata*
Bairdia mediterranea	*Mutilus speyeri*
Bosquetina dentata	*Paracytheridea bovettensis*
Bythocythere ? insignis	*Semicytherura inversa* var.
Callistocythere cf. *C. diffusa*	*Semicytherura punctata*
Carinocythereis carinata var. (fossil)	*Semicytherura ? ruggierii*
Carinocythereis rubra	*Tetracytherura angulosa*
Cytherura rara	*Urocythereis margaritifera*
Loxoconcha mediterranea	*Xestoleberis ? aurantia*

The following near-shore assemblage, dominated by such shallow water forms as *Cytheridea neapolitana*, species of *Callistocythere* and *Leptocythere*, is present in V10–46 (depth 69 m) in the eastern part of the Aegean Sea:

V10–46 (69 m)

Bairdia 'subdeltoidea'	*Carinocythereis quadridentata*
Bosquetina dentata	*Costa edwardsi*
Buntonia sublatissima var.	*Cytherella?* n. sp.
Callistocythere cf. *C. diffusa*	*Cytherella vulgata*
Carinocythereis carinata	*Cytheridea neapolitana*

Cytheropteron latum
Cytheropteron rotundatum
Henryhowella sarsi
Leptocythere aff. *L. castanea*
Leptocythere sp.
Loxoconcha ovulata

Pterygocythereis jonesii
Pterygocythereis n. sp. *1*
Trachyleberis histrix
Xestoleberis decipiens
Xestoleberis dispar

A bathyal 'A' assemblage, which is dominated by species of *Polycope*, *Argilloecia* and *Pterygocythereis* is represented by V10–40, V10–42 and V10–43 which range in depth from 217 to 495 m. The following species are present:

V10–40 (495 m)
Argilloecia acuminata
Argilloecia clavata
Aurila n. sp. *X*
Bairdia 'subdeltoidea'
Buntonia sublatissima
Bythocypris obtusata
Carinocythereis ? quadridentata
? Costa n. sp. *1*
Henryhowella sarsi
Loxoconcha ovulata
Loxoconcha pellucida
Paracytheridea bovettensis
Polycope ? frequens
Polycope n. sp. *4*
Pseudocythere caudata
Sclerochilus contortus
Semicytherura acuticostata
Xestoleberis ? dispar
Xestoleberis ? plana

V10–43 (228 m)
Argilloecia acuminata
Bosquetina dentata
Cytheropteron alatum
Cytheropteron latum
Cytheropteron rotundatum
Erythrocypris cf. *E. serrata*
Henryhowella sarsi
Leptocythere ? rara
Loxoconcha pellucida
Polycope cf. *P. dentata*
Polycope frequens
Polycope n. sp. *3*
Polycope reticulata
Polycopsis n. sp. *1*
Pontocypris sp.
Pterygocythereis jonesii

V10–42 (217 m)
Argilloecia acuminata
Bythocythere ? insignis
Costa edwardsi
?Cytherois sp.
Cytheropteron alatum
Cytheropteron latum
Cytheropteron rotundatum
Loxoconcha pellucida
Polycope cf. *P. dentata*
Polycope n. sp. *3*
Pontocypris sp.
Pterygocythereis jonesii
Pterygocythereis n. sp. *1*

In the central part of the Aegean Sea, the bathyal 'B' assemblage, which is characterized by *Bythocypris obtusata*, species of *Polycope* and *Bairdia*, is present and is represented by V10–44 (depth 801 m), V10–45 (depth 860 m), V10–47 (depth 856 m) and V10–48 (depth 805 m). The following fauna was encountered:

V10–44   (801 m)
Argilloecia acuminata
Aurila sp.
Bairdia sp.
Bairdia 'subdeltoidea'
Bythocythere n. sp. *B*
Bythocypris obtusata
Paracytheridea bovettensis
Polycope n. sp. *4*
Pontocypris n. sp. *B*
Pseudocythere caudata
Urocythereis margaritifera

V10–45   (860 m)
Bairdia 'subdeltoidea'
Bythocypris bosquetiana
Bythocypris obtusata
Polycope maculata (fossil)
Polycope n. sp. *4*
Pontocypris n. sp. *B*
Xestoleberis n. sp. *A*

V10–47   (856 m)
Argilloecia acuminata
Bairdia 'subdeltoidea'
Bythocypris bosquetiana
Cytheropteron ? punctatum
Krithe ? similis
Polycope frequens
Polycope reticulata

V10–48   (805 m)
Argilloecia acuminata
Bairdia 'subdeltoidea'
Polycope cf. *P. dentata*
Pontocypris n. sp. *B*

In this area the greatest depth of about 4700 m is reached in the Rhodes Abyssal Plain, and two cores, V10–60 (depth 3881 m) and V10–61 (depth 4216 m) were obtained. Core V10–61 did not yield any ostracods. *Cytherois* sp., *Polycope* sp., and *Bythocypris* were noted in V10–60.

Sea of Candia. In the sea of Candia, north of Crete, seven cores were studied; the shallowest of these samples, V10–51 (depth 764 m) has yielded the following off-shore assemblage:

V10–51   (764 m)

Argilloecia acuminata
Aurila convexa
Bairdia 'subdeltoidea'
Erythrocypris cf. *E. serrata*
Hemicytherura defiorei
Loxoconcha cf. *L. impressa*
Microcythere inflexa
Occultocythereis dohrni
Polycope cf. *P. dentata*
Polycope frequens

Polycope n. sp. A.
Polycope reticulata
Polycopsis compressa
Polycopsis n. sp. *1*
Pseudocythere caudata
Sclerochilus contortus
Semicytherura paradoxa
Xestoleberis communis
Xestoleberis dispar

The fauna contained in V10–35 (depth 911 m) has yielded a mixture of both Recent and fossil forms; the fossil assemblage is indicative of a shallow-water environment. The following forms were encountered:

V10–35   (911 m)

Argilloecia acuminata
Argilloecia sp. *1*
Aurila sp. (fossil)
Bairdia ? minor

'*Cythereis*' *polygonata*
Cytherelloidea n. sp. *1* (fossil)
Erythrocypris cf. *E. serrata*
Loxoconcha cf. *L. impressa*

Mutilus ? speyeri (fossil)   *Xestoleberis ? dispar*
Polycope n. sp. *3*   *Xestoleberis ? rara*
Semicytherura acuticostata (fossil)

Core V10–36, depth 1226 m, has yielded a rich shallow-water, near-shore assemblage, which consists of the following species:

V10–36   (1226 m)

Argilloecia acuminata   *Microcythere hians*
Argilloecia minor   *Neocytherideis* sp.
Aurila convexa   *Paracytheridea bovettensis*
Bairdia reticulata   *Paracytherois striata*
Bairdia serrata   *Paradoxostoma versicolor*
Bosquetina dentata   *Paradoxostoma* sp.
Bythocypris obtusata   *Polycope* cf. *P. dentata*
Callistocythere cf. *C. diffusa*   *Polycope ? dispar*
Callistocythere lobiancoi   *Polycope* aff. *P. fragilis*
Callistocythere n. sp. *G*   *Polycope* n. sp. *1*
Carinocythereis carinata   *Polycope* n. sp. *A*
Carinocythereis quadridentata   *Polycope reticulata*
'*Cythereis*' *polygonata*   *Pterygocythereis jonesii*
Cytherella vulgata   *Pterygocythereis* sp.
Cytherura n. sp. *X*   *Semicytherura acuticostata*
Erythrocypris sp.   *Semicytherura paradoxa*
Hemicytherura videns gracilicosta   *Semicytherura* sp.
Krithe sp. 1   *Urocythereis margaritifera*
Loxoconcha cf. *L. impressa*   *Xestoleberis communis*
Loxoconcha n. sp. *M*   *Xestoleberis dispar*
Loxoconcha ovulata   *Xestoleberis ? margaritea*
Loxocythere angulosa   *Xestoleberis* sp.

Abundance of near-shore forms like *Aurila convexa, Urocythereis margaritifera*, species of *Carinocythereis, Callistocythere, Neocytherideis* and *Loxocythere angulosa*, would indicate a displaced fauna, perhaps by turbidity currents, at a depth of 1226 m.

Core V10–49 (depth 1130 m) has yielded a bathyal fauna, dominated by species of *Polycope* and *Bairdia* '*subdeltoidea*', which consists of the following assemblage:

V10–49   (1130 m)

Argilloecia acuminata   *Polycope* cf. *P. dentata*
Bairdia '*subdeltoidea*'   *Polycope frequens*
Bythocypris obtusata   *Polycope* n. sp. *1*
Cytherura costata   *Polycope ? striata*
Erythrocypris cf. *E. serrata*   *Sclerochilus contortus*
Paradoxostoma versicolor   *Xestoleberis dispar*

A different bathyal assemblage is represented by V10–37 (depth 1862 m), which contains a mixture of both bathyal and shallow-water forms characterized by the following species:

V10–37   (1862 m)

Costa edwardsi   *Polycope reticulata*
Hemicytherura n. sp. *1*   *Pterygocythereis jonesii*
Loxoconcha ? impressa   *Xestoleberis dispar*
Polycope frequens

A similar bathyal fauna is represented by V10–50 (depth 2443 m) and V10–58 (depth 2194 m) which consists of:

V10–50   (2443 m)
Argilloecia acuminata
Hemicytherura videns
Loxoconcha pellucida
Polycope cf. *P. dentata*
Polycope reticulata

V10–58   (2194 m)
Argilloecia acuminata
Polycope cf. *P. dentata*
Polycope ? fragilis
Xestoleberis sp.

Korfu. The fauna of Korfu has been studied by Klie (1938) and Stephanides (1937, 1948). Klie (1938) described a brackish-water species, *Cytherois stephanidesi* Klie, from Korfu living in a minimum salinity of 2·6‰. This was the first time that a species of a truly marine euryhaline genus, *Cytherois*, had been recorded from such a low salinity. Stephanides (1948, p. 93) later found another specimen of this interesting form in low brackish-water (salinity 4–6‰) from a ditch in Ragusa. The other marine form reported by him is a species of *Leptocythere*, tentatively identified as *L. ilyophila* Hirschmann, a boreal form recorded from Germany and Finland. Other brackish-water forms recorded by Stephanides were *Cyprideis littoralis* Brady, (NaCl 2–3‰ and 20–25‰), *Loxoconcha gauthieri* Klie (NaCl 2–25‰) and *Leucocythere mirabilis* Kaufmann (2–3‰ NaCl) largely cosmopolitan.

The following fresh-water assemblage was reported by Stephanides (1948) from Korfu:

*Ilyocypris gibba Ramdohr, *I. biplicata* '(Koch)', *I. bradyi* Sars, *I. getica* Masi, **I. australiensis* Sars, *Notodromus persica* Gurney, *Cypris pubera O. F. Müller, *C. bispinosa* Lucas, *Eucypris virens* Jurine, *E. kerkyrensis* Klie, *E. elongata* Stephanides, **Herpetocypris reptans* Baird, **H. chevreuxi* Sars, *H. strigata* O. F. Müller, *Ilyodromus olivaceus* Brady and Norman, **Cyprinotus incongruens* Ramdohr, **C. salina* Brady, **C. fretensis* Brady, **C. inaequivalvis* Bronstein, **Cypridopsis vidua* O. F. Müller, **C. parva* Müller, *C. aculeata* Costa, *C. hartwigi* Müller, *C. newtoni* Brady and Robertson, *Potamocypris fulva* Brady, **P. maculata* Alm, *Cyclocypris ovum* Jurine, **Physocypris kerkyrensis* Klie, *Candona neglecta Sars, *C. fabaeformis* Fischer and *Candonopsis kingsleii* Brady and Robertson.

(Forms preceded by * were also recorded by Stephanides from Macedonia, Epirus and Central Greece, together with *Herpetocypris brevicaudata* Kaufmann, *Potamocypris variegata* Brady and *Cyprideis littoralis* (Brady).

Pliny Trench. The name Pliny Trench (Emery, Heezen and Allan, 1966) was proposed for an elongate feature with a depth exceeding 3000 m (see Fig. 2) south-east of Crete, which is landward of the two trenches on the north-western flank of the Mediterranean Ridge.

Core V10–59 (depth 2159 m), contains the following abyssal fauna:

V10–59   (2159 m)
Leptocythere sp.
Loxoconcha ? impressa
Polycope ? fragilis
Semicytherura ? dispar

Strabo Trench. The name Strabo Trench was applied by Emery, Heezen and Allan (1960) to an elongate feature, with depth over 3000 m, which bounds the Mediterranean Ridge on the north-west. Three cores, V10–52 (depth 2488 m), V10–62 (depth

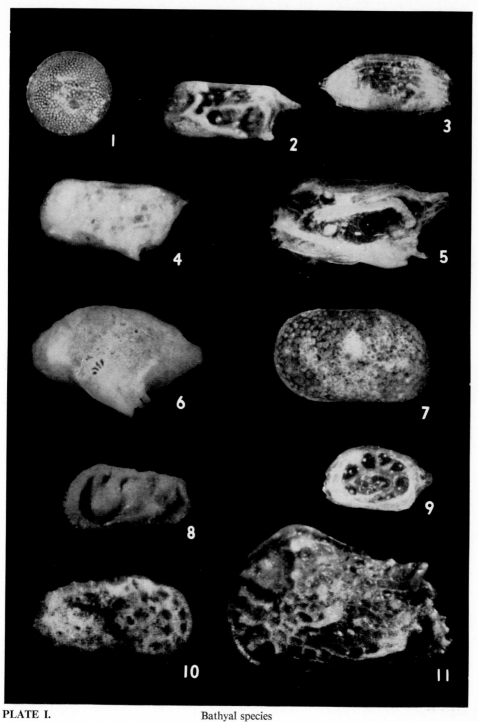

PLATE I.   Bathyal species

1. *Polycope* n. sp. *A*, RV, × 75, 2. *Cytherura dispar* Müller, LV, × 75, 3. *Cytherura* n. sp. *E*, RV.
× 75, 4. *Semicytherura muelleri* Puri, LV, × 60, 5. *Semicytherura paradoxa* (Müller), LV, × 75,
6. *Cytheropteron alatum* Sars, LV, × 75, 7. *Cytherella* n. sp. *2*, LV, × 50, 8. *Occultocythereis dohrni*
Puri, LV, × 50, 9. *Hemicytherura videns* (Müller), LV, × 75, 10. ? *Costa* n. sp. *1*, RV, × 75,
11. '*Cythereis*' *polygonata* Rome, LV, × 75.

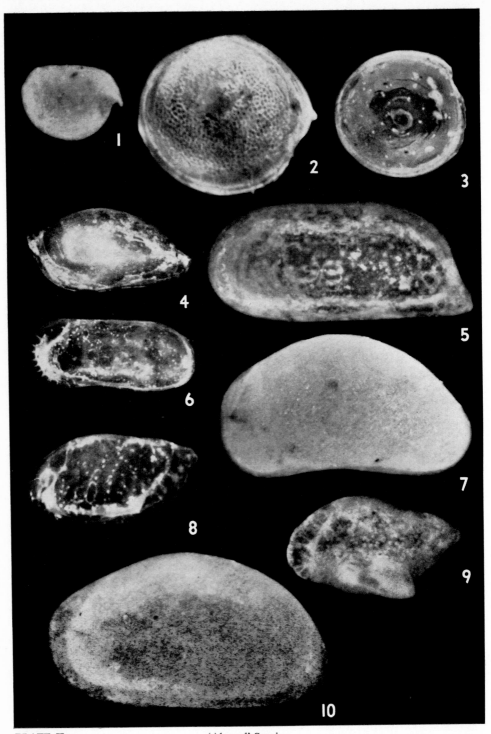

PLATE II. 'Abyssal' Species

1. ?? *Conchoecia* sp. RV × 50, 2. *Polycope* cf. *P. dentata* Brady, RV, × 100, 3. *Polycopsis* n. sp. *1*, RV, × 100, 4. *Paradoxostoma versicolor* Müller, RV, × 50, 5. *Krithe* sp. *1*, LV, × 75, 6. *Krithe similis* Müller, LV, × 75, 7. *Bythocypris bosquetiana* Brady, LV, × 50, 8. *Cytheropteron rotundatum* Müller, LV, × 100, 9. *Cytheropteron latum* Müller, LV, × 100, 10. *Bythocypris obtusata* Sars, LV, × 50.

2384 m) and V10–63 (depth 2818 m), were examined from this area. The following ostracod fauna occurs in these samples:

V10–52   (2488 m)
Argilloecia acuminata
Bairdia 'subdeltoidea'
Bythocypris bosquetiana
Krithe sp. *1*
Polycope frequens
Polycope n. sp. *3*

V10–63   (2818 m)
? *Cytherois* sp.
Paradoxostoma n. sp.
Polycope cf. *P. dentata*
Polycope sp.

V10–62   (2384 m)
Argilloecia sp.
Bythocypris bosquetiana
? ? *Conchoecia* sp.
Cytheropteron aff. *C. alatum*
Cytheropteron rotundatum
Erythrocypris cf. *E. serrata*
Krithe sp. *1*
Monoceratina n. sp. *2*
Mutilus ? *speyeri*
Polycope ? *frequens*
Polycope cf. *P. dentata*
Polycopsis n. sp. *1*
Pseudocythere caudata
Pterygocythereis jonesii

The fauna represented by V10–52 and V10–63 is abyssal while the fauna in V10–62 (depth 2384 m) is mixed with such shallow-water forms as *Mutilus* ? *speyeri*.

Cyprus. Marine ostracods of the Cyprus area are largely unknown, but the following list includes the fresh-water fauna reported by Stephanides from the island of Cyprus:

Ilyocypris divisa Klie, *Eucypris virens* Jurine, *Herpetocypris reptans* Baird, *Cyprinotus* (*Heterocypris*) *incongruens* Ramdohr, *C.* (*H.*) *salina* Brady, ? *C.* (*H.*) *fretensis* Brady and *Cypridopsis* sp.

The Lebanon Sea. Five samples were collected in the Beirut area at depths up to 135 fathoms. The following 18 species are common:

Aurila convexa (Baird)
Bosquetina carinella (Reuss)
Bairdia formosa Brady
Callistocythere diffusa (Müller)
Costa batei (Brady)
'*Cythereis*' *polygonata* Rome
Cytheridea neapolitana Kollmann
Cytheropteron latum Müller
Hemicytherura videns (Müller)

Kangarina abyssicola (Müller)
Loxoconcha impressa Baird
Loxoconcha napoliana Puri
Loxoconcha pellucida Müller
Polycope reticulata Müller
Semicytherura acuticostata (Sars)
Semicytherura alifera Ruggieri
Semicytherura inversa (Seguenza)
Xestoleberis communis Müller

The Levantine Coast. The ostracods of the Levantine coast have been studied by Klie (1935), who reported two species from Alexandria. Later Ruggieri (1953) reported the following assemblage from Port Said:

Cyprideis sp., '*Cythereis*' *emaciata* Brady, *Carinocythereis turbida* (Müller), *C. runcinata* (Baird), *C. scutigera* (Brady), *Hemicythere* sp., *Basslerites teres* (Brady), *Neocytherideis* cf. *N. subulata* (Brady), *Cushmanidea elongata* (Brady), *Caudites* sp., *Paijenborchella* (*Neomonoceratina*) *mediterranea* Ruggieri.

In a more recent study by Lerner-Seggev (1964) on the ostracod fauna from the coast of Israel (from Haifa in the north and Vadi Rubin in the south), nine species were recorded from the eulittoral zone to a depth of 158 m.

2 c

This assemblage consists of:

Cytherella pori Lerner-Seggev, *Bairdia incognite* Lerner-Seggev, *Krithe bartonensis levantina* Lerner-Seggev, *Neocytherideis foveolata* (Brady), *Cytheridea neapolitana* Kollmann, *Buntonia giesbrechti* (Müller), *Loxoconcha stellifera* Müller, and *Paradoxostoma intermedium* Müller.

Except for the new forms reported by Lerner-Seggev, the remainder of the fauna is of a western Mediterranean nature.

Nile Cone. Four cores were studied from the Nile Cone, a submarine part of the Nile delta, which extends from the continental shelf, slope and continental rise towards the Herodotus Abyssal Plain. These cores are located in the north-east part of the delta called the Rosetta Fan by Emery, Heezen and Allan (1960) and vary in depth from 2342 to 2975 m. The following abyssal fauna is contained in V10–54, V10–55 and V10–57:

V10–54   (2347 m)
Polycope cf. *P. dentata*
Polycope frequens
Pseudocythere caudata
Pseudocythere sp. (?n. sp.)

V10–57   (2975 m)
Erythrocypris cf. *E. serrata*
Loxoconcha ? impressa
Polycope cf. *P. dentata*
Polycope ? fragilis

V10–55   (2417 m)

Polycope cf. *P. dentata*
Polycope ? fragilis
Pseudocythere caudata

A displaced shallow-water, near-shore fauna is contained in core V10–53 at depth 2772 m; this fauna could perhaps be drifted by the Mediterranean current or displaced by turbidity currents. This assemblage consists of:

V10–53   (2772 m)
Cytherella vulgata
Cytheridea neapolitana

Cytheropteron latum
Polycope dispar

Beach sand collected at Tripoli by Dr K. McKenzie has yielded the following ostracod fauna:

Aurila convexa (Baird)
Aurila ? woodwardi (Brady)
Bairdia ? mediterranea Müller
Carinocythereis carinata (Roemer)
Carinocythereis rubra (Müller)
Cushmanidea elongata (Brady)
Cyprideis sp.
Cytheretta subradiosa (Roemer)
Heterocythereis albomaculata (Baird)

Loculicytheretta pavonia (Brady)
Loxoconcha ovulata (Costa)
Loxoconcha stellifera Müller
Mutilus speyeri (Brady)
Quadracythere prava (Baird)
Urocythereis margaritifera (Müller)
Urocythereis sp.
Xestoleberis decipiens Müller
Xestoleberis rara Müller

The Black Sea. The modern Black Sea communicates with the Mediterranean through the Bosphorus; in the past, however, it was joined with the Caspian Sea through the Sea of Azov. Consequently the ostracod fauna of the Black Sea has both Mediterranean and Sarmatic elements.

The salinity of the Black Sea is very low (15–18‰ on the surface) as compared with the rest of Mediterranean, (+38‰). The bottom sediments consist of coarse to fine sands, detrital sands, and silt and clays. The ostracod fauna has been studied by Klie

(1937, 1942), Dubowsky (1939), Caraion (1958, 1959, 1960, 1962, 1963) and Marinov (1962, 1963, 1964) and by Cvetkov (1959).

The following three biotopes are recognized on the Bulgarian coast by Marinov (1964):

1. *Phytals*. The fauna commonly encountered among algal vegetation along the coast consists of *Loxoconcha pontica* Klie, *Eucytherura bulgarica* Klie, *Paradoxostoma intermedium* Müller, *Xestoleberis aurantia* (Baird) and *X. decipiens* Müller. Comparatively rare are *Paradoxostoma abbreviatum* Sars, *P. bradyi* (Sars), *Cytheroma karadagiensis* Dubowsky, and *Sclerochilus gewemuelleri dubowskyi* Marinov.

2. *The pelophile biotope*, which inhabits a clayey bottom, consists of *Carinocythereis rubra pontica* Dubowsky, *C. antiquata* (Baird), *Pterygocythereis jonesii* (Baird), *Loxoconcha granulata* Sars and *Callistocythere diffusa* (Müller).

3. *The psammophile biotope*, which prefers a sandy bottom, consists typically of a large assemblage of *Parvocythere hartmanni* Marinov, *Microcythere varnensis* Marinov, *M. longiantennata* Marinov, *Loxoconcha nana* Marinov, *L. bulgarica* Klie, *L. aestuarii* Marinov, *Cytherois pontica* Marinov, *C. pseudovitrea messambriensis* Marinov, *C. pseudovitrea carcinitica* Marinov, *Semicytherura pontica* (Marinov), *S. remanei* Marinov, *Pontocytheroma arenaria* Marinov, *Cushmanidea bacescoi* Caraion, *Pseudocytherura pontica* Dubowsky, *Paracytheridea pauli* Dubowsky, *Cytherois cepa* Klie, *Microcytherura* sp. and *Leptocythere pellucida* (Baird).

Certain species, such as *Loxoconcha minima* Müller and *L. impressa* (Baird), also inhabit algal vegetation while *Carinocythereis rubra pontica* Dubowsky lives also in a muddy bottom. *Cushmanidea tchernjawskii* (Dubowsky) inhabits argillaceous sand. Some of the psammophile species are interstitial forms, such as *Microcythere varnensis* Marinov and *Parvocythere hartmanni* Marinov, which live at a depth of 2 to 5 m, in a salinity of 1–2‰ to 17–18‰ (Marinov, 1964, p. 86).

The fauna in the Bosphorus region consists of forms like *Loxoconcha granulata* Sars, *Carinocythereis antiquata* (Baird), *C. rubra pontica* Dubowsky etc., which have a wide distribution in the Black Sea together with such stenohaline species as *Philomedes intermedia* (Baird), *Pterygocythereis jonesii* (Baird), *Costa hamata* (Müller), *Cytheropteron rotundatum* Müller, *Cytheretta rubra* (?) Müller, *Bythocythere* sp. and some species of Bairdiidae. These typically Mediterranean stenohaline forms have adapted themselves to lower salinity and lower winter temperature of the water.

The distribution of ostracods in the brackish-water Black Sea estuaries and ultra-haline lakes is divided into four categories according to their origin, as follows (Marinov, 1964, p. 83):

1. Fresh-water species which have also been found in the brackish-water basins. Of the 15 species, the most common forms represented are *Ilyocypris biplicata* (Baird), *Candona neglecta* Sars, *C. levanderi* Hirschmann, *Limnocythere inopinata* (Baird), *Darwinula stevensoni* (Brady and Robertson).

2. Typical brackish-water species. Most abundant and common forms are represented by *Cyprideis littoralis* Brady, *C. torosa* Jones, *Hemicythere sicula* (Brady) and *Potamocypris steueri* Klie.

3. Ultrahaline species are represented by *Eucypris inflata* (Sars) which occurs along the Bulgarian coast and in lakes near Schabla, Baltschik, Pomorie and Burgas. Salinity in these lakes and basins during summer months is up to 80‰.

4. Sea-euryhaline species which are open sea forms that have adapted themselves in mesohaline waters. These are represented by *Cytheridea neapolitana* Kollmann, *Callistocythere diffusa* (Müller), *Leptocythere pellucida* (Baird), *Callistocythere*

mediterranea (Müller), *Eucythere declivis* (Norman), *Carinocythereis antiquata* (Baird), *Pterygocythereis jonesii* (Baird), *Costa hamata* (Müller), *Hemicytherura cellulosa* (Norman), *Cytheropteron rotundatum* Müller, *Cytheroma variabile* Müller, *Loxoconcha granulata* Sars, *L. impressa* (Baird), *L. minima* Müller, *L. littoralis* Müller, *Xestoleberis decipiens* Müller, *X. aurantia* (Baird), *Paradoxostoma intermedium* Müller, *P. simile* Müller, *P. variabile* (Baird), *P. abbreviatum* Sars, and *P. bradyi* Sars.

The above assemblage is typically Mediterranean mixed with some European boreal species.

The Sea of Azov. The Sea of Azov is shallow (13 m), and the temperature range of the water is 30° to 0·6°C (mean 25°C), and is very variable. During the Quaternary period, the Mediterranean invaded the Black Sea through the Bosphorus and in the open part of the Black Sea a large percentage of the brackish-water Sarmatic fauna was destroyed, but in the Sea of Azov survivors of the Sarmatic fauna are still living (Ekman, 1953, p. 96). This is also true of ostracods. In the delta parts of rivers in the Azov sea basin, Schornikov (1964) reported eight species of Caspian origin which are still alive. Five of these species *Candona schweyeri* Schornikov (*nom. nov.* pro *Candona elongata* Schweyer, 1949 *non C. elongata* Herrick, 1879 *non C. elongata* Vávra, 1891), *Leptocythere striatocostata* (Schweyer), *L. quinquetuberculata* (Schweyer), *L. longa* (Negadaev) and *Loxoconcha lepida* Stepanaitys, were originally described from the upper Pliocene and post-Pliocene of the South of U.S.S.R. The other three species include *Leptocythere lopatici* Schornikov, *L. relicta* Schornikov and *L. gracilloides* Schornikov.

The Caspian Sea. The Caspian Sea is marked by a very high salinity (up to 170‰). The ostracod fauna was described by Sars (1927) from the littoral region and the two species (*Heterocythereis amnicola* (Sars) and *Loxoconcha umbonata* Sars) were re-described by Elofson (1945). Sars (1888) reported *Heterocythereis amnicola* from the coast of Sicily in the Mediterranean.

The Red Sea. From the material collected by the expedition of the S.M.S. *Pola* in the Red Sea, Graf (1931) described the Cypridinidae. Recently, Hartmann (1964) has described a fauna consisting of 49 species from the eulittoral and supralittoral zones.

The fauna of the Red Sea is most distinct and only two species (*Leptocythere* cf. *L. littoralis* and *Paradoxostoma breve*) are common with the Mediterranean.

Bathyal Species. The following ostracods occur in the Mediterranean from 250 to 880 m:

Species	Depth Range
Polycope fragilis rostrata	150–255
Polycope maculata	860
Polycope n. sp. *A*	265–1226
Polycopsis compressa	764
Argilloecia ? clavata	495–818
Argilloecia ? caudata	1266
Argilloecia bulbifera	265
Argilloecia ? minor	265–1266
Argilloecia sp. *M*	818
Bairdia ? minor	911
Bairdia reticulata	1226
Sclerochilus contortus	495–1330
Sclerochilus aequus	281

Pseudocythere aff. *P. caudata*	598
Cytherura dispar	422–2159
Cytherura rara	495–1033
Cytherura n. sp. *E*	281
Semicytherura muelleri	281
Semicytherura paradoxa	764–1226
Semicytherura n. sp. *B*	281–764
Hemicytherura videns	281–2443
Cytherois n. sp. *A*	911
Paracytherois n. sp. *X*	281
Paradoxostoma cylindricum	265
Microcythere gibba	265
Microcythere hians	265–1226
Microcythere cf. *M. nana*	265
Xestoleberis ? rara	422–1337
Xestoleberis communis	1226
Xestoleberis ? plana	495
Xestoleberis n. sp. *A*	265–860
Loxoconcha ovulata	495–1225
Leptocythere n. sp. *L*	422–764
Mutilus speyeri	281–911
Carinocythereis rubra	495
'*Cythereis*' *polygonata*	911–1226
? Costa n. sp. *1*	495
Occultocythereis dohrni	764
Cytherella n. sp. *2*	281
Cytherelloidea n. sp. *1* (fossil)	911
Bythocythere n. sp. *B*	801

'*Abyssal*' *Fauna*. The following species occur in the Mediterranean from a depth range of 800 to 4285 m:

Species	Depth Range
?? Conchoecia sp.	2049–2499
Polycope n. sp. *1*	265–2723
Polycope n. sp. *3*	217–2488
Polycope cf. *P. dentata*	150–4285
Polycopsis n. sp. *1*	228–2384
Erythrocypris sp.	228–2975
Bairdia formosa	2250
Pseudocythere caudata	150–2574
Hemicytherura videns gracilicosta	1226
Paradoxostoma versicolor	1130–1226
Paradoxostoma maculatum	2049
Loxoconcha sp. *M*	1225
Krithe sp. *1*	1226–2550
Krithe ? similis	885
Caudites n. sp. *A*	1387
Bythocypris bosquetiana	150–3881
Bythocypris obtusata	150–2905

Species	Depth Range
Argilloecia n. sp. *1*	192–1884
Cytheropteron rotundatum	69–3427
Cytheropteron latum	69–2382
Cytherois n. sp. *A*	911
Paracytherois striata	150–1225
Xestoleberis dispar	69–4285

There is no true abyssal fauna in the Mediterranean in as much as the ornate hemicytherids and trachyleberids, which are so typical of the abyssal assemblages elsewhere are absent. Some of the above species are doubtless archibenthic forms (200–1000 m depth) which have acclimatized themselves to live in deeper water since the temperature of the water below 1000 m is about 13°C which is about 1° lower than the bottom temperature at a depth of 300 m.

Nine samples of sediment cores from depths over 3000 m were examined from the Mediterranean Sea. Two of the cores, V10–15, (depth 3434·9 m) from the Tyrrhenian Sea, and V10–61 (depth 4216·3 m) from a small abyssal plain south-east of Rhodes, did not contain ostracods, although Foraminifera were noticed in both of these cores. The substrate at V10–15 consists of a clay, while V10–61 consists of *Globigerina* ooze.

Generally, there is a limit between bathyal and abyssal fauna at a depth of about 3000 m and a temperature gradient of 4°C, which coincides with a great change in the nature of the benthonic fauna. However, this is not supported by the ostracod fauna encountered in the samples at our disposal.

There does not seem to be any great antiquity in the nature of the 'Abyssal' Fauna as it is represented mostly by species of *Polycope* and *Bythocypris bosquetiana*, which are also common constituents of the bathyal fauna.

The following are the species encountered in depths over 3000 m at the various stations:

V10–25   (3101 m) clay
Loxoconcha ? impressa
Polycope sp.

V10–26   (3348 m) *Globigerina* ooze
Polycope cf. *P. dentata*
Polycope fragilis
Polycope reticulata

V10–27   (4285 m) *Globigerina* ooze
Bairdia sp.
Polycope cf. *P. dentata*
Xestoleberis dispar
Xestoleberis sp. *1*

V10–60   (3881 m) *Globigerina* and pteropod ooze
Bythocypris bosquetiana
Cytherois sp.
Polycope sp.

V10–66   (3380 m) Pteropod ooze
Polycope frequens
Polycope n. sp. *4*
Polycope reticulata

V10–68   (3538 m) Pteropods and clay
Polycope cf. *P. dentata*
Polycope frequens
Polycope reticulata

V10–69   (3089 m) Pteropods and clay
Polycope frequens

Nature of the substrate rather than the depth, seems to be controlling the vertical distribution of the ostracod fauna since the presence of *Globigerina* ooze coincides with the occurrence of a stable 'abyssal' fauna. Deep sea abyssal faunas seem to be controlled by the availability of nutrients and consequently abyssal plains closer to the land mass have a higher proportion of ostracod species.

Establishment of the pelagic life of the sea seems to be controlling the ostracod assemblage in the Mediterranean since *Globigerina* ooze substrate, even at depths less than 1000 m (e.g. V10–6, depth 818 m) has a similar fauna as compared with similar ooze at much greater depths except that the number of species progressively decreases from ten species distributed over six genera in V10–6, 818 m, to four species distributed over three genera in deeper waters.

SUMMARY AND CONCLUSIONS

The Mediterranean Sea as a basin, supports a remarkable uniform marine and marginal ostracod fauna. In areas closer to the land, the ostracod assemblages are controlled primarily by salinity and amount of food supply. The near-shore and off-shore marine assemblages are modified by the nature of the substrate and presence or absence of vegetation. There is a stable bathyal fauna in this area but a true abyssal fauna is lacking.

The following ostracod assemblages are recognized in this basin:

1. Limnetic (fresh-water salinity range 0·2 to 0·5‰) consisting primarily of species of *Ilyocypris*, *Notadromus*, *Herpetocypris*, *Ilyodromus*, *Cypridopsis*, *Potamocypris* and *Candona*.

2. (Mixo-) oligohaline (salinity range 0·5 to 3‰) contains, in addition to the limnetic assemblage, forms like *Leptocythere* probably *L. ilyophila*, *Cytherois stephanidesi*, *Cyprideis littoralis*, *Loxoconcha gauthieri* and *Leucocythere mirabilis*.

3. (Mixo-) mesohaline (salinity range 2 to 10‰) *Cyprideis torosa*, *Cytheromorpha fuscata*, *Loxoconcha elliptica*.

4. (Mixo-) polyhaline (salinity range > 17‰) *Cyprideis torosa*, *Xestoleberis aurantia*, *Aurila woodwardi*, *Loxoconcha elliptica*, *L. dorsoinflata*, *Cushmanidea elongata*, *Bairdia corpulenta*, *Carinocythereis carinata*.

5. Euhaline (salinity > 32‰). Locally this assemblage is greatly modified by the nature of substrate and consists of several species of *Bairdia*, *Cytherella*, *Echinocythereis* and *Henryhowella*.

6. Near-shore assemblage (depth up to 100 m) is characterized by large, ornate forms such as *Aurila convexa*, *Buntonia sublatissima*, species of *Carinocythereis*, '*Cythereis*' *polygonata*, *Neocytherideis foveolata*, *Urocythereis margaritifera* and *Eucythere declivis*. Within this assemblage, several sub-assemblages occur depending on the nature of the substrate and vegetation.

7. Off-shore assemblage (depth range 100 to 250 m) is dominated by species of *Argilloecia*, *Polycope*, *Polycopsis* and *Cytheropteron*.

8. Bathyal assemblage (depth range 250 to 800 m) is dominated by several species of *Argilloecia*, *Polycope*, *Pseudocythere*, and *Bythocythere*.

9. 'Abyssal' assemblage (depth range 800 to 4285 m) is dominated by forms like *Bythocypris bosquetiana*, *B. obtusata*, *?? Conchoecia sp.*, *Cytheropteron rotundatum*, *C. latum* and species of *Krithe*.

10. The Mediterranean Ridge as a submarine topographic barrier does not seem to affect the distribution of benthonic ostracods as faunas on both sides of the ridge are remarkably very similar.

ACKNOWLEDGEMENTS

Gratitude is expressed to Dr Maurice Ewing, Director, Lamont Geological Observatory, Columbia University, who made it possible to study sediment cores collected during the cruise 10 of the Research Ship *Vema* in 1956. The sampling programme and storage facilities for housing these cores at the Lamont Geological Observatory were supported by the National Science Foundation Grant GA558 and Office of Naval Research Grant TO–4. We wish to thank Dr G. Hartmann, Hamburg, Germany, who read the manuscript and offered valuable comments. Studies on the Gulf of Naples were supported by National Science Foundation Grant Nos G 14562 and GB 2621; study of the *Challenger* Ostracoda was made possible by National Science Foundation Grant No GB 6706, for which gratitude is expressed.

TABLE 1. Piston cores in the Mediterranean Sea

Core No.	Location		Depth (in m)	Sediment	Shipboard Length (in cm)
	Latitude	Longitude			
1956					
VEMA 10–1	35°41′30″N	03°59′30″W	1365·2	Glob. ooze	830
VEMA 10–2	35°59′30″N	00°40′15″W	(not given)	Glob. ooze	763
VEMA 10–3	36°41′20″N	00°06′30″W	2556·5	Glob. ooze	735
VEMA 10–4	38°41′00″N	04°33′30″E	2260·0	Glob. ooze	735
VEMA 10–5	40°18′30″N	06°47′30″E	2723·0	Glob. ooze	980
VEMA 10–6	41°32′30″N	09°46′40″E	818·0	Glob. ooze	1040
VEMA 10–7	41°18′00″N	11°09′00″E	1387·14	Glob. ooze	870
VEMA 10–8	41°10′00″N	11°06′00″E	1337·73	Glob. ooze	945
VEMA 10–9	40°58′30″N	10°46′00″E	1043·1	Glob. ooze	1073
VEMA 10–10	41°03′30″N	12°47′00″E	543·51	Silt	1165
VEMA 10–11	41°03′00″N	12°42′45″E	192·15	very fine sand	210
VEMA 10–12	40°48′02″N	12°45′30″E	1884·9	Glob. ooze	415
VEMA 10–13	40°18′30″N	12°44′00″E	3422·1		No Core
VEMA 10–14	40°08′13″N	12°19′00″E	3434·9		Fragments
VEMA 10–15	40°07′00″N	12°21′00″E	3434·9	Clay	865
VEMA 10–16	40°28′08″N	14°02′00″E	986·3	Clay	240
VEMA 10–17	40°27′15″N	13°57′45″E	1033·9	Clay	295
VEMA 10–18	40°23′30″N	13°57′40″E	265·3	Silt and shell	275
VEMA 10–19	39°31′00″N	14°49′00″E	150·1	Sand and silt	245
VEMA 10–20	39°30′00″N	14°50′00″E	256·2	Silt and rocks	223
VEMA 10–21	39°19′48″N	14°25′30″E	1266·4	Clay	200
VEMA 10–22	38°12′30″N	18°44′30″E	3670	Clay	724
VEMA 10–23	38°20′40″N	19°35′40″E	3466·0	Clay and shell	1080
VEMA 10–24	38°50′30″N	18°46′30″E	2250·9	Glob. ooze	363
VEMA 10–25	37°45′00″N	20°03′30″E	3101·8	Clay	630
VEMA 10–26	36°16′00″N	21°36′00″E	3348·9	Clay, some shell, (little glob.)	598
VEMA 10–27	36°33′30″N	21°20′00″E	4285·9	Glob. ooze	715
VEMA 10–28	38°20′40″BN	22°03′55″E	347·7	Sand	735
VEMA 10–29	38°15′00″N	22°15′30″E	649·6	Clay, fine sand	221
VEMA 10–30	38°21′40″N	22°25′35″E	73·2	Clay, m. to f. sand, shell	980
VEMA 10–31	38°13′00″N	22°26′30″E	790·60	Clay, f. sand	490
VEMA 10–32	38°06′20″N	22°53′45″E	768·6	Clay	490
VEMA 10–33	38°10′30″N	22°46′15″E	823·5	Clay	490
VEMA 10–34	38°03′00″N	22°51′45″E	422·7	Clay	245
VEMA 10–35	36°20′00″N	23°15′00″E	911·3	Sand	490
VEMA 10–36	35°24′18″N	23°15′24″E	1226·1	Clay, silt, f. sand (few glob.)	326
VEMA 10–37	36°07′50″N	23°44′40″E	1862·9	some glob.—silt, f. sand	160

TABLE 1.(*contd.*)

Core No.	Location		Depth (in m)	Sediment	Shipboard Length (in cm)
	Latitude	Longitude			
VEMA 10–38	36°42'30"N	23°16'50"E	598·4	Glob. ooze	380
VEMA 10–39	36°25'30"N	23°31'30"E	281·8	Glob, silt, f. sand	590
VEMA 10–40	37°15'00"N	23°37'00"E	495·9	Glob. ooze	980
VEMA 10–41	37°30'12"N	23°42'00"E	135·42	Sand (v. coarse v. fine)	80
VEMA 10–42	37°40'48"N	23°45'54"E	217·7	Glob. ooze	895
VEMA 10–43	37°40'00"N	23°43'00"E	228·5	Glob. ooze	735
VEMA 10–44	38°23'00"N	25°31'00"E	801·5	Glob. ooze	510
VEMA 10–45	38°26'45"N	25°38'00"E	860·1	Glob. ooze (few pelecypods)	1092
VEMA 10–46	38°40'45"N	26°38'45"E	69·5	Clay and pelecyopds	240
VEMA 10–47	38°26'30"N	25°36'45"E	856·4	Clay (few glob.)	580
VEMA 10–48	37°44'45"N	25°35'00"E	805·2	Glob. ooze	566
VEMA 10–49	36°05'00"N	26°50'20"E	1130·9	Glob. ooze	703
VEMA 10–50	35°57'45"N	27°04'45"E	2443·0	Glob. ooze	1600
VEMA 10–51	35°55'45"N	27°18'00"E	764·9	Glob. ooze	490
VEMA 10–52	35°00'30"N	27°49'30"E	2488·8	Glob. ooze	490
VEMA 10–53	33°11'30"N	29°06'00"E	2772·4	Glob. ooze, pteropods	735
VEMA 10–54	32°48'00"N	29°37'00"E	2347·9	Glob. ooze	980
VEMA 10–55	32°46'30"N	29°23'00"E	2417·4	Glob. ooze	490
VEMA 10–56	34°01'00"N	30°08'30"E	2574·8	Glob. ooze	980
VEMA 10–57	33°21'30"N	28°25'00"E	2975·6	Glob. ooze	980
VEMA 10–58	35°40'30"N	26°18'00"E	2194·2	Glob. ooze	735
VEMA 10–59	35°07'00"N	27°02'00"E	2159·4	Glob. ooze, Diatom ooze, pteropods	735
VEMA 10–60	35°49'30"N	28°58'00"E	3881·4	as above	730
VEMA 10–61	36°00'00"N	28°47'30"E	4216·3	Glob. ooze	970
VEMA 10–62	34°19'00"N	26°53'30"E	2384·5	Glob. ooze, pteropod	610
VEMA 10–63	34°07'00"N	26°19'00"E	2818·2	Glob. ooze	630
VEMA 10–64	34°23'30"N	24°06'15"E	2049·6	Glob. ooze	682
VEMA 10–65	34°37'00"N	23°25'00"E	2584·0	Glob. ooze	960
VEMA 10–66	35°25'05"N	21°24'00"E	3380	(pteropods) volcanic tephra	515
VEMA 10–67	35°42'00"N	20°43'00"E	2906	as above	846
VEMA 10–68	36°54'05"N	17°57'00"E	3538	pteropods, clay	700
VEMA 10–69	37°13'05"N	17°17'00"E	3089	pteropods, clay	1220
VEMA 10–70	37°33'09"N	16°35'05"E	2499	pteropods, clay	735

TABLE 2. *Ostracoda encountered in*

Depth in m			1365	?	2260	2723	818	1387	1337	1042	543	192	1884	936	1033	265	150	256	1266	3427	2550	3466	3101
Species / Stations	Depth Range		1	2	4	5	6	7	8	9	10	11	12	16	17	18	19	20	21	22	23	24	25
? ? *Conchoecia* sp.	(2049 to 2499 m)	+/θ																					
Polycope fragilis Müller	(265 to 3348 m)	+/θ														5			1				
Polycope frequens Müller	(150 to 3538 m)	+/θ											2/θ			1/θ	1		1		2		
Polycope reticulata Müller	(150 to 3538 m)	+/θ					2/1									3/θ	4/θ					1
Polycope rostrata Müller	(150 to 265 m)	+/θ														1	1						
Polycope dispar Müller	(150 to 2772 m)	+/θ														1	5/θ	3/θ					
Polycope maculata Müller	860 m	+/θ																					
Polycope n. sp. 3	(217 to 2488 m)	+/θ					2																
Polycope n. sp. A	(265 to 1226 m)	+/θ														2							
Polycope n. sp. 1	(265 to 2723 m)	+/θ				1	4	1/θ		1	1/θ				1								
Polycope cf. *P. dentata*	(150 to 4285 m)	+/θ							2/θ							3/θ	1/θ		2	1		1	
Polycopsis compressa (Brady & Robertson)	764 m	+/θ																					
Polycopsis n. sp. 1	(228 to 2384 m)	+/θ			1														2				
Erythrocypris cf. *E. serrata* Müller (all the genus)	(228 to 2975 m)	+/θ			1		1														2		
Pontocypris n. sp. B	(150 to 860 m)	+/θ														1	3						
Argilloecia acuminata Müller	(150 to 2550 m)	+/θ					4/θ	1/θ	2/θ		1/θ		2	2	1	2	2/θ		1/θ		2		
Argilloecia? clavata (Brady)	(435 to 818 m)	+/θ					1																
Argilloecia? caudata Müller	1266 m	+/θ																	1				
Argilloecia bulbifera Müller	265 m	+/θ														2	1						
Argilloecia? minor Müller	(265 to 1266 m)	+/θ														1			1				
Argilloecia sp. M	818 m	+/θ					1																
Argilloecia sp. 1	(192 to 1884 m)	+/θ										1	1										
Bairdia formosa Brady	2250 m	+/θ																					
Bairdia 'subdeltoidea' (Münster)	(69 to 2488 m)	+/θ																					
Bairdia? minor Müller	911 m	+/θ																					
Bairdia reticulata Müller	1226 m	+/θ																					
Bairdia mediterranea Müller	(135 to 256 m)	+/θ									2/θ							1/θ					
Bythocypris bosquetiana (Brady)	(150 to 3881 m)	+/θ													1	46/θ	1/θ	1			3		
Bythocypris obtusata Sars	(150 to 2906 m)	+/θ			1		2/θ	1/θ	2/θ	2			3	2	1	4/θ	1		1				5
Bythocypris n. sp. B	150 m	+/θ															1						
Sclerochilus contortus (Norman)	(495 to 1330 m)	+/θ																					
Sclerochilus aequus Müller	281 m	+/θ																					
Pseudocythere caudata (Sars)	(150 to 2574 m)	+/θ										1				1	5						

Legend: + adults, θ moults F fossil

the Mediterranean Cores

3348	4285	347	649	73	768	790	422	823	911	1226	1862	598	281	495	135	217	228	801	860	69	856	805	1130	2443	764	2488	2772	2347	2417	2574	2975	2194	2159	3881	2384	2818	2049	3380	2906	3538	3089	2499
26	27	28	29	30	31	32	33	34	35	36	37	38	39	40	41	42	43	44	45	46	47	48	49	50	51	52	53	54	55	56	57	58	59	60	62	63	64	66	67	68	69	70
																																			5/0		2/0		.			1
1/0										1																		3/0	1	1/0	1		1			1/0					
											1	1		1			1			1		1		3	1/0		1/0								2			2		1	1	1
1										1/0	1	3	3/0				1			1			1	1															1		1
																											1															
										0																																
																		1																								
									1				1			1/0	1								1																	
										1/0		1/0													1																	
										1		1/0										1																				
1	1									1		2/0	1/0		1	2				1		2/0	1/0	2/0			1/0	1/0		1	1/0		1		1	1		2,				
																								1																		
																		1						1									2									
																		1					2	4						1			1									
																		1	1		3																					
									1	3/0		1/0		1/0			3/0				0				1/0					1											
													1																													
										1		1																														
					1				1																																	
		1											1				1/0	1	4/0	3/0	5/0	1/0		3/0	1																	
									1																																	
													1		1																											
																		1					1										1	1							
										1/0		1				2/0	1/0				1/0															1		1			
												1							1	1																					
										1																															
										2/0	1			2					2	1	1			1			1/0														

TABLE 2. *Ostracoda encountered in*

Legend: + adults, θ moults F fossil

Species	Depth Range	sym	1 / 1365	2 / ?	4 / 2260	5 / 2723	6 / 818	7 / 1387	8 / 1337	9 / 1042	10 / 543	11 / 192	12 / 1884	16 / 986	17 / 1033	18 / 265	19 / 150	20 / 256	21 / 1266	22 / 3427	23 / 3466	24 / 2550	25 / 3101
Pseudocythere aff. *P. caudata* (Sars)	598 m	+/θ																					
Cytherura dispar Müller	(422 to 2159 m)	+/θ											1										
Cytherura costata Müller		+/θ																					
Cytherura rara Müller	(495 to 1033 m)	+													1								
Cytherura n. sp. *X*	1226 m	+/θ																					
Cytherura n. sp. *E*	281 m	+/θ																					
Semicytherura acuticostata (Sars)	(192 to 1226 m) (311)	+/θ										2											
Semicytherura muelleri Puri	281 m	+/θ																					
Semicytherura inversa (Seguenza) var.	135 m	+/θ																					
Semicytherura ruggierii Pucci	135 m	+/θ																					
Semicytherura punctata (Müller)	135 m	+/θ																					
Semicytherura paradoxa (Müller)	(764 to 1226 m)	+/θ																					
Semicytherura n. sp. *B*	(281 to 764 m)	+/θ																					
Cytheropteron alatum Sars	(150 to 2550 m)	+/θ											2 θ				2 θ					1	
Cytheropteron aff. *C. alatum* Sars	150 m	+/θ		1													1 θ						
Cytheropteron rotundatum Müller	(69 to 3427 m)	+/θ																		1			
Cytheropteron latum Müller	(69 to 2384 m)	+/θ																					
Hemicytherura videns (Müller)	(281 to 2443 m)	+/θ		1																			
Hemicytherura videns gracilicosta Ruggieri	1225 m	+/θ																					
Hemicytherura defiorei Ruggieri	(192 to 764 m)	+/θ									1												
Hemicytherura n. sp. 1	(1862 m)	+/θ																					
Eucytherura complexa (Brady)	150 m	+/θ															2						
Eucytherura n. sp. 1	(150 to 823 m)	+/θ															1						
Cytherois n. sp. *A*	911 m	+/θ																					
Paracytherois? oblonga Müller	(150 to 281 m)	+/θ										1				1	1						
Paracytherois striata Müller	(150 to 1226 m)	+/θ															1						
Paracytherois n. sp. *X*	281 m	+/θ																					
Paradoxostoma versicolor Müller	(1130 to 1226 m)	+/θ																					
Paradoxostoma maculatum Müller	2049 m	+/θ																					
Paradoxostoma cylindricum Müller	265 m	+/θ														1							
Paradoxostoma n. sp. *X*	150 m	+/θ															2						
Microcythere gibba Müller	265 m	+/θ														3							
Microcythere hians Müller	(265 to 1226 m)	+/θ														1	1						

the Mediterranean Cores—continued

3348	4285	347	649	73	790	768	823	422	911	1226	1862	598	281	495	135	217	228	801	860	69	856	805	1130	2443	764	2488	2772	2347	2417	2574	2975	2194	2159	3881	2384	2818	2049	3380	2906	3538	3089	2499
26	27	28	29	30	31	32	33	34	35	36	37	38	39	40	41	42	43	44	45	46	47	48	49	50	51	52	53	54	55	56	57	58	59	60	62	63	64	66	67	68	69	70
												1 θ																														
								1 θ																									1									
																							1																			
															1																											
										1																																
													1																													
							1	F		3			4	1																												
													1																													
															4																											
															3																											
															1																											
										2																1																
													2																													
													2 θ			1	1																									
																																		1								
																1	1 θ			2														1								
					Y								1			1	θ			1								1														
													1												1																	
										1																																
													1													1																
												1																														
			1										2																													
								2 θ																																		
													1																													
								2			1																															
													1																													
											1															1																
																																					1					
											1																															

TABLE 2. *Ostracoda encountered in*

Species	Depth Range	+/θ	1 (1365)	2 (?)	4 (2260)	5 (2723)	6 (818)	7 (1387)	8 (1337)	9 (1042)	10 (543)	11 (192)	12 (1884)	16 (986)	17 (1033)	18 (265)	19 (150)	20 (256)	21 (1266)	22 (3427)	23 (3466)	24 (2550)	25 (3101)
Microcythere inflexa Müller	(265 to 764 m)	+/θ														1							
Microcythere cf. *M. nana* Müller	265 m	+/θ														1							
Microcythere n. sp. *X*	150 m	+/θ															1						
Xestoleberis? rara Müller	(422 to 1337 m)	+/θ							1														
Xestoleberis dispar Müller	(63 to 4285 m)	+/θ																					
Xestoleberis? margaritea (Brady)	(73 to 1226 m)	+/θ																					
Xestoleberis fuscomaculata Müller	73 m	+/θ																					
Xestoleberis communis Müller	1226 m	+/θ																					
Xestoleberis decipiens Müller	69 m	+/θ																					
Xestoleberis? plana Müller	495 m	+/θ																					
Xestoleberis? aurantia (Baird)	135 m	+/θ																					
Xestoleberis n. sp. *A*	(265 to 860 m)	+/θ													1								
Paracytheridea bovettensis (Seguenza)		+/θ																					
Loxoconcha decipiens Müller	135 m	+/θ										1				1							
Loxoconcha cf. *L. impressa* (Baird)	(132 to 3101 m)	+/θ										3	1			1						1	1
Loxoconcha mediterranea Müller	135 m	+/θ																					
Loxoconcha pellucida Müller	(217 to 2443 m)	+/θ																					
Loxoconcha sp. 1	192 m	+/θ										1											
Loxoconcha n. sp. *X*	(150 to 281 m)	+/θ														1	2						
Loxoconcha ovulata (Costa)	(135 to 1225 m)	+/θ															1						
Loxoconcha n. sp. *F*	(150 to 265 m)	+/θ														1	1						
Loxoconcha n. sp. *M*	1225 m	+/θ																					
Leptocythere? rara Müller	228 m	+/θ																					
Leptocythere aff. *L. castanea* Sars	69 m	+/θ																					
Leptocythere n. sp. *L*	(422 to 764 m)	+/θ																					
Callistocythere cf. *C. diffusa* Müller	(69 to 1226 m) 281	+/θ										1											
Callistocythere lobiancoi Müller	1226 m	+/θ																					
Callistocythere n. sp. *G*	1226 m	+/θ																					
Krithe? similis Müller	856 m	+/θ																					
Krithe sp. 1	(1226 to 2550 m)	+/θ	1	2									1 / 0								2 / θ	1	
Cytheridea neapolitana Kollmann	(69 to 2772 m)	+/θ																					
Eucythere declivis (Norman)	192 m	+/θ										1											
Cythere sp.	192 m	+/θ										1											

Legend: + adults, θ moults F fossil

the Mediterranean Cores—continued

3348	4285	347	649	73	790	768	823	422	911	1226	1862	598	281	495	135	217	228	801	860	69	856	805	1130	2443	764	2488	2772	2347	2417	2574	2975	2194	2159	3881	2818	2384	3380	2049	3538	2906	2499	3089
26	27	28	29	30	31	32	33	34	35	36	37	38	39	40	41	42	43	44	45	46	47	48	49	50	51	52	53	54	55	56	57	58	59	60	62	63	64	66	67	68	69	70
																									1																	
								1	3																																	
	1	1	2					1	2	1	2			1						4 θ		1			1																	
				1					1																																	
				1																																						
											1											1																				
																				1																						
													1																													
															1																											
																				1																						
										2					1	1		1																								
					1										1																											
									1 θ	1 θ	2	1														1						1	1									
											1																															
													1		1	2								1																		
												2																														
											1			1	3 θ					1 θ																						
											1																															
																		1																								
																				2																						
		1			1			1																																		
											1		1		4					3 θ																						
											1																														
											1																														
																						1																				
											1 θ																1												1 θ			
																				27 θ							1															

TABLE 2. *Ostracoda encountered in*

Species	Depth Range	1365	?	2260	2723	818	1387	1337	1042	543	192	1884	986	1033	265	150	256	1266	3427	3466	2550	3101
Stations		1	2	4	5	6	7	8	9	10	11	12	16	17	18	19	20	21	22	23	24	25
Aurila convexa (Baird)	(192 to 1226 m) 281	+/θ									1/θ	1										
Aurila sp. *X*	(192 to 495 m)	+/θ									1/θ											
Mutilus? speyeri (Brady)	(218 to 311 m)	+/θ																				
Tyrrhenocythere n. sp. 1	73 m	+/θ																				
Buntonia sublatissima dertonensis Ruggieri	(422 to 730 m)	+/θ																				
Buntonia sublatissima (Neviani) var.	(69 to 495 m)	+/θ																				
Carinocythereis laticarina (Brady)	(150 to 265 m)	+/θ													4/θ	3/θ						
Carinocythereis carinata (Roemer)	1225 m	+/θ																				
Carinocythereis quadridentata (Baird)	(69 to 1225 m)	+/θ																				
Carinocythereis rubra (Müller)	495 m	+/θ																				
Carinocythereis carinata (Müller) var.	(69 to 495 m)	+/θ																				
'*Cythereis*' *polygonata* Rome	(911 to 1226 m)	+/θ																				
Pterygocythereis jonesii (Baird)	(73 to 2384 m)	+/θ						1														
Pterygocythereis n. sp. 1 (*jonesii* of Müller)	(73 to 860 m)	+/θ																				
Costa edwardsi (Roemer)	(63 to 1862 m)	+/θ															1					
?Costa n. sp. 1 (= *? Cythereis hartmanni* Caraion)	495 m	+/θ																				
Occultocythereis dohrni Puri	764 m	+/θ																				
Trachyleberis histrix	69 m	+/θ																				
Urocythereis margaritifera (Müller)	(135 to 1226 m)	+/θ																				
Caudites n. sp. *A*	1387 m	+/θ					1															
Henryhowella sarsi (Müller)	(69 to 1042 m)	+/θ							1	1	1				1							
Trachyleberid n.g. *D.* n. sp. *A*	265 m	+/θ													1/θ							
Cytherella vulgata Ruggieri	(69 to 2772 m)	+/θ											1									
Cytherella n. sp. *A*	(138 to 1337 m)	+/θ				1	1	1/θ														
Cytherella? n. sp. (w. punctations)	69 m	+/θ																				
Cytherella n. sp. 2	281 m	+/θ																				
Cytherelloidea n. sp. 1 (fossil)	911 m	+/θ																				
Cyprideis torosa (Jones)	73 m	+/θ																				
Bythocythere? insignis Sars	(69 to 281 m)	+/θ																				
Bythocythere n. sp. *B*	801 m	+/θ																				
Tetracytherura angulosa (Seguenza)	135 m	+/θ																				
Candona candida	73 m	+/θ																				
Candona sp. 1 (or ♀ of *C. candida*)	73 m	+/θ																				

Legend: + adults, θ moults F fossil

the Mediterranean Cores—continued

3348	4285	347	649	73	790	768	823	422	911	1226	1862	598	281	495	135	217	228	801	860	69	856	805	1130	2443	764	2488	2772	2347	2417	2574	2975	2194	2159	3881	2384	2818	2049	3380	2906	3538	3089	2499
26	27	28	29	30	31	32	33	34	35	36	37	38	39	40	41	42	43	44	45	46	47	48	49	50	51	52	53	54	55	56	57	58	59	60	62	63	64	66	67	68	69	70
														1																												
										$\frac{1}{\theta}$				1																												
															1																											
									$\frac{1}{\theta}$					1	1																				1							
			$\frac{13}{\theta}$																																							
						1		$\frac{1}{\theta}$																																		
														1						$\frac{5}{\theta}$																						
												1			F																											
												1			1					$\frac{7}{\theta}$																						
															$\frac{3}{\theta}$																											
												F_1								3																						
									1	2																																
			$\frac{1}{\theta}$							2	1					1	1			$\frac{2}{\theta}$																2						
			1													1				1																						
											1					1				1																						
													1																													
																									1																	
																				$\frac{3}{\theta}$																						
										2					$\frac{5}{\theta}$			1																								
						1	1	1					$\frac{1}{\theta}$	$\frac{1}{\theta}$	1			1		1																						
		1								1			1	1						$\frac{1}{\theta}$						1																
			$\frac{2}{\theta}$																																							
																				$\frac{7}{\theta}$																						
													$\frac{1}{\theta}$																													
								1																																		
			$\frac{5}{\theta}$																																							
													2		1	1																										
																	1																									
													7																													
			$\frac{5}{\theta}$																																							
			$\frac{7}{\theta}$																																							

2 D

REFERENCES

ALLAN, T. D., CHARNOCK, H. AND MORELLI, C. 1964. Magnetic, gravity and depth surveys in the Mediterranean and Red Sea. *Nature, Lond.*, **204**, 1245-8.

ASCOLI, P. 1961. Contributo alla Sistematica degli Ostracodi italiani. *Riv. ital. Paleont.*, **68** (1), 45-52.

——. 1965. Preliminary ecological study on Ostracoda from bottom cores of the Adriatic Sea. *Pubbl. Staz. zool. Napoli* [1964], **33** (suppl.), 213-46, 3 Figs, 4 Pls.

BOULOS, I. 1962. *Carte de reconnaissance des côtes du Liban.* Bassile Frères, Beyrouth, 1:150,000, 2 sheets (2nd ed.).

BRADY, G. S. 1866. On new or imperfectly known species of marine Ostracoda. *Trans. zool. Soc. Lond.*, **5**, 359-93, Pls 57-62.

——. 1880. Report on the Ostracoda dredged by H.M.S. *Challenger* during the years 1873-76. *Report of the Scientific Results of the Voyage of H.M.S. Challenger*, 1873-76, *Zoology*, **1**, Pt. 3, 1-184, Pls 1-XLIV.

—— AND NORMAN, A. M. 1889. A monograph of the marine and freshwater Ostracoda of the North Atlantic and of North-Western Europe. Section 1, Podocopa. *Scient. Trans. R. Dubl. Soc.*, Ser. 2, **4**, 63-270, Pls 8-23.

——. 1896. A monograph of the marine and freshwater Ostracoda of the North Atlantic and of North-Western Europe. Part II, Sections II-IV, Myodocopa, Cladocopa and Platycopa. *Scient. Trans. R. Dubl. Soc.*, Ser. 2, **5** (12), 621-746, Pls 50-68.

BUEN, O. DE 1916. Los Crustaceos de Baleares. *Bol. Soc. esp. Exp. Nat. Hist.*, **16**, 355-67.

CARAION, F. E. 1958. Ostracoda marine din apele rominesti ale Marii negre. *Hidrobiologica*, **1**, 89-101, 6 Figs.

——. 1959. Ostracode noi in Marea Neagra (Apele bosforice). *Comunle Acad. Rep. pop. rom.*, **9**.

——. 1960a. Deux Ostracodes nouveaux pour les eaux du littoral roumain; *Cytheridea tchernjawskii* (Dub.) emend. et *Cytheridea bacescoi* n. sp. *Rapp. P.-v. Réun. Commn int. Explor. scient. Mer Méditerr.*, **15** (2), 113-20, 5 Figs.

——. 1960b. *Loxoconcha bulgarica* n. sp. A new Ostracod collected in the Bulgarian waters of the Black Sea (Sozopol). *Comunle Acad. Rep. pop. rom.*, **11** (1), 21-27, 5 Figs.

——. 1962. Some special problems related with present knowledge of the Black Sea and Azov Ostracoda. *Revue Biol., Buc.*, **3** (7), 45-63.

——. 1963. Reprezentanti noi al Familie Cytheridae (Ostracoda-Podocopa) proveninti di apele pontice rominesti. *Revue Biol., Buc.*, **3** (15), 319-31.

CASPERS, H. 1957. Black Sea and Sea of Azov. Treatise on marine Ecology and Paleoecology. *Mem. geol. Soc. Am.*, **67**, 1, 801-89.

CVETKOV, L. 1959. Materialien über die Süsswasser-ostracodenfauna in Bulgarien. *Mitt. Zool. Inst. B.A.W.*, **8**, 10.

D'ARRIGO, A. 1936. Ricerche sul regime dei litorali nel Mediterraneo. *Ric. Var. Spiagge ital.*, **14**, 1-172.

DOEGLAS, D. J. 1950. Old beaches in the Mediterranean. *18th Internat. Geol. Congress, Great Britain*, 1948, Pt. VIII, Sec. G., 16-20.

DUBOWSKI, N. W. 1939. Zur Kenntnis der ostracodenfauna des Schwarzen Meeres. *Trudỹ karandah. nauch. Sta. T.I. Vyazems'koho*, **5**, 3-68, 68 figs.

DUPLAIX, S. 1958. Étude minéralogique des niveaux sableux des carottes prelevées sur le fond de la Méditerranée. *Rep. Swed. deep Sea Exped.*, **8** (2), 137-66.

EKMAN, S. 1953. *Zoogeography of the Sea.* London, Sidgwick and Jackson. 417 pp.

ELOFSON, O. 1945. On *Cythereis amnicola* (G. O. Sars) and *Loxoconcha umbonata* G. O. Sars, two ostracods from the Caspian Sea. *Ark. Zool.*, **36** (2), 1-5.

EMERY, K. O. AND BENTON, Y. K. 1960. The continental shelf of Israel. *Bull. geol. Surv. Israel*, **26**, 25-41.

——, HEEZEN, B. C. AND ALLAN, T. D. 1966. Bathymetry of the Eastern Mediterranean Sea. *Deep Sea Res.*, **13**, 173-92.

EMILIANI, C. 1955. Pleistocene temperature variations in the Mediterranean. *Quaternaria*, **2**, 87-98.

GONCHAROV, V. P. AND MIKHAILOV, O. V. 1963. New data on the bottom relief of the Mediterranean Sea (In Russian). *Okeanologiia*, **3** (6) 1056-61.

GRAF, H. 1931. Expedition S.M.S. 'Pola' in das Rote Meer. Die Cypridinidae des Roten Meeres. *Denkschr. Akad. Wiss., Wien.*, **102**, 32-46, 10 Figs, 8 tables.

HARRISON, J. C. 1955. An interpretation of gravity anomalies in the Eastern Mediterranean. *Phil. Trans. R. Soc. A*, **248**, 283-325.

HARTMANN, G. 1953. *Iliocythere meyer-abichi* nov. spec., ein Ostracode des Schickwarts von San Salvador. *Zool. Anz.*, **151**, Heft. (11/12), 310-16.

——. 1954. Ostracodes des eaux souterraines littorales de la Méditerranée et de Majorque. *Vie Milieu*, **4** (2) [1953], paru fevrier 1954, 238-53, 6 Figs.

——. 1963. Zur phylogenie und systematik der Ostracoden. *Z. zool. Syst. Evolutionforsch*, **1**, 1-154,

——. 1964. Zur Kenntnis der Ostracoden des Roten Meeres. *Kieler Meeresforsch.*, **20**, 35-127, 62 Pls.

——. 1965. Neontological and paleontological classification of Ostracoda. *Pubbl. Staz. zool. Napoli* [1964] **33** (suppl.), 550-87.

HEEZEN, B. C. AND EWING, M. 1963. The Mid-Oceanic Ridge. In *Seas, Ideas and Observations on Progress in the Study of the Seas*. M. N. Hill, Editor, Interscience, New York, 3, 388-410.

HERSEY, J. B. 1965. Sedimentary basins of the Mediterranean Sea. *Colston Papers, Bristol, England*, 5-9 April 1965, Butterworth, England. No. 17, pp. 75-89.

Hydrographic Department of the Admiralty 1958. *Eastern Mediterranean—Sheet 2, Crete to Port Said. Experimental Depth Contour Chart. Contour interval 100 fathoms*. C. 6192.

IVANOFF, A. 1959. Introduction à une étude des propriétés diffusantes des eaux de la Baie de Naples. *Pubbl. Staz. zool. Napoli*, **31**, 34-43.

JACOBSEN, J. P. 1912. The amount of oxygen in the water of the Mediterranean. *Rep. Dan. oceanogr. Exped. Mediterr.*, **1**, 207.

KLIE, W. 1935. Ostracoda aus dem Tropischen Westafrika. In: Voyage de Ch. Alluaud et P.A. Chappuis en Afrique Occidentale Française (1930-31). *Arch. Hydrobiol.*, **28**, 35-68, 63 Figs.

——. 1937. Ostracoden und Harpacticoiden aus brackigen Gewässern an der bulgarischen Küste der Schwarzen Meeres. *Izv. tsarsk. prirodonauch. Inst. Sof.*, **10**, 1-22.

——. 1938. In DAHL, F. *Die Tierwelt Deutschlands und der angrenzenden Meeresteile. Teil 34, Krebstiere oder Crustacea. III. Ostracoda, Muschelkrebse*. i-iv, 230 pp., 768 Figs. Gustav Fischer, Jena.

——. 1942. Adriatic Ostracoden—III. *Zool. Anz.*, **139**, 67-72, 4 Figs.

LACOMBE, H. AND TCHERNIA, P. 1960. Quelques traits généraux de l'Hidrologie Méditerranéen. *Cah. océanogr.*, **12** (8), 527-47.

LERNER-SEGGEV, R. 1964. Preliminary notes on the Ostracoda of the Mediterranean Coast of Israel. *Israel J. Zool.*, **13**, 145-76, 104 Figs, 2 tables.

McKENZIE, K. G. 1963. A brackish-water ostracod fauna from Lago di Patria, near Napoli. *Ann. Inst. Mus. Zool., Univ. Napoli*, **15** (1), 1-14, 2 Pls, 1 Fig.

MALDURA, C. M. 1952. I fattori chimici nella valutazione dello stock biologico del Mediterraneo. *Boll. Pesca, Piscic. Idrobiol.*, **7**, 117.

MARINOV, T. 1962. Über die Muschelkrebs-Fauna des westlichen Schwarzmeerstrandes. *Bull. Inst. cent. Rech. Sci. pisc. pech. Varna*, **2**, 81-108.

——. 1963a. *Pontocytheroma arenaria* n. g. n. sp. Eine neue Ostracode aus der Sandbiozönose des Schwarzen Meeres. *C.r. Acad. bulg. Sci.*, **16** (5).

——. 1963b. *Cytherois pontica* n. sp. Eine neue Ostracode aus dem Schwarzen Meer. *C.r. Acad. bulg. Sci.*, **16** (7).

——. 1963c. *Loxoconcha aestuarii* n. sp. Eine neue Schwarzmeer-Ostrakode. *C.r. Acad. bulg. Sci.*, **16** (7).

——. 1964a. Untersuchungen über die Ostracodenfauna des Schwarzen Meeres. *Kieler Meeresforsch.*, **20** (1), 82-91.

——. 1964b. Beitrag zur Ostracodenfauna des Schwarzen Meeres. *Bull. Inst. cent. Rech. Sci. pisc. pech. Varna*, **4**, 29-60.

MELLIS, O. 1954. Volcanic ash-horizons in deep-sea sediments from the eastern Mediterranean. *Deep Sea Res.*, **2**, 89-92.

MENZIES, R. J., IMBRIE, J. AND HEEZEN, B. C. 1961. Further considerations regarding the antiquity of the abyssal fauna with evidence for a changing abyssal environment. *Deep Sea Res.*, **8**, 79-94.

MÜLLER, G. W. 1894. Die Ostracoden des Golfes von Neapel und der angrenzenden Meeres-Abschnitte. *Fauna Flora Golf. Neapel*, **21**, Monogr., i-viii, 1-404, Pls 1-40.

NEEV, D. 1960. A pre-Neogene erosion channel in the southern coastal plain of Israel. *Bull. geol. surv. Israel*, **25**, 1-21.

OCHAKOVSKY, Yu E. 1963. The fourth Mediterranean Expedition aboard Research Vessel *Akademik S. Vavilov* (In Russian). *Okeanologiia*, **3** (3), 550-4.

OLAUSSON, E. 1960. Description of sediment cores from the Mediterranean and the Red Sea. *Rep. Swed. deep Sea Exped.*, **8** (3), 285-334.

PARKER, F. L. 1958. Eastern Mediterranean Foraminifera. *Rep. Swed. deep Sea Exped.*, **8** (2), 217-83.

PFANNENSTIEL, M. 1960. Erläuterungen zu den bathymetrischen Karten des Östlichen Mittelmeeres. *Bull. Inst. oceanogr. Monaco*, (1192), 1-60.

PICARD, L. 1943. Structure and evolution of Palestine, with comparative notes on neighbouring countries. *Bull. geol. Dep. Hebrew Univ.*, **84**, 1-187.

PURI, H. S., BONADUCE, G. AND MALLOY, J. 1965. Ecology of the Gulf of Naples. *Pubbl. Staz. zool. Napoli* [1964], **33** (suppl.), 87-199, 67 Figs, 1 table.

——. 1966. Ecologic Distribution of Recent Ostracoda. In *Symposium on Crustacea*, Part I. Marine Biol. Assn India, 457-95, 10 Figs.

REYS, S. 1961b. Recherches sur la systématique et la distribution des Ostracodes de la région de Marseille. *Recl. Trav. Stn mar. Endoume*, **22** (36), 53-109.

——. 1963. Ostracodes des peuplements algaux de l'étage infralittoral de substrat rocheux. *Recl. Trav. Stn mar. Endoume*, **28** (43), 33-47.

ROMANOVSKY, V. 1954. Les courants marins de surface dans le bassin occidental de la Méditerranée. Assn Oc. Phys. U.G.G.I., General Assembly at Rome, Proc. Verb. Nr. 6, Bergen. p. 256-7 (abstract).

ROME, D. R. 1939. Note sur des ostracodes marins des environs de Monaco. *Bull. Inst. océanogr. Monaco*, 768.

——. 1942. Ostracodes marins des environs de Monaco, 2me note. *Bull. Inst. océanogr. Monaco*, 819.

——. 1965. Ostracodes des environs de Monaco, leur distribution en profondeur, nature des fonds marins explorés. *Pubbl. Staz. zool. Napoli* [1964], **33** (suppl.), 200-12.

RUGGIERI, G. 1952. Nota preliminare sugli Ostracodi di alcune spiaggie adriatiche. *Note Lab. Biol. mar. Fano*, **1** (8), 57-63, 4 Figs.

——. 1953. Ostracodi del genere *Paijenborchella* viventi nel Mediterraneo. *Atti Soc. ital. Sci. nat.*, **92**, 1-7, 5 Figs.

——. 1955. *Tyrrhenocythere*, a new recent ostracod genus from the Mediterranean. *J. Paleont.*, **29**, 698-9, 5 Figs.

RYAN, W. B. F. AND HEEZEN, B. C. 1965. Ionian Sea submarine canyons and the 1908 Messina Turbidity Current. *Bull. geol. Soc. Am.*, **76**, 915-32.

——, WORKUM, F. AND HERSEY, J. B. 1965. Sediments on the Tyrrhenian Abyssal Plain. *Bull. geol. Soc. Am.*, **76**, 1261-82.

SACCHI, C. F. 1961. L'évolution récente du milieu dans l'étang saumâtre dit 'Lago Patria' (Naples) analysée par sa macrofaune invertebrée. *Vie Milieu*, **12** (1), 37-65.

SAID, R. 1958. Remarks on the geomorphology of the deltaic coast between Rosetta and Port Said. *Bull. Soc. Géogr. Égypte*, **34**, 115-25.

SARS, G. O. 1888. Nye Bidrag til Kundskaben om middlehavets invertebratfauna. 4. Ostracoda Mediterranea, *Arch. Math. Naturv.*, **12**, 173-324, Pls 1-20.

——. 1899. *An Account of the Crustacea of Norway*, Vol. II. Isopoda. i-x, 1-270, 100 Pls. Bergen Museum, Bergen.

——. 1927. Notes on the Crustacean Fauna of the Caspian Sea. In *Festschrift Knipowitsch*, Leningrad.

SCHORNIKOV, E. I. 1964. An experiment on the distribution of the Caspian elements of the ostracod fauna in the Azov-Black Sea Basin. *Russk. zool. Zh.*, **48** (9), 1276-93, 11 Figs.

SEGUENZA, G. 1883-86. Gli Ostracodi del Porto di Messina. *Il Naturalista Siciliano*; 1884, 39-42, 75-78, 124-8, 149-52, 186-9, 233-57, 319-22; 1885, 44-48, 76-79, 110-12; 1886, 57-61; 2 Pl. Palermo.

SHUKRI, N. M. 1950. The mineralogy of some Nile sediments. *Q. Jl. geol. Soc. Lond.*, **105**, 511-34; **106**, 466-7.

SOLIMAN, S. M. AND FARIS, M. I. 1964. General geologic setting of the Nile Delta Province, and its evaluation for petroleum prospecting. *Fourth Arab Petr. Congr.*, *Beirut*, **23** (B-3), 1-11.

SPARCK, R. 1931. Some quantitative investigations on the bottom fauna at the west coast of Italy in the bay of Algiers, and at the coast of Portugal. *Rep. Dan. oceanogr. Exped. Mediterr.*, **3** (7).

STEPHANIDES, T. 1937. Zwie neue *Eucypris*-Arten (Ostr.) von der Insel Korfu. *Zool. Anz.*, **119**, 269-72.

——. 1948. Korfu and Greece Ostracods. *Prak. Hellenic Hydrobiol. Inst.*, **2**, 70-107, Pls 25-34.

TODD, R. 1958. Foraminifera from western Mediterranean deep-sea cores. *Rep. Swed. deep-Sea Exped.*, **8**(2), 167-215.

UDINTSEV, G. B., Editor 1961. Data of oceanological investigations, Research Vessel *Akademik S. Vavilov*, the First Mediterranean Sea Cruise, 1959. Bottom relief (In Russian). *Mezhd. Geofiz God.* 1957-1958-1959, Inst. Okeanol., Akad. Nauk, S.S.S.R., Moscow, 1-58.

U.S. Hydrographic Office 1959a. *Mediterranean Sea—Antalya Korfazi to Alexandria.* B.C. 3924N (2nd ed.), scale 1:817,635. Contour interval 100 fathoms.

——. 1959b. *Mediterranean Sea—Tobruck to Alexandria.* B.C. 3925N (2nd ed.), scale 1:817,635. Contour interval 100 fathoms.

WALTHER, J. 1910. Die Sedimente der Taubenbank im Golfe von Neapel. *Abh. preuss. Akad. Wiss.* (Anhang).

WENDICKE, F. 1916. Hydrographische Untersuchungen des Golfes von Neapel. *Mitt. zool. Stn Neapel*, **22**, 329.

WÜST, G. 1960. Die Tiefenzirkulation des Mittellandischen Meeres in den Kernschichten des Zwischen- und des Tiefenwassers. *Dt. hydrogr. Z.*, **13**, 105-31, I-VIII.

DISCUSSION

ASCOLI: It appears that what you call *Carinocythereis runcinata* is what Ruggieri called *Costa runcinata*, which is extremely close to *C. edwardsi*, the type species of *Costa*. Could you say why you assigned this species to the genus *Carinocythereis*?

PURI: It has a typical *Carinocythereis* carapace and muscle scars as figured by Ruggieri and we have compared it with some of his types. This is a subjective matter and our interpretations of this genus may well differ.

ASCOLI: Your interpretation of *Carinocythereis quadridentata* corresponds with mine up until a few months ago, when I borrowed Baird's type specimens from the British Museum (Nat. Hist.). The true *C. quadridentata* is a very big species, 1·2 mm long and differs from our so-called *C. quadridentata*, which is about 0·7 to 0·8 mm long, is not so plump, has weaker ornamentation and seems to be another species. This latter has already been described from its type locality of Farnesina near Rome by Namias in his paper of 1900.

PURI: We will look at the material and bear your suggestions in mind, but *C. quadridentata* has been identified on both carapaces and soft parts and is assumed to exist in the Mediterranean.

BENSON: I was intrigued by the maps showing the concentration of samples in the Italian region and areas to the south, and by the absence of correlation of the data with that collected and described by Baird and Brady further east where they had some 15 localities. Would Dr Puri care to comment on this? Secondly, in cores or parts of cores I have examined from the Mediterranean an abyssal fauna has been noticeably absent. I know of no reason for this except possibly the abnormal circulation that occurs in this region. I would like to know more about this part of the fauna. Can Dr Puri throw any light on this?

PURI: Due to lack of time I did not describe the eastern Mediterranean faunas. There is a fauna in the Aegean Province. Brady described some samples from the Levant sponge sand and there are also some papers by Stephanides. We have five cores in shallow water in the eastern Mediterranean off the coast of Greece. One or two of these yielded a fauna which is a mixture of both fresh-water and near-shore forms. The other cores off the Greek coast are full of *Tyrrhenocythere*. West of Greece there is a Bairdiidae assemblage, some 25–30 specimens of *Bairdia*, and about 60 per cent of the fauna is nothing but Bairdiidae.

MCKENZIE: I would be grateful for an opinion from either Dr Puri or Dr Rome on whether or not *Cythereis polygonata* Rome 1942 is a late stage instar of *Cythereis prava* Baird 1850.

ROME: My typical form was an adult form, plus another adult, not larval form, with appendages and so I think that it is a valid species.

KORNICKER: Are these distribution maps based on living forms? If not, especially in view of your statement on tectonic activity in the area, you may be losing a considerable amount of information by not restricting distribution maps of this type to the living ostracods.

PURI: The distribution maps of the Mediterranean are based on both carapace and living material and I assume that all the fauna is Recent. We are just looking at the top one inch of the sediment-water interface.

THE ECOLOGY OF THE MARINE OSTRACODA OF HADLEY HARBOR, MASSACHUSETTS, WITH SPECIAL REFERENCE TO THE LIFE HISTORY OF *PARASTEROPE POLLEX* KORNICKER, 1967[1]

NEIL C. HULINGS
Smithsonian Institution, Washington, D.C., U.S.A.

SUMMARY

A suggested model of the life cycle of *Parasterope pollex* in Hadley Harbor, Massachusetts, based on available data is as follows.

1. The adult populations of summer 1964 and summer 1965 were derived from the spring 1964 and spring 1965 juvenile population, respectively. The percentage of the spring 1964 juvenile population moulting into the summer 1964 adult population was 29 and for the following year, 36.

2. The juvenile population of autumn 1964 was derived from the ovigerous female population of summer 1964.

3. There is periodicity as far as the males and ovigerous females of *P. pollex* are concerned. The male population has a life span of 8 weeks and the ovigerous female population of up to 15 weeks. The female population shows a decline from summer 1964 to spring 1965. However, it never disappears completely but is continuously added to through moulting of the juvenile population during the winter. There is only one generation per year.

4. The juvenile population shows periodicity in occurrence of various size classes. The smaller juveniles appear in late summer and early autumn in the first year and late summer in the second year.

5. The periodicity and reproductive period of *Sphaeronellopsis monothrix* and *Parasterope pollex* are essentially identical.

INTRODUCTION

The present report is an outgrowth of a detailed study of marine benthic invertebrate communities in Hadley Harbor, Massachusetts, initiated by R. H. Parker in September, 1963 (Parker and Williams, 1966; Parker, 1967). By means of weekly quantitative sampling over a two-year period at closely spaced stations in an area of $\frac{1}{3}$ km², Parker hoped to demonstrate the existence of benthic communities based on both macro- and meiofauna with emphasis on the latter. In order to determine seasonal distribution and the relationship of the fauna to ecological factors, he also measured weekly some 15 physical and chemical environmental parameters.

The ostracod and physical-chemical data collected during this study were provided by R. H. Parker through the Systematics-Ecology Program at Woods Hole, Massa-

[1] Systematics-Ecology Program Contribution No. 137.

chusetts. To R. H. Parker and to M. R. Carriker, Director of the Systematics-Ecology Program, I am indebted for having the opportunity to study the material. To numerous other individuals, including L. S. Kornicker, T. E. Bowman, J. H. Baker and J. C. McCain, I am indebted for helpful suggestions and assistance during the course of this preliminary study. I am also indebted to the Texas Christian University Research Foundation for providing initial financial support for the study.

DESCRIPTION OF STUDY AREA

The Hadley Harbor complex is part of the Elizabeth Islands Chain with Buzzards Bay to the north-west and Vineyard Sound to the south-east (Fig. 1). The complex is shallow (depth range, 0·3 to 7 m) and consists of a series of tidal channels and embay-

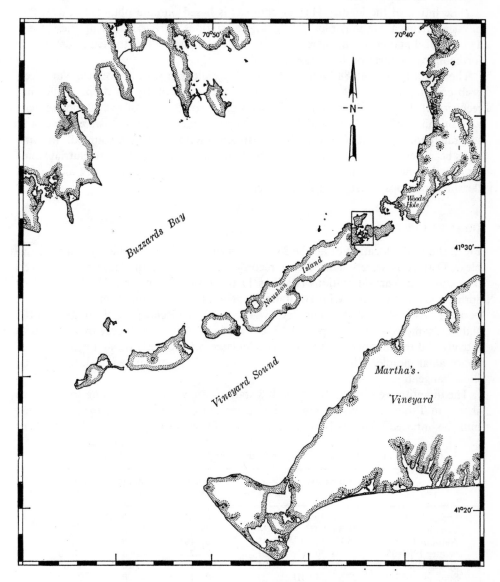

Fig. 1. Location of the Hadley Harbor study area.

ments through which exchange of water occurs between Buzzards Bay and Vineyard Sound. Settling basins and eel grass flats are characteristic of certain portions of the complex.

Bottom salinity throughout the complex is relatively uniform both areally and seasonally (annual range, 31·10 to 32·95‰) as is temperature at any one time. The bottom seasonal temperature range, however, is great, varying from −2·00 to 25·00°C. Other physical-chemical characteristics of the bottom of Hadley Harbor include primarily an oxidizing environment, except in certain areas of the settling basins and, with few exceptions, a neutral to slightly alkaline environment.

The primary sediment types of the complex include sand, silty sand, sandy silt and clayey silt and the median diameter is quite variable. Sorting coefficients range from poorly to very well sorted.

Description of the Hadley Harbor complex is based on data obtained at over 800 physical-chemical stations plus 288 biological stations by Parker during the survey. The physical and chemical characteristics of the complex will be treated in detail in a forthcoming publication by Parker (1967).

The majority of the 288 biological samples were taken with a modified Van Veen grab covering an area of $\frac{1}{25}$ m² and having a bite about 10 cm deep. Following recovery of the grab, the sediment was screened through a set of five 12-inch screens with openings of 4, 2, 1, 0·5 and 0·25 mm. Sorting and faunal counts were carried out separately for each fraction. Only living ostracods were sorted and counted. In some cases because of the great volume of animals present, it was necessary to elutriate and/ or split the finer fractions. Care was exercised, however, to count enough aliquots of the finer fractions to establish confidence.

TREATMENT OF OSTRACOD DATA

Fig. 2 shows the location of all the 288 biological stations in the Hadley Harbor complex. Ostracods were collected and recorded from 199 of the stations. However, ostracod data from 11 stations (stations 17 to 69 sampled prior to 4 May, 1964 or spring 1964) are not treated quantitatively. The 11 stations were sampled during the period 18 August, 1963 to 2 February, 1964. Samples collected from the remaining 188 stations, sampled during the period 4 May, 1964 to 8 August, 1965, are treated quantitatively and the number of ostracods is calculated on the basis of $\frac{1}{25}$ m², the size of the original sample. The terms 'sample station' and '$\frac{1}{25}$ m²' are equivalent and used interchangeably.

The quantitative samples from the 188 stations are grouped according to season as shown in Table 1. A restriction on interpretation of ostracod data for spring 1964 must be imposed because of the small number of samples (5) and the lack of samples

TABLE 1. Seasons and grouping of quantitative samples

Season	Inclusive Dates	No. of Stations	Inclusive Stations	Minimum Temp. (°C)	Maximum Temp. (°C)
Spring 1964	4.V–9.VI	5	69–81	9·70	15·31
Summer 1964	16.VI–8.IX	39	83–150	16·98	22·38
Autumn 1964	15.IX–24.XI	35	151–191	7·48	16·80
Winter 1964–65	2.XII–20.IV	61	192–258	−2·00	6·80
Spring 1965	27.IV–8.VI	21	260–279	9·20	15·80
Summer 1965	22.VI–24.VIII	27	905–2413	17·00	23·50

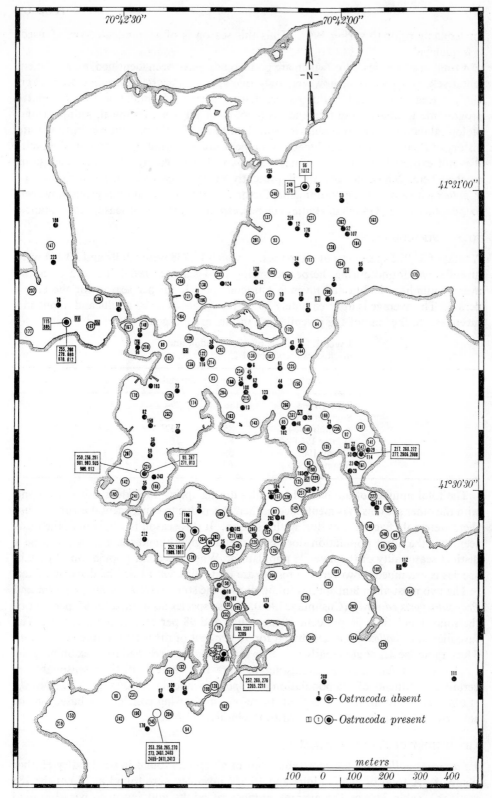

Fig. 2. Location of all biological stations, and station occurrence of ostracods in Hadley Harbor (modified after Parker, 1967).

immediately prior to spring 1964. Thus this season is often eliminated or, if used, it is qualified.

A total of 35 species, 31 of which are podocopids, have been identified, most of them tentatively. Of this total, however, only one species, *Parasterope pollex* Kornicker, 1967, is treated in detail in this paper. Until successful formulation of a computer programme to analyse the tremendous amount of physical, chemical, sediment and biological data that are available, it seems best to restrict the present report to the life history of *P. pollex* and the influence of temperature on various aspects of its life cycle. It is not expected that the computer analysis will significantly alter the conclusions reached here. Thus, no attempt is presently made to correlate the distribution of *P. pollex* with any of the environmental factors. Some consideration is given, however, to the total ostracod fauna of Hadley Harbor, especially in terms of seasonal abundance.

TOTAL OSTRACOD FAUNA

A total of 35,972 specimens of all species, of which 15,758 were adults and 20,214 were juveniles was found during the period spring 1964–summer 1965. Table 2 gives the average number of adult and juvenile ostracods per $\frac{1}{25}$ m² per season for the above period. The average is approximately 203 total ostracods or 90 adults and about 113 juveniles. In the case of the juveniles, all instars are lumped.

TABLE 2. Average number of adult and
juvenile ostracods per 1/25 m² per season

Season	Adults	Juveniles
Spring 1964	186	294
Summer 1964	98	32
Autumn 1964	85	132
Winter 1964–65	85	169
Spring 1965	59	100
Summer 1965	63	24

The total number of adults and juveniles during spring 1964 does not correspond with the other seasons. As mentioned earlier, this is due to experimental error because there were only five stations during this season. If this season is eliminated, then it is seen that the adult population does not fluctuate greatly and there appears to be a distinct seasonality of the juvenile population. When the adult population of certain species is examined, however, the results are quite different as will be discussed later.

The two most abundant species found during the study were *Parasterope pollex* and *Propontocypris edwardsi* (Cushman, 1906). These species accounted for 97 per cent of the total population, 96 per cent of the adults and 98 per cent of the juveniles. The juveniles of *P. pollex* alone accounted for 96 per cent of the total juvenile population. Thus, these species could greatly influence the conclusions reached from examining the total ostracod fauna. It is interesting to note that there is distinct seasonality in certain components of the population of *P. pollex* while no seasonality in *P. edwardsi* is obvious. Care must be exercised, therefore, in drawing conclusions based on the total ostracod fauna without regard to the individual species.

LIFE HISTORY OF *Parasterope pollex*

Parasterope pollex was the most abundant of all species found in the Hadley Harbor complex. A total of 8,209 adults and 19,549 juveniles were found at 170 of the 188 stations. Table 3 gives the average number of adults and juveniles per $\frac{1}{25}$ m² of

P. pollex by season. The average number of adults for all seasons was 48 and for the juveniles 115. Eliminating spring, 1964, there was a decrease in adult population from summer 1964 to spring 1965 and then an increase in summer 1965.

TABLE 3. Average number of total adults, males, non-ovigerous females, ovigerous females and juveniles of *Parasterope pollex* per 1/25 m² by season

Season	Total Adults	Males	Non-Ovigerous Females	Ovigerous Females	Juveniles
Spring 1964	208	55	145	8	258
Summer 1964	74	6	35	33	32
Autumn 1964	52	0	50	2	140
Winter 1964–65	33	0	33	0	183
Spring 1965	14	1	13	0	115
Summer 1965	41	2	17	22	19

When the adult population is broken down by sex and ovigerous against non-ovigerous females, however, the picture is quite different as seen in Table 3. The non-ovigerous females were present throughout the entire period although there was some fluctuation in the number of individuals from one season to the next. The number of juveniles, however, showed more variation with season. Again, if spring 1964 is eliminated and the period summer 1964 to summer 1965 is considered then there was an increase in the number of juveniles with a decrease in temperature. The adult males and ovigerous females exhibited a greater seasonality. The significance of the seasonality of these components of the *P. pollex* population is considered below.

Fig. 3 is a weekly and seasonal plot of the presence of adult non-ovigerous and ovigerous females and adult males plotted against temperature. In spring 1964, the adult males appeared at 15·00°C and in spring 1965 at 15·50°C. In the spring 1964, they were present till early July and disappeared. In spring 1965, the adult males were present till early August. Taking both occurrences into consideration, the temperature range for the males is 15·00 to 22·60°C. Thus the adult males of *P. pollex* exhibit a distinct seasonality.

A similar seasonality is seen in the ovigerous females. In the spring 1964, they appeared at the same time as the males, at a temperature of 15·00°C (9 June, 1964) and were present into early autumn (29 September, 1964). In the spring 1965, however, the ovigerous females were not found until the temperature reached 19·30°C (7 July, 1965) and were present in the summer sampling (17 August, 1965). No data are available for autumn 1965.

The adult male to adult non-ovigerous female ratio for the entire sampling period for *P. pollex* was about 1:12. If ovigerous females are added to the non-ovigerous females, then the ratio is 1:15·5. The ratio of male to ovigerous female is about 1:3·5.

DISCUSSION

There is very little known of the life history of *P. pollex* or other myodocopids. Such questions as the life span, the relationship between adult males and ovigerous females, i.e. whether or not ovigerous females are dependent on the presence of adult males, the time required for brooding and moulting, etc., cannot be answered at present. The field data available does, however, permit hypothesizing on these and other questions. Definitive testing of the hypotheses must await detailed morphological and laboratory studies.

The available data suggest that the adult males of *P. pollex* have a very short life span. In the first year, spring 1964 into summer 1964, they were present for about eight weeks. In the second year, spring 1965 into summer 1965, they were found for about the same number of weeks. Thus the life span of the adult males of *P. pollex* is about eight weeks. It is assumed that the adult males die following the reproductive season. The fact that in both seasons the total life span of the adult male population was about eight weeks supports this assumption.

The source of the adult males is unknown at present. One possibility, as suggested by the well-developed first antenna, is that they are planktonic. Although no plankton samples were available, the method of sampling used would have collected adult males if they are planktonic. The Van Veen grab sampler is open while passing through the water column to the bottom.

Another possibility, and one that is more likely, is that the adult males develop from the juvenile population that is abundant during the winter and early spring.

Whether one male fertilizes more than one female cannot be answered positively but the sex ratio data suggest that each male fertilizes somewhere between 3 and 15 females.

The data on the relationship between adult males and ovigerous females are somewhat conflicting. The data for spring 1965 are reliable because of sampling immediately prior to spring 1965 and the large number of samples. Data for spring 1964, however, are not reliable because of the lack of samples immediately prior to spring 1964 and the small number of samples taken. In the spring 1964, when only five samples were available, both adult males and ovigerous females appeared at the same time and temperature. In the second year, the ovigerous females were not found until about a month after the adult males and at a temperature of 19·30°C. If the assumption is made that the presence of males is necessary for the females to become ovigerous, then this would be the case for spring 1964. It would mean that copulation and deposition of the eggs in the brood chamber would be very rapid. The data for the second season which are more reliable does not support this since the ovigerous females did not appear until one month after the adult males. This suggests that the period of a month is required from the time of copulation to the deposition of the eggs in the brood chamber.

In 1964 and 1965, the ovigerous females were present after the adult males had disappeared. In the first year, they were present into early autumn. No data are available for the second autumn. It appears that the females die soon after the brooding season. This conclusion is based on the fact that very few post-brood chamber females were found in summer and autumn, 1964.

Fig. 4 is a plot of the occurrence of length categories of instars of *P. pollex*. After measurements of the instars had been completed, it was found that the juveniles fell into six size ranges as shown in Fig. 4. Brooks' Principle (Brooks, 1886) which states that the length of the larvae increases uniformly at each moult by one-fourth of its length before the moult, was applied to the size categories and the fit was within an acceptable variation. Poulsen (1965) applied this principle to *Parasterope muelleri* (Skogsberg, 1920) and found a mean increase of 28 per cent. There is, however, a slight discrepancy between what the author of this report used as the lower limit for adults, 1·40 mm, and what Kornicker (in Bowman and Kornicker, 1967) found. Kornicker gives an average length of 1·56 mm (range, 1·41 to 1·68 mm) for adult females and 1·57 mm (range, 1·50 to 1·63 mm) for adult males. Since Kornicker's data were based on appendage morphology and the present author's were not, the present

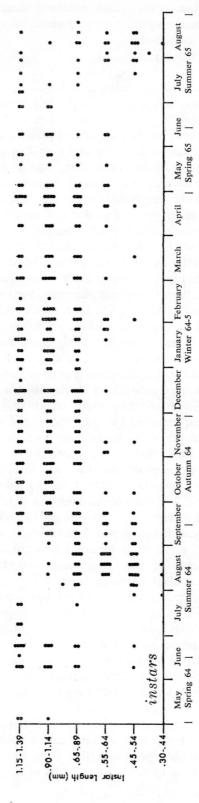

Fig. 4. Weekly and seasonal plot of the presence of lengths of instars of *Parasterope pollex* in Hadley Harbor.

data are subject to revision. Thus certain statements made earlier, such as the presence of adult females throughout the year must be qualified until detailed morphological studies are completed.

In spite of these reservations, it is possible to obtain some idea as to the length of the brooding period and the rate of moulting from the data. The data on the occurrence of early instars and ovigerous females indicate the brooding period ranges from three to four weeks. This is based on the time from the first occurrence of the ovigerous females to the first occurrence of the earliest free instar.

It appears that the early instars have a shorter life than the late instars. If temperature has a significant effect on the rate of moulting in *P. pollex* as it does on other crustaceans, i.e. an increase in the rate of moulting with an increase in temperature, then it would appear that the early instars moult rapidly. The early instars appear during late summer when the temperature is around 20°C. The rate of moulting decreases with a decrease in temperature so that it is drastically reduced during the winter. Some moulting, however, must occur to account for the continuous presence of adult females throughout the year. As the temperature increases in the spring, moulting is accelerated and this would account for the increase in the adult population in the summer. The data in Fig. 4 also show that the reproductive period is seasonal. The data further show that there is only one generation per year. Ideally, the generations last from about August to August. The generation as defined would include the total population including all instars and adults regardless of sex. But it must be remembered that the adult males are present for only a period of eight weeks. Furthermore, it is assumed that the ovigerous females die after the brooding period.

In 1967, Bowman and Kornicker described a parasitic copepod, *Sphaeronellopsis monothrix*, found in *Parasterope pollex* from Hadley Harbor and other areas in the vicinity of Woods Hole. They found the parasites present only in the male and female adults of *P. pollex*. In their most reliable data, they found about 8 per cent of the adults of *P. pollex* infected, 10 of 141 females and 1 of 6 males. Based on several samples, collected from 9 June, 1964 to 17 August, 1965 and containing about 700 adult females including non-ovigerous specimens, an overall infection of 13 per cent was found by the author of this report. Thus, there is good agreement between the results of both investigators. The reference to ovigerous females in this section must, however, be qualified. The presence of the parasite and the ovisacs was learned by the author after having recorded the ovigerous females (those with eggs in the brood chamber). Subsequently, the 'ovigerous females' were separated into those with only ostracod eggs, those with only copepod ovisacs, and those with both. Thus the term ovigerous female refers to a 'filled' brood chamber.

Among non-ovigerous female ostracods, 10 per cent were infected with female parasites and 2 per cent with male parasites. Among the ovigerous females 14 per cent were infected. Of the ovigerous females, 31 out of 336 had only copepod ovisacs, 6 had both ostracod eggs and copepod ovisacs (this agrees with the finding of Bowman and Kornicker), 7 were parasitized by the female but only ostracod eggs were present (Bowman and Kornicker found 1 of 23 like this) and one was parasitized by a male copepod. One other specimen was found with both copepod sexes plus copepod eggs in the brood chamber.

Bowman and Kornicker suggested that the males of *S. monothrix* have a shorter life span than the females and the present author's data support this suggestion. The author found male *S. monothrix* present in 1964 from 19 June to 23 June, over a temperature range of 15·00 to 19·80°C. The females of *S. monothrix* were present, however, from 9 June, 1964 to 29 September, 1964 (a temperature range of 15·00 to 22·38°C) and from 22 June, 1965 to 19 August, 1965 (a temperature range of 19·99 to 23·50°C).

It is interesting to note that the reproductive period of the copepod parasite and that of *P. pollex* coincide almost exactly. The ovigerous females of *P. pollex* were found from 9 June to 29 September in 1964 and from 7 July to 17 August in 1965. Copepod ovisacs were found during the same periods. Thus not only is there egg mimicry in the copepod parasite as reported by Bowman and Kornicker but 'mimicry' in the reproductive period.

REFERENCES

BOWMAN, T. E. AND KORNICKER, L. S. 1967. Two new crustaceans: the parasitic copepod *Sphaeronellopsis monothrix* (Choniostomatidae) and its myodocopid ostracod host *Parasterope pollex* (Cylindroleberidae) from the Southern New England Coast. *Proc. U.S. natn. Mus.*, **123**, (3613), 28 pp., 7 Figs, 2 Pls.

BROOKS, W. K. 1886. Report on the Stomatopoda collected by H.M.S. *Challenger* during the years 1873-76. *Report of the Scientific Results of the Voyage of H.M.S. Challenger, 1873-76*, Zoology **16**, Part 45, 1-116, Pls I-XVI.

PARKER, R. H. 1967. A model study of marine benthic communities, Hadley Harbor, Massachusetts. Prog. in Oceanogr. (in press).

PARKER, R. H. AND WILLIAMS, A. B. 1966. Seasonal study of small-scale variations in benthos and physico-chemical factors, Hadley Harbor, Massachusetts. *Abst. Papers, Second Internat. Oceanogr. Congress*, 278.

POULSEN, E. M. 1965. Ostracoda—Myodocopa, Pt. II: Cypridiniformes—Rutidermatidae, Sarsiellidae and Asteropidae. *Dana Rep.*, **12** (65), 1-484, Figs 1-156.

DISCUSSION

GREKOFF: There is considerable variation in collections of ostracods made at different seasons and you are right to draw attention to the importance of the season in this respect.

DELORME: Did you find a decrease in the number of adults with a combined increase in temperature and salinity?

HULINGS: Salinity is not a factor since it is relatively constant throughout the year. There is no apparent decrease with increase in temperature, the female population tending to remain relatively constant with no significant fluctuation.

FERGUSON: How significant is temperature in relation to the development of young instars? In 1944, working with three species of fresh-water ostracods from St. Louis, Missouri, I found that the rate of development is comparatively faster among young instars than later ones, but I do not think that this has anything to do with the change in temperature. I found this condition both during the winter when we had to collect under ice, and during the summer. I think that it is universally true, even in warm-blooded animals, that the young individual develops at a comparatively greater pace than does the adult.

HULINGS: My data indicate that the earlier instars have a much shorter life span, i.e. the length of time between one instar and the next is much shorter in the early stages of development than in the later stages. Among crustaceans, it has been demonstrated a number of times that temperature does in fact have an effect on the moulting rate; with increase in temperature, the rate of moulting increases and certainly my data indicate a similar relationship between moulting rate and temperature.

SOHN: I am concerned about your lack of instar individuals compared with your number of adults. Since you have about six instar stages the sum of the instar individuals should be much greater than it is. Could migration have taken place?

HULINGS: The average number of eggs per ovigerous female is seven. If you multiply this by the number of ovigerous females present in summer, 1964, you get 231 juveniles. The average number found was 140. I would rather attribute this to mortality than to migration.

BATE: In keeping *Heterocypris incongruens* in the laboratory I find that although the temperature remains fairly constant and the food supply is still there, there is a decline in population during the winter months. It seems probable that the decreasing hours of daylight suggest the approach of winter and that in consequence the ostracods lay resistant eggs. In your work have you noticed that the length of day has any effect?

HULINGS: Not so far, although it is possible that such information will come out of the multivariate factor analysis that is being programmed.

MCKENZIE: Did you notice in your material any microstructures of the type described in the paper presented by Dr Sohn and Dr Kornicker?

HULINGS: In looking at the preserved specimens I do not recall seeing any of the larger, more obvious microstructures.

ROME: Did you find a difference in other species regarding the incidence of males and females which would enable one to assess the rapidity of development?

HULINGS: The only difference I can comment on at present is the difference in the sex ratio among the different species. The average male to female ratio is about 1:4, ranging from 1:2 up to 1:12.

TAXONOMY AND ECOLOGY OF NEAR-SHORE OSTRACODA FROM THE PACIFIC COAST OF NORTH AND CENTRAL AMERICA

F. M. SWAIN
University of Minnesota, Minneapolis, U.S.A.

SUMMARY

The Ostracoda from 40 eulittoral and sublittoral localities along the Pacific Coast of North America from the Strait of Juan de Fuca, Washington to Baja California comprise about 90 species of which 30 or more are believed to be new but are not formally described here.

Two zoogeographic assemblages are recognized, separated more or less at Point Conception, California: (*i*) a northern assemblage having arctic and boreal affinities, and (*ii*) a southern assemblage showing resemblances to faunas from the Gulf of Mexico–Caribbean region.

Most of the living specimens were found on or near attached algae and rock bottom locations; relatively many fewer living specimens occurred in sand-silt environments. The latter are characterized by a high frequency of dead shells, which suggests that ecological analyses based on dead shells from sand-silt deposits are limited in value.

A collection from a small bay at San Juan del Sur, Nicaragua, yielded 51 species from four localities. The assemblage is indicative of the pronounced increase in diversity of assemblages as we approach the equator along the Pacific coast line.

INTRODUCTION

The samples described here were collected by Robert R. Lankford from approximately 125 stations along the Pacific Coast of North America extending from the southern tip of Baja California to Puget Sound, Washington (Fig. 1). The samples were collected mainly from short traverses perpendicular to the shoreline with the use of SCUBA gear from littoral stations to depths of about 130 feet. Lankford (1958) studied the Foraminifera from the samples and generously allowed the writer to examine the Ostracoda.

Ostracoda representing about 90 species were obtained from 40 of the stations. Peter L. Miller separated the specimens from most of the samples. The study was supported by grant GB 4110 of the National Science Foundation for which sincere appreciation is expressed.

LOCATION OF STATIONS AND BOTTOM TYPES

The location of the stations and the characteristics of the bottom at each station are shown in Table 1. The bottom types range from fine silty sand through medium and

2 E

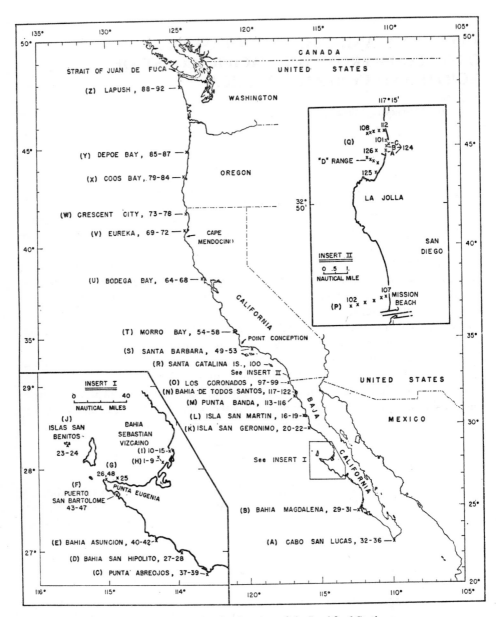

Fig. 1. Index map showing location of the Lankford Stations.

coarse sand and rock cobbles to outcropping bedrock. The sand bottom ranges from smooth to ripple-marked and the rock bottoms are in large part heavily coated with kelp, gorgonians, worm tubes, bryozoans, etc., but some rock outcrops were without much attached life.

OSTRACODA ASSEMBLAGES

The distribution of the Ostracoda species in the samples studied is shown in Figs 2–26. The numbers indicate the total number of specimens of each species and the number of living specimens (in parentheses) at each locality.

Two reasonably distinct assemblages can be recognized in the material studied, with separation of the two groups of species occurring more or less at Point Conception, California.

1. North of Point Conception the following are fairly typical: *Cythere* cf. *C. lutea*, *Heterocythereis dentarticulata*, *Heterocythereis* spp., and *Urocythere* sp. Several of these species have North Atlantic or Arctic affinities. In addition Lucas (1930) and [Lucas-]Smith (1952) recorded 34 Ostracoda species from the Vancouver Island region of which 16 were newly described from the area. Three of her new species are recognized in the present collection.

2. South of Point Conception the following species are reasonably characteristic: *Bairdia phlegeri*, *B. simuvillosa*, *Bradleya* sp., *Caudites* sp., *Cushmanidea guardensis*, *Cytherelloidea californica*, *Cytheropteron? ventrokurtosa*, *Cytherura bajacala*, *C.* cf. *C. gibba*, *C. johnsonoides*, *C. paracostata*, *Loxoconcha lenticulata*, *L.* sp., *L. tamarindoidea*, *Loxocorniculum sculptoides*, *Macrocyprina pacifica*, *Mutilus aurita*, *M. confragosa*, *Orionina pseudovaughani*, *Paracytheridea granti*, *Paracytherois perspicilla*, *Pellucistoma scrippsi*, *Pumilocytheridea vermiculoidea*, *Puriana pacifica*, and *Neocytherideis* cf. *N. subulata*. Many of these species have affinities with Gulf of Mexico–Caribbean forms or actually occur there in strata as old as Upper Miocene.

About seven species of the Lankford localities including some of the more abundant ones occur throughout the area studied.

Relationships of Ostracod Species to Bottom-type and Food Supplies

In nearly all of the samples studied the living specimens (carapaces containing appendages) and generally speaking the most numerous shells of dead specimens were found in association with living attached kelp, other algae, surfgrass, or living invertebrates on coarse sand, gravel and rock bottom. Areas having medium and fine sand and silt bottoms are characterized by a high frequency of dead shells as compared to living specimens. Palaeontological collections from comparable localities would be almost entirely from the sand-silt environments rather than the rock-kelp environments because the latter are mainly areas of erosion or non-deposition. As a result, ecological studies of ostracods based on material from near-shore sand-silt deposits would very likely not be indicative of the true habitats of the species and would have only limited significance.

In the present collection the depth of water and amount of surge seem to have had little effect on the distribution of these living specimens of ostracods.

[*Text continued on* p. 455.]

Table 1. Station Data

Sta. No.	Depth (ft)	Sample Type[1]	Area	Location N. Lat. W. Long.	Bottom Notations
1	157	Sn.	Bahia Sebastian Vizcaino Baja California, Mexico	28°09.80′ 114°19.20′	Medium sand, shelly, gray-brown with apatite grains
3		Sn.	Bahia Sebastian Vizcaino Baja California, Mexico	28°07.00′ 114°16.50′	Medium-fine sand, shelly, gray
5		Sn.	Bahia Sebastian Vizcaino Baja California, Mexico	28°05.60′ 114°14.90′	Medium-fine sand, shelly, gray
12	31	C.	Bahia Sebastian Vizcaino Baja, California, Mexico	28°14.20′ 114°05.68′	Medium sand, lt. gray, rippled, little surge, moderate organic debris, abundant pennatulids (sea pens)
13	51	C.	Bahia Sebastian Vizcaino Baja California, Mexico	28°14.17′ 114°06.00′	Medium sand, gray, egg-crate ripples, abundant organic debris, common pennatulids (sea pens)
19	2½	C.	Bahia Sebastian Vizcaino Baja California, Mexico	30°28.91′ 116°06.31′	Medium sand with small ripples bordered by patches of coarse sand with giant egg-crate ripples, mollusc frags.
21	25	C.	Bahia Sebastian Vizcaino Baja California, Mexico	29°47.33′ 115°47.60′	Medium sand, white, abundant Zostera (eel-grass) beds, faint H_2S odour at base of 4″ core
22	25	C.	Bahia Sebastian Vizcaino Baja California, Mexico	29°47.25′ 115°47.66′	Rock bottom with attached algae and abundant Strongylocentrotus franciascanus (sea urchin)—bottom grades abruptly to medium sand
23	55	C.	Islas San Benitos (Benito del Oeste), B. C., Mexico	28°18.22′ 115°34.03′	Rock bottom, coarse sand, shelly, weak surge, no ripples, abundant algae and Strongylocentrotus spp., gastropods
24	20	C.	Islas San Benitos (Benito del Oeste), B. C., Mexico	28°18.44′ 115°34.08′	Rock bottom, poorly sorted rubble of rock, sand, silt, no ripples or surge, abundant algae, sparse molluscan fauna
25	115	C.	South Bahia Sebastian Vizcaino, B. C., Mexico	28°51.20′ 114°51.30′	Rock bottom, thin (0-4in) veneer of pebbly, silty sand with abundant glauconite, between low strike ridges, no surge, abundant attached kelps and gorgonians
28	20	Scoop	South Bahia Sebastian Vizcaino, B. C., Mexico	26°58.30′ 114°00.30′	Rock bottom, coarse shell sand in pockets and crevices, surge very strong, Corallina, kelp, molluscs
29A	20	Core	Punta Huches, off Bahia Magdalena, B. C., Mexico	24°45.60′ 112°16.70′	Fine sand, white, long-crested ripples, moderate surge, no organisms observed

[1] Abbreviations for sample types: Sn., snapper dredge; C., collected by hand.

TABLE 1 (*contd.*)

Sta. No.	Depth (ft)	Sample Type	Area	Location N. Lat. W. Long.	Bottom Notations
30A	55	C.	Punta Huches, off Bahia Magdalena, B. C., Mexico	24°46.20' 112°16.45'	Rock outcrop, coarse sand with shell and pebbles, giant ripples, moderate surge, sediment grades abruptly into fine sand
32	80	C.	Cabo San Lucas (Canyon Slope), B. C., Mexico	22°52.49' 109°53.58'	Coarse sand, lt. gray, creeping down 30° slope between isolated granitic outcrops, gorgonians on rocks
37	51	Scoop	Whale Rock area, Punta Abreojos, B. C., Mexico	26°41.63' 113°39.35'	Rock bottom, poorly sorted sand and pebbles with shell fragments, very abundant kelp, gorgonians, starfish
38-B	20	Scoop	Whale Rock area, Punta Abreojos, B. C., Mexico	26°41.63' 113°39.35'	Rock bottom, coarse sand in interstices, abundant kelps, *Corallina*, gorgonians, abalone
41	30	C.	Bahia Asuncion, B. C., Mexico	27°08.10' 114°17.43'	Coarse sand near isolated outcrop, rare kelp growth, gorgonians
43	31	C.	Puerto San Bartolome, B. C., Mexico	27°40.83' 114°53.77'	Rock bottom in coarse sand composed of bryozoan fragments near large outcrops with abundant kelps
47	35	C.	Puerto San Bartolome, B. C., Mexico	27°39.65' 114°52.83'	Rock bottom, coarse sand, shelly, between outcrop areas, very abundant worm tubes completely covering the sand surface
48–7	25	Scoop	Puerto San Bartolome, B. C., Mexico	27°39.65' 114°52.83'	Rock bottom, accumulation of debris in bottom of steep-sided rock gully
52	70	C.	Puerto San Bartolome, B. C., Mexico	34°23.82' 119°39.82'	Silty sand, gray-brown, smooth bottom, no surge, *Macrocystis* debris, common pennatulids, worms, molluscs
55	102	C.	Morro Bay, California	35°21.58' 120°53.18'	Medium-fine sand, gray, ripples gentle surge, rare burrowing worms
67	37	C.	Morro Bay, California	38°18.30' 123°02.63'	Medium sand, bottom partially rippled, surge gentle, abundant tube worms stabilizing the sand
75	75	Scoop	Crescent City, California	41°43.21' 124°11.78'	Rock bottom, 15 ft relief, thin veneer of silty sand in rock crevices, no surge, organisms as at Sta. 74
76	30	C.	Crescent City, California	41°43.80' 124°10.41'	Medium sand, gray, rippled surge gentle, rare tube worms and gastropods

TABLE 1 (*contd*)

Sta. No.	Depth (ft)	Sample Type	Area	Location N. Lat. W. Long.	Bottom Notations
81	78	C.	Crescent City, California	43°21.35′ 124°23.10′	Medium sand, dark gray, rippled, no surge, too dark to observe organisms
82	58	C.	Crescent City, California	43°21.21′ 124°22.41′	Medium sand, gray, rippled, gentle surge, common *Dendraster*, rare crabs
83	34	C.	Crescent City, California	43°21.05′ 124°21.40′	Medium sand, gray, rippled, surge moderate to strong, broken shell, pebbles in ripple troughs, abundant wood fragments
85	90	C.	Depoe Bay, Oregon	44°46.60′ 124°05.50′	Coarse sand, brown, rippled, no surge, too dark for bottom observations
86	70	C.	Depoe Bay, Oregon	44°48.60′ 124°04.80′	Coarse sand, brown, abundant shell alternating with sand layers, giant ripples, no surge, no light
87	48	C.	Depoe Bay, Oregon	44°48.45′ 124°04.20′	Coarse sand, brown, near rock outcrops, giant ripples, surge gentle, tube worms, burrowing crustaceans
97	35	C.	Coronados Islands, B. C., Mexico	32°25.30′ 117°15.57′	Rock bottom, detritus from crevice, surge gentle, abundant attached kelp and smaller algae, echinoids, molluscs
100	20	C.	Santa Catalina Island, California	33°26.70′ 118°29.10′	Medium sand, gray to brown, common lignite, not rippled, coll. in small cove, lee side, contains non-living marsh spp.
101	12	C.	La Jolla, California	32°52.52′ 117°15.00′	Fine sand, gray, long-crested ripples, strong surge, near boulder outcrop, no organisms observed
102	132	C.	Mission Beach, San Diego, California	32°45.82′ 117°17.14′	Fine, silty sand, gray, rippled, light surge, abundant benthic organisms including: molluscs, urchins, pennatulids, burrowing worms
105	50	C.	Mission Beach, San Diego, California	32°46.03′ 117°15.83′	Fine sand, gray bits of shell debris, egg-crate to intermediate crested ripples, surge moderate, rare *Astropecten* (sand star)
106	31	C.	Mission Beach, San Diego, California	32°46.15′ 117°15.46′	Fine sand, gray, rare pebbles, shells, long crested ripples, moderate to strong surge causing sheet flow of sand, abundant gastropods, tube worms, clams

TABLE 1 (contd)

Sta. No.	Depth (ft)	Sample Type	Area	Location N. Lat. W. Long.	Bottom Notations
108	119	C.	La Jolla, California	32°52.89' 117°16.04'	Fine silty sand, gray-brown, irregular surface, no ripples or surge, sediment surface covered with brown diatom crust, common pennatulids, dead surf-grass
110	75	C.	La Jolla, California	32°52.96' 117°15.77'	Fine sand, gray, intermediate crested ripples with brown diatom crust, no surge, abundant mysidaceans, burrowing worms, gastropods
111	45	C.	La Jolla, California	32°53.00' 117°15.51'	Fine sand, gray intermediate crested ripples, light surge, no brown diatom crust, abundant heart urchins, rare *Astropecten*, hydroids
114	80	C.	Punta Banda, B.C., Mexico	31°42.01' 116°41.07'	Rock cobble bottom, interstitial coarse sand, dark gray to brown, irregular surface, no ripples or surge, attached kelp, *Heliaster*, abundant dead *Mytilus*
115	50	Scoop	Punta Banda, B.C., Mexico	31°42.01' 116°41.06'	Rock cobble bottom with only thin patches of rippled sand, moderate surge, abundant brittle stars; sample 30' seaward of submarine, vertical rock scarp covered with sea urchins
116	20	C.	Punta Banda, B.C., Mexico	31°42.00' 116°41.02'	Coarse sand, white to tan, between scattered boulders covered with sea urchins, strong surge, rare attached kelp and small algae
121	30	C.	Bahia de Todos Santos, B.C., Mexico	31°43.95' 116°40.22'	Fine sand, light gray, dead lagoon pelecypods (*Chione*), intermediate crested ripples, moderate surge, rare pennatulids, *Astropecten*
122	15	C.	Bahia de Todos Santos, B.C., Mexico	31°43.56' 116°40.00'	Fine sand, gray, intermediate to long crested ripples, strong surge (5 ft/ sec.), bottom sediment thickly suspended in water

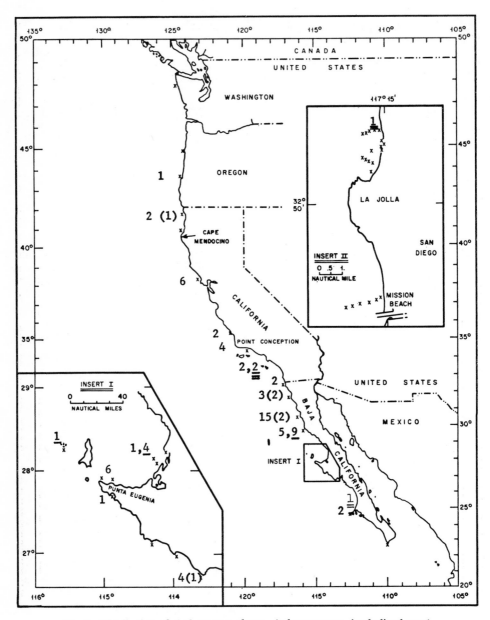

Fig. 2. Distribution of *Ambostracon glauca, Ambostracon* sp. (underlined once), *Anterocythere* sp. (underlined twice), and *Cytherella* cf. *C. vizcainoensis* (underlined three times), in number of specimens at each station; living specimens in parentheses.

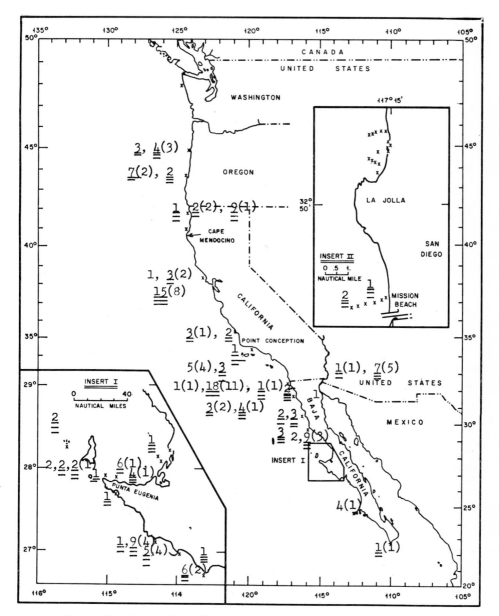

Fig. 3. Distribution of *Aurila conradi californica, Aurila* sp. A. (underlined once), *Aurila jollaensis* (underlined twice) (also 2 at Sta. 22) and *Aurila lincolnensis* (underlined three times), in number of specimens at each Station; living specimens in parentheses.

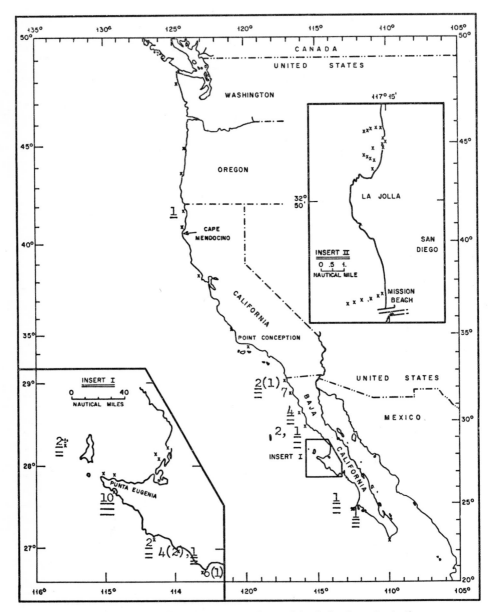

Fig. 4. Distribution of *Aurila* sp. *B*, *Aurila* sp. *C* (underlined once), *Aurila* sp. *D* (underlined twice), and *Bairdia phlegeri* (underlined three times) in number of specimens at each Station; living specimens in parentheses.

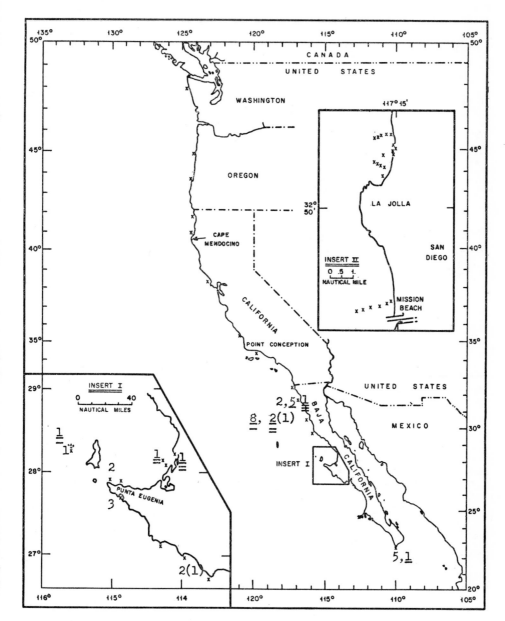

Fig. 5. Distribution of *Bairdia simuvillosa, Bairdia tuberculata* (underlined once),
Basslerites delrayensis (underlined twice) and *Basslerites sonorensis* (underlined
three times), in number of specimens at each Station; living specimens in parentheses.

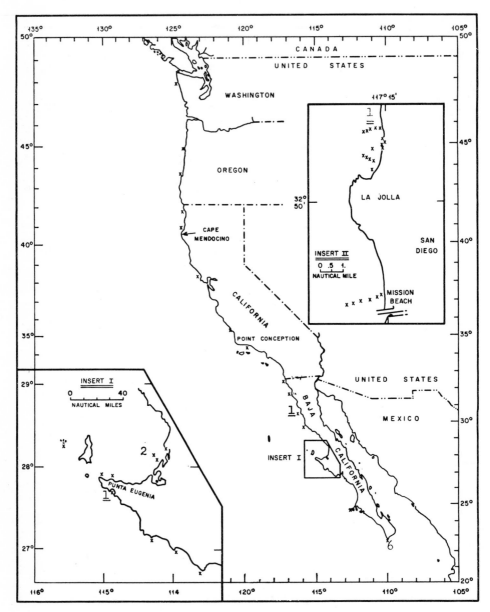

Fig. 6. Distribution of *Basslerites thlipsuroidea*, and *Trachyleberidea*? *henryhowei* (underlined once) in number of specimens at each Station; living specimens in parentheses.

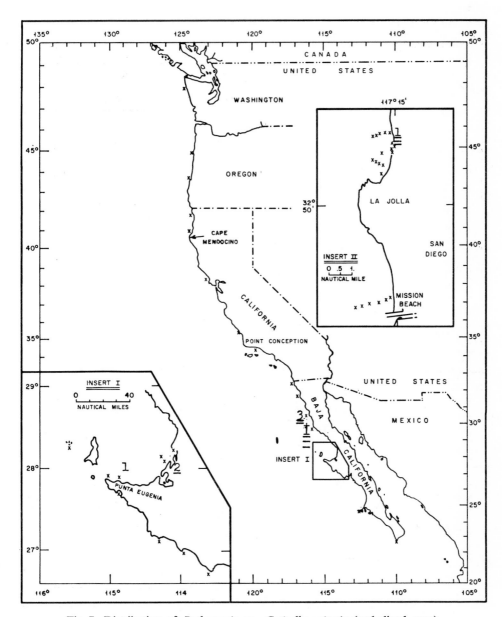

Fig. 7. Distribution of *Bythocypris* sp., *Cativella unitaria* (underlined once),
Caudites sp. A (underlined twice), and *Caudites* aff. *C. leguminosus* (underlined
three times), in number of specimens at each Station; living specimens in parentheses.

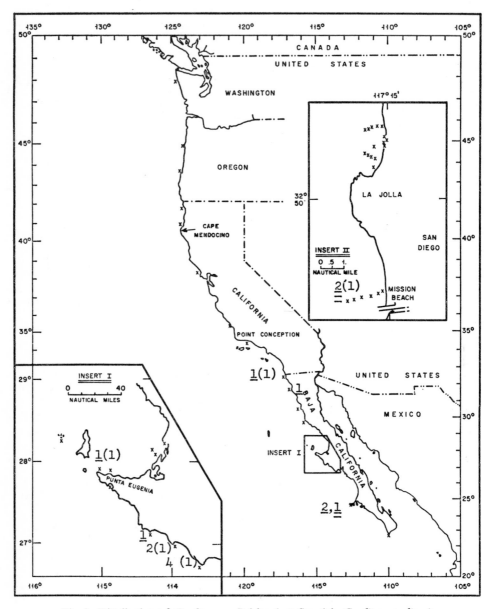

Fig. 8. Distribution of *Caudites* sp. *B* (also 1 at Sta. 1a), *Caudites rosaliensis* (underlined once) (also 4 at Sta. 37), *Caudites* sp. *C* (underlined twice), and *Cushmanidea guardensis* (underlined three times), in number of specimens at each Station; living specimens in parentheses.

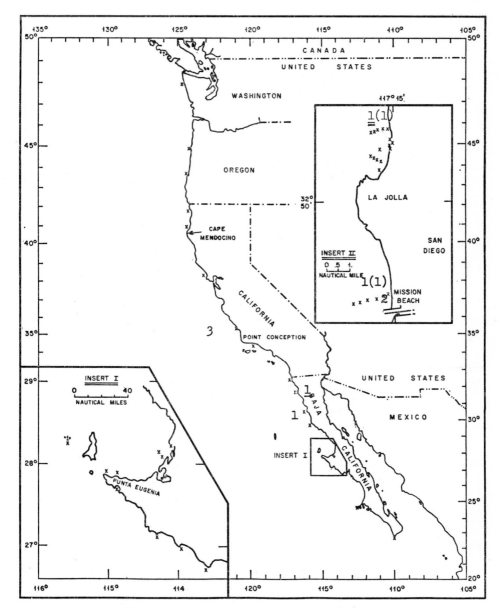

Fig. 9. Distribution of *Cushmanidea pauciradialis, Cushmanidea* sp. (underlined once) and *Pseudophilomedes*? sp. (underlined twice) in number of specimens at each Station; living specimens in parentheses.

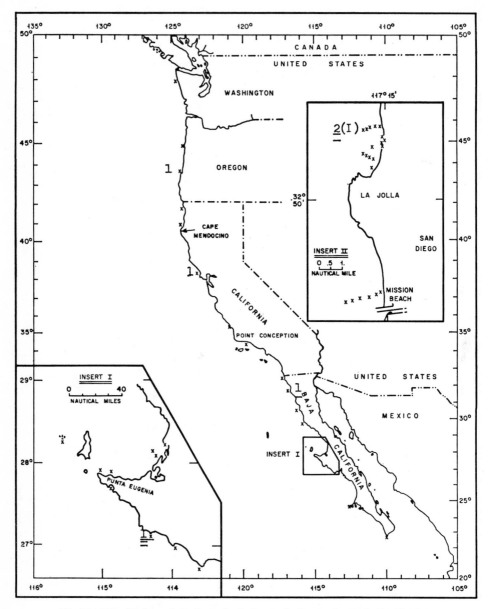

Fig. 10. Distribution of *Cythere* cf. *C. lutea, Cytherois* cf. *C. fischeri* (underlined once), *Cytherella* cf. *C. banda* (underlined twice), and *Cytherelloidea* sp.,
immature (underlined three times) in number of specimens at each Station;
living specimens in parentheses.

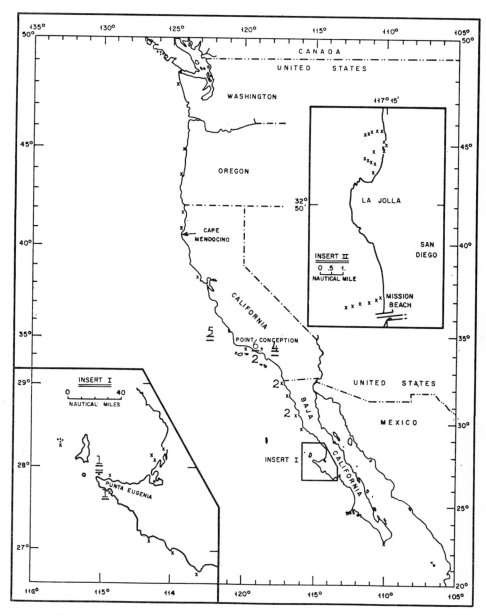

Fig. 11. Distribution of *Cytherelloidea californica*; *Cytheretta* aff. *C. danaiana* (underlined once), *Cytheretta* sp. (underlined twice), and *Cytheroma* sp. aff. '*Microcythere*' *gibba* Müller (underlined three times), in number of specimens at each Station; living specimens in parentheses.

2 F

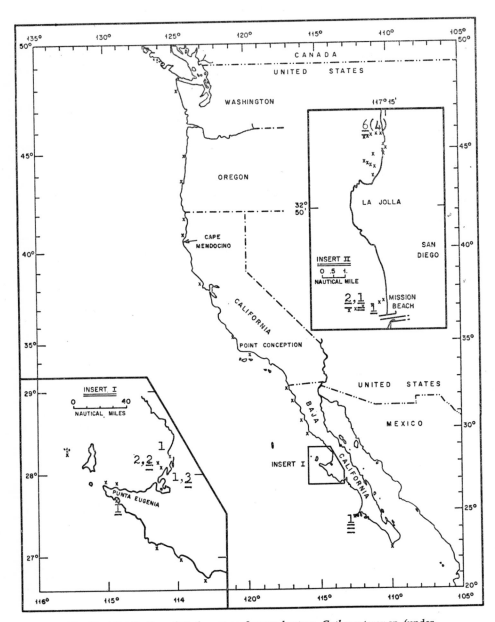

Fig. 12. Distribution of *Cytheropteron*? *ventrokurtosa, Cytheropteron* sp. (underlined once), *Cytherura bajacula* (underlined twice), and *Cytherura* cf. *C. gibba*
(underlined three times), in number of specimens at each Station; living specimens in parentheses.

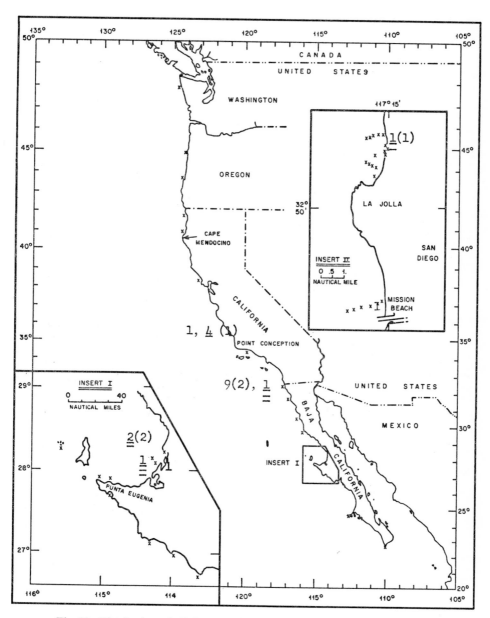

Fig. 13. Distribution of *Cytherura johnsonoides, Cytherura* sp. *A* (underlined once), *Cytherura paracostata* (underlined twice), and *Cytherura* sp. *B,* immature (underlined three times), in number of specimens at each Station; living specimens in parentheses.

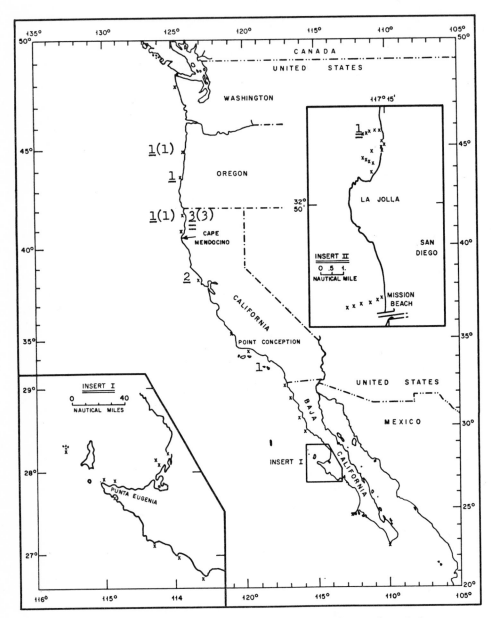

Fig. 14. Distribution of *Echinocythereis* sp. *Heterocythereis dentarticulata* (Smith) (underlined once) (also 1 at Staion 108), *Heterocythereis* sp. *A* (underlined twice) (also 2 at Sta. 67), and *Heterocythereis* sp. *B* (underlined three times), in number of specimens at each Station; living specimens in parentheses.

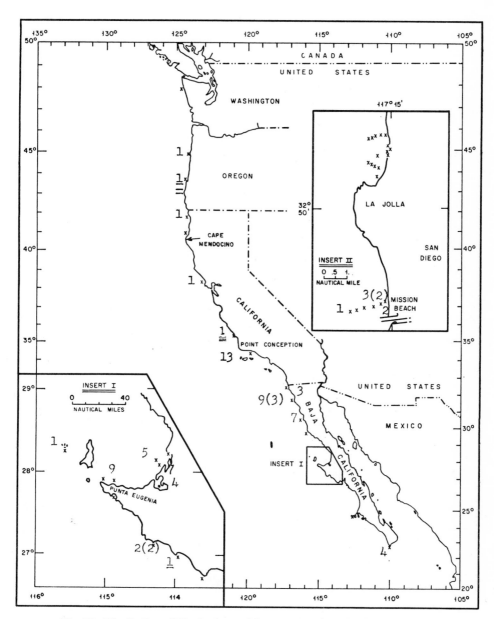

Fig. 15. Distribution of *Hemicythere californiensis, Hemicytherura cranekeyensis* Puri (underlined once), *Hemicytherura* sp. (underlined twice), and *Cythere* sp. *A* (underlined three times), in number of specimens at each Station; living specimens in parentheses.

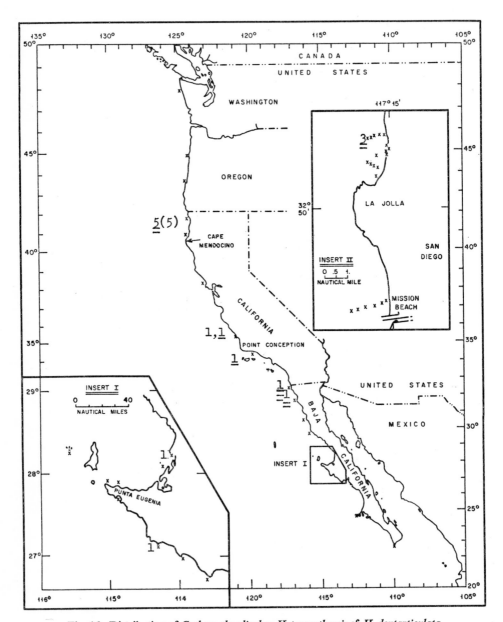

Fig. 16. Distribution of *Cythere alveolivalva, Heterocythereis* cf. *H. dentarticulata* (underlined once), *Jonesia rostrata,* (underlined twice) and *Kingmaina* sp. (underlined three times), in number of specimens at each Station; living specimens in parentheses.

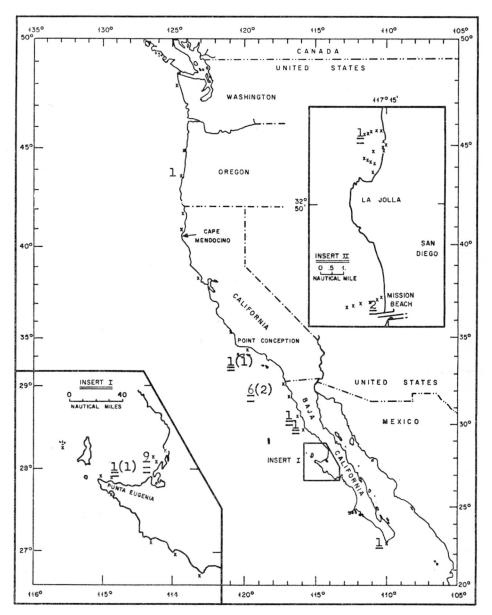

Fig. 17. Distribution of *Leptocythere* sp., *Loxoconcha lapidiscola* (underlined once), *Loxoconcha lenticulata* (underlined twice), and *Loxoconcha* sp. (underlined three times), in numbers of specimens at each Station; living specimens in parentheses.

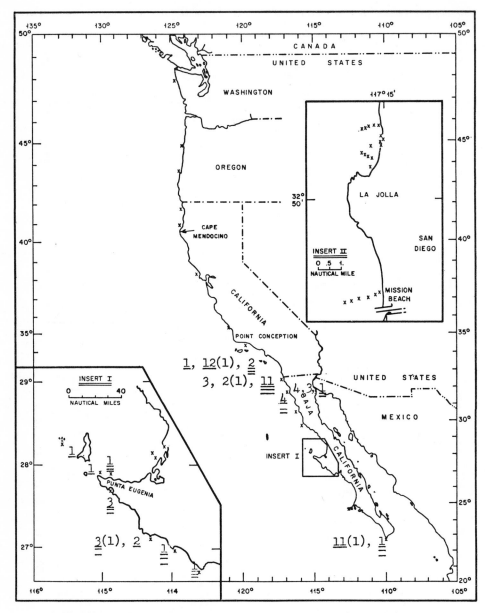

Fig. 18. Distribution of *Loxoconcha tamarindoidea, Loxocorniculum sculptoides* (underlined once), *Loxocorniculum* sp. *B* immature (underlined twice), and *Macrocyprina pacifica* (underlined three times), in number of specimens at each Station; living specimens in parentheses.

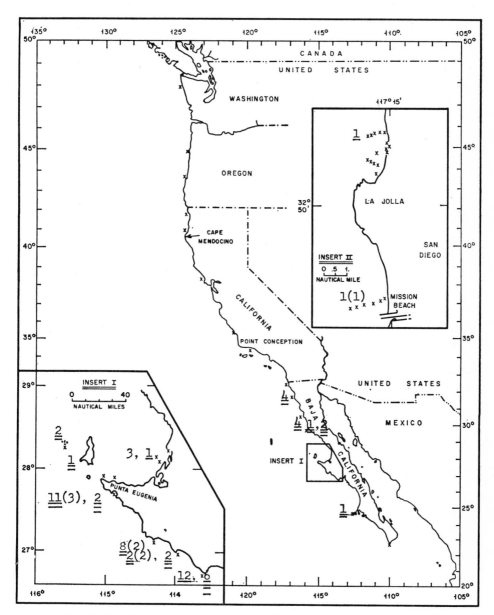

Fig. 19. Distribution of *Miracythere* sp., *Monoceratina* sp. (underlined once), *Mutilus aurita* (underlined twice), and *Mutilus confragosa* (underlined three times), in number of specimens at each Station; living specimens in parentheses.

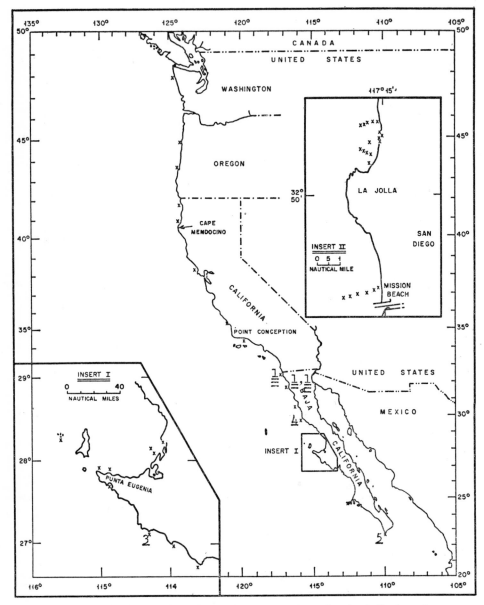

Fig. 20. Distribution of *Mutilus* sp., *Orionina pseudovaughani* (underlined once), *Paracypris franquesoides* (underlined twice), and *Paracypris politella* (underlined three times), in number of specimens at each Station; living specimens in parentheses.

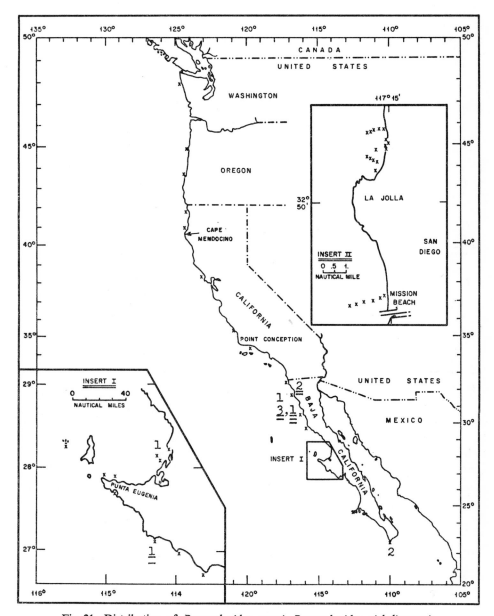

Fig. 21. Distribution of *Paracytheridea granti. Paracytheridea pichelinguensis* (underlined once), *Paracytherois perspicilla* (underlined twice), and *Paradoxostoma* cf. *P. hodgei* (underlined three times), in number of specimens at each Station; living specimens in parentheses.

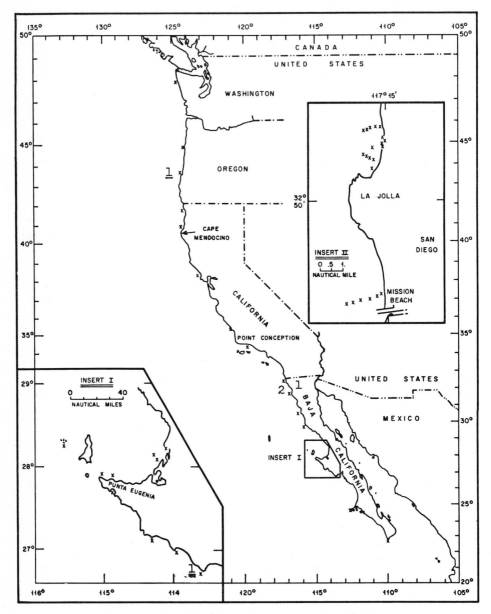

Fig. 22. Distribution of *Pellucistoma scrippsi, Pseudocytthereis* cf. *'P.' simpsonensis* (underlined once), and *Cytherelloidea praecipua* van den Bold (underlined twice), in number of specimens at each Station; living specimens in parentheses.

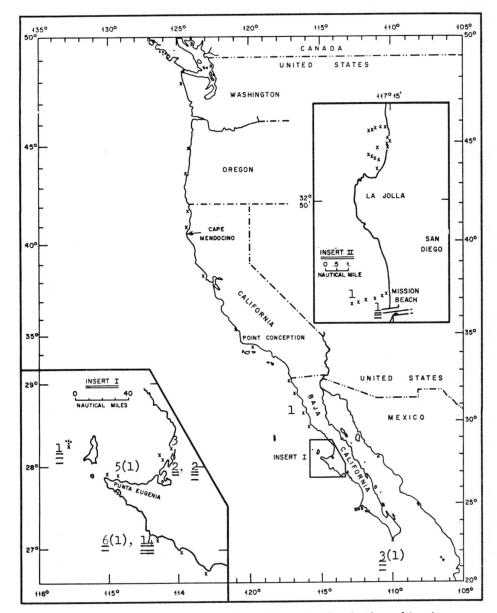

Fig. 23. Distribution of *Pterygocythereis delicata, Pumilocytheridea realejoensis* (underlined once), *Pumilocytheridea vermiculoidea* (underlined twice), and *Puriana pacifica* (underlined three times), in number of specimens at each Station; living specimens in parentheses.

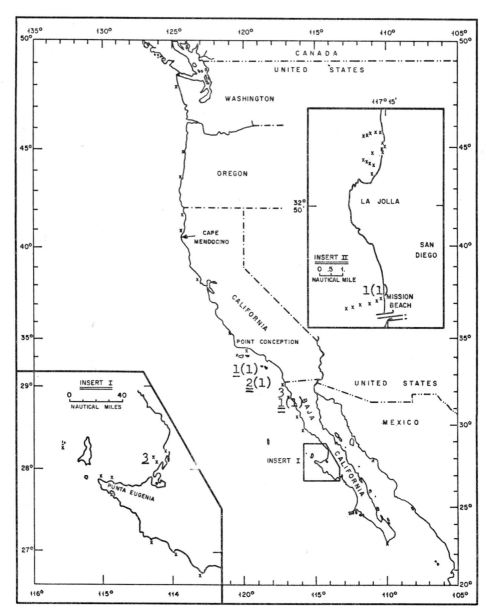

Fig. 24. Distribution of *Neocytherideis* cf. *N. subulata, Sclerochilus contortellus* (underlined once), and *Sclerochilus nasus* (underlined twice), in number of specimens at each Station; living specimens in parentheses.

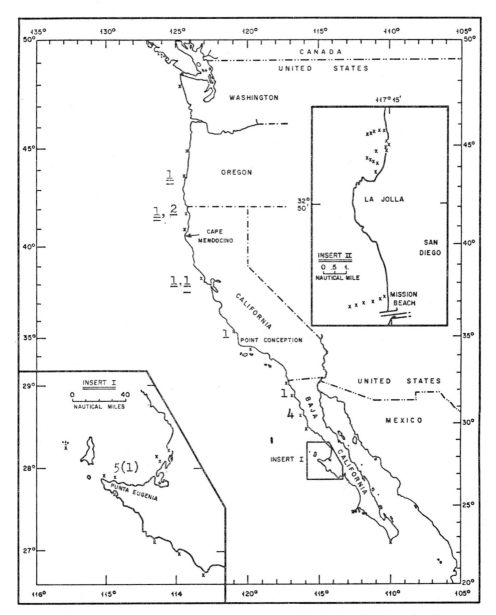

Fig. 25. Distribution of *Trachyleberidea tricornis* (also 1 at Sta. 37), *Urocythere*? sp. *A* (underlined once), and *Urocythere*? sp. *B* (underlined twice) (also 2 at Sta. 75), in number of specimens at each Station; living specimens in parentheses.

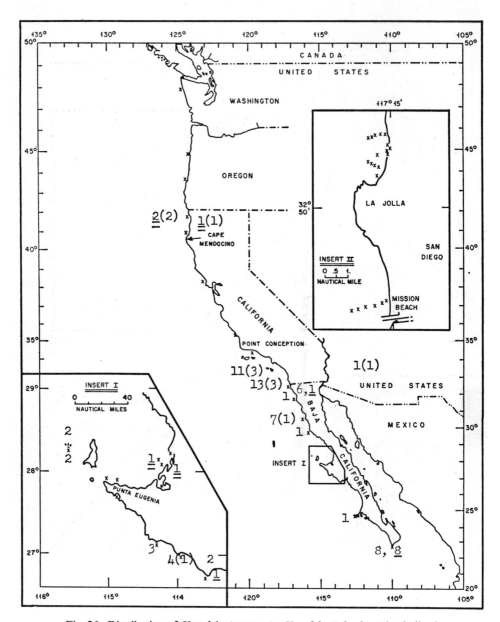

Fig. 26. Distribution of *Xestoleberis aurantia, Xestoleberis hopkinsi* (underlined once), and *Xiphichilus tenuissimoides* (underlined twice), in number of specimens at each Station; living specimens in parentheses.

PLATE I.

Figs 1a, b. *Cytherelloidea praecipua* van den Bold. Left side and dorsal views; Whale Rock area, Punta Abreojas, Baja California, Mexico, Lankford Station 37. × 44. Figs 2a, b. *Cytherelloidea praecipua* van den Bold. (a) Interior of right valve; San Juan del Sur, Nicaragua, Station 4. (b) Left side, San Juan del Sur, Nicaragua, Station 1. × 47. Fig. 3. *Cytherelloidea* sp. *A.* Exterior of holotype left valve; San Juan del Sur, Station 3. × 50. Figs 4a-c. *Cytherelloidea sanlucasensis* Swain. (a), (c) Exterior of two male right valves; San Juan del Sur, Nicaragua, Station 1; (b) exterior of female right valve; San Juan del Sur, Nicaragua, Station 1, × 47. Figs 5a, b. *Cytherelloidea* sp. *B.* Interior of right and left valves; San Juan del Sur, Nicaragua, Station 3. × 45. Figs 6a-c. *Cytherella* cf. *C. vizcainoensis* McKenzie and Swain. (a) Exterior of left valve; (b) left side of shell; (c) dorsal view; Santa Catalina Island, Lankford Station 100. × 50. Figs 7a-e. *Bairdia tuberculata* Brady. (a) Ventral view of complete shell; (b), (d) interior views of two left valves; (c) dorsal view of complete shell; (e) right side of complete shell; Punta Banda, Baja California, Lankford Station 114. × 48. Figs 8a, b. *Triebelina* cf. *T. gierloffi* Hartmann. Exterior views of two right valves; San Juan del Sur, Nicaragua, Station 1. × 48. Fig. 9. *Paracypris politella* Swain. Left side of shell; San Juan del Sur, Nicaragua, Station 1. × 47. Figs 10a, b. *Perissocytheridea meyerabichi* (Hartmann). (a) Exterior of immature right valve; (b) exterior of immature left valve; San Juan del Sur, Nicaragua, Station 5. × 47. Fig. 11. *Cythere* cf. *C. lutea* O. F. Müller. Right side of shell; Bodega Bay, California, Lankford Station 67. × 49. Figs 12a, b. *Cythere* sp. *A.* Dorsal and left side views of shell; Coos Bay, Oregon, Lankford Stations 81-83. × 47. Fig. 13. *Heterocythereis dentarticulata* (Smith). Exterior of right valve; Morro Bat, California, Lankford Station 55. × 49.

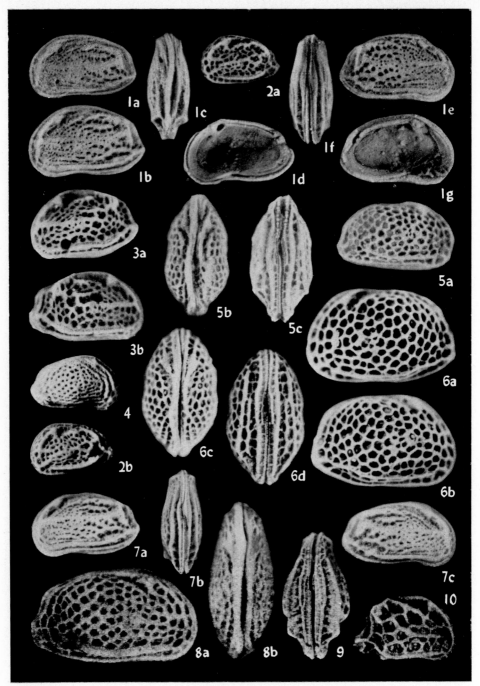

PLATE II.

Figs 1a-g. *Aurila* sp. *A*. (a) Left side of shell; (b) left side of shell; (c) dorsal view of shell; (d) interior of right valve; (e) exterior of right valve; (f) dorsal view of shell; (g) interior of left valve; Punta Banda, Baja California, Lankford Station 114. × 50. Figs 2a, b. *Aurila* sp. *B*. Exterior views of left valves; San Juan del Sur, Nicaragua, Stations 1 and 2 respectively. × 45. Figs 3a, b, 9. *Aurila jollaensis* (LeRoy). 3a, Left side of shell; 3b, 9, right side and ventral views of shell; Isla San Geronimo, Baja California, Lankford Station 22. × 45. Fig. 4. *Aurila lincolnensis* (LeRoy). Exterior of right valve; San Juan del Sur, Nicaragua, Station 2. × 52. Figs 5a-c. *Hemicythere californiensis* LeRoy. (a) Right side of male shell; (b) dorsal view; (c) ventral view; Bahia Sebastian Vizcaino, Baja California, Lankford Station 25. × 50. Figs 6a-d. *Hemicythere californiensis* LeRoy. (a) Left side of female shell; (b) right side of female shell; (c) dorsal view; (d) ventral view; Bahia Sebastian Vizcaino, Baja California, Lankford Station 5. × 55. Figs 7a-c. *Aurila* sp. *A*. (a) Left side of shell; (b), (c) dorsal and right side views of shell; Isla San Geronimo, Baja California, Lankford Station 22. × 52. Figs 8a, b. *Aurila* sp. *C*. Right side and dorsal view of shell; Crescent City, California, Lankford Station 75. × 48. Fig. 10. *Mutilus cardonensis* Swain and Gilby. Exterior of right valve; San Juan del Sur, Nicaragua, Station 2. × 50.

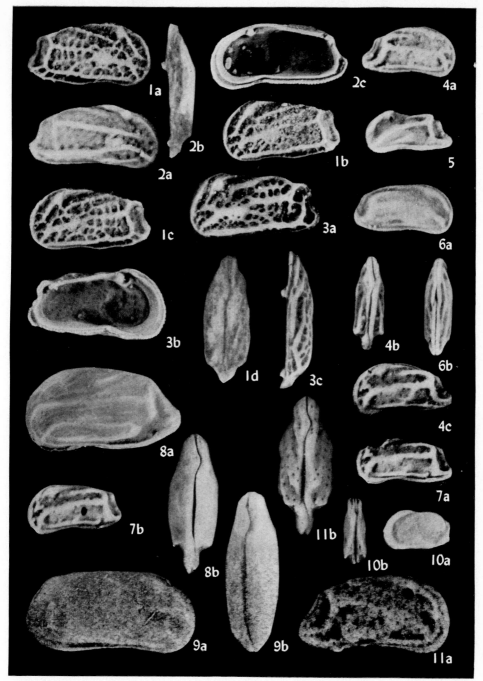

PLATE III.

Figs 1a-d. *Orionina pseudovaughani* Swain. (a) Exterior of right valve; (b) exterior of left valve; (c) exterior of left valve; (d) dorsal view of shell; Isla San Geronimo, Baja California, Lankford Station 21. × 48. Figs 2a-c. *Orionina lienenklausi* (Müller). (a) Exterior of right valve; (b) dorsal view of right valve; (c) interior of right valve; Bahia Asuncion, Baja California, Lankford Station 41. × 47. Figs 3a-c. *Ambostracon glauca* Skogsberg. (a) Exterior of left valve; (b) interior of left valve; (c) dorsal view of right valve; Bodega Bay, California, Lankford Station 67. × 50. Figs 4a-c. *Caudites rosaliensis* Swain. (a), (b) Right side and dorsal views of male shell; Bahia Magdalena, Baja California, Lankford Station 29A. (c) Left side of male shell; Punta Abreojos, Baja California, Lankford Station 37. × 50. Fig. 5. *Anterocythere* sp. Exterior of left valve; Bahia Magdalena, Baja California, Lankford Station 29A. × 49. Figs 6a, b. *Caudites* sp. C. Right side and dorsal views of shell; Bahia Magdalena, Baja California, Lankford Station 29A. × 50. Figs 7a, b. *Caudites rosaliensis* Swain. Left side of male and female shell; Punta Abreojos, Baja California, Lankford Station 29A. × 80. Figs 8a, b, 9a, b. *Caudites* sp. B. Figs 8a, b, Left side and dorsal view of male shell; Bahia San Hipolito, Baja California, Lankford Station 28. Figs 9a, b. Right side and dorsal view of female shell; Isla San Martin, Baja California, Lankford Station 19. × 48. Figs 10a, b. *Miracythere* sp. Left side and dorsal views of shell; Bahia Sebastian Vizcaino, Baja California, Lankford Station 1. × 49. Figs 11a, b. *Caudites* sp. A. Right side and dorsal views of shell; Isla San Geronimo, Baja California, Lankford Station 22. × 50.

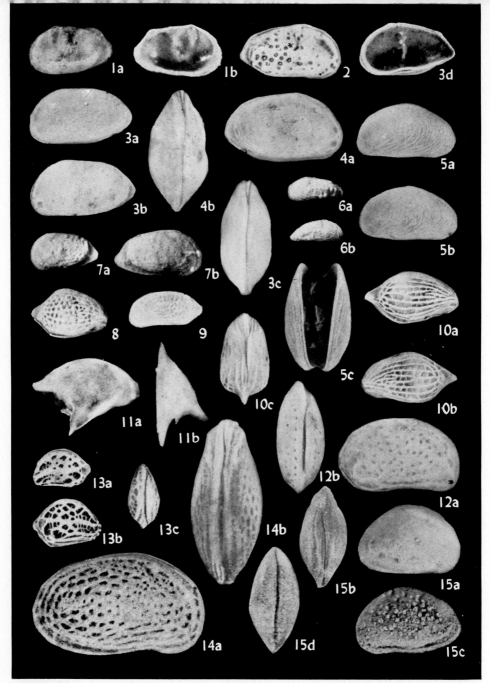

PLATE IV.

Figs 1a, b. *Monoceratina* ? sp. Exterior view and interior view of a left valve; Bahia Sebastian Vizcaıno, Baja California, Lankford Station 1. × 47. Fig. 2. *Urocythere* ? sp. *A.* Right side of shell; Bodega Bay, California, Lankford Station 67. × 48. Figs 3a-d, 4a, b. *Urocythere* ? sp. *B.* Fig. 3a. Left side of a male shell; Crescent City, California, Lankford Station 75; 3b, c, left side and dorsal views of a male shell; Coos Bay, Oregon, Lankford Stations 81-83; 3d, inside of male right valve. Figs 4a, b. Right side and dorsal views of a female shell; Crescent City, California, Lankford Station 75. × 44. Figs 5a-c. *Cushmanidea pauciradialis* Swain. (a) Right side of shell; (b) right side of shell; (c) ventral view of shell; Morro Bay, California, Lankford Station 55. × 73. Figs 6a, b. *Pumilocytheridea realejoensis* Swain and Gilby. (a) Left side of shell; (b) right side of shell; San Juan del Sur, Nicaragua, Station 3. × 56. Figs 7a, b. *Urocythere* ? sp. *C.* (a) Exterior of small left valve; (b) exterior of right valve; San Juan del Sur, Nicaragua Station 1. × 49. Fig. 8. *Cytherura bajacala* Benson. Right side of shell; La Jolla, California, Lankford Station 110. × 47. Fig. 9. *Cytherura johnsonoides* ? Swain. Left side of shell; Morro Bay, California, Lankford Station 55. × 49. Figs 10a-c. *Cytherura* sp. *A.* (a) Right side of shell; (b) left side of shell; (c) ventral view of shell; Lankford Station 55. × 47. Figs 11a, b. *Cytheropteron* sp. Exterior and dorsal views of right valve; Puerto San Bartolome, Baja California, Lankford Station 47. × 55. Figs 12a, b. *Cythere alveolivalva* Smith. Right side and dorsal views of shell; Bahia Asuncion, Baja California, Lankford Station 41. × 48. Figs 13a-c. *Hemi-cytherura* sp. (a) Left side of immature shell; (b) right side of shell; (c) dorsal view of shell; Bahia de Todos Santos, Baja California, Lankford Station 122. × 72. Figs 14a, b. *Aurila lincolnensis* LeRoy. Right side and dorsal views of large shell; Punta Banda, Baja California, Lankford Station 114. × 51. Figs 15a-d. *Aurila* sp. *D.* (a) Right side of shell; (b) dorsal view of shell; (c) left side of shell; (d) ventral view of shell; La Jolla, California, Lankford Station 101. × 51.

List of Species

The following species, alphabetically arranged, were found in the Lankford samples studied:

Ambostracon glauca (Skogsberg)
Ambostracon sp.
Anterocythere sp.
Aurila conradi californica Benson
Aurila jollaensis (LeRoy)
Aurila lincolnensis (LeRoy)
Aurila sp. *A*
Aurila sp. *B*
Aurila sp. *C*
Aurila sp. *D*, nearly smooth
Bairdia phlegeri McKenzie and Swain
Bairdia simuvillosa Swain
Bairdia tuberculata Brady
Basslerites delrayensis LeRoy
Basslerites sonorensis Benson and Kaesler
Basslerites thlipsuroidea Swain
Bythocypris sp.
Cativella unitaria Swain
Caudites sp. *A*
Caudites aff. *C. leguminosus* Bold
Caudites sp. *B*
Caudites rosaliensis Swain
Caudites sp. *C*
Cushmanidea guardensis Swain
Cushmanidea pauciradialis Swain
Cushmanidea sp.
Cythere sp. *A*
Cythere cf. *C. lutea* O. F. Müller
Cythere alveolivalva Smith
Cytherella cf. *C. banda* Benson
Cytherella cf. *C. vizcainoensis* McKenzie and Swain
Cytherelloidea sp., immature
Cytherelloidea californica LeRoy
Cytherelloidea praecipua van den Bold
Cytheretta aff. *C. danaiana* (Brady)
Cytheretta sp.
Cytherois cf. *C fischeri* (Sars)
Cytheroma aff. *Microcythere gibba* Müller
Cytheropteron? *ventrokurtosa* Swain
Cytheropteron sp.
Cytherura bajacala Benson
Cytherura johnsonoides Swain
Cytherura paracostata Swain
Cytherura cf. *C. gibba* Müller
Cytherura sp. *A*

2 G

Cytherura sp. *B.*, immature
Echinocythereis sp.
Hemicythere californiensis LeRoy
Hemicytherura cranekeyensis Puri
Hemicytherura sp.
Heterocythereis sp. *A*
Heterocythereis sp. *B*
Heterocythereis dentarticulata (Smith)
Heterocythereis cf. *H dentarticulata* (Smith)
Jonesia rostrata (Lucas)
Kingmaina sp.
Leptocythere sp.
Loxoconcha lapidiscola Hartmann
Loxoconcha lenticulata (LeRoy)
Loxoconcha sp.
Loxoconcha tamarindoidea Swain
Loxocorniculum sculptoides Swain
Loxocorniculum sp. *B*
Macrocyprina pacifica (LeRoy)
Miracythere sp.
Monoceratina sp.
Mutilis aurita (Skogsberg)
Mutilus confragosa (Edwards)
Mutilus sp., immature
Neocytherideis cf. *N. subulata* (Brady)
Orionina pseudovaughani Swain
Orionina lienenklausi (Müller)
Paracypris franquesoides Swain
Paracypris politella Swain
Paracytheridea granti LeRoy
Paracytheridea pichelinguensis Swain
Paracytherois perspicilla (Benson and Kaesler)
Paradoxostoma cf. *P. hodgei* Brady
Pellucistoma scrippsi Benson
Pseudocytthereis cf. *'P' simpsonensis* Swain
Pseudophilomedes? sp.
Pterygocythereis delicata (Coryell and Fields)
Pumilocytheridea realejoensis Swain
Pumilocytheridea vermiculoidea Swain
Puriana pacifica Benson
Sclerochilus contortellus Swain
Sclerochilus nasus Benson
Trachyleberidea tricornis Swain
Trachyleberidea ? henryhowei McKenzie and Swain
Urocythere? sp. *A*
Urocythere? sp. *B*
Xestoleberis aurantia Baird
Xestoleberis hopkinsi Skogsberg
Xiphichilus tenuissimoides Swain

RECENT OSTRACODA FROM SAN JUAN DEL SUR, NICARAGUA

The bay at San Juan del Sur, south-western Nicaragua is about 0·75 nautical miles in diameter, up to 60 feet deep and has a salinity of 35‰. The bottom is fine sand. The bedrock surrounding the bay is tuffaceous shale, sandstone and graywacke of the Eocene Brito Formation.

The ostracod population of the small bay is somewhat unique in that it is represented by 43 species, a relatively large number compared with other localities in western Central America from which samples have been obtained. Eight of the species are apparently new and three others, totalling 35 per cent of the fauna, are not known outside Central American localities.

Several of the species now living at San Juan del Sur are known to occur only as fossils in the Caribbean region from which they probably migrated in the late Tertiary or Quaternary.

The species identified in the bay are listed in Table 2 together with their known occurrences elsewhere.

The San Juan del Sur assemblage represents one of several along the open western part of North America in which significant numbers of localized species, although apparently free to migrate, have not done so.

CONTINENTAL STABILITY OF REGION AS INDICATED BY OSTRACODA

The provincial distribution of many of the ostracod species from the Miocene to the Recent in the southern Pacific coast of North America, Central America and the Gulf of Mexico–Caribbean region strongly suggests relative stability of these continental blocks in Neogene time. It would be hard to explain the developmental patterns of ostracod distribution under conditions other than that North and South America have had their present location with respect to latitude, longitude, ocean current patterns and general climate at least since the early or middle Miocene.

SYSTEMATIC SECTION

The species and their occurrences are summarized in this section. Synonymies have been omitted to conserve space but are given in related papers: Skogsberg (1928, 1950), Lucas (1930), LeRoy (1943, 1945), Smith (1952), Hartmann (1953, 1956, 1957, 1959), Benson (1959), Benson and Kaesler (1963), McKenzie (1965), McKenzie and Swain (1966), Swain (1967), Swain and Gilby (1967). The presumed new species found in the collections are not described here but will be considered further in later papers.

Subclass OSTRACODA Latreille, 1806
Order PODOCOPIDA Müller, 1894
Suborder PLATYCOPINA Sars, 1866
Family CYTHERELLIDAE Sars, 1866

Genus CYTHERELLA Jones, 1849
Cytherella cf. *C. banda* Benson
Fig. 10

Rare at Station 108, near La Jolla, California. The species was recorded by Benson (1959) from Todos Santos Bay, Baja California.

Cytherella ovularia Swain

This species occurs at San Juan del Sur, Nicaragua. It was described from the Gulf of California (Swain, 1967) and from Corinto Bay, Nicaragua; it is widely distributed (Swain and Gilby, 1967).

TABLE 2. Ostracod species of San Juan del Sur Bay, Nicaragua and their occurrence at other localities

Species	1	2	3	4	5	6	7	8	9
Ambostracon glauca Skogsberg	X	X	X		X				
Aurila convergens Swain	X	X			X				
Aurila sp. E									
Aurila lincolnensis (LeRoy)	X	X	X		X	X			
Bairdia cf. *B. subcircincta* Brady and Norman									
Bairdia simuvillosa Swain	X	X	X		X				
Basslerites sonorensis Benson and Kaesler	X		X		X				
Caudites acosaguensis Swain and Gilby	X								
Caudites aff. *C. nipeensis* Bold									
Cytherella ovularia Swain	X	X			X				
Cytherella parapunctata Swain		X							
Cytherelloidea californica LeRoy	X							X	
Cytherelloidea sanlucasensis Swain	X	X							
Cytherelloidea sp. A	X								
Cytherelloidea sp. B	X	X							
Cytheretta aff. *C. danaiana* (Brady)									
Cytheropteron sp.									
Cytherura bajacala Benson	X		X		X				
Cytherura johnsonoides Swain	X	X			X				
Hemicytherura cf. *H. abyssicola* (Müller)									
Kangarina quellita Coryell and Fields	X	X	X		X				X
Loxoconcha emaciata Swain		X							
Loxoconcha lapidiscola Hartmann			X		X				
Loxocorniculum sculptoides Swain	X	X	X		X				
Loxocorniculum sp. A		X							
Macrocyprina pacifica (LeRoy)		X							
Mutilus cardonensis Swain and Gilby	X								
Neocytherideis cf. *N. subulata* (Brady)									
Paracypris politella Swain		X							
Paracytheridea granti LeRoy	X	X	X		X	X		X	
Paracytheridea pichelinguensis Swain									
Paracytherois perspicilla (Benson and Kaesler)	X	X	X		X				
Pellucistoma scrippsi Benson	X	X	X		X				
Perissocytheridea meyerabichi (Hartmann)	X	X	X	X	X				
Pumilocytheridea realejoensis Swain and Gilby	X								
Puriana pacifica Benson	X	X	X		X				
Sclerochilus contortellus Swain									
Triebelina cf. *T. gierloffi* Hartmann				X					
Urocythere? sp. C									
Xestoleberis hopkinsi Skogsberg	X	X	X		X	X			
Xestoleberis sp. A									
Xestoleberis sp. B									
Xiphichilus tenuissimoides Swain	X	X			X				

1. Corinto Bay, Nicaragua
2. Gulf of California
3. Pacific coast of Southern California and Baja California
4. El Salvador coastal area
5. Gulf of Panama
6. Pacific coast of U.S., north of Point Conception
7. Gulf of Mexico–Caribbean region
8. Fossil, Pacific coast region
9. Fossil, Caribbean–Gulf region

Cytherella parapunctata Swain

The species was obtained at San Juan del Sur, Nicaragua. It was described from the Gulf of California, south-western part, near-shore (Swain, 1967).

Cytherella cf. *C. vizcainoensis* McKenzie and Swain
Plate I, Figs 6a–c; Plate XI, Fig. 10; Fig. 2

This form was obtained from Lankford's Station 100. It was described from Scammon Lagoon, Baja California (McKenzie and Swain, 1967).

Genus CYTHERELLOIDEA Alexander, 1929
Cytherelloidea californica LeRoy
Fig. 11

The species was found at San Juan del Sur, Nicaragua. It also occurs in the Gulf of California (Swain, 1967), on both sides, near-shore, and at Lankford's Stations 19 and 97, Baja California and at Santa Catalina Island. It was described from the Pleistocene Lomita Marl of California (LeRoy, 1943).

Cytherelloidea sanlucasensis Swain
Plate I, Figs 4a–c

This form was obtained from San Juan del Sur, Nicaragua and was described from the Gulf of California (Swain, 1967), on both sides, near-shore.

Cytherelloidea sp. *A*
Plate I, Fig. 3

This species was obtained at San Juan del Sur, Nicaragua and will be described elsewhere.

Cytherelloidea sp. *B*
Plate I, Figs 5a, b

Several specimens of this form, related to *C. tewarii* van den Bold (1963), were found at San Juan del Sur, Nicaragua.

Cytherelloidea praecipua van den Bold
Plate I, Figs 1a, b, 2a, b; Plate XI, Fig. 4; Fig. 22

Representatives of this species were found at San Juan del Sur, Nicaragua, and at Lankford's Station 37, Baja California. It was described from Recent coral sand, Tobago, West Indies (van den Bold, 1963).

Cytherelloidea sp. immature
Fig. 10

This form was found at Lankford's Station 41, Baja California.

Suborder PODOCOPINA Sars, 1866
Superfamily BAIRDIACEA Sars, 1888

Genus TRIEBELINA van den Bold, 1946
Triebelina cf. *T. gierloffi* Hartmann
Plate I, Figs 8a, b

The species was obtained from San Juan del Sur, Nicaragua; it was described from the Pacific coast of El Salvador (Hartmann, 1959) and from Scammon Lagoon, Baja California (McKenzie and Swain, 1967).

Genus BAIRDIA McCoy, 1844
Bairdia cf. *B. subcircincta* Brady and Norman

This species was obtained from San Juan del Sur, Nicaragua. It was described from the mid-Atlantic Ocean (1270–2200 m) and Ceylon (Brady, 1880), as *B. formosa*, (Müller, 1912).

Bairdia simuvillosa Swain
Fig. 5

This occurs at San Juan del Sur and was described from the Gulf of California (Swain, 1967), generally distributed. It also occurs in Scammon Lagoon, Baja California (McKenzie and Swain, 1967) and Lankford's Stations 24, 32, 37, 43, 48 and 116, Baja California.

Bairdia phlegeri McKenzie and Swain
Fig. 4

In the Lankford collection the species was found at Stations 19, 22, 24, 28, 29A, 30A, 41, 43 and 97. It was described from Scammon Lagoon, Baja California (McKenzie and Swain, 1967) where it was found to be characteristic of a lower lagoon euryhaline assemblage.

Bairdia tuberculata Brady
Plate I, Figs 7a–e; Fig. 5

In the Lankford collection the species was found at Station 114; elsewhere the species has been recorded from Torres Strait, Australia (lagoonal) (Brady, 1880).

Genus BYTHOCYPRIS Brady, 1880
Bythocypris? sp.
Fig. 7

A *Bythocypris?* was obtained in the Lankford collection from Station 25, Baja California.

Superfamily CYPRIDACEA Baird, 1845
Family PARACYPRIDIDAE Sars, 1923

Genus MACROCYPRINA Triebel, 1960
Macrocyprina pacifica (LeRoy)
Fig. 18

In the Lankford collection the species occurs at Stations 25, 28, 32, 37, 41, 43, 97, 100, 114 and 122; southern California and Baja California, also at San Juan del Sur, Nicaragua. Previously the species was recorded from the region of the Gulf of California (Swain, 1967), Ensenada and Scammon Lagoon, Baja California (Benson, 1959; McKenzie and Swain, 1967), Catalina Island, California (LeRoy, 1943). It was described from the 'Pleistocene' Timm's Point Formation, southern California. It is one of the more near-shore characteristic species of the southern California–Baja California–western Central America area.

Genus PARACYPRIS Sars, 1866
Paracypris politella Swain
Plate I, Fig. 9; Fig. 20

The species was found at San Juan del Sur, Nicaragua, and at Lankford's Station

122, Baja California. It was described from the Gulf of California where it is rare in occurrence on the western side of the Gulf (Swain, 1967).

Paracypris franquesoides Swain
Fig. 20

In the Lankford collection this species was obtained at Station 97. It was described from the Gulf of California (Swain, 1967) where it is rare in occurrence on the western margin.

Family BRACHYCYTHERIDAE Puri, 1954

Genus PTERYGOCYTHEREIS Blake, 1933
Pterygocythereis delicata (Coryell and Fields)
Fig. 23

The species was found in the Lankford collection at Stations 25, Baja California and 102, southern California. The species was described from the Gatun Formation, Miocene, of Panama (Coryell and Fields, 1937) and has been recorded from Baja California (Crouch, 1949; Benson, 1959) and the Gulf of California (Swain, 1967).

Genus KINGMAINA Keij, 1957
Kingmaina sp.
Plate VII, Figs 5, 9a, b; Fig. 16

In the Lankford collection this form was found at Stations 97 and 115.

Family BYTHOCYTHERIDAE Sars, 1926

Genus MIRACYTHERE Hornibrook, 1952
Miracythere sp.
Plate III, Figs 10a, b; Plate VI, Figs 1a, b; Fig. 19

This species was obtained from Lankford's Stations 1, Baja California and Station 102, southern California.

Genus MONOCERATINA Roth, 1928
Monoceratina sp.
Plate IV, Figs 1a, b; Plate IX, Fig. 9; Fig. 19

The form assigned to this genus was found at Station 1 of the Lankford collection.

Superfamily CYTHERACEA Baird, 1850
Family CYTHERIDAE Baird, 1850
Subfamily CYTHERINAE Baird, 1850

Genus CYTHERE Müller, 1785
Cythere cf. *C. lutea* O. F. Müller
Plate I, Fig. 11; Plate X, Fig. 4; Fig. 10

The species was obtained in the Lankford collection at Stations 67 and 81–83. It is widely distributed in the North Atlantic and North Pacific Oceans (Sars, 1928; R. H. Benson, oral communication).

Cythere sp. *A*
Plate I, Figs 12a, b; Fig. 15

This species was found in the Lankford collection at Coos Bay, Oregon, Station 81–83.

Cythere alveolivalva Smith
Plate IV, Figs 12a, b; Plate VII, Figs. 1a, b; Plate IX, Figs 4a b, Plate X, Figs. 9a, b;
Plate XI, Fig. 7; Fig. 16

In the Lankford collection the species was obtained at Stations 12, 41 and 67. It was
described from the Vancouver Island area (Smith, 1952).

Subfamily PERISSOCYTHERIDEINAE van den Bold, 1963

Genus PUMILOCYTHERIDEA van den Bold, 1963
Pumilocytheridea realejoensis Swain and Gilby
Plate IV, Figs 6a, b; Fig. 23

The species was found at San Juan del Sur, Nicaragua, Station 3, and at Lankford's
Station 19, Baja California. Originally it was described from Corinto Bay, Nicaragua
(Swain and Gilby, 1967).

Pumilocytheridea vermiculoidea Swain
Fig. 23

In the Lankford collection the species occurs at Stations 3 and 41. It was described
from the Gulf of California (Swain, 1967).

Genus PERISSOCYTHERIDEA Stephenson, 1938
Perissocytheridea meyerabichi (Hartmann)
Plate I, Figs 10a, b

The species occurs at San Juan del Sur, Nicaragua. It also occurs at other localities
in Central America (Hartmann 1953, 1957; Swain and Gilby, 1967), the Gulf of
California (Swain, 1967) and Baja California (Benson, 1959; McKenzie and Swain,
1967).

Subfamily NEOCYTHERIDEINAE Puri, 1957

Genus CUSHMANIDEA Blake, 1933
Cushmanidea pauciradialis Swain
Plate IV, Figs 5a–c; Plate X, Fig. 8; Fig. 9

The species occurs at the following stations in the Lankford collection: 55, centra
California, 105, 106, southern California, and 19, Baja California. It was described
from the Gulf of California, generally distributed (Swain, 1967), and also occurs in
Scammon Lagoon, Baja California, where it represents the lower lagoon assemblage
(McKenzie and Swain, 1967).

Cushmanidea guardensis Swain
Fig. 8

This species occurs in the Lankford collection at Station 102, southern California.
It was described from the western side of the Gulf of California (Swain, 1967).

Cushmanidea sp.
Fig. 9

This coarsely-pitted species occurs at Lankford's Station 122, Baja California.

Genus NEOCYTHERIDEIS Puri, 1952
Neocytherideis cf. *N. subulata* (Brady)
Plate VI, Fig. 9; Fig. 24

In the Lankford collection the species occurs at Stations 106 and 121.

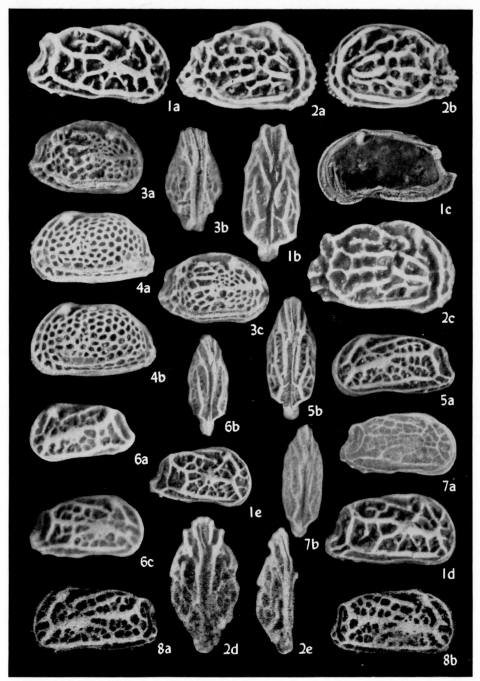

PLATE V.

Figs 1a-e. *Ambostracon glauca* (Skogsberg). (a), (b) Right side and dorsal views of shell; (c) interior of right valve, (d, e) right side of two immature shells; Santa Catalina Island, California, Lankford Station 100. × 49. Figs 2a-e. *Mutilus aurita* Skogsberg. (a), (d) Right and dorsal side of shell; (b), (e) exterior and dorsal views of left valve; Los Coronados, California, Lankford Station 37; (c) Exterior of large right valve; Isla San Geronimo, Baja California, Lankford Station 21. × 49. Figs 3a-c. *Aurila jollaensis* LeRoy. (a) Right side of shell; (b), (c) dorsal view and right side of shell; Santa Catalina Island, California, Lankford Station 100. × 52. Figs 4a, b. *Hemicythere californiensis* LeRoy. Left sides of two shells; Punta Eugenia, Baja California, Lankford Station 25. × 49. Figs 5a, b. *Ambostracon glauca* (Skogsberg). Left side and dorsal view of shell; Isla San Martin. Baja California, Lankford Station 19. × 48. Figs 6a-c. *Ambostracon glauca* (Skogsberg). (a), (b) Left side and dorsal view of shell; (c) right side of shell; Isla San Martin, Baja California, Lankford Station 19. × 47. Figs 7a, b. *Ambostracon* sp. Right side and dorsal view of shell; Islas San Benitos, Baja California, Lankford Station 23. × 45. Figs 8a, b. *Ambostracon glauca* (Skogsberg). (a) Left side of shell; Isla San Geronimo, Baja California, Lankford Station 21; (b) right side of shell; Bodega Bay, California, Lankford Station 67. × 46.

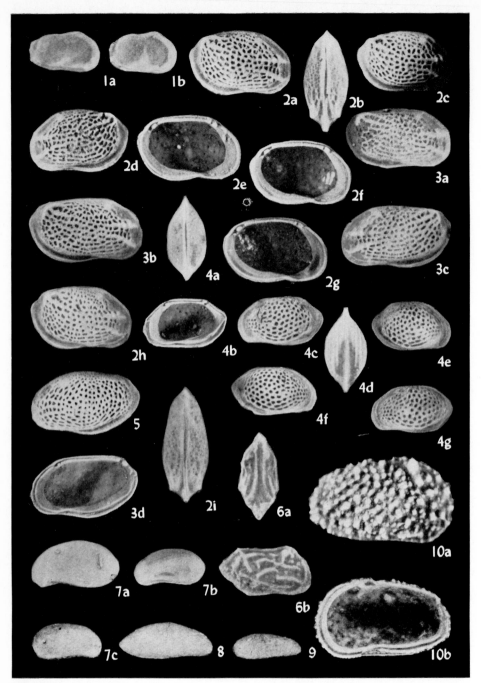

PLATE VI.

Figs 1a, b. *Miracythere* sp. Right sides of two shells; Bahia Sebastian Vizcaino, Baja California, Lankford Station 1. × 52. Figs 2a-i. *Loxocorniculum sculptoides* Swain. (a), (b). Right side and dorsal views of shell; (c) right side of shell; (d) left side of shell; (e), (f), (g) interior view of three left valves; (h), (i) right side and dorsal views of shell; Santa Catalina Island, California, Lankford Station 100. × 41. All are male shells. Figs 3a-d. *Loxocorniculum sculptoides* Swain. (a), (b). Right sides of two shells; (c) left side of shell; (d) interior of left valve; Santa Catalina Island, California, Lankford Station 100. All are female shells. × 46. Figs 4a-g. *Loxoconcha* sp. (a) Dorsal view of shell; (b) interior of right valve; (c) left side of shell; (d) ventral view of shell; (e) right side of shell; (f) right side of shell; (g) left side of shell; Bahia Sebastian Vizcaino, Baja California, Lankford Station 1. × 47. Fig. 5. *Loxoconcha tamarindoidea* Swain. Left side of shell; Santa Catalina Island, California, Lankford Station 100. × 47. Figs 6a, b. *Loxoconcha lapidiscola* Hartmann. Dorsal right side views of shell; Cabo San Lucas, Baja California, Lankford Station 32. × 48. Figs 7a-c. *Sclerochilus contortellus* Swain. (a), (b) Left sides of an adult and an immature shell; (c) right side of an immature shell; San Juan del Sur, Nicaragua, Station 1. × 48. Fig. 8. *Xiphichilus tenuissimoides* Swain. Right side shell; San Juan del Sur, Nicaragua, Station 1. × 48. Fig. 9. *Neocytherideis*, cf. *N. subulata* (Brady). Right side of shell; San Juan del Sur, Nicaragua, Station 1. × 50. Figs 10a, b. *Echinocythereis* sp. Exterior and interior views of right valve; Santa Catalina Island, California, Lankford Station 100. × 49.

Family CYTHERURIDAE Müller, 1894

Genus CYTHERURA Sars, 1866
Cytherura paracostata Swain
Fig. 13

In the Lankford collection, this species was obtained at Stations 111, southern California and 12, Baja California. It was described from marginal stations on both sides of the Gulf of California (Swain, 1967) and also was recorded from Scammon Lagoon, Baja California, inner lagoon assemblage (McKenzie and Swain, 1967).

Cytherura bajacala Benson
Plate IV, Fig. 8; Fig. 12

This species was obtained from San Juan del Sur, Nicaragua, and from Stations 1, 106 and 110 of the Lankford collection. The species was described from Todos Santos Bay, Baja California (Benson, 1959), and has been recorded from the Gulf of California (Swain, 1967) and Corinto Bay, Nicaragua (Swain and Gilby, 1967).

Cytherura cf. *C. gibba* Müller
Fig. 12

In the Lankford collection a form related to *C. gibba* was found at Stations 29A and 102.

Cytherura johnsonoides Swain
Plate IV, Fig. 9?; Fig. 13

The species was found at San Juan del Sur, Nicaragua, and at Stations 3, 55 and 106 in the Lankford collection in central and southern California and in Baja California. It also occurs in the Gulf of California (Swain, 1967) and in Corinto Bay, Nicaragua (Swain and Gilby, 1967).

Cytherura sp. *A*
Plate IV, Figs 10a–c; Fig. 13

This strongly-ribbed, posteriorly acuminate form occurs in the Lankford collection at Station 55, central California.

Cytherura sp. *B*
Fig. 13

Immature specimens from Lankford's Stations 1 and 121, Baja California, are assigned to this species.

Genus HEMICYTHERURA Elofson, 1941
Hemicytherura cranekeyensis Puri
Fig. 15

This species was found in the Lankford collection at Stations 19 and 28, Baja California, and 76, northern California. It was described from the Gulf of Mexico, the western coast of Florida (Puri, 1960).

Hemicytherura cf. *H. abyssicola* (Müller)

A specimen possibly representing this species occurs at San Juan del Sur, Nicaragua. Elsewhere the species occurs in the Mediterranean Sea (Müller, 1894).

Hemicytherura sp.
Plate IV, Figs 13a–c; Fig. 15

The species was obtained from Lankford's Station 122, Baja California.

Genus CYTHEROPTERON Sars, 1866
Cytheropteron? ventrokurtosa Swain, 1967
Fig. 12

In the Lankford collection the species was obtained from stations 1, 3 and 12, Baja California. It was described from the Gulf of California, generally distributed (Swain, 1967), Scammon Lagoon, Baja California (McKenzie and Swain, 1967), near Lankford's Stations 12 and 13, and from Corinto Bay, Nicaragua (Swain and Gilby, 1967).

Cytheropteron sp.
Plate IV, Figs 11a, b; Fig. 12

This form was obtained from San Juan del Sur, Nicaragua and from Lankford's Station 47, Baja California.

Genus KANGARINA Coryell and Fields, 1937
Kangarina quellita Coryell and Fields

The species was obtained from San Juan del Sur, Nicaragua. It was described from the Miocene Gatun Formation of Panama (Coryell and Fields, 1937) and has also been recorded from the Gulf of California (Swain, 1967), Baja California (McKenzie and Swain, 1967), Corinto Bay, Nicaragua (Swain and Gilby, 1967) and from the Miocene of the Gulf of Mexico–Caribbean region (Puri, 1954; van den Bold, 1958, 1963).

Genus PARACYTHERIDEA Müller, 1894
Paracytheridea granti LeRoy
Fig. 21

In the Lankford collection the species was obtained from Stations 32 and 116; also from San Juan del Sur. It was previously recorded from the Pleistocene of California (LeRoy, 1943), Baja California (Benson, 1959), Gulf of California (Swain, 1967), and Corinto Bay, Nicaragua (Swain and Gilby, 1967).

Paracytheridea pichelinguensis Swain
Plate VII, Fig. 7; Fig. 21

The species was obtained from Station 1 in the Lankford collection, and from San Juan del Sur, Nicaragua. It was described from the Gulf of California, western side, near-shore (Swain, 1967).

Family HEMICYTHERIDAE Puri, 1953

Genus HEMICYTHERE Sars, 1925
Hemicythere californiensis LeRoy
Plate II, Figs 5a–c, 6a–d; Plate V, Figs 4a, b; Plate IX, Figs 2a–d; Fig. 15

In the Lankford collection, the species was obtained from Stations 1, 3, 5, 9, 24, 25, 40, 49, 64, 85, 102, 105, 107, 108, 113. It was previously recorded from Southern California (LeRoy, 1943), Baja California (Benson, 1959) and the Gulf of California (Swain, 1967) where it is found near-shore along both margins.

Genus AMBOSTRACON Hazel, 1962

Ambostracon glauca (Skogsberg)

Plate III, Figs 3a–c; Plate V, Figs 1a–e, 5a, b, 6a–c, 8a, b;
Plate IX, Fig. 6; Plate X, Figs 5, 7a, b; Fig. 2

The species is widely distributed in the Lankford collection at Stations 1, 19, 21, 25, 29A, 37, 43, 53, 55, 67, 75, 81–83, 97, 100, 114 and at San Juan del Sur, Nicaragua. The species was recorded previously from southern California (Skogsberg, 1928), the Gulf of California, general distribution (Swain, 1967), lower part of Scammon Lagoon (McKenzie and Swain, 1967) and from Corinto Bay, Nicaragua (Swain and Gilby, 1967).

Ambostracon sp.

Plate V, Figs 7a, b; Plate XI, Figs 6a–d; Fig. 2

Specimens of this form were obtained from Lankford's Stations 1, 21 and 23.

Genus HETEROCYTHEREIS Elofson, 1941

Heterocythereis sp. *A*

Plate VII, Figs 3a, b; 11a, b, Plate X, Figs 2a, b; Fig. 14

This form was obtained from Stations 67 and 108 in the Lankford collection.

Heterocythereis sp. *B*

Plate VII, Figs 4a, b; Fig. 14

The species was found at Stations 81–83 in the Lankford collection.

Heterocythereis dentarticulata (Smith)

Plate I, Fig. 13; Plate VII, Figs 2a, b; Fig. 14, 16?

The muscle scar of this form has the upper middle adductor spot paired and the hinge is amphidont which indicates a relationship to *Heterocythereis* rather than to *Cythere* where it was placed by Smith (1952). In the Lankford collection it was found at Stations 67, 75, 83-85 and 108 (Fig. 14), and questionably at Stations 52 and 55 (Fig. 16). The species was described from the Vancouver Island area (Smith, 1952).

Genus UROCYTHERE Howe, 1951

Urocythere? sp. *A*

Plate IV, Fig. 2; Fig. 25

This coarsely punctate form was found in the Lankford collection at Stations 67 and 76. It is related in shape and caudate posterior end to *Urocythere*, but its hingement appears to be more merodont than amphidont. It may be a new genus of Cytherinae. It superficially resembles the '*Acuticythereis*' group of *Campylocythere* but has different hingement.

Urocythere? sp. *B*

Plate IV, Figs 3a–d, 4a, b; Plate IX, Fig. 10;
Plate XI, Figs 9a–c; Fig. 25

This species is similar to the preceding but nearly smooth; and is evidently dimorphic. In the Lankford collection it occurs at Stations 67, 75, 76 and 81–83.

Urocythere? sp. *C*

Plate IV, Figs 7a, b

This form was obtained from San Juan del Sur, Nicaragua.

Genus AURILA Pokorný, 1955
Aurila conradi californica Benson and Kaesler
Fig. 3

This species was obtained from Lankford's Stations 23, 29A, 67, 97 and 100. It was recorded previously from the Gulf of California (Benson and Kaesler, 1963; Swain, 1967) general distribution and Baja California (McKenzie and Swain, 1967). Along the Pacific Coast of North America it is much more frequent in occurrence south of Point Conception than to the north.

Aurila sp. *A*
Plate II, Figs 1a–g, 7a–c; Plate IX, Figs 5a, b, 8a, b; Fig. 3

This form was obtained from Lankford's Stations 22 and 114.

Aurila convergens Swain

The species was found at San Juan del Sur, Nicaragua. It was described from the Gulf of California (Swain, 1967) and recorded from Corinto Bay, Nicaragua (Swain and Gilby), 1967.

Aurila jollaensis LeRoy
Plate II, Figs 3a, b, 9; Plate V, Figs 3a–c; Fig. 3

In the Lankford collection the species was obtained from Stations 19, 22, 23, 32, 37, 41, 43, 48, 55, 67, 75, 81–83, 85, 97, 100, 115, 122. The species was recorded previously from southern California (LeRoy, 1943) and Baja California (Benson, 1959). It is one of the most widely distributed near-shore species along the Pacific Coast of North America.

Aurila lincolnensis LeRoy
Plate II, Fig. 4; Plate IV, Figs 14a, b; Fig. 3

This species was found in Lankford's collection at Stations 19, 21, 22, 23, 25, 28, 37, 53, 55, 67, 75, 81–83, 85, 97, 102, 107, 114, 115, 122 and at San Juan del Sur, Nicaragua. It has also been obtained from southern California (LeRoy, 1943), Baja California (Benson, 1959; McKenzie and Swain, 1967), Corinto Bay, Nicaragua (Swain and Gilby, 1967), and from the Pleistocene of southern California. Like the preceding species it is one of the commonest near-shore forms of the Pacific Coast region.

Aurila sp. *B*
Plate II, Figs 2a, b; Fig. 4

This form was obtained in the Lankford collection from Stations 22, 28, 37 and 114, and from San Juan del Sur, Nicaragua.

Aurila sp. *C*
Plate II, Figs 8a, b; Fig. 4

This species occurs at Lankford's Station 75.

Aurila sp. *D*
Plate IV, Figs 15a–d; Fig. 4

In the Lankford collection this form occurs at Station 101.

Aurila sp. *E*

This form was obtained from San Juan del Sur, Nicaragua.

Genus CAUDITES Coryell and Fields, 1937
Caudites sp. *A*
Plate III, Figs 11a, b; Plate IX, Fig. 7; Plate X, Fig. 6; Fig. 7

This form is rare in occurrence at Lankford's Stations 19 and 22.

Caudites sp. *B*
Plate III, Figs 8a, b, 9a, b; Plate XI, Fig. 8; Fig. 8

This species was obtained at Lankford's Stations 19, 28 and 37.

Caudites sp. *C*
Plate III, Figs 6a, b; Fig. 8

In the Lankford collection the species was found at Station 29A.

Caudites sp. aff. *C. leguminosus* van den Bold
Fig. 7

In the Lankford collection the species was found at Stations 21 and 111. A similar form was found in Corinto Bay, Nicaragua (Swain and Gilby, 1967).

Caudites acosaguensis Swain and Gilby

This form occurs at San Juan del Sur, Nicaragua. It was described from Corinto Bay, Nicaragua (Swain and Gilby, 1967).

Caudites aff. *C. nipeensis* van den Bold

This species was obtained from San Juan del Sur, Nicaragua. A somewhat similar form was described by van den Bold (1946) from the Caribbean region.

Caudites rosaliensis Swain
Plate III, Figs 4a–c, 7a, b; Plate X, Figs 10a, b; Fig. 8

In the Lankford collection the species was found at Stations 25, 28, 29A, 37, 41, 97 and 122. It was described from the Gulf of California (Swain, 1967) where it is found on both eastern and western margins, near-shore, and from Scammon Lagoon, Baja California (McKenzie and Swain, 1967).

Genus ANTEROCYTHERE McKenzie and Swain, 1967
Anterocythere sp.
Plate III, Fig. 5; Fig. 2

In the Lankford collection this species was found at Stations 29A and 112.

Genus MUTILUS Neviani, 1928
Mutilus cardonensis Swain and Gilby
Plate II,   Fig. 10

The species was obtained from San Juan del Sur, Nicaragua; it was described from Corinto Bay, Nicaragua (Swain and Gilby, 1967).

Mutilus aurita Skogsberg
Plate V, Figs 2a–e; Fig. 19

In the Lankford collection the species was found at stations 21, 22, 23, 24, 28, 30A,

37, 41, 43 and 114. It was previously recorded from southern California and Baja California (Skogsberg, 1928; Benson, 1959; McKenzie and Swain, 1967).

Mutilus confragosa (Edwards)
Fig. 19

In the Lankford collection the species occurs at Stations 22, 28, 37 and 43. In the Gulf of California it was found along both eastern and western margins, near-shore (Swain, 1967). It is known elsewhere in the Upper Miocene and younger deposits of the Atlantic Coastal Plain and Gulf of Mexico–Caribbean regions.

Mutilus sp.
Fig. 20

The species was found at Lankford's Station 115.

Genus ORIONINA Puri, 1954
Orionina pseudovaughani Swain
Plate III, Figs 1a–d; Plate X, Figs 1a, b; Fig. 20

In the Lankford collection the species was obtained from Stations 21, 22, 32, and 41. It was described previously from the Gulf of California.

Orionina lienenklausi Müller
Plate III, Figs 2a–c

The species was found in the Lankford collection at Station 41. In this region it was recorded from Baja California (Benson, 1959).

Family LEGUMINOCYTHERIDEIDAE Howe, 1961 or
CAMPYLOCYTHERIDAE Puri, 1960
Subfamily LEGUMINOCYTHERINAE Howe, 1961

Genus BASSLERITES Howe, 1937
Basslerites delrayensis LeRoy
Fig. 5

In the Lankford collection the species was obtained at Stations 1, 19 and 23. It was described from the Pleistocene of southern California (LeRoy, 1943), and the Recent of Baja California (Benson, 1959), and the Gulf of California (Swain, 1967).

Basslerites sonorensis Benson and Kaesler
Fig. 5

The species was found in the Lankford collection at Stations 5, 19, 75 and 122; also at San Juan del Sur. It was previously recorded from the Gulf of California and Baja California (Benson and Kaesler, 1963; McKenzie and Swain, 1967) and from Corinto Bay, Nicaragua (Swain and Gilby, 1967).

Basslerites thlipsuroidea Swain
Fig. 6

This species was obtained at Stations 1 and 32 in the Lankford collection. It was described from the Gulf of California (Swain, 1967).

Family LEPTOCYTHERIDAE Hanai, 1957

Genus LEPTOCYTHERE Sars, 1925
Leptocythere sp.
Plate VII, Figs 8a, b; Plate IX, Fig. 1; Plate X, Figs. 3a–c; Fig. 17

This form occurs in the Lankford collection from Stations 81–83.

Family LOXOCONCHIDAE Sars, 1925

Genus LOXOCONCHA Sars, 1866
Loxoconcha emaciata Swain

This species was obtained from San Juan del Sur, Nicaragua. It was described from the Gulf of California (Swain, 1967).

Loxoconcha lapidiscola Hartmann
Plate VI, Figs 6a, b; Plate XI, Fig. 1; Fig. 17

The species was found in the Lankford collection at Station 32; and at San Juan del Sur, Nicaragua. It was described from El Salvador (Hartmann, 1959). It apparently does not extend its range into the United States. A closely related species *L. emaciata*, Swain (1967) was obtained from north-western Gulf of California and San Juan del Sur, Nicaragua.

Loxoconcha lenticulata LeRoy
Fig. 17

In the Lankford collection the species was obtained at Stations 19, 22, 25, 100, 105, 108 and 114. The species was described from Pliocene, and Pleistocene deposits in California (LeRoy, 1943) and from the Recent of southern California (Rothwell, 1944), and Baja California (Benson, 1959; McKenzie and Swain, 1967).

Loxoconcha sp.
Plate VI, Figs 4a–g; Plate XI, Fig. 2; Fig. 17

This form was obtained in the Lankford collection at Station 1. A closely similar form from the Gulf of California, was referred to as *Hemicythere* sp. *B* by Swain (1967) but that species has different internal structures from the present one.

Loxoconcha tamarindoidea Swain
Plate VI, Fig. 5; Fig. 18

In the Lankford collection this species was found at Stations 25, 28, 32, 37, 41, 43, 97, 100, 114 and 122. It was previously recorded from the Gulf of California (on margins, near-shore) (Swain, 1967), Scammon Lagoon, Baja California (lower lagoon assemblage) (McKenzie and Swain, 1963), and from Corinto Bay, Nicaragua (Swain and Gilby, 1967).

Genus LOXOCORNICULUM Benson and Coleman, 1963
Loxocorniculum sculptoides Swain
Plate VI, Figs 2a–i, 3a–d; Plate XI, Fig. 3; Fig. 18

The species was obtained in the Lankford collection at Stations 23, 32, 41, 48, 97, 100 and 122. Also from San Juan del Sur, Nicaragua. It is known previously from the Gulf of California on margins, near-shore (Swain, 1967), Scammon Lagoon, Baja

California (McKenzie and Swain, 1967) and Corinto Bay, Nicaragua (Swain and Gilby, 1967).

Loxocorniculum sp. *A*

This form was obtained from San Juan del Sur, Nicaragua.

Loxocorniculum sp. *B*
Fig. 18

Immature specimens of this species were found in Lankford's collection at Station 11.

Family PARADOXOSTOMATIDAE Brady and Norman, 1889
Subfamily PARADOXOSTOMATINAE Brady and Norman, 1889

Genus PARADOXOSTOMA Fischer, 1855
Paradoxostoma cf. *P. hodgei* Brady
Fig. 21

The Lankford collection contained this species at Station 19. In the Pacific Coast region it had been found previously in the Gulf of California (western side, near-shore) (Swain, 1967). It was originally described from the British Isles (Brady, 1870).

Genus CYTHEROIS Müller, 1884
Cytherois cf. *C. fischeri* (Sars)
Fig. 10

This species was found at Lankford's Station 14. It was recorded from the Gulf of California at 488 fathoms (Swain, 1967).

Genus PARACYTHEROIS Müller, 1894
Paracytherois perspicilla (Benson and Kaesler)
Fig. 21

The species was obtained in the Lankford collection at Stations 19, 22 and 41; also at San Juan del Sur, Nicaragua. It was previously found in the Gulf of California (Benson and Kaesler, 1963; Swain, 1967).

Genus SCLEROCHILUS Sars, 1866
Sclerochilus contortellus Swain
Plate VI, Figs 7a–c; Fig. 24

In the Lankford collection the species occurs at Stations 1 and 122; and at San Juan del Sur, Nicaragua. The species was described from marginal shallow water parts of the Gulf of California (Swain, 1967).

Sclerochilus nasus Benson
Fig. 24

In the Lankford collection the species was found at Stations 99 and 114. It was described from Baja California (Benson, 1959).

Genus XIPHICHILUS Sars, 1866
Xiphichilus tenuissimoides Swain
Plate VI, Fig. 8; Fig. 26

The species was obtained in the Lankford collection at Stations 1, 5, 75 and 77; and

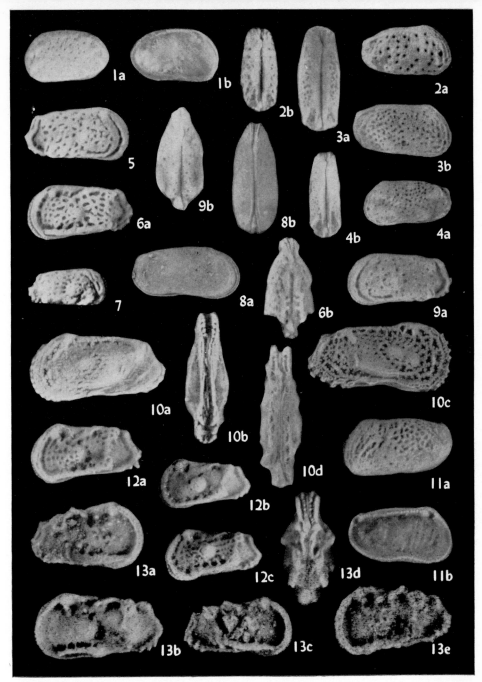

PLATE VII.

Figs 1a, b. *Cythere alveolivalva* Smith. Exterior and interior views of left valve; Bahia Sebastian Vizcaino, Baja California, Lankford Station 12. × 48. Figs 2a, b. *Heterocythereis dentarticulata* (Smith). Right side and dorsal views of shell; La Jolla, California, Lankford Station 108. × 52. Figs 3a, b. *Heterocythereis* sp. *A*. Dorsal and right side view of shell; Bodega Bay, California, Lankford Station 67. × 47. Figs 4a, b. *Heterocythereis* sp. *B*. Left side and dorsal view of shell; Coos Bay, Oregon, Lankford Stations 81-83. × 47. Fig. 5. *Kingmaina* sp. Right side of shell; Los Coronados, Baja California, Lankford Station 97. × 53. Figs 6a, b. *Trachyleberidea tricornis* Swain. Left side and dorsal view of shell; Isla San Martin, Baja California, Lankford Station 19. × 47. Fig. 7. *Paracytheridea pichelinguensis* Swain. Exterior of right valve; San Juan del Sur, Nicaragua, Station 1. × 48. Figs 8a, b. *Leptocythere* sp. Right side and dorsal views of shell; Coos Bay, Oregon, Lankford Stations 81-83. × 48. Figs 9a, b. *Kingmaina* sp. Left side and dorsal views of shell; Punta Banda, Baja California, Lankford Station 115. × 53. Figs 10a-d. *Trachyleberidea ? henryhowei* McKenzie and Swain. (a), (b) Left side and dorsal view of shell; Puerto San Bartoleme, Baja California, Lankford Station 43; (c), (d) right side and dorsal views of shell; Isla San Martin, Baja California, Lankford Station 19. × 50. Figs 11a, b. *Heterocythereis* sp. *A*. Exterior and interior views of left valve; Bodega Bay, California, Lankford Station 67. × 49. Figs 12a-c. *Trachyleberidea tricornis* Swain. (a) Left side of shell; Punta Abreojos, Baja California, Lankford Station 37; (b), (c) Left side of immature shells; Isla San Martin, Baja California, Lankford Station 19. × 52. Figs 13a-e *Trachyleberidea tricornis* Swain. (a) Right side of shell; Punta Eugenia, Baja California, Lankford Station 25; (b) Left side of an elongate and unpitted shell; Isla San Martin, Baja California, Lankford Station 19; (c) Right side of small poorly preserved shell; Punta Eugenia, Baja California, Lankford Station 25; (d), (e) Dorsal view and left side of shell, same locality as preceding. × 47.

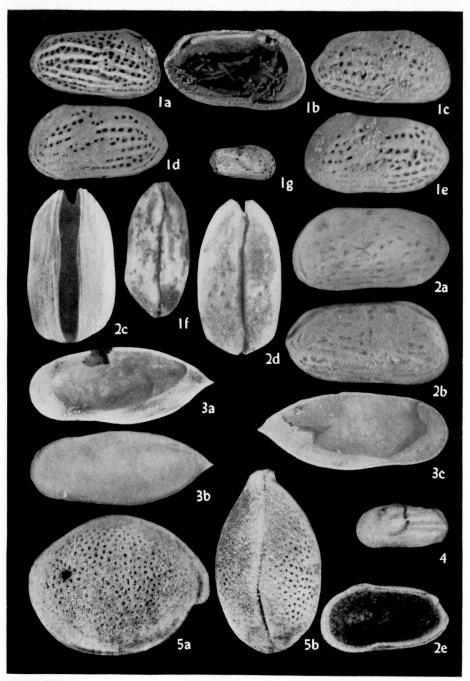

PLATE VIII.

Figs 1a-g. *Cytheretta* aff. *C. danaiana* (Brady). (a) Right side of shell; Puerto San Bartoleme, Baja California, Lankford Station 43; (b) interior of left valve; (c) exterior of right valve; (d), (e) exterior of two left valves; (f) dorsal view of shell; Santa Barbara, California, Lankford Station 53; (g) exterior of immature right valve only questionably assigned to the species; San Juan del Sur, Nicaragua, Station 1. × 50. Figs 2a-e, 4. *Cytheretta* sp. (a) Left side of shell; (b) right side of shell; (c), (d) ventral and dorsal views of shell; (e) interior of immature right valve; Morro Bay, California, Lankford Station 55; 4, Exterior of left valve of immature shell only questionably assigned to the species; San Juan del Sur, Nicaragua, Station 1. × 49. Figs 3a-c. *Jonesia rostrata* (Lucas). (a) Interior of right valve; (b) exterior of left valve; (c) interior of left valve; La Jolla, California, Lankford Station 108. × 49. Figs 5a, b. *Pseudophilomedes*? sp. Right side and dorsal views of shell; La Jolla, California, Lankford Station 111. × 24.

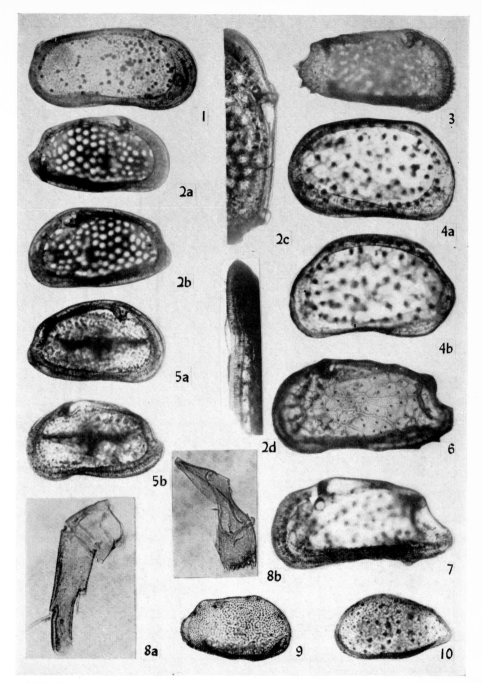

PLATE IX.

Fig. 1. *Leptocythere* sp. Interior of left valve; Coos Bay, Oregon, Lankford Station 81-83. × 59.
Figs 2a-d. *Hemicythere californiensis* LeRoy, (a) Interior of left valve; (b) interior of right valve,
× 59; (c) view of dorsal part of right valve; (d) ventral view of part of left valve, × 95. All from
Bahia Sebastian Vizcaino, Baja California, Lankford Station 5. Fig. 3. *Trachyleberidea tricornis*
Swain. Interior of left valve; Lankford Station 19. × 59. Figs 4a, b. *Cythere alveolivalva* Smith.
Interior views of left and right valves, respectively; Lankford Station 67. × 59. Figs 5a, b. *Aurila*
sp. *A*. Interior views of left and right valves, respectively; Lankford Station 114. × 59. Fig. 6.
Ambostracon glauca (Skogsberg). Interior of right valve; Lankford Station 25. × 59. Fig. 7.
Caudites sp. *A*. Interior of right valve; Lankford Station 22. × 59. Figs 8a, b. *Aurila* sp. *A*. Left
antenna and left mandible, respectively; Lankford Station 114. × 172. Fig. 9. *Monoceratina* sp.
Interior of left valve; Lankford Station 1. × 59. Fig. 10. *Urocythere* ? sp. *B*. Interior of immature
right valve; Lankford Stations 81-83. × 59.

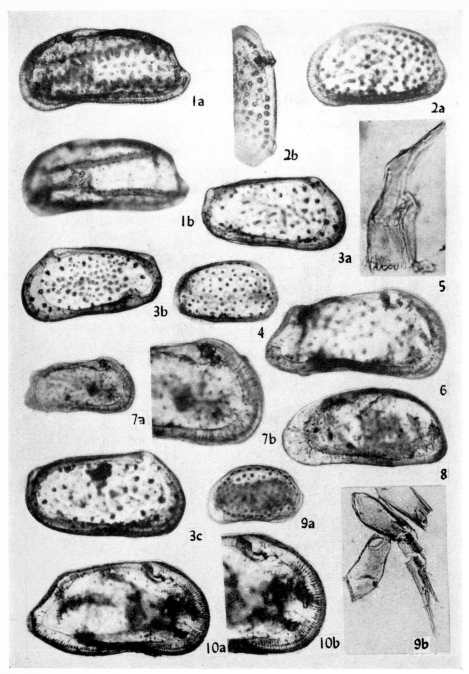

PLATE X.

Figs 1a, b. *Orionina pseudovaughani* Swain. Interior views of right valve at different levels of focus;
Lankford Station 41. × 59. Figs 2a, b. *Heterocythereis* sp. *A.* Interior view of right valve and dorsal
view of left valve; Lankford Station 100. × 59. Figs 3a-c. *Leptocythere* sp. Interior of left valve,
right valve and left valve, respectively; Lankford Stations 81-83. × 59. Fig. 4. *Cythere* cf. *C. lutea* O. F.
Müller. Interior of left valve; Lankford Station 12. × 59. Fig. 5. *Ambostracon glauca* (Skogsberg)
Mandible; Lankford Station 25. × 172. Fig. 6. *Caudites* sp. *A.* Interior of left valve; Lankford
Station 22. × 59. Figs 7a, b. *Ambostracon glauca* (Skogsberg). Interior of left valve, and enlarge-
ment of anterior portion; Lankford Station 67. × 59 and 98. Fig. 8. *Cushmanidea pauciradialis*
Swain. Interior of right valve; Lankford Station 55. × 59. Figs 9a, b. *Cythere alveolivalva* Smith.
Interior of right valve × 59, and antenna × 172; Lankford Station 41. Figs 10a,b. *Caudites rosaliensis*
Swain. Interior of left valve × 59, and enlargement of anterior end × 98; Lankford Station 28.

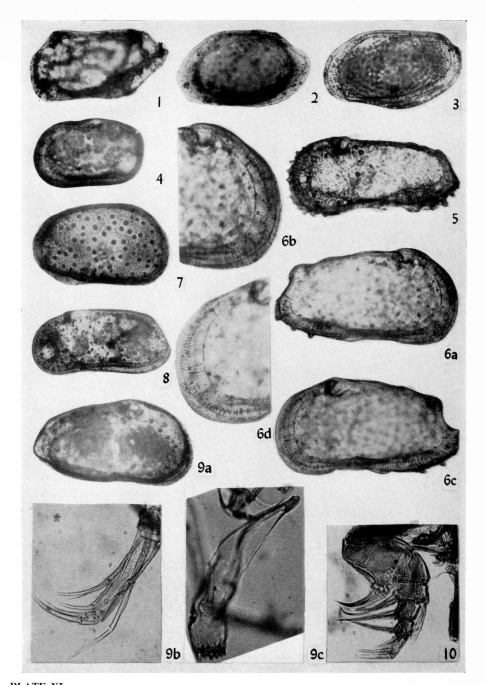

PLATE XI.

Fig. 1. *Loxoconcha lapidiscola* Hartmann. Interior of right valve; Lankford Station 32. × 59.
Fig. 2. *Loxoconcha* sp. Interior of right valve; Lankford Station 1. × 59. Fig. 3. *Loxocorniculum
sculptoides* Swain. Interior of right valve; Lankford Station 100. × 59. Fig. 4. *Cytherelloidea
praecipua* van den Bold. Interior of right valve; Lankford Station 37. × 59. Fig. 5. *Trachyleberidea*
? *henryhowei* McKenzie and Swain. Interior of right valve; Lankford Station 19. × 59. Figs 6a-d.
Ambostracon sp. (a), (b) Interior of left valve and enlargement of anterior end; (c), (d) interior of
right valve and enlargement of anterior end; Lankford Station 23. (a), (c) × 59; (b), (d) × 98.
Fig. 7. *Cythere alveolivalva* Smith. Interior of left valve; Lankford Station 12. × 59. Fig. 8.
Caudites sp. B. Interior of right valve; Lankford Station 19. × 59. Figs 9a-c. *Urocythere* ? sp. B.
Interior of left valve, antenna and mandible; Lankford Station 67. × 59 and × 172. Fig. 10.
Cytherella cf. *C. vizcainoensis* McKenzie and Swain. Antennule; Lankford Station 100. × 172.

at San Juan del Sur, Nicaragua. In previous studies it was found in the Gulf of California, near-shore on the eastern side of the Gulf (Swain, 1967) and in Corinto Bay, Nicaragua (Swain and Gilby, 1967).

Subfamily CYTHEROMATINAE Elofson, 1939

Genus CYTHEROMA Müller, 1894
Cytheroma sp. aff. '*Microcythere*' *gibba* Müller
Fig. 11

In the Lankford collection this form occurs at Station 25. It was previously recorded from the Gulf of California (Swain, 1967).

Genus PELLUCISTOMA Coryell and Fields, 1937
Pellucistoma scrippsi Benson
Fig. 22

This species occurs in the Lankford collection at Stations 114 and 122; and at San Juan del Sur, Nicaragua. Previously the species was found in Baja California (Benson, 1959; McKenzie and Swain, 1967).

Family TRACHYLEBERIDIDAE Sylvester-Bradley 1948

Genus PURIANA Coryell and Fields, 1953
Puriana pacifica Benson
Fig. 23

In the Lankford collection the species was found at Stations 5, 23, 28, 32 and 105; and at San Juan del Sur, Nicaragua. The species was found earlier in Baja California (Benson, 1959; McKenzie and Swain, 1967) and the Gulf of California (Benson and Kaesler, 1963; Swain, 1967).

Genus ECHINOCYTHEREIS Puri, 1954
Echinocythereis sp.
Plate VI, Figs 10a, b; Fig. 14

The species was obtained from Station 100 in the Lankford collection.

Genus PSEUDOCYTHEREIS Skogsberg, 1928
Pseudocythereis cf. '*P*'. *simpsonensis* Swain
Fig. 22

This form was found at Lankford's Stations 81–83. It was previously recorded from the Pleistocene Gubik Formation of Alaska (Swain, 1963). Hazel (1967) has placed the species in *Rabilimis septentrionalis* (Brady), albeit subjectively.

Genus CATIVELLA Coryell and Fields, 1937
Cativella unitaria Swain
Fig. 7

The species was found in the Lankford collection at Station 5. Previously the species was found to have general distribution in the Gulf of California (Swain, 1967).

2 H

Genus TRACHYLEBERIDEA Bowen, 1953
Trachyleberidea tricornis Swain

Plate VII, Figs 6a, b, 12a–c, 13a–e; Plate IX, Fig. 3; Fig. 25

The species occurs in the Lankford collection at Stations 19, 25, 37, 55 and 122. It was recorded earlier from the Gulf of California (Swain, 1967).

Trachyleberidea? henryhowei McKenzie and Swain

Plate VII, Figs 10a–d; Plate XI, Fig. 5

In the Lankford collection this form was obtained at Stations 19 and 43.

Family XESTOLEBERIDIDAE Sars, 1928

Genus XESTOLEBERIS Sars, 1866
Xestoleberis aurantia Baird
Fig. 26

In the Lankford collection the species occurs at Stations 19, 21, 22, 28, 29A, 32, 39, 41, 97, 99, 100, 114 and 122. The species was recorded earlier from Baja California (Benson, 1959) in the Pacific Coast area.

Xestoleberis hopkinsi Skogsberg
Fig. 26

This widespread species was obtained in the Lankford collection from Stations 1, 5, 37, 75 and 78; also from San Juan del Sur, Nicaragua. The species was described previously from southern California (Skogsberg, 1950), Baja California (McKenzie and Swain, 1967), and Corinto Bay, Nicaragua (Swain and Gilby, 1967).

Xestoleberis sp. *A*

This form was obtained from San Juan del Sur, Nicaragua.

Xestoleberis sp. *B*

Examples of this species were also found at San Juan del Sur, Nicaragua.

Genus JONESIA Brady, 1866
Jonesia rostrata (Lucas)
Plate VIII, Figs 3a–c; Fig. 16

The species was obtained in the Lankford collection from Stations 108 and 122. It was described from the Vancouver Island area.

Family CYTHERETTIDAE Triebel, 1951

Genus CYTHERETTA Müller, 1894
Cytheretta aff. *C. danaiana* (Brady)
Plate VIII, Figs 1a–g; Fig. 11

In the Lankford collection the species was found at Stations 43 and 53. It was also found at San Juan del Sur, Nicaragua.

Cytheretta sp.

Plate VIII, Figs 2a–e, 4; Fig. 11

The species was found at Lankford's Station 55, and questionably from San Juan del Sur, Nicaragua.

Order MYODOCOPIDA Sars, 1866
Suborder MYODOCOPINA Sars, 1866
Superfamily CYPRIDINACEA Baird, 1850
Family CYPRIDINIDAE Baird, 1850
Subfamily PHILOMEDINAE Müller, 1912

Genus PSEUDOPHILOMEDES Müller, 1894

Pseudophilomedes? sp.

Plate VIII, Figs 5a, b; Fig. 9

In the Lankford collection this form was found at Station 111.

REFERENCES

BENSON, R. H. 1959. Ecology of Recent ostracodes of the Todos Santos Bay region, Baja California, Mexico. *Paleont. Contr. Univ. Kans.*, Arthropoda, Art. 1, 1-80, 11 Pls.
——. AND KAESLER, R. L. 1963. Recent marine and lagoonal ostracodes from the Estero de Tastiota Region Sonora, Mexico (northeastern Gulf of California). *Paleont. Contr. Univ. Kans.*, Arthropoda, Art. 2, 1-52, 8 Pls, 33 Figs.
BOLD, W. A. VAN DEN. 1963. Upper Miocene and Pliocene Ostracoda of Trinidad. *Micropaleontology*, 9, 361-424, Pls. 1-12.
BRADY, G. S. 1870. Notes on Entomostraca taken chiefly in the Northumberland and Durham District (1869). *Trans. nat. Hist. Soc. Northum.*, 3, 361-73, Pls 12-14.
——. 1880. Report on the Ostracoda dredged by H.M.S. *Challenger* during the years 1873-76. *Report of the Scientific Results of the Voyage of H.M.S. Challenger, 1873-76. Zoology*, 1, Pt. 3, 1-184, Pls. I-XLIV.
CORYELL, H. N. AND FIELDS, S. 1937. A Gatun ostracode fauna from Cativa, Panama. *Am. Mus. Novit.*, No. 956, 1-18, 2 Pls.
CROUCH, R. W. 1949. Pliocene Ostracoda from Southern California. *J. Paleont.*, 23, 594-9, Pl. 96.
HARTMANN, G. 1953. *Iliocythere meyer-abichi* nov. spec. ein Ostracode des Schlickwattes von San Salvador. *Zool. Anz.*, 151, 310-16, 17 Figs.
——. 1956. Zur Kenntnis des Mangrove-Estero-Gebietes von El Salvador und seiner Ostracoden-Fauna I. *Kieler Meerseforsch.*, 12, 219-48.
——. 1957. Zur Kenntnis des Mangrove-Estero-Gebietes von El Salvador und seiner Ostracoden-Fauna II. *Kieler Meeresforsch.*, 13, 134-59, Pls 39-50.
——. 1959. Sur Kenntnis der lotischen Lebensbereiche der pazifischen Küste von El Salvador unter besonderer Berücksichtigung seiner Ostracoden-Fauna (III Beitrag zur Fauna El Salvadors). *Kieler Meeresforsch.*, 15, 187-241, Pls 27-48.
HAZEL, J. E. 1967. Classification and distribution of the Recent Hemicytheridae and Trachyleberididae (Ostracoda) off northeastern North America. *Prof. Pap. U.S. geol. Surv.*, 564, 49 pp. 11 Pls.
LANKFORD, R. H. 1958. Ecology of Recent nearshore Ostracoda of Pacific United States. Unpublished doctoral dissertation, *Univ. Calif. Scripps Inst. Oceanogr.*
LEROY, L. W. 1943. Pleistocene and Pliocene Ostracoda of the coastal region of Southern California. *J. Paleont.*, 17, 354-73, Pls 58-62.
——. 1945. A contribution to ostracodal ontogeny. *J. Paleont.*, 19, 81-86, Pl. 9.
LUCAS, V. Z. 1930. Some Ostracoda of the Vancouver Island region. *Contr. Can. Biol. Fish.*, N. S., 6, 397-416.
MCKENZIE, K. G. 1965. Myodocopid Ostracoda (Cypridinacea) from Scammon Lagoon, Baja California. *Crustaceana*, 9, 57-70, 1 Pl., 6 Figs.
——. AND SWAIN, F. M. 1967. Recent Ostracoda from Scammon Lagoon, Baja California. *J. Paleont.*, 41, 281-305, Pls 29, 30, Figs 1-36.
MÜLLER, G. W. 1894. Die Ostracoden des Golfes von Neapel und der angrenzenden Meeres-Abschnitte. *Fauna Flora Golf. Neapel*, Mongr., 21, i-viii, 1-404, Pls 1-40.
PURI, H. S. 1954. Contribution to the study of the Miocene of the Florida Panhandle. Part 3: Ostracoda. *Bull. Fla St. geol. Surv.*, 36 (1953), 215-345, Pls 1-17, Figs 1-14.

——. 1960. Recent Ostracoda from the west coast of Florida. *Trans. Gulf-Cst Ass. geol. Socs.*, **10**, 107-49.

SARS, G. O. 1928. *An account of the Crustacea of Norway. Vol. IX. Ostracoda.* Parts 15 and 16, Cytheridae (concluded) (241-77, Pls CXIII-CXIX). Bergen, Bergen Museum, 1-277, 119 Pls.

SKOGSBERG, T. 1928. Studies on marine ostracods Part II. External morphology of the genus *Cythereis* with descriptions of twenty-one new species. *Occ. Pap. Calif. Acad. Sci.*, **15**, 1-155.

——. 1950. Two new species of marine Ostracoda (Podocopa) from California. *Occ. Pap. Calif. Acad. Sci.*, ser. 4, **26**, 483-505.

SMITH, V. L. 1952. Further Ostracoda of the Vancouver Island region. *J. Fish. Res. Bd Can.*, **9**, 16-41.

SWAIN, F. M. 1963. Pleistocene Ostracoda from the Gubik Formation, Arctic Coastal Plain, Alaska. *J. Paleont.*, **37**, 798-834, Pls 95-99.

——. 1967. Ostracoda from the Gulf of California. *Mem. geol. Soc. Am.*, **101**, 139 pp, 9 Pls, 58 Figs.

——. AND GILBY, J. M. 1967. *J. Paleont.*, **41**, 306-34, Pls 31-34, Figs 1-23.

DISCUSSION

GREKOFF: In order to establish ecological differences or boundaries we always have to distinguish between the dead and living fauna, a point well brought out in your paper. In addition I think that these relationships between North and South America during the Miocene are extremely interesting and will prove of great interest during future research in this area.

HOWE: When I did some consulting work for one of the major oil companies on the Pacific Coast, based on foraminifera not ostracods, I could find very little resemblance between the Cretaceous of northern California and the Cretaceous of the Gulf Coast. But when you come to the Middle and Upper Eocene, I think that the resemblance of the faunas in the two areas is greater than it is in Miocene and younger faunas. Those in the section from central California on up into southern Oregon are greater in the Middle and Upper Eocene than later, so that the connections you are talking about are not Cretaceous connections but started earlier than the Miocene.

BATE: Is there any known similarity between the faunas of the east and west coasts of the United States, north of the latitude of Point Conception?

SWAIN: From the little that we know about them, there is a similarity in some of the Cytheracea on both coasts of the United States north of Cape Hatteras in the east and north of Point Conception in the west. The amount of information we have is much less than for the areas to the south because the number of ostracod species is smaller, and so far, the number of collections has been less.

BENSON: First, in regard to biogeographic similarity, I found in a study of Todos Santos Bay, which covered most of the environments in that particular region, very little similarity between this fauna and that of the Gulf Coast. With the work of Dr Kaesler and myself on the Estero de Tastiota (Gulf of California), there was a very much greater similarity. In that paper we suggested the possibility of migration and the trapping of the faunas in shallower water by the northern migration of isotherms in the Pleistocene. Secondly, concerning the forms referred to *Heterocytpereis* and *Elofsonella*, I question their identification in the Pacific. Having looked at some of these northern faunas I find them notably lacking; although it is possible that we have not sampled the right environment. Thirdly, I would like to ask about the size of the samples you used and also if you would say more about the Rose Bengal staining technique which I think several of us have found difficult to use.

SWAIN: The size of the samples was generally rather large, of the order of 100 g. As to the Rose Bengal staining method, certainly there are many problems. I am reasonably sure, however, that all or nearly all the specimens listed as living, were living, i.e. they had a dark reddish-brown stain. In many cases they also had appendages or some body material inside that could be seen. Many of these specimens were soaked in detergent, made flexible, opened up and dissected and so we have the appendages.

BENSON: Did you find the stain an advantage? I have found that it also stains many things that did not have soft parts.

SWAIN: It is true that living specimens can generally be noted without staining, but staining does offer an advantage over the normal picking process. It is a distinct disadvantage, however, in dissection. If the ostracod is stained it is very difficult to clean it and obtain a good mount for making drawings.

PRELIMINARY REPORT ON THE STUDY OF ABYSSAL OSTRACODS

RICHARD H. BENSON
Smithsonian Institution, Washington, D.C., U.S.A.

The following report, is a brief summary of the progress being made toward the study of the podocopid ostracod fauna of the deep sea. This report was presented informally at the Hull Symposium with illustrations and much more discussion than will be given here, to solicit ideas and criticism from the colleagues assembled. Its purpose here is to make the fact known that such a study is being undertaken and to indicate something of its size, intent, and problems.

First, it should be stated that the term 'abyssal' has been used in various ways in the literature and that its inclusion in the title is not truly definitive. All podocopid ostracods which were found in samples collected at depths below about 400 m were considered. This includes the bathyal as well as the abyssal environments.

After having examined samples from many regions containing specimens from depths as great as 5000 m, it is apparent that the ostracods from the greater depths are diverse in kind and different from those assemblages described from shallow waters, in particular shelf faunas. The assumption that one fauna would necessarily be confined to depths greater than from 2000 to 4000 m or that ostracods would be different in bathyal regions did not seem reasonable in the beginning of the study. Because the faunas vary from station to station and sampling was difficult and scattered, a method was needed to show in what degree the faunas of the different stations were similar and if they could be grouped according to depth or some other environmental parameter. There is evidence now to indicate that, at least in one well sampled region, those forms living in depths greater than 3000 m tend to be different from those of depths less than 2000 m. It can also be stated that assemblages of the same depths but living in different parts of the world are somewhat different although individual species may be closely related. In short, the pattern is quite polythetic and ranges of species appear to be discontinuous. A clustering method using coefficients of association was employed to examine a portion of this pattern.

There are two past studies of importance whose main interest was deep-sea ostracods; Brady (1880)–the *Challenger* Report, and the paper of Tressler (1941) whose study was based on a series of cores across the North Atlantic along the track of the transatlantic cable. So that the present study will be strongly tied to these two previous ones, both collections were obtained and studied. The designation of lectotypes and photographic study of the *Challenger* collection by H. S. Puri, N. C. Hulings and myself, will appear elsewhere.

The present study includes samples from all of the oceans with concentrated attention on the Mozambique Channel, the North American Basin, the Gulf of Mexico, and off western South America. These are regions from which many large samples were obtained. Some moderately large samples containing deep-water forms have been obtained from New Zealand and from north of Scotland. Ostracods from numerous cores and scattered dredge samples have been obtained from over a wide area of the south Pacific Ocean and more sparsely from the Atlantic and Indian Oceans. Obtaining samples of sufficient size has been difficult and not all samples contained ostracods. A few samples from the Arctic and from the Mediterranean suggest that neither place has a normal deep-water fauna at the present time. The deeper parts of the Red Sea seem to be devoid of any ostracods. Samples collected from below the carbonate compensation depth as yet have contained no ostracods. Over-all about 200 samples have been examined ranging in size from 10 cc to about one-half cubic metre. Some fossil material has been examined, but the present effort has been concentrated on Recent forms.

Specimens in foraminiferal, pteropod and other deep-sea oozes are very rare. A Recent sample with six or seven dead specimens per gramme of 20–40 mesh (420–840 microns openings) sieved and dried sample would be considered to have an abundant population. Living specimens are extremely rare. Even the Menzies dredge failed to collect more than five or ten live forms in any of the samples thus far examined. In spite of these low population counts, a few cores containing Tertiary ostracods had as many as 15 specimens in a standard 10 cc sample collected for foraminifers.

The kinds of ostracods found include many forms of *Krithe, Bairdia, Bythocypris, Echinocythereis, Henryhowella, Quasibuntonia, Ambocythere, Bythoceratina*, plus many variations of forms previously described by Brady as *Cythere dictyon, C. suhmi, C. ericea, C. radula, C. viminea, C. dasyderma, C. circumdentata, Cytheropteron mucronalatum* and *Cytheropteron fenestratum*. As one would expect, many new forms have been found and some of the older described ones seem to have much regional and local variation.

General properties especially characteristic of the deep-sea fauna include large size adults (commonly from 1·0 to 1·3 mm in length), exaggerated ornamentation (reticulation, spinosity, alateness), blindness, and long and delicate appendages. Compared with a shallow open shelf fauna, those of the deep are less diverse with seldom more than 15 species represented in a sample.

The stability, bearing strength, and other mechanical parameters of the sediment–water interface at great depths on a scale significant to an ostracod have not been described. One can only guess at the specialized functions of the adaptive features of the carapace for the present; nevertheless several possibilities suggest themselves. Broad ventral surfaces seem to allow the ostracod to remain upright on a soft or irregular substrate. If spines contain sensory setae, extension outward of these organs may reflect a greater sensitivity to the surrounding environment in compensation for blindness. A heavily reticulate carapace may be relatively stronger, although the presence of fragile spines on one form and heavy reticulation on another one found in the same sample seems to be contradictory. Another possibility may be the added weight needed to remain on the bottom as was suggested earlier by Neale (1965). Yet another possibility is the inheritance of characteristics from shallow-water ancestral forms which subsequently have become exaggerated simply because the bizarre has not been eliminated by the rigorous selection that was active in past habitats.

During the oral presentation many examples of unexpected combinations of taxonomic characters and forms whose phyletic relationship seems either old or doubtful were given. These will be dealt with more fully in the future and are only touched upon here. In this category are: divided adductor scars in association with V-shaped frontal scars; presence of the knee apparatus in trachyleberids; a brachycytherid-like form without modified adductors; and a living *Cythereis*-like form with an entomodont hinge. Many specimens of a form very similar to *Phacorhabdotus* have been found in Recent sediments. The species described as *Cythere sulcatoperforata* by Brady seems to have a primitive hinge, and except for its tumid proportions it is similar to *Veenia*. Consequently, one cannot help but be impressed with the possible Cretaceous aspect of the deep-sea ostracod fauna. I believe, however, that it is premature to make a strong statement about this affinity until this or other studies have progressed further.

The usefulness of ostracods in the study of the history of the ocean floor stems from their benthic habitus and the fact that other kinds of benthic animals which leave a microfossil record are extremely rare. If ostracods are restricted in kind to specific depth zones and structural basins, and especially if they have long stratigraphic ranges, they could compliment the planktonic foraminifers as tools for showing changes of climatological and geological significance. This is basically an extension of their potential as recognized for shallow-water forms, only on a much greater scale.

To test the segregation of assemblages of species according to depth, samples from stations along a traverse from the vicinity of Durban, South Africa, across the Mozambique Channel to Tulea, Madagascar, were selected. Depths from about 400 to 4000 m were represented by 30 samples containing 25 species. The coefficients of association (Jaccard and Simple Matching) based on the distribution of the presence and absence of the species in the various samples were calculated to represent the linkage of their occurrence together. These were then clustered to show the overall similarity of stations according to the species shared, and secondly the natural assemblages of species based on their tendency to occur together in the same samples. The method has been described by Kaesler (1966) and is also discussed elsewhere in this volume. This method has several assumptions which may not be entirely accept- able, such as the exclusive selection of a species to join only one cluster; and of the representiveness of presence and absence as a measure of species distribution. None- theless, the results are interesting and fulfil at least a large part of the heuristic intent of the model. In short, the stations (or samples) tended to cluster into two major depth zones, from 400 to 1350 m and from 2200 to 3750 m, with some noise in the 1500 to 2000 m range. This tends to support the suspicion that the bathyal fauna can be recognized even though a deeper fauna seems more evident from simple inspection. The assemblages of species also were segregated by this method demonstrating the varying degrees of reliability of any species or group of species for identification of these depth zones in displaced sediments. Further tests are required of the data and the method, but it looks promising.

Geographic variation within such forms as '*Cythere*' *dictyon*, *C. viminea* and *C. radula* are especially interesting in that certain aspects of carapace ornamentation seem to be typical of particular basins. The fauna of the Gulf of Mexico seems to have endemic expressions of massiveness, such as seen in *C. radula*. The ventro-lateral ridge of *C. dictyon* becomes very pronounced and even detached in the New Zealand region. Efforts have been made to avoid assignment of many of these forms to formal taxa until these variations within the ornament are better understood. They may be of specific rank, and if so the formation of new taxa will be necessary to receive them.

In summary, the 'abyssal' ostracods seem to be well adapted to life in the deep-sea and occur in sufficient numbers to be potentially useful for palaeoecological studies. For the most part they are morphologically distinctive and much more diverse than previously thought. As a fauna, they may range backwards in the geologic record at least into the Cretaceous. At present, elements tend to be segregated by depth and consequently they are isolated into large oceanic, basinal provinces, at least on the species level. On higher levels the fauna seems cosmopolitan. Examination of older sediments in cores reveals ostracods as old as Turonian (eastern Indian Ocean). Not enough material has been available for study for further statements now about this older fossil fauna. It is hoped that the present study can be extended to include a comprehensive study of the fossils.

REFERENCES

BRADY, G. S. 1880. Report on the Ostracoda dredged by H.M.S. *Challenger* during the years 1873-76. *Report of the Scientific Results of the Voyage of H.M.S. Challenger. Zoology*, **1**, Pt. 3, 1-184, Pls I-XLIV.

KAESLER, R. L. 1966. Quantitative re-evaluation of ecology and distribution of Recent Foraminifera and Ostracoda of Todos Santos Bay, Baja California, Mexico. *Paleont. Contr. Univ. Kans.*, Arthropoda, Paper 10, 50 pp., 14 tables.

NEALE, J. W. 1965. Some factors influencing the distribution of Recent British Ostracoda. *Pubbl. Staz. zool. Napoli* [1964] **33** (suppl.), 247-307, 11 Figs, 1 Pl., 5 tables.

TRESSLER, W. L. 1941. Geology and biology of North Atlantic deep-sea cores between Newfoundland and Ireland; Part IV, Ostracoda. *Prof. Pap. U.S. geol. Surv.*, **196C**, 95-106, 4 Pls.

DISCUSSION

WHATLEY: I was struck by the considerable amount of homoeomorphy exhibited by these abyssal faunas. You have shown us many trachyleberids and hemicytherids which bear a very strong resemblance to *Echinocythereis*. Could you explain why this particular adaptation should be apparently so advantageous in the abyssal realm? Secondly, are you implying that there is a continuation of forms living in the abyssal environment from the original *Cythereis* and *Veenia* stocks of the Cretaceous up to the present day? If so, what evidence have you that *Veenia* and *Cythereis* inhabited an abyssal or sub-abyssal environment in the Mesozoic?

BENSON: Taking the last question first, I have no evidence that *Veenia* and *Cythereis* did live in the abyssal environment. I do have comparative forms of Cretaceous age from the abyssal area which are very similar. With regard to the question of homoeomorphy, the original title of my paper was 'Adaptive Convergence in Abyssal Faunas' and I believe that it is so striking that it serves to demonstrate itself. I have found several common characteristics in abyssal forms. First, they are very large, secondly they are exaggerated to some extent in their ornamentation whether this be smooth with a very well-developed, platform-like venter, or be it spinose. For example, there is a trachyleberid from off South America which has secondary spines even within the reticulations. The function of this is still elusive because when they come up from the bottom most of them, even the living ones, are very dirty. Since they are blind for the most part, I don't think that they use this for species recognition. I assume, therefore, that this is to support them on the rather soft ooze, analagous to the very spinose planktonic forms which use surface tension. Lastly, a characteristic which I think is universal is the long and thin appendages; these are not only characteristic of abyssal ostracods but of all abyssal crustaceans.

It looks very much as though we had the basic fauna inhabiting this area in the Cretaceous and that, while there have been some subsequent invasions by other forms, the present fauna represents adaptation and geographic isolation of the basic fauna.

KAESLER: Do you have any evidence that the depth zonation is actually controlled by depth or could the ostracod distribution be controlled by other factors that are highly correlated with depth?

BENSON: I think that the action of depth as an ecological barrier is an integral function of a number of factors, including change of sediment type, temperature, light intensity, etc. I cannot yet find any direct correlation with temperature; one would expect some change to take place between the Polar regions and the equatorial regions along the 2–4°C isotherm but I find no evidence of this as yet. This may be obscured by lack of data that I have at some critical interfaces. The most difficult places to collect are those which are of the most interest in the upper bathyal zone. As yet we simply don't know what causes depth zonation.

NEALE: With regard to the question of size, there is now considerable evidence in both ostracods and other groups to suggest that, within limits, this is temperature controlled. As regards depth zonation, experience with Antarctic and South American faunas suggests that temperature is the principal control. Am I right in thinking that at these abyssal depths the temperature must be fairly constant, in which case other differentiating factors must influence the distributions? Since light intensity at these depths is negligible the main considerations would seem to be type of bottom and perhaps hydrostatic pressure.

ASCOLI: I would also like to ask if Dr Benson thinks that the distribution of these ostracods is controlled by temperature. I was interested to see that a species you place in *Cytheropteroides* we, in Italy, put in *Bosquetina*. I think that this species, or a very close relative, is present in our upper Pleistocene and lower Pliocene. In the type Pliocene this kind of species was found in a neritic, not abyssal environment. As to upper Pleistocene (type Sicilian), this is known to carry a very cold fauna. Thus, having found this species in a cold, but not deep environment, I would suggest that temperature might also exercise a strong control on the distribution of this species as well as depth. Did you find this species at great depths or not?

BENSON: I should point out that it is the species *cytheropteroides* and not the genus of that name. Brady found this particular species of '*Cythere*' and if I recall correctly it is the shallowest of all this group. It comes from 150 fathoms off the Cape of Good Hope and is obviously shallow. But let me comment on the occurrence of your so-called cold-water species in the Plio-Pleistocene of Italy. You have many deep forms that occur through the upper part of the La Castella section which I believe to be reworked material, simply because they occur in a pure assemblage in the Pliocene or lower section and in a more scattered and diluted state in the upper part of the section. To what extent do you know they were living in place in the deposits containing this mixed assemblage? The La Castella section is extremely steep, tectonically active, and is noted for the mixture of cold-water (or deep-water as I would prefer to call them) forms with the shallow-water ones. In other words the record is that of a continually rising sea floor with no reversability or later depression. This would seem to be a strong argument for postulating the fauna to be deep water.

ASCOLI: Possibly, I have not examined the La Castella samples yet, but as far as the type Sicilian is concerned, it is not a tectonically disturbed zone.

   Another question arises in connection with my specimens of *Quasibuntonia* which show a very peculiar feature in having a festoon-like outer margin. Did you find this feature in your specimens?

BENSON: Not as far as I recall.

ASCOLI: Might your specimens be a transitional form between *Quasibuntonia* and something else? This is interesting, because Professor van den Bold and I found that a *Quasibuntonia* ranging from lower Miocene to lower Pleistocene in Italy was also found in Gabon (Africa). Consequently, this deep-water species appears to be a quite good ecological indicator and widespread all over the world.

BENSON: In your studies of the Tortonian fauna have you found any of these deep-water elements?

ASCOLI: No, not in the type Tortonian because it is a shallow water to neritic deposit.

SWAIN: With regard to the use of muscle scars to distinguish hemicytherids and trachyleberids, it seems to me that your work shows something that some of us have noticed for some time. That is that we must not depend too much on the muscle scar pattern to distinguish members of these two families. The variation from moult to adult and also the variation in the anterior V-shaped scar is something we should use with considerable caution and I would depend on other shell features as well as the muscle scars in distinguishing these families. Secondly, concerning the species *sulcatoperforata*. Although the external morphology is similar to *Veenia* the hinge structure, and in fact the whole internal shell structure is so different from *Veenia* that I do not see how *sulcatoperforata* can be placed in that genus. Apparently *sulcatoperforata* has a superstructure or doubling of the shell wall very similar to others from the South

Pacific described by Hornibrook and in fact is very similar to some Ordovician Ostracoda. It seems to me that the species may be a descendent of much more primitive forms than any that we are talking about and has a hinge more like *Cythere* than *Veenia*.

BENSON: Taking the second question first—I alluded to the similarity of this form (from what we could see in the photographs) to *Veenia*, but I make no statement that we should classify this as *Veenia*. It is *Veenia*-like in some characteristics and we have much to learn about it. It is very rare, one finds perhaps one or two specimens in a sample, and there is a fair amount of variation. This variation is apparently sufficient to include some of the forms you have just described and others that I have seen that have a more advanced type of development than *Cythere*.

I think that one has to use muscle scars with some degree of caution and consider all the characteristics in classification. I also think that depth favours an increase in size of the species and a division of the scars, perhaps so that the muscles are more functionally useful. Simple division or non-division of scars as a classificatory criterion tends to be an over-simplification. At least in fresh-water ostracods, it has to be tempered with the kind of homological study carried out by Pokorný.

McKENZIE: Could you please discuss further the dendrograms which you obtained?

BENSON: The dendrograms (or ecograms as I prefer to call them) show clusters of species that tend to occur together in the same stations, or the associations of stations sharing the same or common species. They are subsequently interpreted as assemblages of species which compose natural faunas or assemblages of stations which are reflecting similar environments.

McKENZIE: Is it just one specimen in your populations which shows variation in muscle scar pattern or do you find it in many specimens?

BENSON: No, I found them to be reliable and continuous, and I think the ones I showed here occurred in almost all specimens. They are fairly consistent.

McKENZIE: *Quasibuntonia* occurs in the Atlantic (*Cythere sulcifera*) as was recognized by Ruggieri when he established the genus in 1958. It is living off the coast of West Africa in deep water. Furthermore, I feel that the muscle scar is quite reliable in common with Hazel who has used it as a good character in what I think is an excellent paper. It may not be so useful in the subfamily Orionininae, but these can be separated out if we use the pillar structures used by Professor van den Bold. Thus, while I think that the muscle scar pattern is a useful character, it should be used in conjunction with other characters.

NEALE: I should like to associate myself with Dr McKenzie's remarks on the value of the muscle scar pattern and on the value of Dr Hazel's work. In my own experience the muscle scar pattern is a good guide to differentiating the Hemicytherinae and Trachyleberidinae. At the same time I would agree that it should be used in conjunction with other characters, and it seems significant to me that in the two subfamilies quoted, where the evidence is available, the differences in the muscle scar patterns of the two groups are reinforced by the evidence of the first antenna and mandibular exopodite.

ECOLOGY – FACTORS, APPROACHES, TECHNIQUES, DATA HANDLING

CHAIRMAN: DR H. S. PURI

This account edited and prepared for publication by the Chairman

PURI: Let us first discuss factors.

KILENYI: In plotting my various maps showing the proportion of biocoenosis to thanatocoenosis I found that this increased very rapidly down the river. In the inner estuary 50 per cent of the fauna was actually living there and 50 per cent was dead. Going down river the percentage of dead carapaces or valves decreased to about 10 per cent at about 50 miles from London Bridge and then about one hundred miles out it was down to $0.3–0.5$ per cent, really significantly low. But this is a very wide interpretation of the term biocoenosis—anything that lived or could have lived there.

SWAIN: Professor Howe remarked that the height of the tides along the Pacific Coast of North America might have an effect on the number of specimens per unit area collected in our sub-tidal materials. I have been considering this and it seems to me that along the parts of the Californian coast where these samples were taken, many localities of which I have visited later, the tides are more or less normal. At these localities the tidal range is three or four feet.

BENSON: It seems that in my experience the total ostracod count in the amount of sediment that one obtains depends primarily on the rate of sedimentation combined with productivity. The very high productivity along the western coast of both North and South America seems to be reflected in a very high population of living ostracods in these regions in contrast to regions such as the eastern coast of Africa. Perhaps others may have also noted the effects of upwelling and in general the effects of productivity upon their faunas.

BATE: Dr Benson has mentioned rate of sedimentation as a factor. Recently I investigated the Grey Limestone Series which a number of people here sampled yesterday. During Middle Jurassic times a very large delta lay to the north of this series. On approaching this delta the rate of sedimentation from the land was so considerable that I feel no organism could have survived in that area. Nor is this just a question of decalcification; even the macrofossils are absent in that region. I do feel that the rapid rate of sedimentation provides an environment in which no larval state can live long enough to produce a slightly larger organism which might be able to survive.

KILENYI: I agree that this is a very important factor. In the Thames for instance the rate of sedimentation is practically nil. You wouldn't expect this, but what happens is that the sediments move into the estuary and up into the inner estuary from which they are constantly dredged. This material is taken out, dumped and then returns again. The amount of sediment entering the estuary is in the region of 750 tons a day, which entails a lot of dredging. The second point concerns the mobility of the sediment. This is an ill-defined term and I do not know what it means exactly. If taken to mean that the sediment is being transported at that particular point, I agree. In fact the sand banks which I showed on the slide produce a very high rate of biocoenosis because they represent a closed circulation. Usually on one side of the sand bank there is a flood channel, on the other side there is an ebb channel. The material goes up the flood channel and then goes down the ebb channel. The same material stays in one place and therefore the valves are not transported they are just simply moved round.

KORNICKER: I think that shifting sands seldom have many living ostracods; this has been my experience in the Bimini area of the Bahamas. Secondly, I think it should be kept in mind that rate of sedimentation is especially important in the thanatocoenosis. If you have the same numbers of living specimens present in different areas, you will have many more specimens present in a thanatocoenosis where sedimentation is very slow than where sedimentation is rapid, that is in samples of the same volume or weight of sediment.

PURI: A recent paper by Ryan and Heezen (1965, *Bull. geol. Soc. Am.*) examined the effects of turbidity currents produced by earthquakes and their effect on the oceanography and on the fauna of the area. This is something right in the area of the various palaeoecologists here.

KAESLER: An approach that has not been discussed here and one that has been largely neglected by palaeontologists is the study of geographic variation of morphology. By making a quantitative study of relationships between geographic variation of morphology and various environmental factors, we can begin to understand the effects of environmental factors not on the distribution of forms, but on their morphology. A geographic variation study of *Cythere lutea* from the North Atlantic, across the Arctic Ocean, through the Bering Strait, and down the west coast of North America would be extremely interesting.

HART: We have noticed considerable variation in the copulatory appendages of certain of the entocytherids. For instance, we first found what we now call *Sagittocythere barri* in Alabama caves and later found similar, but distinct, animals in Indiana caves. Had we not then found intergradations in the cave systems between these two localities we would, no doubt, have described the two forms as different species.

LEVINSON: Speaking of turbidites, we see this quite often in the fossil record in wells. They can be recognized by a mixed assemblage; a near-shore assemblage and a 'deep-water' assemblage. In the study of the facies sequence in cores, this interpretation can be supported. I have used both ostracods and foraminifera to recognize turbidites.

HOWE: Continuing Levinson's comments about turbidites, the migration of ostracod carapaces can come from deep water inland as well as moving from the shore out. Louisiana State University used to maintain a biological station on Grand Isle and they were very concerned because they couldn't find any ostracods. They brought in two different ostracod workers in succeeding summers, but these men were only able to find ostracods in ponds on the Island. They couldn't find them on the beach, or by dredging. I happened to drop down there and instead of looking at the swash-line, I picked up some sea-weed that had been brought up by one of the very large storms we get on the Gulf Coast. Shaking it I got more ostracods than had been collected by other ostracod workers from all other Gulf beaches combined. I therefore suggest that where you have storms bringing in sea-weed you will very frequently get ostracod faunas brought inland from deep water. This was shown both by the ostracod content of the sample and by the foraminifera present. I think the weed had been brought from a depth of approximately 100 feet. Both detached sea-weeds and sponges are capable of transporting micro-organisms a very considerable distance.

KAESLER: I agree with Professor Howe that sampling is a very important factor that needs to be worked out. Eventually we will probably be forced to use different sampling techniques for different environments. Such practice will invalidate many quantitative approaches that might otherwise be useful. It is for this reason and because, as Dr Kornicker mentioned, different sedimentation rates have different effects that I am inclined to consider presence and absence of species rather than number of specimens of a species. I see no way of interpreting number of specimens, and I believe it introduces false precision.

KORNICKER: I believe that in reference to the living fauna the proper sample is one based on bottom area, and included in the area should be everything living there, regardless of whether the animals are on algae, crawling on the sediment, or burrowing into it. Now, if you are dealing with dead forms, the results of the study will be of most use to the palaeoecologist if dry weight is used because the dry weight of sediment is closest to the palaeontologist's sample of rock, though naturally there will be some difference because of cementational processes.

KAESLER: I wonder with all the burrowing instars that Mr Williams mentioned if we are really justified now in using 'area' for Recent or for soft part studies?

KORNICKER: When I say 'area' I mean surface area of bottom regardless of how deep you have to go to get all the living ostracods.

KILENYI: I suppose you have to go down to the anaerobic layer because below that you won't find anything in any case, so if you can get a core or a sample that brings up a relatively undisturbed portion then you have to slice off where the colour changes to dark and anything above should be counted.

SYLVESTER-BRADLEY: There is another approach which I feel that we have rather neglected so far in ecological studies of ostracods and that is the *autecological* approach, the ecology of single ostracod species. If the chairman will excuse me from slightly confusing approaches with techniques, I think that in order to study single ostracod species and the various physiological effects of the environment, we need to use the experimental method and to develop in the laboratory aquaria in which the conditions can be controlled and in which all the natural conditions can be mimicked. Botanists who have been studying ecological conditions like this have built what they call 'phytotrons', which have sophisticated control of light, humidity and everything else that you need for a growing plant. For our experiments we would need similarly sophisticated aquaria for the growth, particularly, of marine species. Have any investigators here been working along these lines of controlled experiments on single species, in which one can see how the effect of salinity, rate of sedimentation, movement of sediment and so on control the growth and habits of ostracods? Such work would be of value both for ostracods with a variable morphology like *Cyprideis* and for those which show less obvious changes of morphology with environment.

HAGERMAN: I quite agree with Professor Sylvester-Bradley especially as I have done a lot of experiments with controlled environmental conditions. Most of the shallow-water forms are extremely easy to keep in cultures. Just put them in a petri-dish with a little bottom material and they will grow and reproduce very easily.

LEVINSON: For the past 15 years, I have been raising fresh-water and marine ostracods in tanks, and have varied bottom conditions, temperature and salinity. I am

considering buying a new tank which is sold by Wards which contains both a re-
frigerant unit and a heating unit to obtain a very close control of temperature. There
have been many interesting results from my studies which I shall publish one of these
days. I was mentioning to one of our group here the other night that I had a tank of
Chlamydotheca unispinosa and periodically I would find a female in which all of her
eggs produced young with a particular mutation. Sometimes the mutation affected
the shell and in other cases appendages (usually setae) were affected. Some of the
mutations of the shell were so unusual that if it was found in nature, it certainly would
be described as a new genus. In no case did I find that mutant forms could reproduce.

McGREGOR: I agree with Professor Sylvester-Bradley and suggest that we begin
using some of the radioactive isotopes in studies of feeding, predation, activity cycles,
or other population characteristics in laboratory or field experiments with fresh-water
and marine ostracods. At our laboratory, Dr Robert G. Wetzel and I are currently
measuring self-absorption of C^{14} radiation by fresh-water ostracods for use in preliminary studies of nutrition and secondary production rates in this group.

McKENZIE: I would like to put on record the work that has been done by workers
on foraminifera in culturing and maintaining populations under conditions which
approximate closely to those in the field. A group at the American Museum have been
doing this now for some years; Jack Bradshaw at Scripps has another approach in
which he takes an ecology meter into the field and studies the animal in its natural
habitat; and at the University of Queensland, Dr Stephenson and a colleague have
been observing a rather large endemic local foraminifer in the laboratory under varied
ecologic conditions with some considerable success. Perhaps we could modify such
techniques to our work with ostracods.

The slides illustrate some points we need to consider in making palaeoecological
syntheses.

Slide 1. Illustrates the strong radiation which has taken place in the group since the
beginning of the Cenozoic, e.g., Cytheridae which comprised about 50 extant genera
at the beginning of the Tertiary now includes about 200 genera.

Slide 2. Shows the latitudinal variation in species diversity for Halocyprididae.
This obviously has been contaminated by over-sampling in the region of 40°S Lat.
The slide records data for the Atlantic and Indian Oceans and emphasizes the rather
restricted known fauna of the Mediterranean.

Slide 3. Deals with variation in species diversity with depth. I think we are all
familiar with this and so I will merely draw your attention to the strong decrease in
species number with depth which is a factor to be considered in ecologic and palaeo-
ecologic syntheses.

Slide 4. Records the importance of the sediment substrate on ostracod distribution
in an area where the taxonomy is ironclad (the Mediterranean). The data has been
taken from various workers, namely Puri, Bonaduce and Malloy (1965), Dom Rome
(1965) and from several of Madame Reys' papers. You see that there are hardly any
species which are common to the different substrates (sand, mud and organogenic fine
sand) over a considerable range of ostracod families and subfamilies.

Slide 5. Illustrates another substrate effect in the Mediterranean, this time with
respect to the phytobenthos (marine angiosperms, inter-tidal and infra-littoral algae,
and calcareous algae). You can see what a considerable part these angiosperms and
seaweeds play in the ecology of ostracods.

Slide 6. Illustrates sweepstakes dispersal in the Pacific Islands. It shows that
certain genera cut out progressively as we go east in the Pacific, from New Caledonia,

to Fiji, to Samoa, and finally to Bora Bora near Tahiti. I appreciate that the data may not be wholly accurate because our sampling is poor in this region, but the slide is based on samples of mine from all these areas.

Slide 7. Records the distribution of large cypridid ostracods and lists the number of species, the number of endemic species, number of genera and endemic genera in this group. When I say large I am defining this arbitrarily as 3 mm or greater in length. You see a strong tendency for clustering in the southern hemisphere in these large ostracods. Extending this a bit further, I might point out that there are correlations between the forms of the Australian region and those of the Ethiopian region which are in my opinion of biogeographic interest.

I hope that these various approaches are of interest to this discussion.

KAESLER: It is important to consider these ideas and their ecological and palaeo-ecological implications. It is also important that we keep in mind, as you have done, the difficulties introduced by unequal sampling.

BATE: I have always been told that the key to the past is a study of the present. So far I am getting very little help from zoologists for the simple reason that they never seem to find the same species in more than one sediment. In my work in the Middle Jurassic I can correlate the Lincolnshire Limestone with shales quite far removed and not really in open contact with Lincolnshire. The same species are present in oolitic limestones in an area which could be say equated with the present day Bahamas, these same species are found in the Yorkshire Basin in sandy beds and shale beds. It isn't just a question of mixing because the Yorkshire area is almost completely isolated from the Lincolnshire Basin and I am wondering just what the zoologists are doing. How is it that they seem to have such a nice pattern of isolation between species of muds, silts, sands, calcareous beds and isolated niches which just did not seem to exist geologically?

PURI: Let us now discuss data handling.

ASCOLI: I remember that during the last Symposium at Naples a whole session was devoted to the problem of tabulating data by means of IBM cards. After 4 years I would like to ask how this worked out and the results that were obtained using IBM punch cards?

PURI: In answer to your question we put all our data on the IBM cards and we had a 60 page index to the last Symposium Volume. To my surprise, and to everyone else's surprise, we had what Professor Howe pointed out on the first day, multiple entries, one species referred to four or five different genera and we were trying to give one number to a species. This work was done through F.A.O. in Rome and with the IBM people in Milan. We could not resolve the basic classification of Pokorný and Hartmann on the sub-ordinal level and we reached a stage where we could not convince the various authors involved that they should work on the common taxonomic and nomenclatural problems in their respective papers. Consequently we decided to leave the index out since we could not give four different numbers to the same species which appeared under four different names. We have been able to put our entire data on the Gulf of Naples species on IBM cards; the stations are keys to latitude and longitude and all the species are numbered. This consists of the Gulf of Naples data; the rest of it is all keyed but still unpublished.

SYLVESTER-BRADLEY: If punched cards are to be used for data retrieval, it is likely that the pack will grow to such a size that only automatic machinery can deal with it. In ecology what we are looking for is not so much *data retrieval* as *data association*— we are looking for correlation effects of species (or characters) with environments. In such a case there is an alternative to a large pack of cards. I am a man with very little mathematical ability and I am terrified of computers, partly because they always use a language I cannot understand, but mostly because of the extreme cost of the few seconds that one is allowed on computers. I have been trying to work out a system in which I can hold preliminary conversations with my cards without any expense. I believe this is now possible by using the 'peek-a-bo' or 'feature-card' method. This is a reversal of the IBM method, in that the items are the holes in the card and the characters are the cards themselves. This reduces the number of cards to something which can easily be handled; correlations can be made visually in one's own laboratory. Subsequently these cards can, if desired, be fed into a special device which reverses the holes to cards and the cards to holes, or which records the data on tape. This can then be fed to a computer. Such a method should be of great assistance to the less mathematically equipped palaeontologist.

Puri: Maybe you are aware of the McLean IBM card system. He has been able to put a tremendous amount of data, both for foraminifera and ostracods on IBM cards and he has published a brochure on data handling which could be obtained from his address in Alexandria, Virginia.

Kaesler: To reduce costs, IBM punched cards may be sorted with a card sorter rather than a computer.

Levinson: I started to develop a manual punch card system but now it is longer than my arms can reach and I have given up the whole scheme. I have come to the conclusion that the only hope is to use an E.D.P. data retrieval system which now is coming into general use. I think that this is our only solution because ostracod bibliographic references are becoming so voluminous that one person just cannot handle them. A joint project among ostracod specialists is needed.

Bate: I feel here that there are two problems, one for the individual research worker and one for the institutions whether they are big oil companies, Universities or Museums such as the one to which I am attached. For our purpose I feel that we are now moving towards punched tape with the data fed into a computer and also retrievable from the tape in the system of card indexes which are convenient for rapid individual use whilst computer storage is useful for a much larger purpose at a later date.

CATALOGUES AND BIBLIOGRAPHIES. VALUE, PUBLICATION, COMPILATION.

CHAIRMAN: DR I. G. SOHN

These minutes prepared for publication by the chairman.

Professor Sylvester-Bradley asked whether Dr Levinson would resume the annual bibliography published in *Micropaleontology*, up to Volume 8, 1962. The meeting supported this unanimously and Dr Levinson agreed to resume the preparation of the bibliography. Dr McKenzie suggested that Professor Howe prepare an updated version of *Ostracod Taxonomy* because many new genera had been proposed since the 1962 edition. Professor Howe agreed to prepare a supplement as Volume 2 of *Ostracod Taxonomy*. He stated that he knew of at least 1,500 references to publications on Ostracoda that are not available to him. Dr Oertli suggested that a list of these references be circulated in order that copies of those books might be sent to Professor Howe. It was further agreed that Professor Howe would circulate a list of genera and families proposed after the 1962 edition of *Ostracod Taxonomy* with the object of adding published taxa to that list as a start for the proposed supplement.

Professor Howe described his card index of Ostracoda which was discussed by Drs Oertli, Sylvester-Bradley, Swain, Puri, Sohn, Levinson and Benson. Professor Howe indicated that his card index is available for consultation. The chairman announced that the cost of the bibliographic cards issued monthly by the Centre Scientifique et Technique, B.R.G.M. was increased in 1967, and that the operation was moved from Paris to Orleans. The cost of the Ostracoda cards is now $10 a year, and the new address is B.P.555, 45 Orleans, La Source, France.

IV. LOCALISATION OF MATERIAL

DESIGNATION OF LECTOTYPES OF FRESH-WATER SPECIES DESCRIBED BY DOBBIN, 1941 (OSTRACODA, CRUSTACEA)

CATHERINE DOBBIN EVENSON
Lewis and Clarke College, Portland, Oregon, U.S.A.

Recently type specimens have been designated for some species of fresh-water Ostracoda from the State of Washington and other western localities. These species had been named and described in the publication by Dobbin in 1941. The specimens now have become a part of the collection of the Thomas Burke Memorial Washington State Museum, University of Washington, Seattle, Washington.

The slides had been the property of Professor Trevor Kincaid, who, at his retirement, presented them to the museum. Most of the specimens had been collected and the slides had been prepared by Professor Kincaid. The author was allowed the use of these for the preparation of a master's thesis. Due to her inexperience and inadequate foundation in the principles of taxonomy she failed to designate type specimens at that time and committed some other errors. After a lapse of many years and with an opportunity to renew her study of ostracods, she has gone over the entire collection and has indicated lectotypes. Although some of these specimens are really inadequate for such a designation, they will be more adequate, it is hoped, than nothing at all having been set apart.

In addition to indicating type specimens, the author has chosen representative specimens, to which museum catalogue numbers have been given, of the other species included in her paper. A few specimens have been added to the original collection for the sake of clarification, but these have been labelled to that effect.

TYPE SPECIMENS

diss. = dissection, w.m. = whole mount, F. = Female, M. = Male.

Entocythere columbia Dobbin, 1941
  Type locality: indefinite.
    Specimens taken from unrecorded species of crayfish, from undesignated place in Columbia R., probably in SW Wash. No date.
  (Note: = *Uncinocythere columbia* (Dobbin, 1941) Hart, 1962.)

No. 26050 Lectotype: w.m. M.
           Paralectotypes: w.m. 4M. 4F.
26051 Paralectotypes: w.m. 2F.
26052 Paralectotype: diss. F.

Limnocythere glypta Dobbin, 1941
  Type locality: Granite L., near Spokane, Wash., 24/VII/36

No. 26065 Lectotype: diss. F.
26066 Paralectotype: diss. M.

Cypris (Cyprinotus) scytoda Dobbin, 1941
   Type locality: Portola Valley, Cal.,
     1/III/36
   (Note: = *Cyprinotus scytoda* Dobbin,
    1941 in Tressler, 1959.)

                        No. 26082 Lectotype: diss. F.
                          26083 Paralectotype: w.m. M.
                          26084 Paralectotype:
                                 immature F.

Cypricercus dentifera Dobbin, 1941
   Type locality: Plantation Pond, near
    Seattle, King Co., Wash., 26/II/28

         No. 26090 Lectotype: diss. M.
            26091 Paralectotype: diss. F.
            26092 Paralectotype: w.m. F.

Cypricercus columbiensis Dobbin, 1941
   Type locality: indefinite.
    Pond near Columbia R., probably
    SW Wash., 24/IV/23

         No. 26097 Lectotype: diss. M.
            26098 Paralectotype: diss. M.
            26099 Paralectotype: diss. F.

Cypricercus elongata Dobbin, 1941
   Type locality: Pond, Ballard, Seattle,
    King Co., Wash.
   (Note: believed by author to be a late
    instar of *Cypricercus dentifera* Dobbin,
    1941.)

         No. 26100 Lectotype: diss. F.
            26101 Paralectotype: diss. F.
            26102 Paralectotype: w.m. F.
            26103 Paralectotype: w.m. M.

Prionocypris longiforma Dobbin, 1941
   Type locality: stream, Bald Hill Lake,
    Washington, 18/IV/36

         No. 26104 Lectotype: diss. M.
            26105 Paralectotype: diss. F.
            26106 Paralectotypes: w.m. M.
                                F.
                    1 immature.

Cypriconcha alba Dobbin, 1941
   Type locality: Dry Alkali Lake, Grand
    Coulee, Wash., 26/IV/36

         No. 26111 Lectotype: diss. M.
            26112 Paralectotype: diss. M.
            26113 Paralectotypes: w.m. M.
                                2F.

Cypriconcha gigantea Dobbin, 1941
   Type locality: St Michael's, Alaska. No
    date.
   (Note: = *Cypriconcha macra* Blake, 1931.)

         No. 26115 Lectotype: diss. M.
            26116 Paralectotype: w.m. M.
            26117 Paralectotype: diss. F.

Potamocypris hyboforma Dobbin, 1941
   Type locality: Alkali Lake, Chelan Co.,
    Washington, No date.

         No. 26126 Lectotype: diss. F.
            26127 Paralectotype: diss. F.
            26128 Paralectotype: w.m. F.

Cyclocypris washingtoniensis Dobbin, 1941
   Type locality: Pond, Ballard, near Seattle,
    King Co., Wash., 16/II/32
   Culture in Laboratory, Seattle,
    Washington, 22/V/22

         No. 26129 Lectotype: diss. M.

              26130 Paralectotype: diss. M.
              26131 Paralectotype: w.m. M.
              26132 Paralectotype: diss. F.

Cyclocypris nahcotta Dobbin, 1941
   Type locality: lake near Nahcotta,
    Pacific Co., Wash., 16/VIII/22

         No. 26133 Lectotype: diss. M.
            26134 Paralectotype: diss. F.
            26135 Paralectotype: diss. F.

Cyclocypria kincaidia Dobbin, 1941	No. 26139 Lectotype:	diss.	F.
Type locality: Lake Washington, King Co.,	26140 Paralectotype:	diss.	F.
Seattle, Wash., May, 1922	26141 Paralectotype:	diss.	F.
(Note: believed by author to be a late	26142 Paralectotype:	diss.	F.
instar of *Cyclocypris kincaidia* (Dobbin,			
1941).)			
Candona foviolata Dobbin, 1941	No. 26154 Lectotype:	diss.	M.
Type locality: pool, Deception Pass,	26155 Paralectotype:	diss.	F.
Washington, 3/II/34	26156 Paralectotype:	w.m.	M.

The following representative specimens have been added to the collection in order to augment and clarify certain features seen in the type specimens:

Entocythere columbia Dobbin, 1941			
From *Pacifastacus klamathensis*	No. 26053	w.m.	M.
(Stimpson), Mt. Scott Cr., Milwaukie,	26054	w.m.	F.
Clackamas Co., Oregon.	26055	w.m.	F.
Coll. C. D. Evenson, 4/IX/65			
(Note:= *Uncinocythere columbia* (Dobbin,			
1941) Hart, 1962.)			
Cypricercus dentifera Dobbin, 1941	No. 26093	diss.	M.
Pond, Kuehn Rd., Milwaukie,	26094	w.m.	M.
Clackamas Co., Oregon.	26095	diss.	F.
Coll. C. D. Evenson, 24/III/60	26096	w.m.	F.
Cypriconcha alba Dobbin, 1941	No. 26114	diss.	F.
Pond, Tule Lake, 2/V/37			

REFERENCES

BLAKE, C. H. 1931. Two fresh-water ostracods from North America. *Bull. Mus. comp. Zool. Harv.*, **72**, 279-92.
COLE, G. A. 1960. The cyprid ostracod genus, *Cypriconcha* Sars. *Trans. Am. microsc. Soc.*, **79** (3) 333-9.
DOBBIN, C. N. 1941. Fresh-water Ostracoda from Washington and other western localities. *Univ. Wash. Publs Biol.*, **4**, 174-246.
HART, C. W., JR. 1962. A revision of the ostracods of the family Entocytheridae. *Proc. Acad. nat. Sci. Philad.*, **114** (3), 121-47.
TRESSLER, W. L. 1959. Ostracoda. In Edmondson, W. T. *Fresh-water Biology*. pp. 657-734. Wiley, New York.

DISCUSSION

BATE: I was interested to hear Dr Evenson say that sexual dimorphic characters appeared in different sized specimens. This is not uncommon in fossil ostracods and has resulted in a difference of opinion between myself and Dr Malz over a Middle Jurassic *Praeschuleridea* species from this country. He has divided my single species into a smaller species which is morphologically indistinguishable apart from the fact thats exual characters appear at a smaller size. It is my opinion that, as you said in your paper, sexual characters can become apparent at different instars or at least at different sizes. How common is this in Recent ostracods and how many people have recorded this appearance of 'pre-adult' male dimorphism within their ostracod populations?
EVENSON: I know of nothing published on the development of males. As far as I know, most of the life histories that have been worked out are for the parthenogenetic forms which have only the females.

DELORME: I think you are quite correct in your present belief that both you and Gerald Cole were working with immature forms of *Cypriconcha alba*. In the interior plains of Canada I have found both males and females of *Cypriconcha alba* to be very abundant and with very well developed Zenker's organs. The same applies in the case of *Cypriconcha macra* and *C. barbata*, as well as for the new species *C. ingens* on which I have just published.

HART: Concerning Dr Bate's remarks, we are in the process of describing an entocytherid ostracod that is commensal on a marine isopod from India. We believe that this ostracod has six instars before reaching adulthood, and that the male genital apparatus is present in the sixth instar as a not-fully-developed, yet recognizable structure—an *anlagen*.

BATE: Is there a difference in carapace morphology?

HART: In the Indian species the valves of the sixth instar and adult males show a characteristic concavity in the postero-dorsal margin. However, such differences are rare among entocytherids, and most have thin, delicate shells with virtually no distinguishing features such as muscle scars or protuberances. Adult males may be slightly larger or slightly smaller than adult females.

BATE: But the carapace does have a permanent shape, the outline doesn't vary?

HART: No. As far as we know there is little variation in shell shape among adults of a given species. As I pointed out, however, there may be slight differences in shell shapes of different instars as can be seen in the case of the Indian entocytherids. There is little other available information on entocytherid instars.

SWAIN: On the question of pre-adult sexual dimorphism, I believe that Hoff identified mature males in about the fifth instar in *Candona*.

KORNICKER: Concerning Dr Bate's original comment, I have in press a study of a collection of myodocopid ostracods from Hadley Harbor, Massachusetts, in which three males have some, but not all, morphological characters of the adult. The stage of development of their appendages is part way between the normal N–1 instar and the adult. Two of the specimens have larger carapaces than adult males, the other is in the same size range as the adult. Apparently these specimens moulted from N–1 stage instars, but did not become fully developed males. Adult males which might have been produced by a further moulting of the undeveloped males were not present in the collection.

SYLVESTER-BRADLEY: I think Dr Kornicker's observation is extremely interesting and important in relation to fossils, because it has been my experience that just occasionally one finds a gigantic male, much bigger than all the others, in a fossil species that is showing dimorphism. Thus a cluster of normal mature males is a whole instar behind these very few rare ones. Am I right in thinking that you are suggesting that very occasionally the size of a male is one whole instar above those which are normally produced in the species?

KORNICKER: Carapace lengths of the three partly developed males are 1·56, 1·70 and 1·72 mm. The range of lengths of fully developed males is 1·50–1·63 mm ($N = 8$). Two of the partly developed males are considerably larger than the largest adult male in the collection, but the difference is not equivalent to a whole instar (the range of lengths of the normal N–1 instar is 1·38–1·43 mm ($N = 5$)). If the partly developed males should moult again in order to reach full maturity, their carapaces would be longer than the normal adult by at least one instar length. In my paper I suggest that unusually large specimens occasionally reported in the literature may be caused by delay of sexual maturity.

SOME TYPE SPECIES OF FRESH-WATER OSTRACODA IN THE BRITISH MUSEUM (NATURAL HISTORY)

F. M. SWAIN and J. M. GILBY[1]
University of Minnesota, Minneapolis, U.S.A., and 21, The Crescent, Ashford, Middlesex, England

INTRODUCTION

Illustrations are given of the shells of specimens of several type species of fresh-water Ostracoda housed in the British Museum (Natural History), London. The following brief remarks about each of the species give the writers' opinion as to the status of each genus represented.

Amphicypris nobilis Sars, 1901 (Cyprididae-Cypridinae) (= *Eucypris* by Daday, 1905) differs from typical *Eucypris* in its more elliptical shell outline, more compressed valves, more numerous adductor muscle scars, narrower inner lamellae, and slightly different furca. *Amphicypris* possibly should be retained as a valid genus.

Eurycypris latissima G. W. Müller, 1898 (Cyprididae-Cypridinae) (= *Cypris* by Müller, 1912) has its shell somewhat more convex than, and lacks the anterior flange-like extension of *Cypris pubera* O. F. Müller, but is otherwise similar in appendages and shell to *Cypris* as recognized by Müller and by Sars and probably should remain suppressed.

The muscle scar pattern is so variable in type specimens of *Megalocypris princeps* Sars (Cyprididae-Cypridinae), particularly in the dorsal field, that little generic characterization seems possible on the basis of the muscle scar pattern. The shell form of *Megalocypris* is closely similar to that of primitive brackish-water genera such as *Fabanella* (Cytheridae) of the Jurassic and Cretaceous. Such shell types may be useful as environmental indicators.

Stenocypris cylindrica major (Baird) (Cyprididae-Cypridinae), recently studied by Edward Ferguson (1967), has a characteristic pattern of numerous, evenly spaced, large-diameter radial canals that possibly are unique in this subfamily.

Paracypridopsis zschokkei Kaufmann 1900 was placed in *Potamocypris* by Müller (1912) (Cyprididae-Cypridopsinae) and by Sars (1928), but that genus is typified by strong dorsal overlap of the right valve by the left that is not present in *P. zschokkei*. Retention of *Paracypridopsis* as was recommended in Moore (1961) seems justified.

The shell of *Herpetocypris reptans* (Baird, 1835) (Cyprididae-Herpetocypridinae) has a complex pattern of radial canals which, however, differs from one specimen to another in the British Museum material in more than incidental detail, perhaps owing to differences in shell maturity. The shell structure of *Microcypris reptans* Kaufmann 1900 (= *Eucypris* by Müller, 1912) is so different from that of *Herpetocypris* that it seems unlikely they belong in the same subfamily as was held by Sharpe (1903). Whether or not *Microcypris* should be suppressed remains unsettled.

[1] Now at: 38 Henry Street, Darien, Conn., U.S.A.

495

Prionocypris serrata (Norman) (Cyprididae-Herpetocypridinae) (= *Eucypris* by Müller, 1912) does not have the flange-like anterior expansion of the shell and other shell features of *Eucypris*, and although similar anatomically to *Herpetocypris* has a different furca than in that genus as pointed out by Sars. Retention of *Prionocypris* as indicated in Moore (1961) is favoured.

Pontocypris mytiloides (Norman) (Pontocyprididae) was treated by Sylvester-Bradley (1947). Type specimens were illustrated by us to provide additional documentation of the type material.

Mecynocypria obtusa Sars. A few additional shell features are shown which supplement the original diagnosis by Rome.

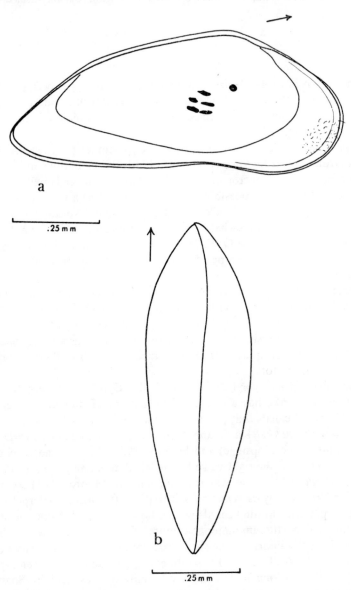

Fig. 1. *Pontocypris mytiloides* (Norman). × 96. B.M. No. 1911.11.18.M 3012. (a) Left valve. Internal lateral view; (b) Carapace. Dorsal view.

SYSTEMATIC DESCRIPTIONS

Family PONTOCYPRIDIDAE

Genus PONTOCYPRIS Sars, 1866

Type species. *Cypris serrulata* Sars, 1863, [= *Cythere* (*Bairdia*) *mytiloides* Norman, 1862], North Atlantic.

Geologic Range. Devonian—Recent.

Pontocypris Sars, 1866. *Forh. VidenskSelsk. Krist.*, 1865, p. 13.

Erythrocypris, G. W. Müller, 1894, *Fauna Flora Golf. Neapel*, **21**, 256 (same type species as *Pontocypris*).

Pontocypris mytiloides (Norman)
Figs 1a, b

Cythere (*Bairdia*) *mytiloides* Norman, 1862, *Ann. Mag. nat. Hist.*, (3) **9**, 50 pp., Pl. 3, Figs 1–3.

Cypris serrulata Sars, 1863, *Nyt Mag. Naturvid.*, **12**, 250.

Pontocypris serrulata (Sars), Sars, 1866, *Forh. VidenskSelsk. Krist.*, 1865, p. 15.

Pontocypris mytiloides (Norman), Brady, 1868, *Trans Linn. Soc. London*, **26**, 385, Pl. 25, Figs 26–30; Pl. 37, Fig. 4; Pl. 38, Fig. 1.

Erythrocypris serrata Müller, 1894, *Fauna Flora Golf. Neapel*, **21**, 258, Pl. 11, Figs 10, 11, 30–38.

Shell elongate, subtriangular in side view, highest about one-third from anterior end; dorsal margin moderately convex, straightened on either side of position of greatest height; ventral margin concave medially; anterior margin broadly curved, extended below, slightly truncate above; posterior margin acuminately extended medially, subtruncate above. Valves rather compressed, greatest convexity anterior in position; left valve slightly larger than and overlaps right. Surface smooth and glossy except for very small normal canals and hairs.

Hingement consists of weak groove in dorsal margin of left valve into which fits edge of right. Inner lamellae very broad terminally, narrower ventrally, vestibule broad; radial canals few, short and widely spaced anteriorly, indistinct elsewhere. Narrow free marginal zone hyaline.

Length of figured left valve 0·96 mm, height 0·37 mm, convexity 0·30 mm.

The principal features of the appendages are listed in Table 1. The appendages were described in detail by previous authors.

Occurrence

The type locality for *Cythere* (*Bairdia*) *mytiloides* Norman is Lamlash Bay, Firth of Clyde, western Scotland; 'found among shell sand dredged in Lamlash Bay'; collected 1854; types are in the British Museum (Natural History); approximately three dozen dry specimens labelled types; most have appendages. All specimens appear to have the same shape.

Family CYPRIDIDAE
Subfamily HERPETOCYPRIDINAE

Genus HERPETOCYPRIS Brady and Norman, 1889

Type species. *Cypris reptans* Baird, 1835, near London, England.

Geologic Range. Tertiary to Recent.

Erpetocypris Brady and Norman, 1889, *Sci. Trans. R. Dubl. Soc.*, ser 2, **4**, p. 84.

Herpetocypris, G. O. Sars, 1890, *Forh. VidenskSelsk. Krist.*, 1889, no. 8, p. 34.

Herpetocypris Brady and Norman, 1889, International Commission of Zoological Nomenclature, Opinion 533, 1958.

Herpetocypris reptans (Baird)
Figs 2a–f

Cypris reptans Baird, 1835, *Hist. Berwicksh. Nat. Club*, **1**, 99, Pl. 3, Fig. 11; Lilljeborg, 1853, *De Crustaces ex ordinibus tribus; Cladocera Ostracoda et Copepoda, in Scania occurrentibus.* Figs 7–9; Müller, 1900, *Zoologica, Stuttg.*, **30**, 58, P. 14, Figs 4, 6, 12, 13, 17.

Candona virescens (moult), Brady, 1864, *Ann. Mag. nat. Hist.*, (3) **13**, 61, Pl. 4, Figs 1–5.

Cypris ornata Fric, 1872, *Arch. naturw. LandDurchforsch. Böhm.*, **2** sect. iv, 226.

Erpetocypris reptans Brady and Norman, 1889, *Sci. Trans. R. Dubl. Soc.*, ser. 2, **4**, 84, Pl. 13, Fig. 27.

Herpetocypris reptans (Baird), Sars, 1891, *Forh. VidenskSelsk. Krist.*, 1890, no. 1, p. 17.

Description of shell

Shell large elongate elliptical sub-reniform, highest postero-medially; dorsal margin slightly convex, straightened medially; ventral margin sinuous, slightly concave

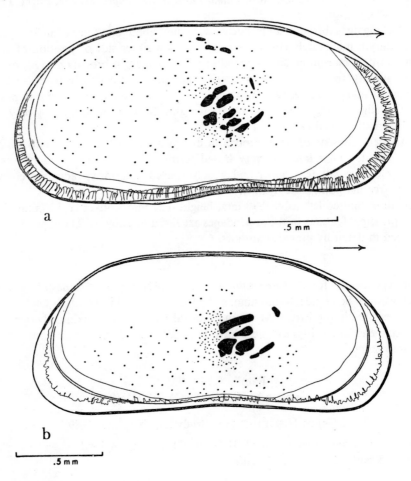

a

.5 mm

b

.5 mm

antero-medially; anterior margin broadly curved, slightly truncate above; posterior margin also broadly curved, somewhat extended below, sub-truncate above; anterior margin extends much farther beyond end of hinge surface than does posterior margin. Left valve slightly larger than and overlapping right; valves compressed, tapering more gradually toward anterior than toward posterior margin. Surface of valves smooth.

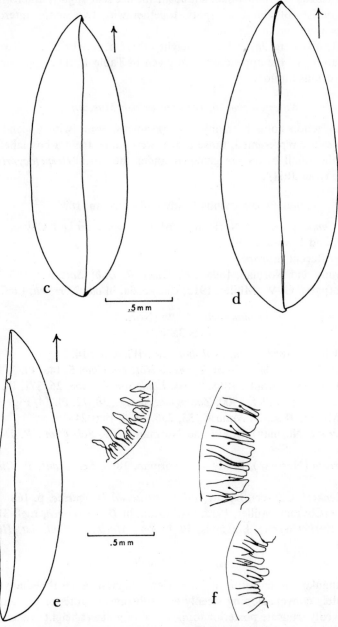

Fig. 2. *Herpetocypris reptans* (Baird). B.M. No. 1966.6.16.4. (a) Left valve. × 48. Internal lateral view; (b) Left valve. × 48. Internal lateral view; (c) Carapace. × 36. Dorsal view; (d) Carapace. × 36. Dorsal view; (e) Right valve. × 36. Dorsal view; Nature of marginal area, × *c.* 120; (f) Nature of marginal area showing branching and funnel-like inner terminations of radial pore canals. × *c.* 170.

Hinge margin of left valve slightly grooved for reception of edge of right valve. Inner lamellae moderately broad anteriorly, narrower posteriorly; line of concrescence and inner margin separate; inner lamellae with slight ridge on surface near mid-width; radial canals numerous, closely spaced and in part branching toward outer margin as illustrated in Fig. 2a; mid-ventrally canals are shorter and narrower; anteriorly, inner terminations of canals are funnel-like, adductor muscle scar slightly antero-median in position and consists of about six spots, together with two, more antero-ventral, 'antennal' spots as shown in Fig. 2a.

Length of left valve (Fig. 2a) 2·19 mm, height 1·00 mm, convexity of valve 0·68 mm.

The main features of the appendages are given in Table 1. The appendages were described by previous authors.

Notes on specimens in the British Museum

The Baird collection contains many dry specimens, some with appendages; six different localities are represented, those drawn were taken from a box labelled 'near London'. All the small boxes are grouped under the label 'Herpetocypris reptans (Baird) (not the types JPH)'.

Genus PRIONOCYPRIS Brady and Norman, 1896

Type species. Candona serrata Norman, 1861 (= *Cypris zenkeri* Chyzer, 1858 by Müller, 1912, and Lowndes, 1931).

Geologic Range; Recent, Europe.

Prionocypris Brady and Norman, 1896, *Sci. Trans. R. Dubl. Soc.*, ser. 2, **5**, 724.

Eucypris Vávra (part), G. W. Müller, 1912, Ostracoda, in *das Tierreich*, Lief. 31, 168.

Prionocypris serrata (Norman)
Figs 3a–c

Cypris zenkeri Chyzer, 1858, *Verh. zool.-bot. Ges. Wien*, **8**, 514.

Candona serrata Norman, 1861, *Trans. Tyneside Nat. Fld Club*, **5**, 148, Pl. 3, Figs 1–6.

Cypris serrata (Norman), Brady, 1868, *Trans. Linn. Soc. London*, **26**, 371, Pl. 25, Figs 15–19; Pl. 36, Fig. 3; Müller, 1900, *Zoologica, Stuttg.*, **30**, 72, Pl. 14, Figs 3, 11, 14.

Cypris bicolor Müller, 1880, *Z. Naturw.*, **53**, 236, Pl. 4, Figs 24–26.

Erpetocypris serrata (Norman), Brady and Norman, 1889, *Sci. Trans. R. Dubl. Soc.*, ser. 2, **4**, 87.

Prionocypris serrata (Norman), Brady and Norman, 1896, *Sci. Trans. R. Dubl. Soc.*, ser. 2, **5**, 724.

Herpetocypris zenkeri (Chyzer), Daday, 1900, *Ostracoda Hungariae*, p. 168.

Eucypris zenkeri (Chyzer), Müller, 1912, Ostracoda, in *Das Tierreich*, Lief. 31, 174.

Prionocypris zenkeri (Chyzer), Lowndes, 1931, *Rep. Marlboro. Coll. nat. Hist. Soc.*, No. 79, p. 93.

Description of shell

Shell subtriangular-subreniform in side view, highest antero-medially; dorsal margin moderately convex, sloping steeply and subtruncate anterior to, and sloping gently and markedly truncate posterior to, position of greatest height; ventral margin nearly straight to slightly concave medially; ventral slope slightly overhangs valve margin; anterior margin broadly curved, slightly extended and with about ten upward-curved marginal spines, below; posterior margin narrowly curved, extended medially and bears about five upward-curved marginal spines. Valves moderately convex, greatest convexity median, valves taper more or less uniformly to ends. Left valve slightly larger than and overlaps right. Surface evenly and very finely pitted.

Hinge of left valve bears a shallow rabbet groove for reception of edge of right. Adductor muscle scar consists of four large spots in a group together with two more antero-ventral, 'mandibular' spots. Inner lamellae of moderate width anteriorly, narrower posteriorly, line of concrescence and inner margin widely separated; radial canals numerous, irregular, short and closely spaced, not all extending to outer margin.

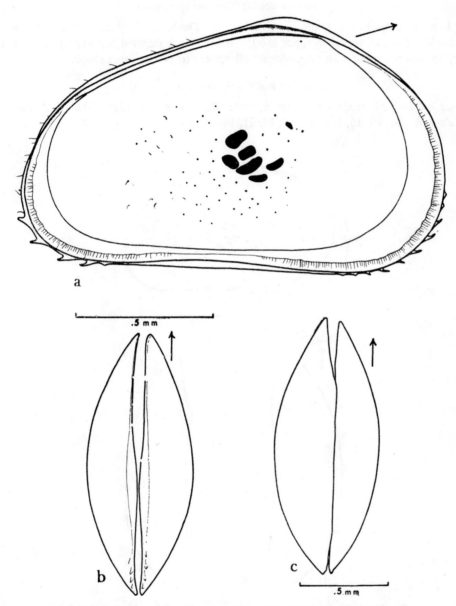

Fig. 3. *Prionocypris serrata* (Norman). B.M. No. 1900.3.6.66. (a) Left valve. × 76. Internal lateral view; (b) Carapace. × 76. Ventral view; (c) Carapace. × 48. Dorsal view.

Length of figured left valve (Fig. 3a) 1·56 mm, height 0·94 mm, convexity 0·60 mm.

The main features of the appendages are given in Table 1. The appendages were described by Brady and Norman.

Remarks

Prionocypris serrata does not have the flange-like anterior expansion of the shell and other features of *Eucypris*, and although similar anatomically to *Herpetocypris* has a different furca than in that genus as was pointed out by Sars.

Notes on British Museum specimens

Type locality Forge Dam, Sedgefield, Co. Durham, England, habitat 'stagna limpida'. Date collected October 1860; specimens are labelled topotypes; about ten dry specimens, some with appendages; all appear to be the same shape.

Genus MICROCYPRIS Kaufmann, 1900

Type species. M. reptans Kaufmann, 1900, *Zool. Anz.*, **23**, 132; 1900, *Revue suisse Zool.* **8**, 301, Pl. 16, Figs 16–18; Pl. 21, Figs 14–16.

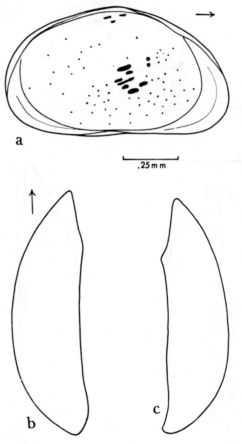

.25 mm

Fig. 4. *Microcypris reptans* Kaufmann.   × 60.   B.M.   No.   1911.11.8. 32480-481.   (a) Left valve. Internal lateral view; (b) Left valve. Dorsal view; (c) Right valve. Dorsal view.

Geologic Range. Recent, Switzerland.
Microcypris Kaufmann 1900, *Revue suisse Zool.*, **8**, 264.
Eucypris Vávra (part) G. W. Müller, 1912, Ostracoda, in *Das Tierreich*, Lief. 31, 168.

Microcypris reptans Kaufmann
Figs 4a–c

Microcypris reptans Kaufmann, 1900, *Zool. Anz.*, **23**, 132; *Revue suisse Zool.*, **8**, 301,
    Pl. 16, Figs 16–18; Pl. 21, Figs 14–16.
Cypris kaufmanni Masi, 1905, *Boll. Soc. zool. ital.*, ser. 2, **6**, 123, (fide Müller, 1912).
Eucypris reptans (Kaufmann), Müller, 1912, Ostracoda, in *Das Tierreich*, Lief, 31, 178.

Description of shell

Shell subovate-subtriangular in lateral view; highest medially; dorsal margin moderately convex, slightly truncated on posterior slope; ventral margin nearly straight, slightly concave medially; anterior margin rounded, extended below; posterior margin more broadly curved than anterior, slightly extended below. Valves subequal, the left slightly larger and overlapping right; moderately convex, curving uniformly to terminal margins. Surface smooth and glossy.

Hinge margin of right valve fits into a shallow furrow in edge of left valve. Inner lamellae broad anteriorly, somewhat narrower posteriorly, line of concrescence and inner margin widely separated; a slight ridge lies near middle of anterior inner lamella; radial canals only weakly developed. Adductor muscle scar consists of a slightly antero-median group of six oblique spots as illustrated, two oblique more antero-ventral 'antennal' spots and two small rounded more antero-dorsal mandibular spots.

Length 1·02 mm, height 0·57 mm, convexity of left valve 0·31 mm.

The main characteristics of the appendages are given in Table 1. Kaufmann has described the appendages in *M. reptans*.

Remarks

The shell structure of *Microcypris reptans* is so different from that of *Herpetocypris* that it seems unlikely they should be placed in the same subfamily as was done by Sharpe (1903).

Subfamily CYPRIDINAE

Genus AMPHICYPRIS Sars, 1901

Type Species. A. nobilis Sars.
Geologic Range. Recent, Argentina, Paraguay.
Amphicypris Sars, 1901. *Arch. Math. Naturv.*, **24** (1), 18.
Eucypris Vávra (part), Daday, 1905, *Zoologica, Stuttg.*, **44**, 243.

Amphicypris nobilis Sars
Figs 5a–e

Amphicypris nobilis Sars, 1901, *Arch. Math. Naturv.*, **24** (1), 18, Pl. 4.
Eucypris nobilis (Sars), Daday, 1905, *Zoologica, Stuttg.*, **44**, 243.

Description of shell

Shell large, subelliptical; highest medially in male shell; highest postero-medially in female shell; dorsal margin gently convex; ventral margin slightly concave antero-

2 K

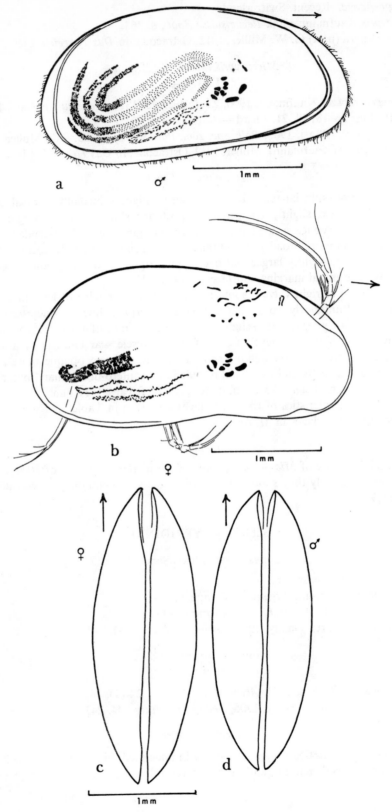

Fig. 5. (*Caption opposite*)

medially; anterior margin rounded, slightly extended below; posterior margin nearly equally rounded but slightly broader in females, extended below. Valves compressed, greatest convexity median; left valve slightly larger than right overlapping dorsally. Surface smooth except for tiny normal canals.

Hinge consists of simple rabbet groove in left valve into which fits edge of right valve. Muscle scars are as illustrated. Inner lamellae narrow and sinuous, vestibule present. Radial canals indistinct, very short or absent.

Length of male shell 2·86 mm, height 1·32 mm, convexity 0·92 mm; female shell 2·80 mm, 1·44 mm, and 0·96 mm.

Occurrence

The illustrated specimens are from Argentina. Type reference: Sars, G. O., 1901, *Arch. Math. Naturv.*, **24** (1), 16.

Genus EURYCYPRIS G. W. Müller, 1898

Type species. E. latissima Müller, 1898.
Geologic Range. Recent, Madagascar.
Eurycypris, G. W. Müller, 1898, *Abh. senckenb. naturf. Ges.*, **21**, 264.
Cypris O. F. Müller (part) G. W. Müller, 1912, Ostracoda, in *Das Tierreich*, Lief, 31, 178.

Eurycypris latissima Müller
Figs 6a, b

Eurycypris latissima G. W. Müller, 1898, *Abh. senckenb. naturf. Ges.*, **21**, 264, Pl. 13, Figs 15–21.
Cypris latissima (Müller), Müller, 1912, Ostracoda, in *Das Tierreich*, Lief, 31, 179.

Description of female shell

Shell ovate in side view, highest medially; dorsal margin evenly and moderately convex, sloping nearly uniformly from position of greatest height; ventral margin

Fig. 5. *Amphicypris nobilis* Sars. × 30. B.M. No. 1901.12.12.326-336. (a) Male carapace. Lateral view from right; (b) Female carapace. External lateral view from right; (c) Female. Dorsal view; (d) Male. Dorsal view; (e) Muscle scar patterns showing variation within the species, × c. 45.

nearly straight, slightly concave medially, anterior portion of concavity more abrupt than posterior part; anterior margin broadly, and nearly uniformly curved, slightly extended below; posterior margin a little more broadly curved than anterior, also slightly extended below, faintly truncate above. Valve surfaces smooth. Valves strongly convex, subequal, the left slightly larger than, and overlapping right valve except along hinge where right valve fits over edge of left.

Hinge surface of left valve with narrow rabbet groove for reception of edge of right valve. Inner lamellae moderately broad anteriorly, much narrower posteriorly; line of concrescence and inner margin separate, line of concrescence smoothly curved and lies close to outer margin; radial canals not clearly observed, but apparently are short and

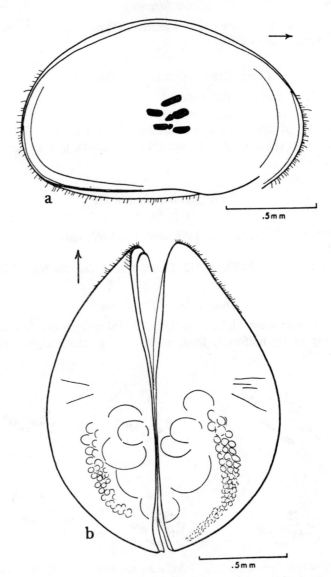

Fig. 6. *Eurycypris latissima* Müller. × 48. B.M. No. 1911.11.8.31635-637.
(a) Female left valve. Internal lateral view from right; (b) Female carapace.
Dorsal view.

closely spaced. Adductor muscle scar consists of five spots as illustrated, lying slightly anterior to middle.

Length of figured female specimen 1·60 mm; height 1·00 mm, convexity 1·33 mm.

The main appendage features are given in Table 1. Müller originally gave detailed descriptions of the appendages.

Notes on specimens in the British Museum

There are two female specimens labelled types. One is complete and one consists of the left valve with entire soft parts; the valves are soft and pliable, probably decalcified; drawings were made in a glycerine-water mixture.

Genus MEGALOCYPRIS Sars, 1898

Type species. M. princeps Sars.
Geologic Range. Recent. South Africa.
Megalocypris Sars, 1898, *Arch. Math. Naturv.*, **20** (8), Pl. 4.

Megalocypris princeps Sars
Figs 7a–c

Megalocypris princeps Sars, 1898, *Arch. Math. Naturv.* **20** (8), 5, Pl. 1.

Description of male shell

Very large, subelliptical to subquadrate in lateral view, highest about one-third from posterior end; dorsal margin nearly straight, somewhat sinuous, with broadly obtuse cardinal angles; ventral margin slightly concave antero-medially; anterior margin broadly and nearly uniformly curved; posterior margin broader than anterior but extended ventro-medially, subtruncate above. Left valve slightly larger than right, overlapping and extending beyond edge of right; in dorsal view valves moderately convex, terminal margins tapering, somewhat compressed. Male and female shells similar in dorsal view.

Antero-dorsally is a shallow rounded sulcus; general surface smooth medially becoming very finely reticulate towards anterior and posterior ends.

Hingement consists of faint groove in dorsal margin of left valve into which fits edge of right valve. Inner lamellae very narrow; line of concrescence and inner margin nearly if not actually coinciding. Radial canals only weakly developed. Adductor muscle scar lies anterior to mid-length, and consists of a group of five or six spots and two more antero-ventral antennal spots as illustrated. Dorsal to main group is a small rounded mandible pivot scar; additional shell scars in antero-dorsal region as shown. Trace of testes marks inner surface of valve.

Length of figured specimen (Fig. 7a) 6·00 mm, height 3·06 mm, convexity 2·40 mm.

Appendages

The appendages of *Megalocypris princeps* were described by Sars (1898) and Müller (1912). Principal features of the appendages are given in Table 1.

Occurrence

Port Elizabeth, South Africa, shallow lake.

Genus STENOCYPRIS Sars, 1889

Type species. Cypris malcolmsoni Brady, 1886.

Geologic Range. Tertiary to Recent, Africa, Australia, East Indies, India.
Stenocypris Sars, 1889, *Forh. VidenskSelsk. Krist.*, no. 8, 27.

Stenocypris cylindrica major Baird
Figs 8a–d

The species has recently been re-studied by Ferguson (1967). The illustrations given here supplement those of Professor Ferguson.

Genus MECYNOCYPRIA Rome, 1962

Type species. Paracypria obtusa Sars.
Geologic Range. Recent, Africa, Lake Tanganyika.
Mecynocypria Rome, 1962, *Résult. scient. Explor. hydrobiol. Lac Tanganika*, **3** (8), 26.

Mecynocypria obtusa Sars
Figs 9a–c

The accompanying drawings are supplementary to the illustrations given by Dom Rome.

Subfamily CYPRIDOPSINAE

Genus PARACYPRIDOPSIS Kaufmann, 1900

Type species. P. zschokkei Kaufmann, 1900.
Geologic Range. Recent, Switzerland.
Paracypridopsis Kaufmann, 1900, *Zool. Anz.*, **23**, 131.
Poracypridopsis Kaufmann, 1900, *Revue suisse Zool.*, **8**, 316 (error).
Potamocypris Brady (part), G. W. Müller, 1912, Ostracoda, in *Das Tierreich*, Lief, 31, 215.

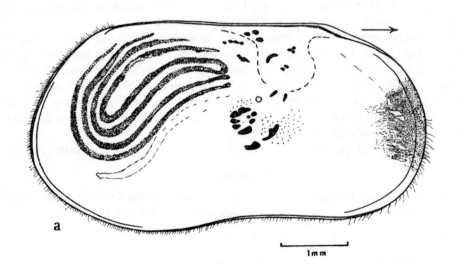

a

1 mm

Paracypridopsis zschokkei Kaufmann
Figs 10a–c

Paracypridopsis zschokkei Kaufmann, 1900, *Zool. Anz.* **23**, 131; *Revue suisse Zool.* **8**, 317, Pl. 19, Figs 18–20; Pl. 22, Figs 26–30.
Potamocypris zschokkei (Kaufmann), Müller, 1912, Ostracoda, In *Das Tierreich*, Lief, 31, 215.

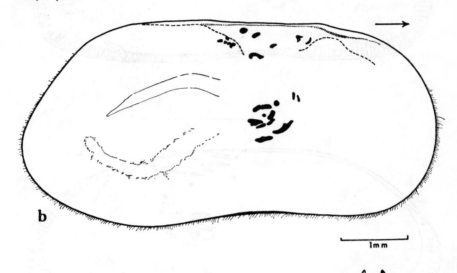

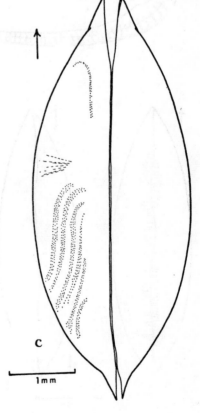

Fig. 7. *Megalocypris princeps* Sars. ×18. B.M. No. 1901.12.12.121. (a) Male carapace. External lateral view from right; (b) Female carapace. External lateral view from right; (c) Male carapace. Dorsal view.

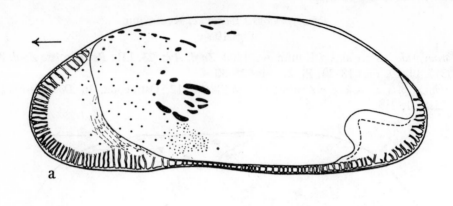

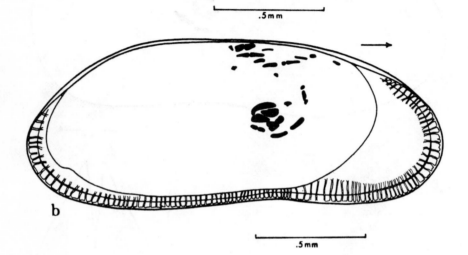

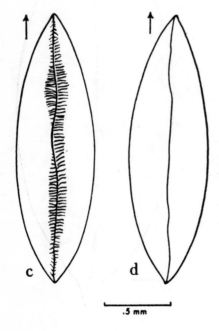

Fig. 8. *Stenocypris cylindrica major* (Baird). B.M. No. 1945.9.26.211. (a) Right valve. ×60. Internal view; (b) Left valve. ×60. Internal view; (c) Carapace. ×36. Ventral view; (d) Carapace. ×36. Dorsal view.

510

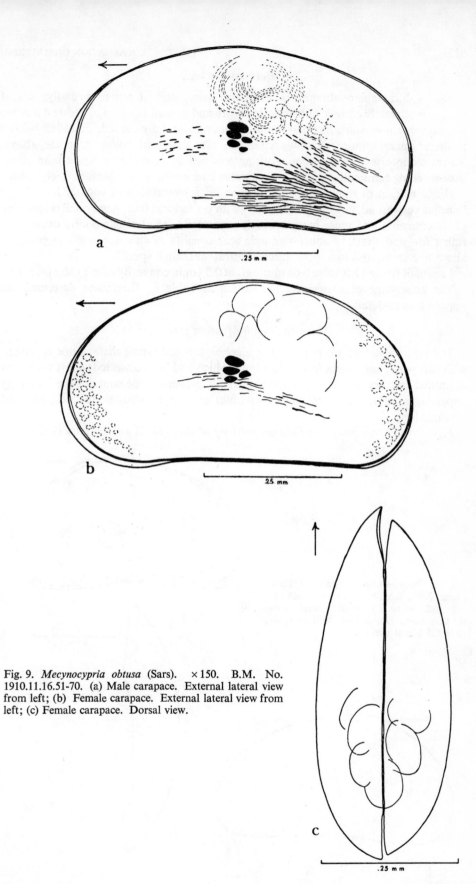

Fig. 9. *Mecynocypria obtusa* (Sars). ×150. B.M. No. 1910.11.16.51-70. (a) Male carapace. External lateral view from left; (b) Female carapace. External lateral view from left; (c) Female carapace. Dorsal view.

511

Description of shell

Shell subtriangular-subreniform in lateral view, highest antero-medially; dorsal margin moderately convex, subtruncate before and behind position of greatest height; ventral margin concave medially; anterior margin broadly curved, extended below; posterior margin more narrowly rounded, also extended below, truncate above. Valves compressed, greatest convexity antero-median. Left valve larger than right, over-reaching terminally; dorsally right valve over-reaches left. Surface finely pitted.

Hinge margin of right valve slightly grooved for reception of edge of left. Inner lamellae very broad, especially anteriorly (with posterior submarginal small ridge); line of concrescence and inner margin widely separated; radial canals poorly developed, rather few and short. Adductor muscle scar consists of an antero-median group of about five spots, and two more antero-ventral 'antennal' spots.

Length of figured left valve 0·68 mm, height 0·36 mm, convexity of left valve 0·13 mm.

The appendage characteristics are shown in Table 1. Kaufmann described the appendages in detail.

Notes on British Museum specimens

The type specimens are preserved in alcohol, labelled types; there is one specimen with soft parts, and valves from a specimen which had been dissected; a left valve was examined in transmitted and reflected light; a prepared slide contains the following appendages: antennules, antennae, mandibles and palps, maxillae, second and third thoracic legs.

P. zschokkei was placed in *Potamocypris* by Müller (1912) and by Sars (1928), but

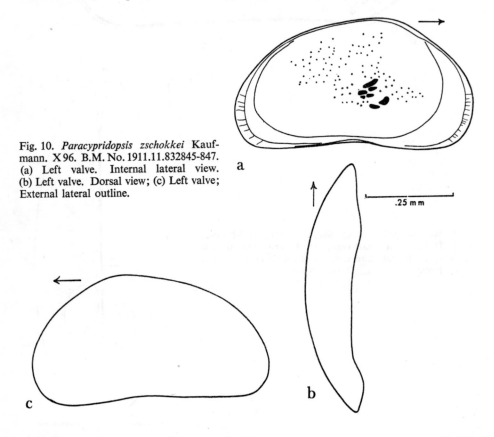

Fig. 10. *Paracypridopsis zschokkei* Kaufmann. X96. B.M. No. 1911.11.832845-847. (a) Left valve. Internal lateral view. (b) Left valve. Dorsal view; (c) Left valve; External lateral outline.

.25 m m

that genus is typified by strong dorsal overlap of the right valve by the left that is not present in *P. zschokkei*. Retention of *Paracypridopsis* as was recommended in Moore (1961) seems justified.

TABLE 1. Fresh-water Ostracoda types in the British Museum: summary of appendage characteristics

Species	Antennule	Antenna	Mandible	Maxilla	Thoracic Legs & Furca
Pontocypris mytiloides (Norman)	7-jointed, terminal part 5-jointed, long terminal setae.	6-jointed, penultimate joint short, with claw; sensory appendage of 1st joint large and club-shaped.	Branchial appendage well defined.	No diagnostic features.	P_1 with 2 terminal unequal claws, P_2 with strong terminal joint; furca well developed, with 2 terminal claws.
Herpetocypris reptans (Baird)	7-jointed, terminal part 5-jointed, long terminal setae.	6-jointed, joints poorly defined, natatory setae reduced.	No diagnostic features.	Masticatory lobes produced; spines on outer lobe denticulate.	P_1 with small vibratory plate, P_2, P_3 elongated, furca with distal claw, both denticulated.
Prionocypris serrata (Norman)	7-jointed, terminal part 5-jointed, long terminal setae.	6-jointed, joints poorly defined, natatory setae reduced.	No diagnostic features.	No diagnostic features.	Furca with special claws nearly equal.
Microcypris reptans (Kaufmann)	No diagnostic features.	Natatory setae short.	No diagnostic features.	No diagnostic features.	Furca with proximal claw $\frac{2}{3}$ length of distal claw.
Amphicypris nobilis (Sars)	No diagnostic features.	No diagnostic features.	No diagnostic features.	No diagnostic features.	Furca slender, nearly straight; posterior margin with row of 5 tubercles reaching nearly to basis; anterior claw $\frac{1}{2}$ length of anterior margin; posterior claw $\frac{2}{3}$ length of anterior claw.
Eurycypris latissima (Müller)	No diagnostic features.	No diagnostic features.	No diagnostic features.	No diagnostic features.	Furca nearly straight, anterior claw strong, distinctly curved: the anterior bristle about $\frac{1}{5}$ length of anterior claw.
Megalocypris princeps (Sars)	No diagnostic features.	No diagnostic features.	No diagnostic features.	No diagnostic features.	Furca weakly curved; anterior claw only $\frac{1}{3}$ length of anterior bristle, $\frac{1}{4}$ length of anterior claw; ejaculatory duct rudimentary.
Paracypridopsis zschokkei (Kaufmann)	No diagnostic features.	Short natatory setae.	No diagnostic features.	No diagnostic features.	Furca single clawed; with small posterior bristle.

REFERENCES

BAIRD, W. 1835. List of Entomostraca found in Berwickshire. *Hist. Berwicksh. Nat. Club.*, **1**, 95-100, Pl. 3.

DADAY, E. 1905. Untersuchungen über die Süsswassermikrofauna Paraguays. IV. Ostracoda. *Zoologica, Stuttg.*, **44**, 234-70, Pls 15-19.

FERGUSON, E. 1967. The type species of the Genus *Stenocypris* Sars, 1889, with descriptions of two new species. *Programme of Symposium on Recent Ostracoda, Hull, England, July* 10-14, 1967.

KAUFMANN, A. 1900. Neue Ostracoden aus der Schweiz. *Zool. Anz.*, **23**, 131-33.

MOORE, R. C. (ed.). 1961. *Treatise on Invertebrate Paleontology, Pt. Q. Arthropoda. 3. Crustacea: Ostracoda*. 442 pp., 334 Figs, Lawrence, Kansas, University of Kansas and Geological Society of America.

MÜLLER, G. W. 1898. Die Ostracoden in Wissenschaftliche ergebnisse der Reisen in Madagaskar und Ostafrica in den Jahren 1889-95 von Dr A. Voeltzkow. *Abh. senkenb. naturf. Ges.*, **21**, 257-96, Pls 13-19.

——. 1912. Ostracoda. In *Das Tierreich*. Eine Zusammenstellung und Kennzeichnung der rezenten Tierformen. *Auftr. k. preuss. Akad. Wiss.*, Leif 31, i-xxxiii, 1-434, 92 Figs.

ROME, D. R. 1962. Ostracodes. *Résult. scient. Explor. hydrobiol. Lac Tanganika*, **3** (8), 305 pp, 34 Figs, 49 Pls.

SARS, G. O. 1901. Contributions to the knowledge of the fresh-water Entomostraca of South America as shown by artificial hatching from dried material. *Arch. Math. Naturv.*, **24** (1), 1-52, Pls 1-8.

——. 1922-1928. *An account of the Crustacea of Norway, Vol. IX, Ostracoda*. 1-277, 119 Pls. Bergen Museum, Bergen.

SHARPE, R. W. 1903. Report on the fresh-water Ostracoda of the United States National Museum, including a revision of the subfamilies and genera of the family Cyprididae. *Proc. U.S. natn Mus.*, **26**, 969-1001, Pls 64-69.

SYLVESTER-BRADLEY, P. C. 1947. Some Ostracod Genotypes. *Ann. Mag. nat. Hist.*, (11) **13**, 192-9.

DISCUSSION

BENSON: It goes without saying that the types are the last recourse for accurate naming of species. I think that they have limited use in defining the characteristics that perhaps define the concept of the species. It seems to me in looking at these drawings that there is a great danger that the preservation of some of these forms might make it impossible to express adequately some of the more difficult features such as the muscle scars. I noticed the absence of some scars in the drawings that might have been caused by inadequate preservation. When working out the muscle scars of a large fauna one can only find relatively few specimens that show the scars and all of the details well. I differ to some extent from Professor Swain, in that I have found a consistency within the general patterns of muscle scars. Like any other variable feature, these take some interpretation and I think a considerable effort is needed to try and work out the homologies among these features. In the original designation of the types, I am sure that the workers were not aware of the possible importance of the muscle scars and therefore did not select specimens that might have had these features well displayed. I hope that we shall be aware of this in future studies.

BATE: I agree with Dr Benson. In 1963 I pointed out that in using muscle scars as diagnostic of a genus one should beware of the variation that is bound to be found within the various species which make up that genus. You may have the fusion of two closely connected scars giving a slightly different pattern and you must work out the variation so that you are not confused when you next see a slight difference and led into making an erroneous designation.

SYLVESTER-BRADLEY: I am concerned that we may be misled by this useful atlas if we fail to understand the worth of the type material being discussed. Am I right in saying that none of these specimens are type specimens; that they are all taken from a collection which has been identified, probably by somebody other than the original author, with the type species of these ostracods and that they may very well be in part misidentified?

In some cases where we were seeing two specimens of one species, I am not happy that we were looking at one species unless this has come from one collection and we have inter-gradations. Nor am I altogether happy that we are always dealing with the type species. For example, I am surprised to see that *Eurycypris latissima* is taken as the type species of *Eurycypris* because I was under the impression that the type species of this genus was *Cypris pubera*.

SWAIN: This species was placed in *Eurycypris* by Müller in 1898 but in 1912 he decided that the genus was a synonym of *Cypris*. Both *latissima* and *pubera* were among the species originally assigned to *Eurycypris*, and *latissima* should be considered as the type. The specimens are from the material labelled as types in the British Museum (Nat. Hist.). In this particular case the specimens were those of G. W. Müller.

HARDING: The type species of *Eurycypris* is *E. latissima* Müller, a distinct species later put into the genus *Cypris* by Müller. *Eurycypris* and *Cypris* have different type species and their synonymy is subjective. The specimens in the British Museum are from Müller's original material from Madagascar. We have been trying to let Professor Swain have material that came from the original person and which, as far as we can make out, are the original type specimens collected from the type locality before the date of publication. We have a number of pill boxes of ostracods left by Baird who more than a century ago was on the British Museum staff and it is this collection that Professor Swain is concentrating upon.

SYLVESTER-BRADLEY: These are syntypes?

HARDING: They are as near as we can get to syntypes. Professor Swain is not selecting lecto-types, this would be dangerous at this stage. I, myself, am particularly concerned about *Herpetocypris reptans* because this is a very common species and Dom Rome has described a number of new species showing that what we have called *Herpetocypris reptans* is, in fact, several species. It would be rather important if there was a choice of which to select as lectotype, to choose a specimen that was what is generally understood as *Herpetocypris reptans*. I wonder whether all Baird's specimens have the type of shell that Dom Rome would recognize as *Herpetocypris reptans*.

PETKOVSKI: Mir scheint, daß die Wiederverwendung mancher von Ihnen vorgeschlagenen Genera nicht genügend begründet ist. Es fehlt ein breiterer Vergleich mit den ähnlichen Arten der verwandten Genera. So sollte *Eurycypris latissima* mit *Cypris pubera*, *C. bispinosa* und *C. subglobosa* sowie mit mancher anderen, außereuropäischen *Cypris*-Arten verglichen werden. Das gilt auch für *Paracypridopsis zschokkei*, *Prionocypris serrata* und *Microcypris reptans*.

Potamocypris steueri ähnelt im Habitus viel mehr einer *Cypridopsis* als *P. zschokkei*, ihr Maxillartaster ist aber für das Genus *Potamocypris* sehr charakteristisch: das Endglied ist kurz und am Ende verbreitert. Daher steht diese Art im Genus *Potamocypris*. Weiter sind die Beziehungen *P. zschokkei* zu *Potamocypris wolfi* Brehm ganz unklar. Klie hat sich darüber nicht kritisch geäußert, sondern hat *P. wolfi* einfach in seine Fauna von Deutschland (1938) übernommen.

SWAIN: Did you say that there was a difference in the mandibular palp in *Paracypridopsis* as compared with *Potamocypris*?

PETKOVSKI: Ja, ich sagte: zwischen *Potamocypris* und *Cypridopsis*. Weiter möchte ich sagen, daß lange, radiäle Porenkanäle am Vorderrande der linken Schale besonders für *Herpetocypris chevreuxi* und nicht für *H. reptans* charakteristisch sind. Daß die radiären Striae erst bei den reifen Individuen zum Ausdruck kommen, bzw. daß sie bei den jugendlichen fast völlig fehlen, ist für *Stenocypris malcolmsoni* seit langem bekannt.

Es gibt Hinweise, die die Ausscheidung der Arten: *serrata, clavata, lutaria* und *zenkeri* aus dem Genus *Eucypris* rechtfertigen. Das Genus *Prionocypris* konnte also wieder aufgestellt werden. Demgegenüber soll das Genus *Microcypris* als Synonym zu *Heterocypris* gestellt werden.

GENERA AND SPECIES INDEX

References to taxa which appear in the running text or lists are shown in Roman sic. 123
References to taxa which appear in distribution maps, diagrams, histograms etc., are shown in Italic sic. 123
References to taxa which are figured in whole or part appear in Bold Face sic. **123**

abbreviatum (Paradoxostoma), 280, 303, *319*, *324*, 393-4
abbreviatus (Sclerochilus), 332, f. 333, 333, 359, 373, 381
aberrans (Allocypria), **178**, 179
abyssicola (Hemicytherura cf. *H.)*, 458, 463
abyssicola (Kangarina), 377, 391
Acocypris, 74
—*hirsuta*, **186**, 187
—*longiuscula*, **186**, 187
acosaguensis (Caudites), 458, 467
aculeata (Cypridopsis), 261, 390
acuminata (Argilloecia), 378-81, 384-5, 387-91, *400-1*
acuminata (Candona), 79, 136-7, 145, 147, 224-5, 233, 236
acuminata (Paracytherois), 332, f. 333, 333
acuminata (Semicytherura), 377, 384
acuticosta (Cytherura), 381
acuticosta (Semicytherura), 278, 283-4, 302, *316*, 335, *337*, 338, 340, 344, 348, 355, 377-8, 382, 384, 386-7, 389, 391, *402-3*
'*Acuticythereis*', 465
adriatica (Callistocythere), 338, 340, 344, 348, 352
aequus (Sclerochilus), 330, 332, f. 333, 333, 359, **369**, 373, 394, *400-1*
aestuarii (Loxoconcha), 393
affinis (Argilloecia), 7
affinis (Cypricercus), 181, **182**
affinis (Loxoconcha), 385
africana (Candonopsis), **174**, 175
agilis (Herpetocypris), 185, **186**
agilis (Loxoconcha), 338, 340, 344, 348
?*Aglaiella setigera*, 5, 11
alatum (Cytheropteron), 278, 377-82, 384-5, 387, **I**, *402-3*
alatum (Cytheropteron aff. *C.)*, 374, 378, 391, *402-3*
alba (Cypriconcha), 492-4
albomaculata (Cythere), 381
albomaculata (Heterocythereis), 26, *29*, *33-41*, *4* (f. 255), 261-3, *13* (f. 264), *14* (f. 265), 273, 278, 280, *281*, 283-4, 290, 294, *295*, 302, 304, *305-7*, 309, *310-11*, 312, *313*, *316*, *320*, 321, **326**, 392

alifera (Semicytherura), 335, 338, *339*, 340, 344, 348, 384, 391
Allocypria aberrans, **178**, 179
— *claviformis*, **178**, 179
— *inclinata*, **178**, 179
— *navicula*, **178**, 179
almasyi (Potamocypris), 78
altoides (Candona), 76-9
alveolivalva (Cythere), *444*, **IV**, 445, 462, **VII**, **IX-XI**
amberii (Elofsonella), 85-90, **I**, 91
Ambocythere, 476
ambophora (Uncinocythere), 157
Ambostracon glauca, *430*, **III**, 455, 458, **V**, 465, **IX, X**
— *pumila*, 5, 12
'—' *pumila*, 12
— sp., *430*, 455, **V**, 465, **XI**
americana (Cyclasterope), 105
Ammonia beccarii, 303
amnicola (Heterocythereis), 394
amphiakis (Dactylocythere), 154
Amphibolocypris exigua, 183, **184**
Amphicypris, 495, 503
— *nobilis*, 495, 503, **504-5**, *513*
angulata (Bairdia), 11
angulata (Bairdia cf. *B.)*, 11
angulata (Candona), 78, *376*
angulata (Hemicythere), 302, *306-7*
angulata (Pseudophilomedes), 377
angulata (Semicytherura), 278, 302, *318*, *324*
angulosa (Cytheridea), 6
angulosa (Loxocythere), 6, 384, 389
angulosa (Tetracytherura), 377, 386, *406-7*
angustum (Paradoxostoma), 332, *f. 333*, 333
anisoacantha (Stenocypris), **186**, 187
Ankylocythere, 158
— *heterodonta*, 160, 164
Anomocytheridea, 89
antarcticum (Paradoxostoma), 8
Anterocythere, 467
— sp., *430*, **III**, 455
Anthrapalaemon parki, 106
antiquata (Carinocythereis), 375, 393-4
antiquata (Pterygocythereis), 278, *281*, 284, *295*
Apatelocypris brevis, **188**, 189

517

GENERAL INDEX

abyssal environment, 475, 478-9
abyssal fauna, assemblage, 356, 373, **379-80**, 381, **382**, **390**, **392**, **II** (f. 395), **395-7**, 411, 478
abyssal ostracods, 373, 379-80, **II** (f. 395), **475-80**
abyssal plain(s), 379, 385, 396, 397
acanthocephalans, 200
Acarina, 305, 310, 313, 317, 322
acid bog lakes, 216
Adamczak, F. J., ix, **93-8**
adaptation, 393, 397; (advantage of) 478; (for grappling) 62
adaptive convergence, 66, 478
adaptive features, 476
Adriatic Sea, **334-55**, 357, 373, **380-1**
Aegean Province, **385-8**, 411
Aegean Sea, 357, **358**, 385-7
Africa, 3, 192, 197-8, 242-4, 477, 479-81, 507-8
age classes, 206, 208-9, 211, 214
air circulation (causing dessication), 309
Alabama, 482
alae, 49, 57
Alaska, 471, 492
alateness, 476
Albania, 5, 380
Alexandria, Egypt, 358, 360, 362-72, 391
algae, 106, 268, 276, 280, 281, 289, 292, 294, 298, 300-24, 328, 330, 332, f. 333, 338, 340, 344, 375, 393, 423, 425-9, 483-4
algal assemblages (populations of ostracods), **295**, 299, 304, **305**, **310**, **313**, 314, **318-9**, 321, **323**, 325, f. **333**
algal samples, 294, 301, 302, 328
algal zones, **295**, 299, 301, 304, 309, 312, 325
alkaline bogs, 216
alkaline environment, 414
alkalinity, 200
Allen, F. H., 253
Alm, G., 197-8, 233-4
Alnmouth, England, 7
American Museum, 484
amiculum, 167
amorphous phase of $CaCo_3$, 99, 103, 104
Amphipoda, 66, 80, 105-6, 305, 310, 313, 317, 322
anaerobic condition, 255
anaerobic layer, 312, 483
anaesthetic effect (Mg.), 108
anaesthetisation, 153, 207, 302
ancestral genus, 48, 58, 65
Anglesey, 256, **269**, **299-329**
anlagen, 167, 199
Antarctic Realm, 7
Antarctic species, 6, 57, 479

antennal muscle scar, 27, 46, 224, 235
antennal 'spots', 500, 503, 512
anterior rounding trend (paradoxostomatids), 62
anterior tooth, (development of), 135
antislip tooth, 52
appendages, 245-6 (in classification), 513 (table of characteristics)
appendages (mutation in), 484
approaches to ecology (discussion), **481-6**
aquaria, 483-4
aragonite, 100, 101, 103, 108
archibenthic forms, 396
arctic affinities, 423, 425
Arctic fauna, **249-50**
Arctic Ocean, 476, 482
'area' use in ecology. 414, 484
Argentina, 503-4
Aristotle, 357
Arkansas, 8
artifacts, 99-102, **I-III**, 103-8, 238-9; (possible) 166
artificial holdfasts, 283, 296
Ascoli, P., ix, 354-5, 380, 382, 410-11, 479, 485
assemblage of species groups, 249
assemblages (increasing diversity), 423
Athens, 385-6
Atlantic Coastal Plain, 468
Atlantic Fauna, 249-50, 359
Atlantic Ocean, 6, 110, 131, 244, 356, 359, 460-1, 476, 480, 482, 484, 497
Atlantic Shelf, 110, 119
attache de la furca (value in taxonomy), **168-93**, 246
attached margin, **109-35**
attenuation of the oral cones trend, 62
Australia, 52, 55, 59, 61, 68, 75, 244, 460, 485, 508
Australian species (entocytherids), 154, **157**, 158, 164
autecological approach, 483
automatic data collection, 21, 45
average distance coefficient, 21, 42-3
average linkage method, 32
axial line, **111**, **113**, 116, 118, 122, 126, 129, 132
axial zone, **116**
Azan process, 111
Azerbaijan, 59
Azores, 294

bacteria, 105
Baer, J. G., 200
Bahamas, 110, 256, 482

539